Lecture Notes in Computer Science 12432

More information about this series at http://www.springer.com/series/7409

Guojun Wang · Xuemin Lin ·
James Hendler · Wei Song ·
Zhuoming Xu · Genggeng Liu (Eds.)

Web Information Systems and Applications

17th International Conference, WISA 2020
Guangzhou, China, September 23–25, 2020
Proceedings

 Springer

Editors
Guojun Wang 🆔
Guangzhou University
Guangzhou, China

Xuemin Lin 🆔
The University of New South Wales
Sydney, NSW, Australia

James Hendler 🆔
Rensselaer Polytechnic Institute
Troy, NY, USA

Wei Song 🆔
Wuhan University
Wuhan, China

Zhuoming Xu 🆔
Hohai University
Nanjing, China

Genggeng Liu 🆔
Fuzhou University
Fuzhou, China

ISSN 0302-9743 ISSN 1611-3349 (electronic)
Lecture Notes in Computer Science
ISBN 978-3-030-60028-0 ISBN 978-3-030-60029-7 (eBook)
https://doi.org/10.1007/978-3-030-60029-7

LNCS Sublibrary: SL3 – Information Systems and Applications, incl. Internet/Web, and HCI

This Springer imprint is published by the registered company Springer Nature Switzerland AG
The registered company address is: Gewerbestrasse 11, 6330 Cham, Switzerland

Preface

It is our great pleasure to present the proceedings of the 17th Web Information Systems and Applications Conference (WISA 2020). WISA 2020 was organized by the China Computer Federation Technical Committee on Information Systems (CCF TCIS), Guangzhou University, and Guizhou University. WISA 2020 provided a premium forum for researchers, professionals, practitioners, and officers closely related to information systems and applications, to discuss the theme of "Artificial Intelligence and Information Systems," focusing on difficult and critical issues, and promoting innovative technology for new application areas of information systems.

WISA 2020 was held in Guangzhou, Guangdong, China, during September 23–25, 2020. Given the theme of WISA 2020, the focus was on intelligent cities, government information systems, intelligent medical care, fintech, and network security, emphasizing the technology used to solve the difficult and critical problems in data sharing, data governance, knowledge graph, and blockchains.

This year we received 165 submissions, each of which was assigned to at least three Program Committee (PC) members to review. The peer-review process was double-blind. The thoughtful discussions on each paper by the PC resulted in the selection of 42 full research papers (an acceptance rate of 25.45%) and 16 short papers. The program of WISA 2020 included keynote speeches and topic-specific invited talks by famous experts in various areas of artificial intelligence and information systems to share their cutting-edge technologies and views about the academic and industrial hotspots. The other events included industrial forums, CCF TCIS salon, and PhD forum.

We are grateful to the general chairs Prof. Baowen Xu (Nanjing University), Prof. Yanming Sun (Guangzhou University), and Prof. Ge Yu (Northeastern University), as well as all the PC members and external reviewers who contributed their time and expertise to the paper reviewing process. We would like to thank all the members of the Organizing Committee, and many volunteers, for their great support in the conference organization. Especially, we would also like to thank publication chairs Prof. Wei Song (Wuhan University), Prof. Zhuoming Xu (Hohai University), and Prof. Genggeng Liu (Fuzhou University) for their efforts on the publication of the conference proceedings. Many thanks to all the authors who submitted their papers to the conference.

August 2020

Guojun Wang
Xuemin Lin
James Hendler

Organization

Steering Committee

Baowen Xu	Nanjing University, China
Ge Yu	Northeastern University, China
Xiaofeng Meng	Renmin University of China, China
Yong Qi	Xi'an Jiaotong University, China
Chunxiao Xing	Tsinghua University, China
Ruixuan Li	Huazhong University of Science and Technology, China
Lizhen Xu	Southeast University, China
Xin Wang	Tianjin University, China

General Chairs

Baowen Xu	Nanjing University, China
Yanming Sun	Guangzhou University, China
Ge Yu	Northeastern University, China

Program Committee Chairs

Guojun Wang	Guangzhou University, China
Xuemin Lin	The University of New South Wales, Australia
James Hendler	Rensselaer Polytechnic Institute, USA

Local Chairs

Hanpin Wang	Guangzhou University, China
Yongbin Qin	Guizhou University, China

Forum Chairs

Chunxiao Xing	Tsinghua University, China
Ruixuan Li	Huazhong University of Science and Technology, China

Publicity Chair

Xin Wang	Tianjin University, China

Publication Chairs

Wei Song	Wuhan University, China
Zhuoming Xu	Hohai University, China
Genggeng Liu	Fuzhou University, China

Website Chairs

Weifeng Zhang	Nanjing University of Posts and Telecommunications, China
Guanghui Feng	Guangzhou University, China

Program Committee

Dezhi An	Gansu University of Political Science and Law, China
Zhifeng Bao	RMIT University, Australia
Yu Cao	University of Massachusetts Lowell, USA
Xingong Chang	Shanxi University of Finance and Economics, China
Lemen Chao	Renmin University of China, China
Guolong Chen	Fuzhou University, China
Lin Chen	Nanjing University, China
Ling Chen	Yangzhou University, China
Yanhui Ding	Shandong Normal University, China
Zhiming Ding	Beijing University of Technology, China
Yongquan Dong	Jiangsu Normal University, China
Kening Gao	Northeastern University, China
Wenzhong Guo	Fuzhou University, China
Yanbo Han	North China University of Technology, China
Tieke He	Nanjing University, China
Xin He	Henan University, China
Mengxing Huang	Hainan University, China
Fangjiao Jiang	Jiangsu Normal University, China
Shujuan Jiang	China University of Mining and Technology, China
Weijin Jiang	Xiangtan University, China
Wang-Chien Lee	Penn State University, USA
Bin Li	Yangzhou University, China
Chunying Li	Guangdong Polytechnic Normal University, China
Feifei Li	The University of Utah, USA
Jianxin Li	Deakin University, Australia
Juanzi Li	Tsinghua University, China
Qingzhong Li	Shandong University, China
Ruixuan Li	Huazhong University of Science and Technology, China
Zhenxing Li	Agile Century, China
Ye Liang	Beijing Foreign Studies University, China
Xuemin Lin	The University of New South Wales, Australia

Bingxiang Liu	Jingdezhen Ceramic Institute, China
Chen Liu	North China University of Technology, China
Genggeng Liu	Fuzhou University, China
Huan Liu	Arizona State University, USA
Qing Liu	Renmin University of China, China
Youzhong Ma	Luoyang Normal University, China
Weiyi Meng	SUNY Binghamton University, USA
Weiwei Ni	Southeast University, China
Baoning Niu	Taiyuan University of Technology, China
Zhiyong Peng	Wuhan University, China
Yong Qi	Xi'an Jiaotong University, China
Weiguang Qu	Nanjing Normal University, China
Jiadong Ren	Yanshan University, China
Yonggong Ren	Liaoning Normal University, China
Derong Shen	Northeast University, China
Xiaohua Shi	Shanghai Jiao Tong University, China
Baoyan Song	Liaoning University, China
Wei Song	Wuhan University, China
Chenchen Sun	Tianjin University of Technology, China
Haojun Sun	Shantou University, China
Yong Tang	South China Normal University, China
Guojun Wang	Guangzhou University, China
Haofen Wang	Tongji University, China
Mingyan Wang	Nanchang University, China
Xiaoguang Wang	Virginia Tech University, USA
Xin Wang	Tianjin University, China
Xingce Wang	Beijing Normal University, China
Yanlong Wen	Nankai University, China
Feng Xia	Dalian University of Technology, China
Shixiong Xia	China University of Mining and Technology, China
Xing Xie	Microsoft Research Asia, China
Chunxiao Xing	Tsinghua University, China
Li Xiong	Emery University, USA
Guandong Xu	University of Technology Sydney, Australia
Bin Xu	Northeastern University, China
Lei Xu	Nanjing University, China
Lizhen Xu	Southeast University, China
Zhuoming Xu	Hohai University, China
Zhongmin Yan	Shandong University, China
Nan Yang	Renmin University of China, China
Jianming Yong	University of Southern Queensland, Australia
Ge Yu	Northeast University, China
Hong Yu	Dalian Ocean University, China
Mei Yu	Tianjin University, China
Shui Yu	University of Technology Sydney, Australia
Fang Yuan	Hebei University, China

Guan Yuan	China University of Mining and Technology, China
Xiaojie Yuan	Nankai University, China
Jeffrey Xu Yu	Chinese University of Hong Kong, Hong Kong, China
Guigang Zhang	Institute of Automation, Chinese Academy of Sciences, China
Mingxin Zhang	Changshu Institute of Technology, China
Rui Zhang	The University of Melbourne, Australia
Weifeng Zhang	Nanjing University of Posts and Telecommunications, China
Ying Zhang	Nankai University, China
Yong Zhang	Tsinghua University, China
Yongxin Zhang	Shandong Normal University, China
Feng Zhao	Huazhong University of Science and Technology, China
Minghui Zheng	Hubei Minzu University, China
Jiantao Zhou	Inner Mongolia University, China
Qiaoming Zhu	Soochow University, China
Fang Zuo	Henan University, China

External Reviewers

Tiecheng Bai	Tarim University, China
Xiangjie Kong	Dalian University of Technology, China
Bohan Li	Nanjing University of Aeronautics and Astronautics, China
Zhilei Liu	Tianjin University, China
Jun Pang	Wuhan University of Science and Technology, China
Zhongbin Sun	Xi'an Jiaotong University, China
Buyu Wang	Inner Mongolia Agricultural University, China
Dong Wang	Henan University, China
Shiyu Yang	East China Normal University, China
Xiang Zhao	National University of Defense Technology, China

Contents

Query Processing and Algorithm

Natural Language Processing

Machine Learning

Edge Computing and Data Mining

Data Privacy and Security

Blockchain

World Wide Web

Influence of Periodic Role Switching Intervals on Pair Programming Effectiveness

Bin Xu, Sheng Yan$^{(\boxtimes)}$, Kening Gao, Yu Zhang, and Ge Yu

School of Computer Science and Engineering, Northeastern University,
Shenyang 110189, China
{xubin,gkn,zhangyu,yuge}@mail.neu.edu.cn, yansheng1117@foxmail.com

Abstract. Pair programming has been widely used in programming experiment teaching in programming courses. One of the important factors affecting the successful completion of pair programming is the timing of periodic role switching. We organized an experiment in the course of Python Programming for 102 freshmen who did not major in computer science. By comparing the accuracy of code submitted by students in the online judge system, we evaluated the influence of pairing programming on students' programming ability under three different periodic role switching intervals of 15 min, 20 min and 30 min, and collected students' perception towards pairing programming under different modes. We also made a standard to judge the normalization of code, to study the influence of pair programming on the normalization of code written by students. The results show that when the periodic role switching interval is 30 min, pairing programming is helpful for students to solve difficult problems, and it has a positive impact on the solution of subsequent problems after experiencing the process of solving difficult problems. When the periodic role switching interval is 20 min, students have a positive attitude towards pair programming. Therefore, the best switching interval can be set between 20 min and 30 min. However, in terms of code normalization, there is no significant relationship between the standard degree of student code and the switching interval of pair programming. We gave some explanations for this in this paper.

Keywords: Programming course · Pair programming · Assessment

1 Introduction

In the past few decades, computers have become a basic tool in people's daily life. Programming is also required by all walks of life as an important skill. Therefore, more and more people choose to study computer science courses in colleges and universities [1]. Pair programming has been proved that students' programming levels can be improved in programming courses [13,14,16]. In pair programming,

Supported by the National Natural Science Foundation of China (U1811261).

G. Wang et al. (Eds.): WISA 2020, LNCS 12432, pp. 3–14, 2020.
https://doi.org/10.1007/978-3-030-60029-7_1

two students coordinate to build codes by playing different roles, the driver is responsible for writing codes, and the navigator is responsible for finding errors and providing feedback. Periodic role switching allows pairs of programming students to share their ideas and collaborate in writing higher-quality programs [5,17].

Previous studies have shown that there are many factors that have a significant impact on the effectiveness of pair programming [6,8,19], including whether students have been exposed to programming before and the differences between teachers' pair programming methods. We have studied the correct rate and standard degree of codes written by students at different switching intervals in pair programming to explore the influence of different times on the effectiveness of pair programming.

2   Background

Pair programming is a programming technique in which two programmers, usually from the same workstation, work together on a task. With the rise of extreme programming [2], it has been widely used in professional applications for the first time and has since become a common programming method in the software industry. Two programmers work side by side in front of the same computer. One enters the code, while the other reviews every line of code he enters. The person who enters the code is called the "driver" and the person who reviews the code is called the "navigator". Usually, two programmers switch roles regularly.

Pair programming, as a well-known learning programming strategy, has proved to be a more effective method than individual programming methods [7,12,20], so pair programming is also used as part of or in conjunction with various teaching methods [9,15]. One advantage of pair programming is accessibility. Pair programming enables students who have just come into contact with programming to have confidence in programming, which essentially lowers the threshold for students to learn to program [4,18].

Under the educational background, there have been many kinds of research on pair programming, such as exploring the correlation between self-reporting preferences [10] and curriculum performance [3] and the core contributors in projects [11]. Pair programming also enables students to constantly check their views on programming and exchange learning strategies. This expands students' learning programming strategies and helps to open up students' thinking mode. However, the factors that affect pair programming are still largely unexplored, so this paper makes a quantitative analysis of the periodic switching interval of the two roles in pair programming.

3   Method

3.1   Experimental Design

The data for this study come from the Python programming introductory course of the Northeastern University of China in autumn 2019. Students have seven weeks of computer programming assignments throughout the semester. We selected the last four weeks of the semester to conduct pairing programming experiments to ensure that students have a certain programming foundation. Students can choose whether to do pair programming or not, and the matching partners of pair programming are chosen voluntarily. In the experiment, our class time is usually 120 min, and we provide 15 min, 20 min, and 30 min for students to switch drivers and navigators. Students can choose the role switching interval of their group and the role switching interval chosen by each group of students also remains the same, that is, the students use the same role switching interval in the whole process of the experiment. There are eight programming problems in the assignments arranged in four weeks. All students do not adopt pair programming in the first week, and the students in the next three weeks will carry out group pair programming. In the whole experiment, once random matching partners are selected, the matching partners of each student are fixed. Figure 1 shows the design pattern of the experiment. We counted the behavior records of the code submission of students in the online judge system, and the homework was only rated as passed or failed. In the following chapters, we also give a standard method to evaluate the normalization of code written by students.

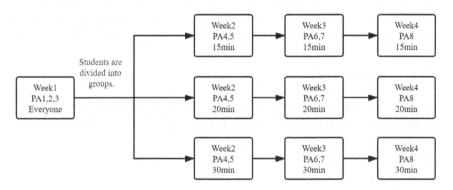

Fig. 1. Design of experiments for comparing three different switching times in pair programming.

3.2 Data

The data in this study were collected by two methods: (1) Homework records submitted by students in the online judge system; (2) Voluntary Pair Programming Strategy Questionnaire. The homework records submitted by the students in the online judge system include the time the students completed, the number of exercises completed, and whether they passed the evaluation. The questionnaire includes questions about pair programming activities. A total of 102 students participated in the survey, of which 96 students were grouped into pairs for programming, with a total inclusion rate of 94.1%. We selected 12 pairs of students from three groups respectively and took the remaining 12 students who did not adopt pair programming as the control group for experiments. The information we got from the questionnaire is shown in Table 1.

Table 1. Comparison of the number of people participating in the survey.

	15 min	20 min	30 min	Solo
Total number	24	24	24	12
Male	18	14	13	7
Female	6	10	11	5
Have programming experience	8	11	8	10
Heard of pair programming	8	9	4	6

In the experiment, each group of pair programming students completed an assignment together, and the students who programmed independently completed an assignment. We can see from Table 1 that some students have some programming experience. Therefore, before the experiment began, we conducted a test on the students to verify their programming level. We conducted two-sample T-tests on the scores of each group of students in different modes and the control group respectively, and the results are shown in Table 2.

Table 2. Differences in programming ability between groups.

Group	t	Degree of freedom (df)	p	Homogeneity of variance test (Sig)
15 min	−1.42	212.38	0.16	No
20 min	1.18	249.36	0.23	No
30 min	−1.13	253.00	0.26	No

The results showed that there was no significant statistical difference between the test results of each group and the control group. Therefore, the level of students' programming ability in our experiment is similar and will not affect the results of the experiment.

3.3   Research Questions

The research questions in this study are as follows.

- RQ1: What is the influence of different rotation intervals on the passing rate of students' pair programming code writing?
- RQ2: Will pair programming improve students' coding normalization?
- RQ3: What are the students' perceptions of pair programming under different switching intervals?

4 Result

4.1 RQ1: Code Pass Rate

To ensure that the students in each mode can accept the same amount of homework, we collected the passing rate of 8 program questions in the programming task and scored the passing rate of each question equally. The code passing rate of each homework in each mode can be calculated by the following formula:

$$R(G_i, P_j) = \frac{1}{N_i} \sum_i^{N_i} \frac{A(\theta_k^i, P_j)}{S(\theta_k^i, P_j)} \tag{1}$$

Where G_i represents the i-th mode; N_i represents the number of people in i-th mode; P_j represents the j-th homework; θ_k^i represents the k-th person in i-th mode; $A(\theta_k^i, P_j)$ represents the correct number of homework submissions in P_j; $S(\theta_k^i, P_j)$ represents the total number of homework submissions in P_j.

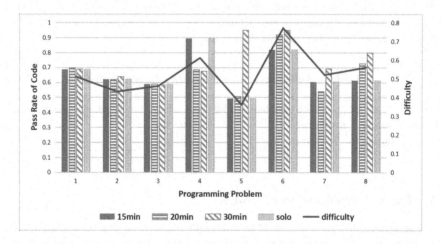

Fig. 2. Illustration of the passing rate of each group of codes and the difficulty of homework. (Color figure online)

According to the students' feedback and the weekly teaching content, combined with the code submitted by all the students, the change curve of homework difficulty can be obtained. As shown in Fig. 2, the yellow curve indicates the difficulty of the homework, and the smaller the ordinate indicates the greater the difficulty of the homework. The difficulty of homework is calculated from the records submitted by everyone on Online Judge, and the values of each point on the curve indicate the passing rate of each homework submission. The lower the pass rate, the more difficult the homework is, and the more difficult it is for the students to solve the homework. The cylindrical graph represents the passing rate of students' code submission in the online judge system under the switching interval of 15 min, 20 min, 30 min, and control group respectively. From the picture, we can find that the passing rate of students' codes is positively related to the difficulty of homework, which accords with our cognition.

In the phase of no pair programming, the three groups of students all adopt the single programming mode, and there is no obvious difference in the passing rate of homework submission. In the phase of pair programming, all the students begin pair programming. When solving the first homework, the group with a switching interval of 15 min has great advantages, can quickly solve some problems, and improve the passing rate of homework, while the other two groups have no obvious difference at the beginning. However, when the difficulty of homework increases, the group with a 30-min switching leads the other two groups by more than 40%.

Two-sample T-test was conducted with the pass rate data of codes with a switching interval of 15 min and 20 min and the data with a switching interval of 30 min respectively, and the difference between them is shown in Table 3. It can be seen that the group with a switching interval of 30 min has obvious statistical differences with the other two groups. Generally speaking, the difference in group scores is negative, which indicates that the other two groups are worse than the students whose switching interval is 30 min.

Table 3. Differences in performance between groups.

Group	t	Degree of freedom (df)	p	Homogeneity of variance test (Sig)
15 min & 30 min	−6.89	212.38	0.00	Yes
20 min & 30 min	−7.34	187.42	0.00	Yes

4.2   RQ2: Normalization of Codes

To determine whether pair programming can make students write codes more standardized, we have formulated a standard to grade students' codes with a standard degree, and detailed standard information is as follows:

1. Items with a weight of 3
   (a) Indentation: Use 4 spaces per indentation level. Continuation lines should align wrapped elements either vertically using Python's implicit line joining inside parentheses, brackets, and braces, or using a hanging indent.
   (b) Tabs or Spaces: Spaces are the preferred indentation method. Tabs should be used solely to remain consistent with code that is already indented with tabs.
   (c) Maximum Line Length: Limit all lines to a maximum of 79 characters.
   (d) Line Break: A-Line Break should before a Binary Operator.
   (e) Blank Lines: Surround top-level function and class definitions with two blank lines. Method definitions inside a class are surrounded by a single

blank line. Extra blank lines may be used (sparingly) to separate groups of related functions. Blank lines may be omitted between a bunch of related one-liners (e.g. a set of dummy implementations). Use blank lines in functions, sparingly, to indicate logical sections.
2. Items with a weight of 2
   (a) Source File Encoding: Code in the core Python distribution should always use UTF-8.
   (b) Imports: Imports should usually be on separate lines. Imports are always put at the top of the file, just after any module comments and docstrings, and before module globals and constants.
   (c) String Quotes: When a string contains single or double quote characters, however, use the other one to avoid backslashes in the string. It improves readability.
   (d) Whitespace in Expressions and Statements: Avoid extraneous whitespace.
3. Items with a weight of 1
   (a) Comments: Good code should have wonderful comments.
   (b) Naming Conventions: Naming in python code should conform to the naming specification.

The total score of the standard score is 30 points. If there is any nonconforming item in the code, the corresponding points will be deducted. Students have no professional foundation of programming before, so we pay more attention to the standardization of students' code layout. So when we calculate the deducted score, we give different weights to different items. For example, we give smaller weights to two items that enhance code readability (variable naming and code comments). The specific calculation method is as follows:

$$S = \beta \sum_{i=1}^{11} nI_i \tag{2}$$

Where I_i represent the item i in the table, β represents the weighting coefficient of each term, and the basic score of each item is 1 point.

We collected the codes of all students participating in pair programming and then calculated the standard rate according to the code length of each program. The standard rate of each group of modes takes the average value of the standard rate of all codes in that group. The formula of code normalization is calculated as follows:

$$N(G_i, P_j) = \frac{1}{N_i} \sum_i^{N_i} \frac{S_{sum} - S\left(\theta_k^i, P_j\right)}{L\left(\theta_k^i, P_j\right)} \tag{3}$$

Where $N(G_i, P_j)$ represents the standard degree of the code of the i-th mode, N_i represents the number of people in the i-th mode, P_j represents the j-th topic, S_{sum} represents the total score of each program, $S\left(\theta_k^i, P_j\right)$ represents the score deducted by each program according to the standard items, and $L\left(\theta_k^i, P_j\right)$ represents the length of the code.

Figure 3 shows the changing trend of standard rates for each group of codes. It can be seen from the figure that there is no correlation between the standard

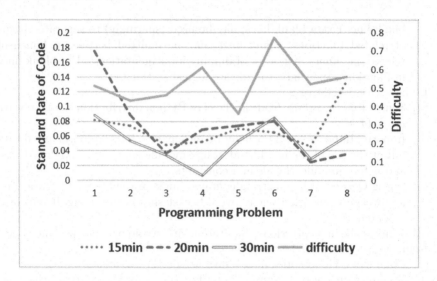

Fig. 3. A comparative diagram of the changing trend of the standard rate of codes in each group and the difficulty of homework.

rate and the difficulty of the homework, and the standard rate of each group has no obvious upward trend with the pair programming. We analyze that the reason why this may happen is that students learn and do while learning, which leads to the students' programming foundation not reaching the standard level. Therefore, we analyzed the codes written by students in the final examination and selected two questions to calculate the standard rate of codes written by students in the examination. The significance test of the regression coefficient for each group of data shows that there is no obvious linear correlation between the standard rate of codes and the time of pair programming.

4.3   RQ3: Students' Attitude to Pair Programming

After the experiment, to measure the students' experience in pair programming, a short questionnaire was distributed to the students.

- Q1: Are you willing to continue pair programming?
- Q2: Do you think pair programming is more efficient?
- Q3: Do you think pair programming is helpful to improve the programming level?

The results in Fig. 4 show that the students' attitude towards pair programming is positive. Most students are in favor of pair programming mode and are willing to continue pair programming. Figures show that students with a 20-min switching interval for pair programming feel more able to benefit from pair programming, and they feel that pair programming is more effective and can improve their programming level.

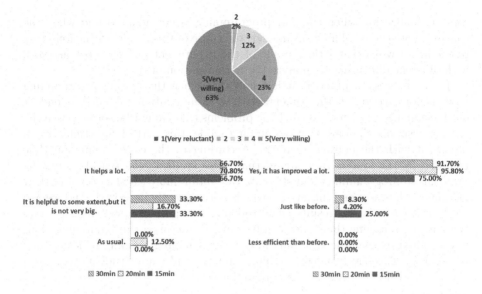

Fig. 4. Results of the questionnaire.

5 Discussion

The results show that the longer the pairing time is, the better it is for students to overcome difficult homework. Moreover, judging from the submission of homework after solving difficult problems, the passing rate of the submitted codes of this group of students is higher than that of the other two groups after experiencing cooperation in solving difficult problems. This shows that students benefit more from pair programming within 30 min.

In order to verify the validity of this result, we have a test for students after the pair programming course. It includes a simple programming problem and a programming problem beyond the syllabus. We counted the passing rate of each group of students. The results show that whether it is a simple question or a more difficult question, the students with a 30-min rotation time for pair programming have a higher pass rate. This shows that our previous results are valid.

As far as the normalization of codes is concerned, from the data obtained from the overall statistics, pairing programming has no effective influence. There is no significant linear correlation between the increase of pairing time and the standard degree of codes. We analyze the possible reasons for this situation include students' lack of programming experience, a small number of samples, and difficulty in homework. In order to verify whether it is related to the students' programming experience, we also selected the codes submitted by the students at the end of the final exam and calculated the standard scores of each group. The results show that after the end of the course, the code normalization has not been greatly improved, and there is no significant difference between the code normalization when the pair programming is not adopted and when the course is not studied. This reason is not the cause of this result. In the following research, we will expand the sample of the experiment and set up homework with different difficulties to analyze the normalization of the code.

In the process of practice, it is found that when the pairing programming mode is not carried out, the students with weak foundation are more dependent on the guidance teachers and treat the problems that cannot be solved passively. After developing the pairing programming mode, this part of the students can discuss it with the pairing members. According to the code submission time recorded by the system, it is found that these students can quickly solve the problems in programming, and students can obtain more satisfaction, and their learning enthusiasm and initiative have also been greatly improved. And we can see from the statistical results that students with a 20-min switching interval for pair programming think they can benefit more from pair programming. This also inspired us that the switching interval of pair programming is not as long as possible. Excessively long switching intervals will cause students to have a negative attitude towards pair programming.

6   Conclusion

In this study, we discussed the influence of periodic role switching interval on the effectiveness of pair programming and analyzed the pass rate, normalization, and students' cognitive attitude towards pair programming under different switching intervals. Our results show that students can benefit from pair programming when the switching interval for pair programming is 30 min. However, when we investigate students' attitudes, we find that students prefer to set the switching interval at 20 min.

Therefore, in the actual programming language teaching environment, we suggest that more attention should be paid to the pairing scheme of pairing programming. However, in the switching interval, we suggest to give priority to work continuity, and there is no need to specify specific time rigidly. According to the experimental results, to balance students' enthusiasm for pair programming and the benefits of pair programming in normal teaching, we suggest that the switching interval of pair programming in class can be set to 20–30 min.

When studying the influence of pair programming on the normalization of students' code writing, we got the opposite result. Our results show that there is no obvious linear correlation between the standard degree of code and the time of pair programming. We analyzed and verified that the reason for this result has nothing to do with whether the students study the whole course. In the following research, we will further expand the sample size and set up different difficult assignments to verify the remaining reasons.

References

1. Computer Research Association., et al.: Generation CS: computer science under-graduate enrollments surge since 2006 (2017). http://cra.org/data/Generation-CS/
2. Beck, K., Andres, C.: Extreme Programming Explained: Embrace Change. Addison-Wesley Professional, USA (2004). https://doi.org/10.5555/318762
3. Braught, G., Wahls, T., Eby, L.M.: The case for pair programming in the computer science classroom. ACM Trans. Comput. Educ. (TOCE) **11**(1), 1–21 (2011). https://doi.org/10.1145/1921607.1921609
4. Carver, J.C., Henderson, L., He, L., Hodges, J., Reese, D.: Increased retention of early computer science and software engineering students using pair programming. In: 20th Conference on Software Engineering Education & Training (CSEET 2007), pp. 115–122. IEEE (2007). https://doi.org/10.1109/CSEET.2007.29
5. Celepkolu, M., Boyer, K.E.: Thematic analysis of students' reflections on pair programming in cs1. In: Proceedings of the 49th ACM Technical Symposium on Computer Science Education, pp. 771–776 (2018). https://doi.org/10.1145/3159450.3159516
6. Chaparro, E.A., Yuksel, A., Romero, P., Bryant, S.: Factors affecting the perceived effectiveness of pair programming in higher education. In: Proceedings of PPIG, pp. 5–18 (2005)
7. Declue, T.: Pair programming and pair trading: effects on learning and motivation in a CS2 course. J. Comput. Sci. Coll. JCSC **18** (2003). https://doi.org/10.5555/771832.771843
8. Hanks, B.: Student attitudes toward pair programming. In: Proceedings of the 11th Annual SIGCSE Conference on Innovation and Technology in Computer Science Education, pp. 113–117 (2006). https://doi.org/10.1145/1140123.1140156
9. Heinonen, K., Hirvikoski, K., Luukkainen, M., Vihavainen, A.: Learning agile software engineering practices using coding dojo. In: Proceedings of the 14th Annual ACM SIGITE Conference on Information Technology Education, pp. 97–102 (2013). https://doi.org/10.1145/2512276.2512306
10. Khan, S., Ray, L., Smith, A., Kongmunvattana, A.: A pair programming trial in the CS1 lab. In: Proceeding Annual International Conference on Computer Science Education: Innovation and Technology (CSEIT), pp. 6–7 (2010)
11. Liu, X., Bai, J., Liu, L., Ouyang, H., Zhou, H., Xu, L.: Mining core contributors in open-source projects. In: Ni, W., Wang, X., Song, W., Li, Y. (eds.) WISA 2019. LNCS, vol. 11817, pp. 690–703. Springer, Cham (2019). https://doi.org/10.1007/978-3-030-30952-7_70
12. McDowell, C., Hanks, B., Werner, L.: Experimenting with pair programming in the classroom. In: Proceedings of the 8th Annual Conference on Innovation and Technology in Computer Science Education, pp. 60–64 (2003). https://doi.org/10.1145/961511.961531
13. McDowell, C., Werner, L., Bullock, H., Fernald, J.: The effects of pair-programming on performance in an introductory programming course. In: Proceedings of the 33rd SIGCSE Technical Symposium on Computer Science Education, pp. 38–42 (2002). https://doi.org/10.1145/563340.563353
14. Nagappan, N., et al.: Improving the CS1 experience with pair programming. vol. 35, pp. 359–362. ACM New York (2003). https://doi.org/10.1145/792548.612006
15. Porter, L., Simon, B.: Retaining nearly one-third more majors with a trio of instructional best practices in CS1. In: Proceeding of the 44th ACM Technical Symposium on Computer Science Education, pp. 165–170 (2013). https://doi.org/10.1145/2445196.2445248

16. Quintana, H., Grados, B.: Applying pair programming practice in the improvement of software design skills, in an undergraduate course. In: Proceedings of the 2020 ACM Conference on Innovation and Technology in Computer Science Education, pp. 543–544 (2020). https://doi.org/10.1145/3341525.3393985

17. Reckinger, S., Hughes, B.: Strategies for implementing in-class, active, programming assessments: a multi-level model. In: Proceedings of the 51st ACM Technical Symposium on Computer Science Education, pp. 454–460 (2020). https://doi.org/10.1145/3328778.3366850

18. Venkatesan, V., Sankar, A.: Adoption of pair programming in the academic environment with different degree of complexity in students perspective-an empirical study. Int. J. Eng. Sci. Technol. **2**(9), 4791–4800 (2010)

19. Xinogalos, S., Satratzemi, M., Chatzigeorgiou, A., Tsompanoudi, D.: Factors affecting students' performance in distributed pair programming. J. Educ. Comput. Res. **57**(2), 513–544 (2019). https://doi.org/10.1177/0735633117749432

20. Zhong, B., Wang, Q., Chen, J.: The impact of social factors on pair programming in a primary school. Comput. Hum. Behav. **64**, 423–431 (2016). https://doi.org/10.1016/j.chb.2016.07.017

Test Case Minimization for Regression Testing of Composite Service Based on Modification Impact Analysis

Xintang Lin[1], Haibo Zhang[1], Hui Xia[1], Liangjiang Yu[1], Xiangyan Fang[1], Xuan Chen[1], and Zhikai Wang[2(✉)]

[1] Wuhan Digital Engineering Institute, Wuhan 430074, China
work_lxt@126.com
[2] Department of Computer Science and Technology, Nanjing University, Nanjing 210023, China
mg1933064@smail.nju.edu.cn

Abstract. Composite service implements complex functions by combing different web services. It evolves a lot during its life cycle, and external web service may evolve without prior notice. To ensure the correctness of each evolved version, regression testing must be performed. In this article, an approach is proposed to automatically minimize test cases for regression testing of WS-BPEL (Web Services Business Process Execution Language) compositions based on modification impact analysis. The proposed approach can detect interfacial, structural, and variable changes of WS-BPEL compositions on the basis of WP (WSDL Parsing), BH (BPEL Heading), and BAST (BPEL Abstract Tree). Afterward, we perform a dependency analysis by extracting def-use pairs of modified variables to cover all affected paths. We conducted experiments with 8 WS-BPEL compositions to evaluate the efficiency of our approach. Experimental results show that our approach can recognize the modifications in most cases, and can reduce the original test suite to 51.75% on average.

Keywords: Test case minimization · Modification impact analysis · Regression testing

1 Introduction

With the rapid development of cloud computing, web services have gained significant attention due to their characteristics of well-encapsulation and loose-coupled. Web services are usually described by WSDL (Web Services Description Language) [1], a language based on XML.

Since single web service provides limited functions, multiple web services have been composited to achieve complex tasks [2]. WS-BPEL [3] has become a standard for service compositions. To adapt to various of requirements in daily business, the structure of WS-BPEL has to keep evolving. Apart from structural

© Springer Nature Switzerland AG 2020
G. Wang et al. (Eds.): WISA 2020, LNCS 12432, pp. 15–26, 2020.
https://doi.org/10.1007/978-3-030-60029-7_2

changes, external web services also evolve without prior notification. Given these circumstances, any change in the WS-BPEL compositions should be fully tested before and during its deployment.

Currently, many tools exist for testing WS-BPEL to ensure correctness by automating test case generation [4–8]. However, the execution of a great number of test cases is time-consuming yet ineffective. Therefore, regression testing [9] has gained growing concerns. Although frequent execution of regression tests can maximize the discovery of errors in WS-BPEL compositions, the process itself consumes too much time. To reduce the cost of regression tests, researchers have conducted in-depth research on automated regression testing techniques. Typical regression testing techniques include test suite minimization [10–12], test case selection [13–15], and test case prioritization [16–19].

In this work, we present an approach to the minimization of test cases based on modification impact analysis. Since WS-BPEL is a kind of programming language, we can define an AST (abstract syntax tree) to extract necessary information. Then with the generated BASTs (BPEL Abstract Syntax Tree), we can perform modification analysis. Specifically, we compare two versions of BASTs and discover interface, structural or variable changes, and use dependency analysis to locate influenced nodes. With modified nodes and influenced nodes, we can then minimize test suite to exclude redundant test cases.

The rest of the paper is organized as follows: Sect. 2 gives an motivating example. Section 3 introduces the methodology. Section 4 presents our implementation. Section 5 discusses our experiment and evaluation, Sect. 6 is about related work, and Sect. 7 is about the conclusion.

2   Motivating Example

When an external service of a WS-BPEL composition is deemed unsuitable, the service integrators may modify the binding address to a replacement service. The changes are known by testers if is notified. However, there could be some other modifications that involve without prior notice.

We motivate our work via a model of WS-BPEL evolution to illustrate the different types of change. Figure 1(a) shows an original version of a composition that contains three services A, B and C.

Consider the next evolved version in Fig. 1(b), although the interface of service B remains unchanged, its internal functions have been edited. This situation occurs when the service provider finds that some operations in the original service B need to be changed to meet current requirement better. Since changes to these external services will not notify the service integrator, we need to actively retrieve the WSDL documents of these services.

In Fig. 1(c), suppose that when the edited Service B' is deemed unsuitable, it is more practical to replace it. In this case, we should ensure that the newly added service is compatible with the original WS-BPEL composition and that no new errors will be introduced. Therefore, our tool should automatically minimize the original test suite and not perform redundant tests on parts that are not affected.

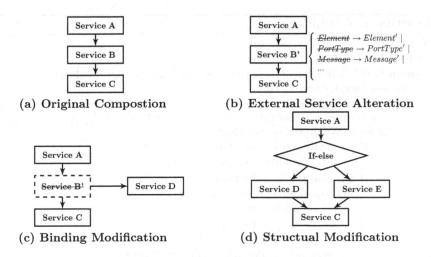

Fig. 1. Illustrations of WS-BPEL evolution

Suppose that there is an incoming requirement that makes the original <SEQUENCE> structure unsatisfactory. To make the WS-BPEL composition meet the requirement, as Fig. 1(d) shows, the service integrator decides to introduce service E and add the control structure of <IF-ELSE>. This complex type of modification could introduce new errors.

How to discover errors while reducing test overhead is an urgent problem we need to solve. We will introduce our method in Sect. 4.

3   Methodology

This section introduces the methods we used in modification extracting. The key idea is using DOM4J to parse the information of WSDL, XSD, and BPEL documents and transforms them into the structure we need.

3.1   WSDL Parsing

In WP, we implement WSDLParser and XSDParser to retrieve service information. WSDLParser is shown in Algorithm 1. XSDParser is not shown for space concerns, since besides enumerated types, XSDParser is generally the same as WSDLParser.

For WSDL, (1) we first parse the <PORTTYPE> nodes (Line 6 to 7 of Function WSDLParser()) to obtain the operations and each <OPERATION> contains input and output <MESSAGE>. (2) Then we parse the <MESSAGE> nodes to obtain the parameters of each operation (Line 8 to 10). (3) Parameter information is described in <TYPES> of the WSDL or XSD document (Line 11 to 17). (4) Also, we parse the <SERVICE> and <BINDING> in the WSDL to obtain the target address and the protocol type (SOAP GET POST). This article only

---

Algorithm 1. WSDLParser()

Input: *file* //XML file(*.wsdl,*.xsd)
Output: List<service> *serviceInfo*

1: **enum** WSDLNodeTypes {portType, message, types, service, binding}
2: **if** file isInstanceOf(*.wsdl) **then**
3:    **for** element in elements **do**
4:        nn=element.getName())
5:        **if** nn in WSDLNodeTypes **then**
6:            **if** nn.equals("portType") **then**
7:                portType.add(createNode(element))
8:            **if** nn.equals("message") **then**
9:                message.add(createNode(element))
10:                portType.addOperation(message);
11:            **if** nn.equals("Types") **then**
12:                **if** hasAtrribute("Import") **then** //means has a *.XSD file
13:                    parameters.add(createNode(XSDParser()))
14:                    message.add(parameters);
15:                **else** parameters.append(createNode(element))
16:                    message.add(parameters)
17:            **if** nn.equals("binding") **then**
18:                binding.add(createNode(element))
19:    **for** element in elements **do**
20:        **if** nn.equals("service") **then**
21:            service.add(portType)
22:            service.setAddress()
23:            serviceInfo.append(service)
24: **return** *serviceInfo*

---

considers the SOAP protocol (Line 18 to 24). The output of WSDLParser is an array list of service information. Each service information has a top-down tree structure, and the structure follows the parsing order.

3.2 BPEL Parsing

BPEL parsing takes BPEL documents as inputs, BAST(BPEL abstract syntax tree) and BH(BPEL headers) as outputs. The parsing method is similar to Algorithm 1, except for some details. To distinguish different types of modifications, we traverse BPEL documents and then split them into BH and BAST.

Definition (BH). BH is defined as a triad (I, PL, V), where I is a set of imports, PL is a set of defined partnerLinks, V is a set of defined variables.

$I = I_n \cup I_l$. The information about external services is included in node I, where I_n denotes external services' *namespace*, and I_l denotes the *location*.

PL records related information of each partnerLink, namely partnerLink *name*, *namepspace* and *role*.

V records related information of each variable, namely variable *name*, *namespace* and *messageType*. It should be noted that *messageType* corresponds to *<message>* in WSDL documents.

Definition (BAST). BAST is defined as a quintuple (N, C, PL, V, F), where N is the BAST node, C is a set of pointers to children nodes, PL is a set of partnerLinks, V is a set of variables and F is the field of each BAST node.

$N = N_B \cup N_S \cup N_P \cup N_L \cup N_C$. In BPEL, basic and structural activities can be transformed into the five kinds of BAST nodes.

- N_B is created for basic activities, i.e. *<receive>*, *<reply>*, *<assign>*, *<invoke>*.
- N_S is created sequential activities, i.e. *<sequence>*.
- N_P is created for selective activities, i.e. *<if>*, *<elseif>*, *<condition>*.
- N_L is created for loop activities, i.e. *<while>*, *<for>*.
- N_C is created for concurrency activities, i.e. *<flow>*.

C stores the pointers to the address of children nodes. The root node is *<sequnce>*. If current node is a leaf node, the value of C is empty.

PL and V are sets of partnerLinks and variables defined in BH and used in current node.

F records related information of each BAST node.

- *id:* records the identification of a node with a unique natural number.
- *name:* records the name of a node in XML documents.
- *type:* records the type of a node.
- *Gen(n):* records the set of variables defined in a node n.
- *Kill(n):* records the set of definitions killed in a node n.
- *Use(n):* records the set of variable used in a node n.

$Gen(n), Kill(n)$ and $Use(n)$ in F field are used for dependency analysis. With $Gen(n)$ and $Kill(n)$, we can further compute $In(n)$ and $Out(n)$ sets. $In(n)$ and $Out(n)$ denote the reaching definitions at the point before and after each node n. The computation formulas are as follows [21]:

$$Out(n) = Gen(n) \cup (In(n) - Kill(n)) \qquad (1)$$

$$In(n) = \bigcup_{p \in pred(n)} Out(p) \qquad (2)$$

where $Gen(n)$ is the set of definitions generated by the statement in n, $Kill(n)$ is the set of all other definitions of the defined variable in the program, and $pred(n)$ is the set of predecessor nodes of n. In our approach, we consider concurrent activities as asynchronous to simplify dependency analysis.

$In(n)$ and $Out(n)$ will be applied to data dependency analysis in Sect. 4.

4   Modification Impact Analysis

In this section, we propose a method of modification impact analysis. In Sect. 4.1, we present the framework of our approach. In Sect. 4.2, we give a detailed explanation of extracting rules.

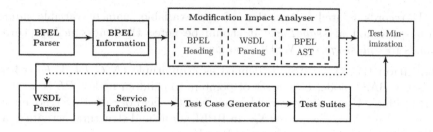

Fig. 2. Framework of our approach

4.1   Framework

Figure 2 is the general framework of our approach. We utilize BPEL Parser and WSDL parser to conduct modification impact analysis. The BPEL information contains two elements, namely BH and BAST. BH is for variable inspection, and BAST is for BPEL comparison. After analyzing of modifications, we minimize test suites with the results. To support new services, we develop a tool for test case generation (Sect. 5.1).

4.2   Extracting Modification Rules

Our extracting modification rules focus on the revision of variables. Since most changes all involved in the modification of variables, directly or indirectly, i.e., in Fig. 1(d), the predicate of <IF-ELSE> contains constraints with variables, and the newly introduced service E must also include new variables. We use portType as additional information since in BPEL activities, portType either appears with variable or does not appear.

A modification rule, denoted r, represents how a variable of WS-BPEL composition is changed to a new one. The rule can be expressed as a 3 tuple, i.e., $r = (C, L, R)$, where C represents the composition, L the outdated variable, and R the new variable. For example, service B's old variable, which is *Message* in Fig. 1(b), can be expressed as follows: $(C, Message \rightarrow Message')$.

A change rule r can have three different cases:

- $(C, \emptyset \rightarrow R)$
- $(C, L \rightarrow R)$
- $(C, L \rightarrow \emptyset)$

These three cases are associated with corresponding actions. For example, when a variable is added or invoked, r has the change case $(C, \emptyset \rightarrow R)$. If the variable is edited or replaced, r has the change case $(C, L \rightarrow R)$. If the variable is deleted or spared, r has the change case $(C, L \rightarrow \emptyset)$. We extract modification rules according to three cases and six actions shown in Table 1. To extract change rules, we use BH, WP and BAST in order, with regard to a modification set $M_{var}(k)$. We explain the details in the following subsections.

Table 1. Extracting modification rules

Rules	Action	Description	Method	Change patterns
$(C, \emptyset \to R)$	ADD	R is added	BH	Additional External Service
	INVOKE	R is invoked	$BAST$	Idle Functionality of Composition
$(C, L \to R)$	EDIT	L is changed into R	WP	Edited External Service
	REPLACE	L is replaced by R	$BAST$	Replace of Existing External Service
			BH/BAST	Replace of Additional External Service
$(C, L \to \emptyset)$	DELETE	L is deleted	$BAST$	Removed External Service
	SPARE	L is spared	$BAST$	Decreasing Functionality of Composition

4.2.1 Using BPEL Headings
Since BH contains the $<import>$, $<portType>$ and $<variable>$, we can compare two versions of headings to extract modifications. If a new variable appears in the new revision, the modification action may be ADD or REPLACE and can be further distinguished with BAST. The new variables will be added into set M_{var}.

4.2.2 Using WSDL Parsing
Even if some variables may remain the same in both BHs, there may be modifications in the WSDL. We locate the codes in WSDL with the information provided by $<variable>$ in BH. According to the previous introduction, we know that $<variable>$ has two attributes: namespace and messageType. namespace can link a variable to corresponding $<import>$ and $<import>$ stores the location of WSDL. messageType name is matched to the $<message>$ in WSDL. With location and $<message>$, we use WP to retrieve the information from two versions and compare them to determine if something has changed. The edited variables will be added into set M_{var}.

4.2.3 Comparing BPEL AST
We compare two BASTs through pre-order traversal. Figure 3 shows two simplified BASTs, circle boxes represent structural nodes, and rectangle boxes represent basic nodes, e.g. $<receive>$. The numbers are unique for different nodes. As is shown in Fig. 3, traversing the v1.0 in pre-order yields $1 - 2 - 3 - 4$ and v1.1 yields $1 - 2 - 3' - 4 - 6 - 7 - 8$. Through comparing two traces, we can extract 2 kinds of modifications (EDIT and ADD) and 4 changed variables($3', 6 - 7 - 8$). Changed variables will be added into set M_{var}.

4.2.4   Data Dependence Analysis

We have obtained the changed variable set M_{var} through previous steps. In the regression testing of WS-BPEL compositions, it is not sufficient to consider the variables that are directly changed. To ensure that there is no error in the changed service composition, we need to extract the nodes with variables that depend on the variables in set M_{var}. The method of extraction is data dependency analysis.

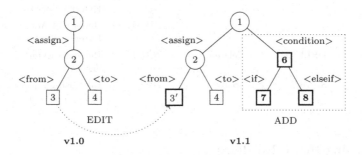

Fig. 3. BAST modification example

In Sect. 3, we store $Gen(n), Use(n), Kill(n)$ of node n during parsing, and compute $In(n)$ and $Out(n)$. If a node A's $Use(A)$ contains a definition defined in node B, i.e., in $Gen(B)$, and other nodes between the two nodes do not kill the definition, then we infer that node A data depends on node B.

For each variable var in M_{var}, we look for other variables that depend on var and add these variables into affected nodes set E_{var}. With the changed variable set and the affected node set, we reduce the test suite by removing test cases that do not cover these two node sets.

5   Evaluations

In this section, we first introduce some implementation issues of our prototype tool in Sect. 5.1, and then conduct several experiments in Sect. 5.2.

5.1   Implementation

According to the extracting rules (Table 1), we have implemented a prototype tool. The inputs of this tool include composition documents, namely *.bpel, .wsdl* and *.xsd* files and the outputs are minimized test suite.

We have also developed a tool for test case generation considering that there may not be test cases for newly added services. This tool retrieves information by parsing WSDL and XSD documents. Then, it generates values for the input parameters of services with constraints and random values.

We chose eight benchmarks to evaluate our work. These compositions come from some popular BPEL engines and BPEL specification (Table 2). They are also used in other BPEL test case generation [19] or regression test prioritization [20]. A to H are used to represent the corresponding benchmarks. The columns "Benchmark," "Source," "Element," and "LOC" represent the name of benchmarks, the source, the number of XML elements, and the number of lines of code of each composition, respectively.

Table 2. Relevant information of the experimental application.

Ref.	Benchmark	Source	Elements	LOC	Versions
A	ATM	ActiveBPEL	94	180	4
B	Gymlocker	BPWS4J	23	52	3
C	Dslservice	BPWS4J	50	123	3
D	LoanApproval	ActiveBPEL	41	102	7
E	MarketPlace	BPWS4J	31	68	3
F	Purchase	BPWS4J	41	125	5
G	Loan	Oracle BPEL Process Manager	55	147	3
H	LoanProcess	WSO2 BPS	33	161	5

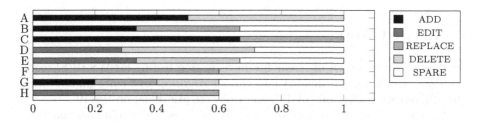

Fig. 4. The statics of extracting modifications for WS-BPEL compositions.

For evaluation, we need some modified versions of these WS-BPEL compositions. Therefore, we invited some research partners (non-authors) to make some modifications to the 8 original compositions. To guarantee the independence of modification, each version only involves one modification.

5.2 Results and Analysis

We apply the 3 rules in Table 1 to the original and modified versions respectively.

Figure 4 shows the results using stacked bar chart, the vertical axis represents the benchmarks' reference, and the horizontal axis represents the proportion of different actions to all modifications. ADD, EDIT, REPLACE, DELETE, SPARE are the actions of three extracting modification rules.

The graph shows that for 7 of these compositions, our approach can detect all the modifications. For example, composition B has 3 versions of modifications, namely ADD, REPLACE, SPARE, and the amount of each modification is 1(1/3 of all versions). The graph also shows that 2 (40%) of the modified version of H has not been detected. After checking the documents, we find that 2 modified versions are called external services through URLs, which caused our tool to failed in obtaining the WSDL documents.

Figure 5 presents performance of test suite minimization, the value of each column equals the mean of corresponding modified versions' number of minimized test cases divided by the number of test cases in the initial test suite. The graph shows that the average minimization rate is 51.75%, our tool works best for D's test suite minimization, and the worst for F's.

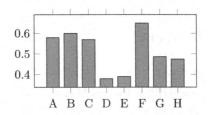

Fig. 5. The statics of test suite minimization.

6   Related Work

At present, there are lot of researches in regression testing for WS-BPEL compositions, which can be divided into three categories: test suite minimization [10–12], test case selection [13–15], and test case prioritization [16–19].

Francisco et al. [10] proposed a search-based test suite minimization technique. The approach generates mutants for original composition and compares their behaviors with original composition under test suite. If the mutant cannot be killed, the test case is deemed redundant.

Singal et al. [14] conducted test case selection based on the extended BPEL flow graph (XBFG). It compares the XBFGs of compositions to extract modifications, select test cases, and generate test cases. However, the method assumes that the system always needs to be fully tested, which is redundant.

Wang et al. [17] proposed a test case prioritization technique based on modification impact analysis. They build the BPEL activity dependency graph to calculate the modification impact and priority. Compared with the coverage-based method, this method can achieve a higher precision. The drawback is that it is not suitable for large-scale compositions due to heavy work load.

7   Conclusion

In this paper, we have proposed a modification impact analysis based regression testing approach to capture the modification caused by interfacial, structural, and variable changes. Our highlight is designing rules to comprehensively cover possible changes. Compared to most technologies, we not only detect changes in external services coarsely at the interface level but more fine-grained. It significantly complements the inadequacy of existing work in theoretical studies of TSM. We have verified our proposal through an experiment.

Our current work focuses on test suite minimization problem of service compositions in static testing environment. In the future, we will broaden our work to solve regression test prioritization problem in dynamic testing environment.

References

1. Web Services Description Language (WSDL) Version 2.0 Part 1: Core Language, W3C (2007). http://www.w3.org/TR/wsdl20/
2. Zhao, H., Chen, J., Xu, L.: Semantic web service discovery based on LDA clustering. In: Ni, W., Wang, X., Song, W., Li, Y. (eds.) WISA 2019. LNCS, vol. 11817, pp. 239–250. Springer, Cham (2019). https://doi.org/10.1007/978-3-030-30952-7_25
3. Jordan, D., et al.: Web services business process execution language version 2.0. *OASIS Stdandard* 11(120), p. 5 (2007)
4. Nakngern, P., Suwannasart, T.: A design of WS-BPEL test case generation tool based on path conditions. In: International MultiConference of Engineers and Computer Scientists (2017)
5. Jahan, H., Rao, S., Liu, D.: Test case generation for BPEL-based web service composition using Colored Petri Nets. In: 2016 International Conference on Progress in Informatics and Computing (PIC), pp. 623–628. IEEE, December 2016. https://doi.org/10.1109/pic.2016.7949575
6. Jehan, S., Pill, I., Wotawa, F.: SOA testing via random paths in BPEL models. In: 2014 IEEE 7th International Conference on Software Testing, Verification and Validation Workshops, pp. 260–263. IEEE, March 2014. https://doi.org/10.1109/icstw.2014.28
7. Sun, C.A., Zhao, Y., Pan, L., Liu, H., Chen, T.Y.: Automated testing of WS-BPEL service compositions: a scenario-oriented approach. IEEE Trans. Serv. Comput. 11(4), 616–629 (2015). https://doi.org/10.1109/tsc.2015.2466572
8. Estero-Botaro, A., García-Domínguez, A., Domínguez-Jiménez, J.J., Palomo-Lozano, F., Medina-Bulo, I.: A framework for genetic test-case generation for WS-BPEL compositions. In: Merayo, M.G., de Oca, E.M. (eds.) ICTSS 2014. LNCS, vol. 8763, pp. 1–16. Springer, Heidelberg (2014). https://doi.org/10.1007/978-3-662-44857-1_1
9. Yoo, S., Harman, M.: Regression testing minimization, selection and prioritization: a survey. Softw. Test. Verif. Reliab. 22(2), 67–120 (2012). https://doi.org/10.1002/stv.430
10. Palomo-Lozano, F., Estero-Botaro, A., Medina-Bulo, I., Núñez, M.: Test suite minimization for mutation testing of WS-BPEL compositions. In: Proceedings of the Genetic and Evolutionary Computation Conference, pp. 1427–1434, July 2018. https://doi.org/10.1145/3205455.3205533

11. Bozkurt, M.: Cost-aware pareto optimal test suite minimisation for service-centric systems. In: Proceedings of the 15th Annual Conference on Genetic and Evolutionary Computation, pp. 1429–1436, July 2013. https://doi.org/10.1145/2463372. 2463551

12. Nabuco, M., Paiva, A.C.R.: Model-based test case generation for web applications. In: Murgante, B. (ed.) ICCSA 2014. LNCS, vol. 8584, pp. 248–262. Springer, Cham (2014). https://doi.org/10.1007/978-3-319-09153-2_19

13. Böhmer, K., Rinderle-Ma, S.: A genetic algorithm for automatic business process test case selection. In: Debruyne, C. (ed.) OTM 2015. LNCS, vol. 9415, pp. 166–184. Springer, Cham (2015). https://doi.org/10.1007/978-3-319-26148-5_10

14. Singal, P., Mishra, A. K., Singh, L.: Test case selection for regression testing of applications using web services based on wsdl specification changes. In: International Conference on Computing, Communication and Automation, pp. 908–913. IEEE, May 2015 https://doi.org/10.1109/CCAA.2015.7148505

15. Li, B., Qiu, D., Leung, H., Wang, D.: Automatic test case selection for regression testing of composite service based on extensible BPEL flow graph. J. Syst. Softw. 85(6), 1300–1324 (2012). https://doi.org/10.1016/j.jss.2012.01.036

16. Mei, L., et al.: A subsumption hierarchy of test case prioritization for composite services. IEEE Trans. Serv. Comput. 8(5), 658–673 (2014). https://doi.org/10.1109/TSC.2014.2331683

17. Wang, H., Xing, J., Yang, Q., Han, D., Zhang, X.: Modification impact analysis based test case prioritization for regression testing of service-oriented workflow applications. In: 2015 IEEE 39th Annual Computer Software and Applications Conference, vol. 2, pp. 288–297. IEEE, July 2015. https://doi.org/10.1109/COMPSAC.2015.11

18. Mei, L., Chan, W.K., Tse, T.H., Jiang, B., Zhai, K.: Preemptive regression testing of workflow-based web services. IEEE Trans. Serv. Comput. 8(5), 740–754 (2014). https://doi.org/10.1109/TSC.2014.2322621

19. Mei, L., Chan, W.K., Tse, T.H., Merkel, R.G.: Tag-based techniques for black-box test case prioritization for service testing. In: 2009 9th International Conference on Quality Software, pp. 21–30. IEEE, August 2009. https://doi.org/10.1109/QSIC.2009.12

20. Song, W., Ma, X., Cheung, S.C., Hu, H., Yang, Q., Lü, J.: Refactoring and publishing WS-BPEL processes to obtain more partners. In: 2011 IEEE International Conference on Web Services, pp. 129–136. IEEE, July 2011. https://doi.org/10.1109/ICWS.2011.12

21. Aho, A.V., Sethi, R., Ullman, J.D.: Compilers, principles, techniques. Addison Wesley 7(8), 9 (1986)

Test Case Generation of Composite Web Services Based on Semantic Matching and Condition Recognition

Haibo Zhang[1], Hui Xia[1], Xintang Lin[1], Liangjiang Yu[1],
Xiangyan Fang[1], Xuan Chen[1], and Hongquan Zhu[2(✉)]

[1] Wuhan Digital Engineering Institute, Wuhan 430074, China
work_lxt@126.com
[2] Department of Computer Science and Technology, Nanjing University,
Nanjing 210023, China
zhq98@foxmail.com

Abstract. With the rapid development of Internet technology, the application of Web service and their combinations is spreading widely as a vital role. However, due to the black box feature of Web services, test cases of Web service can be obtained only from the perspective of users, which is more challenging than traditional tests. We propose a method based on semantic matching and condition recognition, combined with document parsing to generate test cases of composite Web services. In this method, the first step is to parse the relevant XML documents, then to obtain the parameter types, keywords, as well as conditions and orders in the control flow. The following step is to match the parameter instances according to the conditions in the process or the semantic knowledge base. The last is to encapsulate them into test cases. Experimental results show that our method can generate test cases with low redundancy and high coverage, which can cover more than 85% nodes and paths in BPEL.

Keywords: Composite Web services · Test case generation · Document parsing · Semantic matching

1 Introduction

Web services [1] adopt a Service-Oriented Architecture (SOA) [2] to achieve service invocation through the interaction among entities such as service providers, registration centers and service requesters, without depending on language, platform and protocol. With the rapid development of Internet, cloud computing technology and service-oriented technology tend to mature, Web services have also been put in wide utilization. Developers combine several atomic Web services in accordance with a certain process to provide users with more comprehensive value-added services. It is necessary to test them fully to guarantee the accuracy and stability of the combined Web services.

The generation [3] of test cases is the first stage of composite Web service testing. This paper aims to achieve automated generation of test cases at a relative low cost and high coverage. In this paper, we propose a method to generate test cases for composite

© Springer Nature Switzerland AG 2020
G. Wang et al. (Eds.): WISA 2020, LNCS 12432, pp. 27–35, 2020.
https://doi.org/10.1007/978-3-030-60029-7_3

Web services based on semantic matching and condition recognition, and verify the effectiveness of this method through comparative experiments.

The rest of this paper is as follows. Related work is discussed in Sect. 2. Parsing XML documents related to Web services is in Sect. 3. How to generate test cases based on semantic matching and condition recognition is presented in Sect. 4. Experiments and evaluation are in Sect. 5. The last part is conclusion and future work.

2 Related Work

Test case generation is crucial for software testing, model-based test case generation is the most popular method [1, 4–8]. This method improves the accuracy of test cases, but few of them are fully automated without the testing of atomic Web services.

Zhou et al. [9] proposed a test case generation method based on parsing document and solving constraint, which parses XML documents related to composite Web Services. They obtain and encode constraint conditions, and then use the Z3-Solver [10] to solve the encoded constraints, and finally combine the SOAP protocol to generate test cases. However, the speed that Z3 processes string type variable is relatively slow.

Estero-Botaro et al. [11] proposed a genetic algorithm with some bacterial algorithm characteristics to generate a test suite for mutation testing. Experiment shows that the error detection rate and coverage of test cases rise greatly, but costs a lot of time.

In addition, Sun C et al. [12] and Mei et al. [13] studied BPEL runtime testing from the perspective of workflow, and proposed a scenario oriented testing framework for the runtime binding characteristic of composite Web services.

3 Document Parsing

Since service integrators cannot obtain the source code of Web service, test cases can only be generated with the information from interface documents [14]. This section will introduce the parsing method of XML documents in composite Web services in detail.

3.1 XML Documents in Web Services

XML documents related to Web services include BPEL, WSDL and XSD documents, which are respectively used to describe the business processes, interface information, and variable format information. The relationship is shown in Fig. 1 [9].

The basic unit of BPEL [15] is activity, which can be divided into basic activities, structured activities and fault handling activities. In addition to the above three types of activities, there are some nodes in the BPEL documents that describe the external service addresses, partner links, and variable declarations of Web service references.

WSDL document contains nodes: <service>, <binding>, <portType>, <message> and <types>, which are respectively used to describe the information of interface address, communication protocol, operation, message and variable type.

XSD document makes formal definition of the variables with XML format, whose composition structure is in the XSD part in Fig. 1.

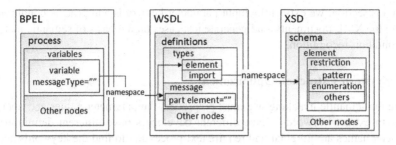

Fig. 1. Relationship of documents related to Web services

3.2 Analytical Method

When parsing the composite Web service document, we start with the BPEL document, where all the atomic Web services and business processes are defined. To construct the controlling flow graph, it starts from the root node <process> to all the nodes in turn, and stores the relationship between nodes and information of themselves. We obtain node information such as loop conditions and branch conditions, and then obtain child nodes iteratively.

Then, we parse WSDL documents. We first retrieve all the <service> elements, for which we jump to <binding> and the elements that match the service. Thus, we get a set of operations (<operation>), and then jump to <portType> to find the input and output <message> corresponding to each operation. Next, we parse the <message> and <types> nodes to obtain the constraint information. Further, it needs to obtain the namespace of the WSDL file for parsing the nodes and constructing test cases.

When parsing XSD documents, we still need to traverse all the nodes to obtain constraint conditions, such as the ordering relationship, values, types, lengths, value ranges, and regular expressions of variables. In addition, the variable definitions in the WSDL file need to parse the XSD file with the same namespace.

4 Test Case Generation Based on Semantic Matching and Condition Recognition

Former test cases generated based on interface parameter types have high redundancy, mainly because many invalid cases are generated based on the type information instead of taking the actual meaning of the parameters into account. For example, in the Web service of attribution query of domestic mobile phone numbers, when we input parameter named "mobileCode" with type "string", it is likely to generate a bunch of meaningless strings if we only judge by the type "string".

Likewise, we also need to generate test cases in accordance with various process conditions. For example, a loan service needs to deal with business processes according to the user's loan amount.

In this section, we extract the input parameter keywords and process conditions by parsing the XML documents, and generate test cases on the basis of semantic matching and condition recognition.

4.1 Keyword Extraction Optimization

Before the semantic matching, we must preprocess the parameter names and operation names to extract keywords. Since different developers have different definitions, the parameter names are irregular and require text processing to find the keywords that can explain the parameters most clearly. We propose a method to deal with the parameter names by analyzing the thousand parameter names in WSDL documents:

(1) Consider IO (input/output): Many inputs and outputs of the same service share the same forms, such as input in the form of "getAbyB". Obviously, B is the input keyword and A is the output keyword. For example, we can extract the keyword "Currency" in the service "currencyService" with an operation "getCurrencyRate".
(2) Separate Words: Considering the definition of parameter names without spaces, we separate these words according to the capitalized first letter.
(3) Filter meaningless words: After separation, some high-frequency but meaningless words should be ignored. The frequency of occurrences of such words in WSDL exceeds 90%, which are saved in a table of our database.

4.2 Building a Local Semantic Library

Building a local semantic library is to facilitate the semantic matching method. We combine both the automatic and manual method to construct the local semantic library:

(1) Automatic method: Present Web services sometimes provide interfaces for the access of specific values allowed by them. For example, a weather forecast service [16] offers weather forecast of 400 cities through the interface "getWeatherby-CityName". We call these interfaces for the legal input parameter value, tag them (the name of a city) with labels like weather, city, and then store into the database.
(2) Manual method: For various parameter types, we have designed a table with artificial maintenance. For example, for the type "double", 3.2, −983.2 and types like "dateTime", "int" etc. are stored in the database. With this table, the program only needs to parse the parameter type for the corresponding input instance.

4.3 Parameter Matching Based on the Semantic Library and DBpedia Domain Knowledge

As the example of currencyService (Sect. 4.1), an operation named "getCurrencyRate" queries the exchange rate between two currencies by inputting two currency codes with an output named "getCurrencyRateReturn". Through keyword extraction, the keyword is "Currency". Then we utilize the local library and DBpedia to match the keywords in the service elements. The parameter matching mechanism is as follows:

(1) Query the local semantic library according to the input parameter type for the suitable preset input instance, which should be listed to the instance table.

(2) Add the matching record to the instance table if it is in the local semantic library, according to the "keyword" after parameter optimization.

(3) If there's no instance to match, we call the DBpediaSpotlight service. We filter parameters by configuring and input the keyword and its description. Spotlight will respond content, which contains the DBpediaURI that matches the keyword, then extracts the string that can be used as an input parameter and adds it to the instance table.

4.4   Conditional Recognition

As mentioned in the previous section, only a part of the input values is obtained according to semantic matching, while others involve in controlling the BPEL process. Thus, we also need to consider the conditions (<condition>). For example, a parameter "amount" of a loan approval service requests users to input the loan amount. The program will estimate the loan risk based on the amount. For low-risk customers whose request is less than 10,000, the loan will be automatically approved. If it is more than 10,000, the approver will check the loan with an <if> node in the process. Based on the analysis result of the BPEL documents, we will identify the corresponding node for condition judgment and match it with the input parameters.

4.5   Test Case Packaging

SOAP, as a protocol specification for exchanging data, is often used to exchange structured and fixed information in a distributed environment. We need to encapsulate the test data in a SOAP message to generate test cases. A test case for mobile phone number location query service [17] is shown in Fig. 2.

```
1 ⊟ <soapenv:Envelope
       xmlns:soapenv="http://schemas.xmlsoap.org/soap/envelope/"
       xmlns:web="http://WebXml.com.cn/">
2       <soapenv:Header/>
3 ⊟     <soapenv:Body>
4 ⊟        <web:getMobileCodeInfo>
5              <web:mobileCode>18111111111</web:mobileCode>
6              <web:userID>12345</web:userID>
7           </web:getMobileCodeInfo>
8        </soapenv:Body>
9 </soapenv:Envelope>
```

Fig. 2. Mobile phone number location query service.

5 Experiment and Evaluation

5.1 Experimental Data and Environment

The hardware environment of the experiment is an Intel Core i5-6400 quad-core processor, 8G memory and the operating system is Windows 10. The development environment is jdk1.8 and Eclipse Oxygen.

We selected some typical composite Web services, including Loan Approval Service (LAS), BookLoan, FlowLinks, and CaculatorProj. These services involve various of activities in Sect. 3.1. The scale of the program is shown in Table 1. LOC represents the number of lines of BPEL, and AN (activities number) is the number of activities.

Table 1. Composite service information and experimental results.

Composite Web service	BPEL LOC	AN	Test case evaluation				
			Total	Redundancy	Node coverage	Path coverage	Time (ms)
CombineUrl	38	4	2	0	100%	100%	93
HelloWorld	45	4	2	0	100%	100%	107
While	82	8	2	0	100%	100%	120
Alarm	86	11	3	33%	100%	100%	142
CaculatorProj	151	13	3	0	100%	100%	97
LAS	204	14	7	40%	100%	100%	280
FlowLinks	92	22	4	33%	90%	86%	236
BookLoan	251	37	8	20%	89%	91%	302

5.2 Analysis of Results

In order to evaluate our method, we chose some common evaluation indicators, including coverage, redundancy, generation time, etc. Since the composite Web service was fully tested before the release, and there is no publicly known bug set, we have not considered the evaluation of the ability to search for bug of the test case.

In Table 1, the evaluation indicators of test case are as follows:

(1) Total: the number of all test cases generated in one execution;
(2) Redundancy: the proportion of test cases that overlap with nodes or paths covered by others;
(3) Node coverage: the proportion of nodes covered by the generated test case set;
(4) Path coverage: the proportion of paths covered by the generated test case set;
(5) Time (unit: milliseconds): the average time required to generate a test case.

The experimental results are shown in Table 1. We found that for small-scale composite Web services (the first four in Table 1), we can achieve 100% coverage with fewer test cases but almost no redundancy. For more complex services (the following four in Table 1), the coverage may decrease, but still with little redundancy and low time cost. Compared with other work, the number of test cases generated by our method is less, because generating by semantic matching is more practical, and the scope of parameters is constrained by condition recognition.

5.3 Comparative Experiment

To verify the effectiveness, we set three experiments: (a) remove semantic matching; (b) remove condition recognition; (c) remove both. The results are in Table 2.

Table 2. Comparison of node coverage.

Composite Web service	Result	Control group A	Control group B	Control group C
CombineUrl	100%	100%	100%	100%
HelloWorld	100%	100%	100%	100%
While	100%	50%	50%	0
Alarm	100%	100%	36%	36%
CaculatorProj	100%	100%	38%	38%
LAS	100%	60%	57%	35%
FlowLinks	90%	32%	41%	23%
BookLoan	89%	41%	32%	27%

By comparison, we found that when any one of the core points "semantic matching" and "conditional recognition" is removed, the effect of test case generation will shrink. When both are removed, similar to the traditional test case generation method based on interface parameter types, the coverage of test cases is sharply reduced. Therefore, using semantic matching and conditional recognition techniques is necessary (Table 3).

Table 3. Comparison of path coverage.

Composite Web service	Result	Control group A	Control group B	Control group C
CombineUrl	100%	100%	100%	100%
HelloWorld	100%	100%	100%	100%
While	100%	50%	0	0
Alarm	100%	100%	0	0
CaculatorProj	100%	100%	0	0
LAS	100%	66%	66%	33%
FlowLinks	86%	25%	50%	25%
BookLoan	91%	40%	40%	20%

6 Conclusion and Future Work

In this paper, we introduce a method for test case generation of composite Web service. This method parses XML documents to obtain semantics, constraints, and other information related to input parameters, overcoming the problem of high redundancy of test cases used to be merely based on interface parameter type.

In the future, we plan to incorporate mutation strategies into our test case generation, combined with fuzzing technology to strengthen the detection of service vulnerabilities and to improve coverage.

References

1. Yan, J., Li, Z., Yuan, Y., Sun, W., Zhang, J.: BPEL4WS unit testing: test case generation using a concurrent path analysis approach. In: 2006 17th International Symposium on Software Reliability Engineering, pp. 75–84. IEEE (2006). https://doi.org/10.1109/issre. 2006.16
2. Krafzig, D., Banke, K., Slama, D.: Enterprise S.O.A. Service-Oriented Architecture Best Practices. Prentice Hall PTR, Indiana (2004)
3. Wurth, E., Delpiroux, J.: Web service testing (Google Patents, 2018). http://www. freepatentsonline.com/y2016/0127409.html
4. Mei, L., Chan, W., Tse, T.: Data flow testing of service choreography. In: Proceedings of the 7th Joint Meeting of the European Software Engineering Conference and the ACM SIGSOFT Symposium on the Foundations of Software Engineering, pp. 151–160 (2009). https://doi.org/10.1145/1595696.1595720
5. Mei, L., Chan, W., Tse, T.: Data flow testing of service-oriented workflow applications. In: Proceedings of the 30th International Conference on Software Engineering, pp. 371–380 (2008). https://doi.org/10.1145/1368088.1368139
6. Hou, S.-S., Zhang, L., Lan, Q., Mei, H., Sun, J.-S.: Generating effective test sequences for BPEL testing. In: 2009 Ninth International Conference on Quality Software, pp. 331–340. IEEE (2009). https://doi.org/10.1109/qsic.2009.50
7. Wu, C.-S., Huang, C.-H.: The web services composition testing based on extended finite state machine and UML model. In: 2013 Fifth International Conference on Service Science and Innovation, pp. 215–222. IEEE (2013). https://doi.org/10.1109/icssi.2013.46
8. Ni, Y., et al.: Effective message-sequence generation for testing BPEL programs. IEEE Trans. Serv. Comput. **6**, 7–19 (2013). https://doi.org/10.1109/TSC.2011.22
9. Zhou, L., Xu, L., Xu, B., Yang, H.: Generating test cases for composite web services by parsing XML documents and solving constraints. In: 2015 IEEE 39th Annual Computer Software and Applications Conference (2015). https://doi.org/10.1109/compsac.2015.51
10. Zheng, Y., Zhang, X., Ganesh, V.: Z3-str: a z3-based string solver for web application analysis. In: Proceedings of the 2013 9th Joint Meeting on Foundations of Software Engineering, pp. 114–124 (2013). https://doi.org/10.1145/2491411.2491456
11. Estero-Botaro, A., García-Domínguez, A., Domínguez-Jiménez, J.J., Palomo-Lozano, F., Medina-Bulo, I.: A framework for genetic test-case generation for WS-BPEL compositions. In: Merayo, M.G., de Oca, E.M. (eds.) ICTSS 2014. LNCS, vol. 8763, pp. 1–16. Springer, Heidelberg (2014). https://doi.org/10.1007/978-3-662-44857-1_1

12. Sun, C., Zhao, Y., Pan, L., Liu, H., Chen, T.Y.: Automated testing of WS-BPEL service compositions: a scenario-oriented approach. IEEE Trans. Serv. Comput. **11**, 616–629 (2018). https://doi.org/10.1109/TSC.2015.2466572
13. Mei, L., et al.: A subsumption hierarchy of test case prioritization for composite services. IEEE Trans. Serv. Comput. **8**, 658–673 (2015). https://doi.org/10.1109/TSC.2014.2331683
14. Zhao, H., Chen, J., Xu, L.: Semantic web service discovery based on LDA clustering. In: Ni, W., Wang, X., Song, W., Li, Y. (eds.) WISA 2019. LNCS, vol. 11817, pp. 239–250. Springer, Cham (2019). https://doi.org/10.1007/978-3-030-30952-7_25
15. Jordan, D., et al.: Web services business process execution language version 2.0. OASIS Stand. **11**, 5 (2007). http://docs.oasis-open.org/wsbpel/2.0/OS/wsbpel-v2.0-OS.html
16. Weather Web Service. http://ws.webxml.com.cn/WebServices/WeatherWebService.asmx
17. MobileCode Web Service. http://ws.webxml.com.cn/WebServices/MobileCodeWS.asmx

An Integrated Optimization Approach for Production-Distribution Planning in Supply Chain

Lingjuan Hou[1(✉)], Chenchen Sun[2], and Zhijiang Hou[3]

[1] School of Management, Tianjin Normal University, Tianjin, China
lingjuan258@163.com
[2] School of Computer Science and Engineering,
Tianjin University of Technology, Tianjin, China
[3] Library, Tianjin University of Technology, Tianjin, China

Abstract. Under the new mode of green supply chain management, in order to overcome the trade-off problem caused by independent optimization, from the global perspective, coordinating forward logistics and reverse logistics effectively, integrating optimization the closed-loop structure of production and distribution planning is very important. In this paper, aiming at the integrated Production–Distribution–Recycle problem which consists of lot sizing, vehicle routing and recycling simultaneously, a mathematical model is established. In view of the complexity of the problem studied, a hybrid optimization algorithm with a combination of harmony search algorithm and genetic algorithm is proposed to solve the problem. Experiments are conducted on four different scale problems adopting integrated approach and decoupled approach respectively. Results show the superiority of integrated optimization approach.

Keywords: Supply chain management · Production–distribution–recycle · Stochastic programming · Harmony search · Genetic algorithm

1 Introduction

Supply chain management (SCM) is very important to the operation of enterprises. It includes a series of optimization topics [1]. Production and distribution are two very important parts in the supply chain system. In the new mode of supply chain management, in order to effectively coordinate production planning and distribution planning, from the overall point of view, the integrated optimization of production and distribution planning has attracted the attention of many researchers and enterprise managers. In particular, some research has been done on the model establishment and algorithm design of the integrated optimization of production and distribution in uncertain environment. An integrated optimization model coordinating production scheduling, demand allocation and transportation was established [2]. The production-distribution problem between suppliers and producers under the mode of centralized distribution center was studied, without considering the random factors [3]. The optimization of closed-loop supply chain network was studied including production,

© Springer Nature Switzerland AG 2020
G. Wang et al. (Eds.): WISA 2020, LNCS 12432, pp. 36–45, 2020.
https://doi.org/10.1007/978-3-030-60029-7_4

distribution and recycling, and a cultural gene algorithm was proposed, but without considering both the random factors and distribution routing [4]. Integrated production-distribution model was studied [5].

In the existing literature, either the random factors are ignored, or the distribution process is simplified too much, and the vehicle routing problem (VRP) in the distribution plan is not considered [6]. Some research is limited to a single enterprise or single production system [7]. Especially the impact of the recycling link in reverse logistics on the original production distribution system is not considered. Therefore, the research on integration of forward logistics and reverse logistics, the integrated optimization of production, distribution and recycling is still limited.

In view of the above situation, this paper considers a supply chain system with multiple periods, single product, single factory and multiple distributors. Taking minimizing the total cost of the system as the optimization objective, a stochastic programming model with chance constraints is established. In the model, the production lot size problem and the vehicle routing problem of pickup and delivery are integrated as a whole. A hybrid optimization algorithm based on harmony search (HS) and genetic algorithm (GA) is designed to solve the problem. The simulation analysis highlights the superiority of the integrated optimization strategy.

2 Problem Description and Model Establishment

This paper aims at the integrated optimization of SC system under the structure of production, distribution, recycling and redistribution under the uncertain demand. The assumption is as follows. Products are produced in batches in the factory, and delivered to the distributors after the completion of production in each period, regardless of the inventory holding cost of the factory. The factory has a group of vehicles of the same size, and the capacity of the vehicles can serve at least one distributor. The product demand of each distributor in any period is random and subject to normal distribution. Products in each period are allowed to be out of stock. The reworked products of the distributors are not allowed to be used directly, and can only be returned to the market after being transported back to the factory for maintenance. Each vehicle delivers goods to each distributor and pickups goods from the distributor at the same time. Each distributor only accepts one service in each period. The factory has production capacity constraints, and the distributors have inventory capacity constraints.

For the convenience of expression, the following symbol system is established: N denotes the number of distributors. I denotes the set of factories and distributors, 0 denotes the set of factories. K denotes the number of vehicles owned by factories, C denotes the capacity of each vehicle. T denotes the set of periods in the planning cycle. DC_i denotes the maximum inventory handling capacity of distributors i in each time period. α denotes the level of customer demand satisfaction. Pc denotes the production cost of unit product. r denotes the maintenance cost rate of unit product, in which, $0 < r < 1$. Mc denotes the maintenance cost of unit product, in which, $Mc = rPc$. Sc_i denotes the unit inventory cost of distributor i in each period, Lc_i denotes the unit out of stock cost of distributor i in each period. C_{ij} denotes the driving cost from position i to position j, and d_{it} denotes the demand of distributor i in t period, in which, $d_{it} \sim N(\mu_{it}, \sigma_{it}^2)$, λ is the product failure rate. p_{it} denotes the quantity of products to be repaired at distributor i in period t, in which: $p_{it} = \lambda q_{i(t-1)}$. u_{it} denotes the total quantity of products to be distributed to distributor i in period t, in which, $u_{it} = q_{it} + p_{i(t-1)}$. S_{it} denotes the product quantity at the end of period t of distributor i. AC denotes the maximum production capacity of the factory in each period, and ac is the production capacity of unit product consumption. $r \times ac$ denotes the production capacity of unit product maintenance consumption, which is consistent with the cost relationship.

Decision variables include:

x_t: Production batch in period t
q_{it}: Number of new products delivered to distributor i in period t
z_{ijkt}: Whether the vehicle k passes through the arc(i, j) in period t
y_{ikt}: Whether the vehicle k serves the distributor i in period t.

Objective function

$$C = Total\ cost = \sum_{t=1}^{T} [x_t Pc + Mc \cdot \sum_{i=1}^{N} p_{it} + \sum_{k=1}^{K} \sum_{i=1}^{N} \sum_{j=1}^{N} C_{ij} z_{ijkt}]$$
$$+ \sum_{t=1}^{T} [\tfrac{1}{2} \sum_{i=1}^{N} Sc_i \cdot (\boldsymbol{max}(S_{i(t-1)}, 0) + \boldsymbol{max}(S_{it}, 0)) + \sum_{i=1}^{N} Lc_i \cdot |\boldsymbol{min}(S_{it}, 0)|]$$

The opportunity constraint model of this problem is as follows:

$$Min\ Total\ cost \tag{1}$$

$$s.t. \quad \sum_{i=1}^{N} q_{it} = x_t \ \forall\ t \tag{2}$$

$$x_t + r \cdot \lambda \cdot x_{(t-1)} \leq \frac{AC}{ac} \quad \forall\ t \tag{3}$$

$$S_{it} = S_{i(t-1)} + q_{it} - d_{it} - p_{it} \quad \forall\ t, i \neq 0 \tag{4}$$

$$S_{i(t-1)} + q_{it} \leq DC_i \ \forall i \neq 0 \tag{5}$$

$$prob\{\sum_{t=1}^{T} q_{it} - \sum_{t=1}^{T} d_{it} \geq 0\} \geq \alpha \quad \forall i \neq 0 \tag{6}$$

$$\sum_{k=1}^{K} y_{ikt} = 1 \quad \forall i, t \neq 0 \tag{7}$$

$$\sum_{i=0}^{N} z_{ilkt} = \sum_{j=0}^{N} z_{ljkt} \quad \forall l, k, t \neq 0 \tag{8}$$

$$\sum_{i=1}^{N} z_{0ikt} + \sum_{i=1}^{N} z_{i0kt} = 2y_{0kt} \quad \forall k, t \neq 0 \tag{9}$$

$$\sum_{i=0}^{N} z_{ilkt} + \sum_{j=0}^{N} z_{ljkt} = 2y_{lkt} \quad \forall l, k, t \neq 0 \tag{10}$$

$$\sum_{i=0}^{\theta} p_{it} y_{ikt} + \sum_{i=\theta+1}^{N} u_{it} y_{ikt} \leq C \quad \forall \theta = 0, 1, \cdots, N; k, t \neq 0 \tag{11}$$

$$y_{ikt}, z_{ijkt} \in \{0, 1\}, q_{it} \geq 0, x_t \geq 0 \quad \forall i, j, k, t \tag{12}$$

The objective is to minimize the total cost. (2) denotes the balance formula of production and sales of the factory. (3) denotes the constraint of production capacity. (4) denotes the inventory balance formula of the distributor. (5) denotes the constraint of the inventory capacity. (6) ensures that the demand meets the level. (7–11) are the constraints of VRPSPD. (12) gives the range of the decision variables.

3 Algorithm Design

The model includes the continuous production batch problem and the discrete vehicle routing problem. In view of the successful application of GA and HS in various optimization problems [8, 9], in particular, HS is suitable for solving production batch problems; GA is suitable for optimizing vehicle distribution path. Therefore, this paper proposes a hybrid intelligent optimization algorithm based on HS and GA. The algorithm flow is shown in Fig. 1. Each harmony solution obtained by HS represents a production batch plan. For each batch plan, GA is used to get the best distribution scheme. This cycle continues until the algorithm is terminated.

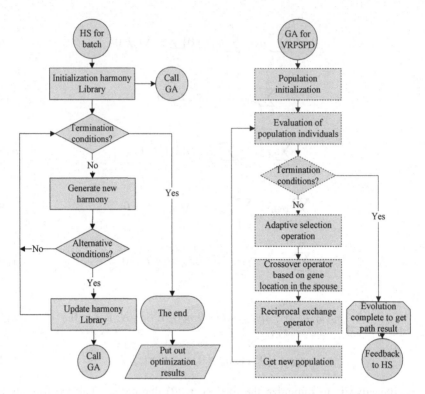

Fig. 1. Flow chart of HS-GA algorithm

3.1   Encoding Method of Solution

The solution of batch part can be expressed as $S = [s_t]_T$, T is the number of time periods, and s_t is the production batch of the period t. Generation method of initial batch:

Step 1. Let μ_{it} represent the mathematical expectation of d_{it}. S_t is randomly generated in $[0.75 \sum_i \mu_{it}, 1.25 \sum_i \mu_{it}]$

Step 2. Repair S to make the total planned batch equal to the total expected demand. Repair the batch plan of each period according to formula (13), in which $\delta = \sum_{t=1}^{T} \left(s_t - \sum_{i=1}^{N} \mu_{it} \right) \bigg/ T$.

$$s_t^* = \begin{cases} max\{s_t - \delta, 0\}, & t < T \\ max\{\sum_t \sum_i \mu_{it} - \sum_{t=1}^{T-1} s_t^*, 0\}, & t = T \end{cases} \quad (13)$$

A route can be encoded as a chromosome $(i_{11}, i_{12}, ..., i_{1s}; 0, i_{21}, ..., i_{2t}; ...; 0, i_{l1}, ..., i_{lw})$, in which 0 denotes the depot, i_{lk} denotes the distributor number in route k.

3.2 Fitness Function

Constraint (6) is an opportunity constraint condition with random variables. Firstly, transform it into the deterministic equivalent representation as formula (14). Using the penalty function method, the penalty coefficient c is introduced, and constraint (6) is included in the objective function Z as formula (15).

$$\sum_{t=1}^{T} q_{it} \geq \Phi^{-1}(\alpha) \sqrt{\sum_{t=1}^{T} \sigma_{it}^2 + \sum_{t=1}^{T} \mu_{it}}, \quad \forall i \neq 0 \tag{14}$$

$$Z = Total \ \cos t + c \sum_{i=1}^{N} \max \left\{ \Phi^{-1}(\alpha) \sqrt{\sum_{t=1}^{T} \sigma_{it}^2 + \sum_{t=1}^{T} \mu_{it}} - \sum_{t=1}^{T} q_{it}, \ 0 \right\} \tag{15}$$

At last, transform Z into fitness function: $f_l = bZ'/Z_l$, where, f_l denotes the fitness value of chromosome l, Z' denotes the best objective value corresponding to chromosome in initial population, Z_l denotes the objective value of chromosome l.

3.3 HS for Batch Optimization

The steps of HS in this paper are as follows:

Step 1. Select the algorithm parameters and generate the initial harmony memory randomly. The harmony memory is expressed as the set $HM = [x_n^i]_{HMS \times N}$, the element x_n^i represents the nth component of the ith harmony solution. HMS represents the scale of the memory. N is the dimension of the solution, in this paper, the dimension of the batch solution is the number of production planning periods.

Step 2. Generate a new harmonic solution $(x_1', x_2', \cdots, x_N')$, in which each component x_n' is obtained according to the formula (16).

$$x_n' = \begin{cases} x_n^{rand} & 1 - P_{HM} \\ x_n^{HM} \in \{x_n^1, x_n^2, \cdots x_n^{HMS}\} & P_{HM} \times (1 - P_{pitch}) \\ x_n^{HM} + b_{range} \times \varepsilon & P_{HM} \times P_{pitch} \end{cases} \tag{16}$$

In the formula, P_{HM} is the probability of selecting a component from the harmony memory. $1 - P_{HM}$ represents the probability of randomly generating the component. P_{pitch} is a probability of variation of components obtained from harmony memory, i.e. harmony adjustment rate. b_{range} represents the harmonic adjustment bandwidth. ε is a random variable that obeys the uniform distribution on $[-1, 1]$.

Step 3. Update the harmony memory.

Step 4. Repeat step 2 and 3 until termination conditions are met. End algorithm.

3.4 GA for VRPSPD

In this paper, an improved genetic algorithm with adaptive selection mechanism is used to optimize the vehicle routing.

Adaptive Selection Mechanism. In order to overcome the premature convergence of the algorithm, the adaptive selection mechanism is introduced to adjust the individual selection probability [8].

Crossover Operator. A crossover operator based on gene location in the spouse [8] presented in this paper is as follows: firstly, two parents are selected randomly from groups; secondly, two crossover points are selected randomly from the first parent and the segment between two points copied to the first offspring; thirdly, other genes are written to the first offspring according to the order in the second parent. Similarly can the second offspring be obtained. An example of the crossover method used is the following:

Before   A : 5 - 3 - **4** - **2** - **1** - **9** - **10** - 7 - 6 - 8   B : 4 - 2 - **7** - **6** - **5** - **1** - **8** - 3 - 10 - 9
After   A : 7 - 6 - **4** - **2** - **1** - **9** - **10** - 5 - 8 - 3   B : 3 - 4 - **7** - **6** - **5** - **1** - **8** - 2 - 9 - 10

Reciprocal Exchange Operator. Two positions are selected randomly and the genes on the position are swapped.

4 Computational Results and Analysis

In order to test the algorithm, four groups data with different scales are designed, they (distributors × periods) are 10×4, 20×4, 10×8, 20×8 respectively. The model parameters are set as follows: The location coordinates of factories and distributors are randomly generated in the area $(-50, 50) \times (-50, 50)$. $d_{it} \sim N(\mu_{it}, \sigma_{it}^2)$, the expectation and variance are generated between 50–100 and 5–10, respectively. Pc is generated between 0.3–0.5. Sc_i of distributor i in each period is generated between 1.5–2.0. Lc_i of distributor i is generated between 1.5–2.0. λ is 10%. Customer demand satisfaction level is 90%. The vehicle capacity is $C = 1.5 \times \left(\sum_{i \neq 0} \sum_{t > 0} \mu_{it} \middle/ T \times K \right)$.

The algorithm parameter settings of HS and GA are shown in Table 1, evolution generation (G_{max}) is 100, population size (NP) is 100.

Table 1. The algorithm parameters.

MaxTrial	MaxImp	HMS	P_{HM}	P_{pitch}	b_{range}	G_{max}	NP	$P_{crossover}$	$P_{mutation}$
50	100	5	0.85	0.3	5	100	20	0.855	0.055

Experiments are carried out with two strategies of independent optimization and integrated optimization respectively. The independent optimization strategy is to optimize the production batch first, then optimize the distribution path according to the final batch plan. The integrated optimization strategy is to optimize the production batch and distribution path in coordination. GA optimizes each path solution, and then the path results are fed back to HS. The optimization results of 4 different scale problems under two optimization strategies are shown in Table 2.

Table 2. Results comparison of different scale problems under two optimization strategies.

		Total cost	Production cost	Distribution cost	Inventory cost	Shortage cost
10 × 4	Independent optimization	141491.6	51033.6	72722.7	7001.5	10733.8
	Integrated optimization	132105.5	48754.8	68308.2	6600	8442.5
	Improvement rate	6.63%	4.47%	6.07%	5.73%	**21.35%**
20 × 4	Independent optimization	262977.4	100994.4	126802.8	13176.9	22003.3
	Integrated optimization	249329.7	98480.4	118935	13076.8	18837.5
	Improvement rate	5.19%	2.49%	6.20%	0.76%	**14.39%**
10 × 8	Independent optimization	317988	104797.2	159173.1	21267.4	32750.3
	Integrated optimization	301628.9	102868.8	152123.4	20716.3	25920.4
	Improvement rate	5.14%	1.84%	4.43%	2.59%	**20.85%**
20 × 8	Independent optimization	590292.7	206106	276459.3	41108.1	66619.3
	Integrated optimization	565378.8	204723.6	262566	41218.1	56871.1
	Improvement rate	4.22%	0.67%	5.03%	-0.27%	**14.63%**

From Table 2, we can see that, compared with the independent optimization algorithm, the integrated optimization algorithm reduces the total system cost by 6.63%, 5.19%, 5.14% and 4.22% respectively. The improvement of the distribution cost and the shortage cost is very significant. Shortage cost has been reduced by 21.35%, 14.39%, 20.85% and 14.63% respectively, which shows that though the improvement of the total cost of the system is not great, shortage cost has been greatly reduced, that is, the customer's demand has been met to a greater extent, which can greatly improve customer satisfaction. Distribution cost decreased by 6.07%, 6.20%, 4.43% and 5.03% respectively. With the increase of the number of distributors, the improvement rate of distribution cost also increased. This is because, through coordination production batch and distribution plan, the distribution plan can be reasonably arranged according to production batch in real time, the vehicle loading rate can be improved.

To sum up, the integrated optimization can meet the customer demand to the greatest extent and improve the logistics service level of the enterprise on the basis of reducing the total cost of the system. This is the value of integrated management of logistics system: through optimizing the allocation of logistics resources, integrating the logistics system, strengthening the coordination among the elements of the system, and achieving the overall benefits of logistics system integration.

5   Conclusion

In this paper, the integrated optimization problem of production distribution recycling and distribution under uncertain demand is studied. A stochastic programming model considering both forward logistics and reverse logistics with opportunity constraints is established. A hybrid optimization algorithm based on HS and GA is proposed. Results of numerical experiments show that the integrated optimization can effectively reduce the operating costs of enterprises. With the increase of the problem scale and the improvement of service level, the value of integration optimization strategy increases correspondingly.

Acknowledgement. This work is supported by the Tianjin Philosophy Social Science Research General Project (No. TJGL16-032).

References

1. Maravelias, C.T., Sung, C.: Integration of production planning and scheduling: overview, challenges and opportunities. Comput. Chem. Eng. **33**(12), 1919–1930 (2009)
2. Chan, F.T.S., Chung, S.H.: Multi criterion genetic optimization for due date assigned distribution network problems. Decis. Support Syst. **39**, 661–675 (2005)
3. Ma, S.-H., Gong, F.-M., Liu, F.-H.: Production and distribution collaborative decision-making based on supply hub. Comput. Integr. Manuf. Syst. **14**(12), 2421–2430 (2008)
4. Pishvaee, M.S., Farahani, R.Z., Dullaert, W.: A memetic algorithm for bi-objective integrated forward/reverse logistics network design. Comput. Oper. Res. **37**, 1100–1112 (2010)
5. Li, Q.-P., Liu, Y.: An optimal policy for a single-vendor-single-buyer integrated production-distribution model for deteriotating items. Oper. Res. Manag. Sci. **28**(01), 90–97 (2019)
6. Chen, C.-P., Han, S.-J., Lu, J.-S., Chen, Q.-F.: A multi-chromosome genetic algorithm for multi-depot and multi-type vehicle routing problems. China Mech. Eng. **29**(02), 218–223 (2018)
7. Grossmann, I.E.: Challenges in the application of mathematical programming in the enterprise-wide optimization of process industries. Theor. Found. Chem. Eng. **48**(5), 555–573 (2014)
8. Hou, L.-J., Zhou, H., Liang, C.-H.: Vehicle routing problem with uncertain demand and travel time. Comput. Integr. Manuf. Syst. **17**(1), 101–108 (2011)

9. Wang, J.-F., Yu, J.-Q.: Order picking optimization of warehouses based on insect intelligent algorithm. Ind. Control Comput. **32**(03), 61–63 (2019)
10. Yu, J., An, Y., Xu, T., Gao, J., Zhao, M., Yu, M.: Product recommendation method based on sentiment analysis. In: Meng, X., Li, R., Wang, K., Niu, B., Wang, X., Zhao, G. (eds.) Web Information Systems and Applications. WISA 2018. Lecture Notes in Computer Science, vol. 11242. Springer, Cham (2018). https://doi.org/10.1007/978-3-030-02934-0_45

Extraction and Portrait of Knowledge Points for Open Learning Resources

Jian Yu[1,2,3], Tingxu Jiang[1,2,3], Tianyi Xu[1,2,3], Jie Gao[1,2,3], Jun Chen[4],
Mei Yu[1,2,3], and Mankun Zhao[1,2,3](✉)

[1] College of Intelligence and Computing, Tianjin University, Tianjin, China
{yujian,jiangtingxu,tianyi.xu,gaojie,yumei,zmk}@tju.edu.cn
[2] Tianjin Key Laboratory of Advanced Networking (TANK Lab), Tianjin, China
[3] Tianjin Key Laboratory of Cognitive Computing and Application, Tianjin, China
[4] School of Information Science and Technology,
University of International Relations, Beijing, China
chen_jun05@sina.com

Abstract. This article explores how to use the technology of text summarization and keyword extraction to automatically extract key knowledge points from massive educational resources and use open resources to generate feature portraits of relevant knowledge points. Specifically, this article takes the field of programming competitions as an example, firstly, crawl the problem solution resources of program design related issues, use data preprocessing to clean the data, then, use unsupervised extraction models based on Bert and centrality to summarize the documents of the resources, the LDA model is used to extract keywords from the generated document summary to identify relevant knowledge points in the resource. Finally, crawl and analyze resources based on knowledge points to establish relevant feature portraits for knowledge points. Unlike manual analysis of resources, this method can automatically select candidate knowledge points, greatly reduce labor costs.

Keywords: Open educational resources · Text summarization · Key words · Knowledge points · Feature portraits

1 Introduction

With the advent of the information age and the rapid popularization of network technology, learning and communication on the Internet has become an indispensable and important part of daily life. News, blogs, and forums are all very mainstream sources of information. The acquisition of a large number of learning resources on the Internet is simple, but the sharp contrast is that the sorting of related knowledge points relies on manual sorting, and often requires people with rich knowledge in the field to sort out. Such work is undoubtedly time-consuming and laborious.

© Springer Nature Switzerland AG 2020
G. Wang et al. (Eds.): WISA 2020, LNCS 12432, pp. 46–56, 2020.
https://doi.org/10.1007/978-3-030-60029-7_5

How to extract useful information from a large number of lengthy open resources, the text summarization is very suitable for this scenario. The text summarization is to keep the most important information in the source document as possible while generating a short version of the source document [1]. Relying on the gradual enrichment of corpora and the continuous advancement of neural network technology [2], the effect of text summarization is increasing year by year. The main methods of text summarization can be roughly divided into two categories, extractive and generative [3]. Extractive text summarization extract the sentences or words in the original document according to a given number and through a certain algorithm, and extract them. Generative summaries are often obtained through neural networks to obtain deep information of the source document, generate new documents, and can generate new words to replace the words in the original document [1]. At present, the extractive and generative methods cannot distinguish between the better and the worse. The results of the extractive method are often more coherent and easy to read, but the disadvantage is that the information compression may not be comprehensive, which makes the abstract not concise. The possible shortcoming of the generative text summary is that the content is not smooth enough. When faced with OOV (out-of-vocabulary words) words, the processing of the generative model is often not ideal, and there will be cases of ignorance or inaccurate replacement, resulting in practical application is more difficult. It also depends on the data set, and different scenarios of the data set have their own efficiency.

Keyword extraction technology is also an important means to obtain key information. If the keywords in the document can be well extracted, then there is no doubt that the efficiency of reading the document can be improved. Keyword extraction techniques can be roughly divided into three categories, including supervised, unsupervised, and semi-supervised methods [4]. Some commonly used algorithms are TF-IDF, Text-rank, and Topic model, which TF-IDF [5] has good results in most application scenarios.

In this paper, Our goal is to automatically extract knowledge points from a large number of learning resources. In terms of models, consider using abstract and keyword extraction techniques to obtain knowledge points. Summarizing open learning resources can extract important information in the text, thereby facilitate the extraction of keywords and reduce interference. We carefully design a set of processes to extract knowledge points from open learning resources. First, we use the crawler architecture to crawl network resources, and use the PacSum model [2] with the versatility advantage to extract the part of the important information of network resources. Finally, we use the LDA model to extract keywords to generate knowledge points. According to the generated knowledge points, we use the obtained information to generate customized feature portraits, and excavate a rich information chain.

2 Background

Scrapy is a crawler framework for crawling data and extracting structured data written based on pure Python. It uses the Twisted framework to implement automatic asynchronous processing in the background. It does not require users to write asynchronous processing code. Several important components, the components are engine, scheduler, downloader, crawler, physical pipeline, crawler middleware and download middleware. Scrapy-splash is to encapsulate the splash in the scrapy framework, users can easily use splash in the scrapy framework. Why use splash? Some simple web pages can be processed in the scrapy framework, but it seems to be unable to deal with complex front-end web pages. Because many web pages use Javascript rendering, ordinary crawlers cannot normally obtain the content that the page actually sees, and splash can return the rendered page.

The methods of extractive summarization are achieved by selecting important parts (usually sentences) in the article [6]. Before the population of deep learning, traditional methods use word frequency and corpus information, linguistic rhetorical structure, traditional machine learning, etc., and achieved good results. The simpler method is TF-IDF. This method was proposed by Salon, using the information provided by the corpus to consider word frequency and inverse document frequency as elements [5], Textrank is another graph-based sorting algorithm [7], which treats all sentences in an article as nodes, the method of calculating the weight of the edge is to calculate the similarity between the node and the node, and iteratively calculate the importance of each node by referring to the method of Pagerank [8]. An important help of deep learning in summarization is to obtain word vectors in sentences. Sentence representation based on deep learning can better mine word-level information. In the extractive model using neural network, most of the text summaries are converted into sentence classification problems. The SummaRuNNer model can be visualized due to multiple features such as novelty and saliency [9]. It has simple and interpretable characteristics. It is the earliest batch of models that use RNN as a model encoder. It incorporates the training methods of the generative model and provides ideas for training the model. Rouge [10] is a commonly used scoring rule in text summary evaluation. The REFRESH model optimizes the Rouge score globally through reinforcement learning to optimize the summary model [11], the BERTSUMEXT [12] model uses the latest Bert [13] technology. Bert can grasp the semantics of the document and express the sentences in the document, this article emphasizes the importance of Bert model encoding for text summarization, and the paper applies Bert to both extractive and abstract tasks at the same time, and has achieved very good results. In the paper of NEUSUM [14], the abstract selection and scoring methods in the article are merged together to prevent the separation of the two in the previous work, and the goal of the training function is to maximize the Rouge score. In addition to the extractive model for label prediction, there are also extractive models that incorporate deep learning Bert and traditional graph-based sorting algorithms, which is PacSum [2], under the influence of Textrank and other graph-based sorting algorithms, reconsidering the calculation method of sentence centrality.

The extractive summarization focuses on the selection of important sentences in the text, while the abstractive summarization is more similar to human thinking. After reading the full text, it is summarized according to the semantics of the document. It has the characteristics of abstract overview and new word replacement. The PTGEN [15] paper is originally the Seq2Seq model is improved, and a pointer generation network is designed. The pointer part allows the network to directly copy words from the original document, while the generation part retains the function of replacing new words with the network. BERTSUMABS [12] is different from the previous work using RNN, the model uses Transformer to solve the problem of generative summary. The encoder part is the same as the encoder in the previously described BERTSUMEXT, using the improved version of Bert as the encoder.

Keyword extraction technology relied on humans in the early, requiring manual annotation by people with rich domain knowledge to expand the keyword library. TF-IDF has an important meaning in keyword extraction technology, and it relies on word frequency and inverse document frequency to sort the importance of words. This method relies on the construction of a corpus and has good results in most common. Because the method is simple and fast, it can be used as a baseline to judge the performance of other models. At the same time, the shortcomings of this method are also obvious, relying too much on statistics, ignoring the situation that the word frequency is less but equally important. The other direction of keyword extraction is subject-oriented, and the notable representatives are the PLSA [16] and LDA [17] models. The idea of this type of model is based on the consideration of people when writing an article. When writing an article, you need to determine several topics. After determining the topic, consider the relevant words based on each topic.

3   Related Work

The automatic summarization of single documents has gained extensive attention in recent years due to the availability of large-scale corpora and the continuous renovation of deep learning models, but it is unrealistic to expect high quality corpus training sets for different languages and different summary types [2].

Therefore, PacSum [2] attempts to solve the above problems from the traditional graph-based sorting abstraction method. First of all, in view of the problem that traditional graph-based sorting algorithms treat graphs as undirected graphs, the paper was inspired from the viewpoint of rhetorical structure theory RST and modified undirected graphs in traditional algorithms to directed graphs, which means that the centrality of the two sentences before and after will be affected by the position, and the similarity between each other is different. The centrality score can be defined as follows:

$$centrality(s_i) = \lambda_1 \sum_{j<i} e_{ij} + \lambda_2 \sum_{j>i} e_{ij}. \tag{1}$$

λ_1, λ_2 are different weights for forward-looking and backward-looking directed edges. e_{ij} is the similarity score between sentences s_i and s_j.

How to express the sentence well, this problem is chosen to be solved with Bert in PacSum. As the best neural network representation learning model, Bert has a very good effect on obtaining sentence semantics.

The LDA topic generation model is a document topic generation model proposed by Blei et al. [17], which is essentially a three-layer probability model based on the Bayesian principle. The LDA model can use the co-occurrence characteristics of words to mine text topics without knowing the relevant background knowledge of the text. The premise of the LDA model is that an article is composed of several topics, and a topic is composed of several words. In this way, the process of generating a document can be divided into two parts, from the document to the topic generation process and from the topic to the word generation process, both of which are based on probability distribution.

4 Resource Crawling

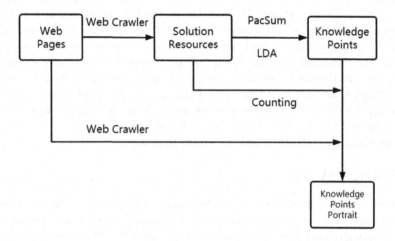

Fig. 1. Process for generating feature portraits of knowledge points.

The crawling goal of this article is problem-solving resources. As POJ (pku judgeonline) is an important domestic problem-solving website, it has a large influence, a large number of people who solve problems, and related problem-solving resources are also relatively rich (Fig. 1 shows how to obtain a feature portrait of knowledge points through network resources).

Perform Baidu search based on the POJ title number, select the content in the specified web page to crawl. From the question bank according to POJ, from question number 1000 to question number 4053, using "POJ" and the question number as a keyword, crawling the web resource returned by Baidu, because the time to render the page and obtain the page resource is not stable, it depends on when the actual crawling code is running, the network congestion is good or bad, so need to set the wait time so that the splash has enough time to render the page. If the time is too short, the content of the returned page may be incorrect. In the returned rendered page, we filter the page according to the url in the response, and reserve the website of a specific domain name. The content structure of each type of web page is different, so we choose those more important sites and crawl them according to the site structure.

5   Knowledge Point Extraction

For the crawled open learning resources, preprocessing is important, because the crawled content may not meet the expected requirements, such as the crawled text content is empty, or there is no solution information about the relevant topic in the text content, so in the document it will interfere with the extraction of important sentences and knowledge points. At the same time, because the method used in this article is an extractive text summary, it is necessary to perform sentence processing on the content. After preprocessing the resources, the PacSum method will be used to extract the document sentences.

According to the content of the article it is cleaned, the content with empty text or meaningless part is removed, and the article is sentence-processed, each article is processed into a list, and each element in the list is a sentence of the article. Then process each sentence and divide it into units of characters.

The first step of PacSum is to get the Bert pre-trained model. In the paper, the author uses the news training set as an input to fine-tune the Bert model, and then uses this model as a pre-trained language model to express sentences, but this method is not applicable in web blogs, because the structure of web blog resources is more scattered, the content is written more casually, and because it is a solution, the blog is full of code and comments. Therefore, if the method in PacSum is used, Bert cannot grasp the deep semantics well. In this paper, the Bert model [18] using the whole word masking technology and a large amount of data training is used. This technology is based on an upgraded version of Bert released by Google. This technology uses the training generation sample generation strategy to change to train the model, the original Bert training strategy is to mask based on some Chinese characters in the word, and this method performs Mask on all words in a word, so that the pre-trained language model can better express the deep semantics of the sentence. The Chinese document data used by the model is more extensive, including Chinese Wikipedia, and other data sources, such as encyclopedia, news, Q&A, etc. The total size of other data sources is about 10G, and the number of words is about 5.4B. Experiments show that this type of model has an improvement on the downstream natural language processing tasks compared to the original Bert model, so this model is more suitable for the task requirements of this paper.

After obtaining the Bert pre-trained model, import the text resources and first calculate the most suitable hyper-parameters. Because the crawled blog resources do not have relevant abstracts as a match, the NLPCC training set with a text style different from the general news writing style is selected. Conventional news is more inclined to put abstract sentences in the first few sentences of the article, the first three sentences are often used as baseline to get a good score. However, due to the influence of its own structure, the sentence that appears in the front of the text is often not the body of the article, so it does not have the writing characteristics of general news, so the hyper-parameters learned from the NLPCC data set are used as parameters when extracting blog resources. After performing the Pre-training processing on the NLPCC data set, the PacSum is used to search for the optimal hyper-parameter settings, and the resulting parameter $(\beta, \lambda_1, \lambda_2)$ is $(0.8, 0.2, 0.8)$, which is in agreement with the conclusion of the original paper that λ_1 tends to be negative. Apply this parameter as a hyper-parameter to the crawl resource dataset.

After completing the document summary of the resource, the LDA theme model is used to extract keywords using the generated document summary. During the keyword extraction process of the LDA theme model, a stop word list is first introduced to filter a large number of meaningless words. Secondly, jieba word segmentation is used to perform word segmentation on the document content. Since the processing of LDA is based on word units, the quality of the word segmentation tool directly affects the effect of document keyword extraction. We introduced the latest vocabulary as a dictionary supplement to make to achieve better results in word segmentation, run the LDA model multiple times. Because the results of each run are different, the union of the set of run results is selected as the final judged set.

6   Knowledge Point Feature Portrait

In view of the selected knowledge points, this article designed several attributes to describe the feature portraits of knowledge points: commonness, related knowledge points, encyclopedia abstracts, and encyclopedia features, of which the last two points use knowledge points as keywords from the encyclopedia Crawl relevant content as attributes of knowledge points.

According to the presence or absence of knowledge points in the original resource, we can link resources to knowledge points, and count the frequency of a single knowledge point in all resources, we divide the commonness of knowledge points according to the threshold. This article chooses to divide the commonness of knowledge points into four categories, then use the knowledge points as keywords to crawl the top three pages of Baidu search. Similarly, perform domain name filtering, leaving certain reserved web pages such as Baidu Encyclopedia, extract the text content of the web pages, and count whether these web pages have the knowledge points, in this way, the remaining knowledge points associated with the current knowledge point can be counted. The source of the encyclopedia abstract and encyclopedia features is the content part of Baidu encyclopedia, and it also shows a brief introduction related to the knowledge points collected manually and some of its own attributes, such as the English name and the subject area to which it belongs.

7   Experiment

7.1   Datasets

This article uses the NYT, CNN/DM, NLPCC competition 2017 text summary data set and the blog data set we obtained by crawling blog resources. The NYT dataset is called the New York Times Labeling Training Set. The content is the New York Times content and the corresponding artificial summary. CNN/DM is extracted from the US Daily Mail and Cable News Network. The content is also for news. NYT and CNN/DM datasets are English datasets, which are reproduced using the method in the article to view the results and verify the summary effect of the PacSum method.

The NLPCC data set is used to approximate the setting of hyperparameters. Finally, the blog resource data set crawled in this article is used to extract the text summary of the data set using the PacSum method. More than 15000 blog resources are crawled using scrapy and splash. The storage format is stored in json format. The resource includes four keywords: id, content, number, and url.

7.2   Experimental Results

Table 1. Part of the knowledge point extraction results and corresponding related knowledge points

Problem number	Knowledge points	Related points
1007	Sort	Heap, tree, queue, list
2020	Dynamic programming	Knapsack problem, recursion
2102	bfs	stack, dfs

The summarization model filters the disturbing sentences in front of the document. Comparing the original text and the summary, sentences such as "topic link" will be filtered. In addition, it can be seen from the generated summary that the generated summary contains some knowledge related to the Knowledge point, such as keywords such as "Eulerian Path", "Eulerian Ring" and "network flows" (Table 1).

After using the LDA model to extract keywords, we can find many knowledge points, such as dictionary trees, dfs, bfs, etc., and the keywords in each type of theme are in line with people's common sense, such as search, dichotomy, bfs, dfs, etc. Among the categories of topics, keywords like graph theory, network flows, maximum flow, and minimum spanning tree are aggregated in another type of topic. After multiple rounds of selection of knowledge points, a knowledge base under all problems can be established.

In the construction of feature portraits of knowledge points, taking topological sorting as an example, the first attribute is associated knowledge points, including ranking, partial order relationship, and queue. The commonness is considered an attribute in this paper, because the keyword of topological sorting is the highest in the number of occurrences of all knowledge points, the topological sorting is classified as the most common type of knowledge points, namely, type A.

7.3 Analysis

This article uses a variety of technical methods to try to mine knowledge points and corresponding feature portraits from open learning resources and achieve good results, but due to the limitations of extractive data sets, it is impossible to use problem-solving blog resources for hyperparameter training. In the future, we can consider constructing a text summary data set of problem-solving resources to further improve the accuracy of extraction. Because the content of blog resources is relatively fragmented, and the articles are filled with a lot of code, comments, and other websites' own content, it is difficult for clauses and further Bert pre-training. If the document resources can be filtered, only the body text is left by analyzing part of the topic, we can perform pre-training in Bert. Due to the lack of problem-solving resource summaries, the effect of extraction cannot be evaluated, and only the extraction of keywords can be used to determine whether the extracted content is reliable. Therefore, it is very important to manually construct the relevant text and corresponding abstract of the problem-solving resources, which can be used as the focus of future work to further improve the effect.

7.4 Conclusion

This article describes how to extract knowledge points and feature portraits for open learning resources. At the beginning, the domestic and foreign research process of text summarization and keyword extraction is introduced, and the significance and value of open learning resources developed on the Internet are analyzed. First of all, for the knowledge point extraction module, crawling data for problem-solving blog resources is taken as the research object, and the text summary and keyword extraction methods are used to obtain knowledge points. Due to the strong applicability of the PacSum method, it can adapt to text summaries in different languages and different writing styles. In this paper, the PacSum technology is applied to the problem-solving blog resources to perform text summaries and extract key sentences. For the obtained key sentences, LDA technology is used to extract the key words of the key sentences. Since the blog resources contain a large number of classified knowledge points, it is suitable to use the topic model to extract keywords. Only need to manually filter the keywords extracted by LDA, you can get the knowledge point database. Such a convenient and efficient method greatly improves the efficiency compared to manual, making the correspondence between the topic and the knowledge point quick and efficient.

Faced with the extracted knowledge base, this paper designs knowledge point attributes to describe its characteristics. The attributes of the feature portraits are derived from Baidu Encyclopedia resources, blog resources, and knowledge crawling resources. Focusing on a knowledge point, it comprehensively depicts the characteristics of all aspects. Using network resources, a chain of problem, knowledge point, related knowledge point, knowledge point summary is formed, and a wealth of information is mined.

Acknowledgments. This work is supported in part by National Natural Science Foundation of China (No. 61877043) and the National Natural Science Foundation of China (No. 61877044).

References

1. Nenkova, A., Maskey, S., Liu, Y.: Automatic summarization. In: Proceedings of the 49th Annual Meeting of the Association for Computational Linguistics: Tutorial Abstracts of ACL 2011, vol. 3. Association for Computational Linguistics (2011)
2. Zheng, H., Lapata, M.: Sentence centrality revisited for unsupervised summarization. arXiv preprint arXiv:1906.03508 (2019)
3. Pan, H.-X., Liu, H., Tang, Y.: A sequence-to-sequence text summarization model with topic based attention mechanism. In: Ni, W., Wang, X., Song, W., Li, Y. (eds.) WISA 2019. LNCS, vol. 11817, pp. 285–297. Springer, Cham (2019). https://doi.org/10.1007/978-3-030-30952-7_29
4. Jingsheng, Z., Qiaoming, Z., Guodong, Z., et al.: Summary of research on automatic keyword extraction. J. Softw. **9**, 2431–2449 (2017). (in Chinese)
5. Salton, G., Buckley, C.: Term-weighting approaches in automatic text retrieval. Inf. Process. Manag. **24**(5), 513–523 (1988)
6. Zhang, X., Lapata, M., Wei, F., et al.: Neural latent extractive document summarization. arXiv preprint arXiv:1808.07187 (2018)
7. Mihalcea, R., Tarau, P.: Textrank: bringing order into text. InL Proceedings of the 2004 Conference on Empirical Methods in Natural Language Processing, pp. 404–411 (2004)
8. Brin, S., Page, L.: The anatomy of a large-scale hypertextual web search engine (1998)
9. Nallapati, R., Zhai, F., Zhou, B.: SummaRuNNer: a recurrent neural network based sequence model for extractive summarization of documents. In: Thirty-First AAAI Conference on Artificial Intelligence (2017)
10. Lin, C.Y., Hovy, E.: Automatic evaluation of summaries using n-gram co-occurrence statistics. In: Proceedings of the. Human Language Technology Conference of the North American Chapter of the Association for Computational Linguistics, vol. 2003, pp. 150–157 (2003)
11. Narayan, S., Cohen, S.B., Lapata, M.: Ranking sentences for extractive summarization with reinforcement learning. arXiv preprint arXiv:1802.08636 (2018)
12. Liu, Y., Mirella, L.: Text Summarization with pretrained encoders, pp. 3728–3738 (2019)
13. Devlin, J., Chang, M.W., Lee, K., et al.: Bert: pre-training of deep bidirectional transformers for language understanding. arXiv preprint arXiv:1810.04805 (2018)
14. Zhou, Q., Yang, N., Wei, F., et al.: Neural document summarization by jointly learning to score and select sentences. arXiv preprint arXiv:1807.02305 (2018)

15. See, A., Liu, P.J., Manning, C.D.: Get to the point: summarization with pointer-generator network. arXiv preprint arXiv:1704.04368 (2017)
16. Hofmann, T.: Probabilistic latent semantic analysis. arXiv preprint arXiv:1301.6705 (2013)
17. Blei, D.M., Ng, A.Y., Jordan, M.I.: Latent Dirichlet allocation. J. Mach. Learn. Res. **3**(Jan), 993–1022 (2003)
18. Cui, Y., Che, W., Liu, T., et al.: Pre-training with whole word masking for Chinese Bert. arXiv preprint arXiv:1906.08101 (2019)

Recommendation

Deep Hybrid Knowledge Graph Embedding for Top-N Recommendation

Jian Li, Zhuoming Xu$^{(\boxtimes)}$, Yan Tang, Bo Zhao, and Haimei Tian

College of Computer and Information, Hohai University, Nanjing 210098, China
{jli,zmxu,tangyan,bzhao,hmtian}@hhu.edu.cn

Abstract. In knowledge graph (KG) based recommender systems, path-based methods make recommendations by building user-item graphs and exploiting connectivity patterns between the entities in the graph. To overcome the limitations of traditional meta-path based methods that rely heavily on handcrafted meta-paths, recent deep neural network based methods, such as the Recurrent Knowledge Graph Embedding (RKGE) approach, can automatically mine the connectivity patterns between entities in the KG, thereby improving recommendation performance. However, these methods usually use only one type of neural network to encode path embeddings, which cannot fully extract path features, limiting performance improvement of the recommender system. In this paper, we propose a Deep Hybrid Knowledge Graph Embedding (DHKGE) method for top-N recommendation. DHKGE encodes embeddings of paths between users and items by combining convolutional neural network (CNN) and the long short-term memory (LSTM) network. Furthermore, it uses an attention mechanism to aggregate the encoded path representations and generate a final hidden state vector, which is used to calculate the proximity between the target user and candidate items, thus generating top-N recommendation. Experiments on the MovieLens 100K and Yelp datasets show that DHKGE overall outperforms RKGE and several typical recommendation methods in terms of Precision@N, MRR@N, and NDCG@N.

Keywords: Top-N recommendation · Knowledge graph · Deep hybrid model · CNN · LSTM · Attention mechanism

1 Introduction

Knowledge graphs (KGs) have proven to be effective in improving recommendation performance [7, 16]. According to [7], there are three categories of KG-based recommendation methods: path-based methods, embedding-based methods, and unified methods. Path-based methods make recommendations by building a KG which contains users, items, and user-item interactions, and then exploiting connectivity patterns between the entities (users or items) in the KG. The traditional meta-path based methods use the semantic similarity of entities in different meta-paths [18] as graph regularization to refine representations of users and items [7]. However, such methods rely heavily on handcrafted meta-paths, which further rely on domain knowledge [14].

G. Wang et al. (Eds.): WISA 2020, LNCS 12432, pp. 59–70, 2020.
https://doi.org/10.1007/978-3-030-60029-7_6

To overcome the limitations of meta-path based methods, deep neural network based methods have recently been devised to automatically mine the connectivity patterns between entities (i.e., path embeddings) in the KG. Path representations are learned by extracting path features from connectivity patterns to characterize user preferences towards items, which are finally used to generate recommendation.

However, existing deep neural network based methods, such as the Recurrent Knowledge Graph Embedding (RKGE) approach [14], usually use only one type of neural network to encode path embeddings. But this cannot fully extract path features, which limits performance improvement of the recommender system. Recently proposed deep hybrid models, such as [12], can combine several neural building blocks to form a more powerful recommendation model. To the best of our knowledge, existing deep hybrid models seldom use KGs for recommendation.

To overcome the weaknesses of existing methods, in this paper we propose a Deep Hybrid Knowledge Graph Embedding (DHKGE) method for top-N recommendation. DHKGE encodes embeddings of paths between users and items that are involved in the recommender system by combining convolutional neural network (CNN) and the long short-term memory (LSTM) network. It further uses an attention mechanism to aggregate the encoded path representations and generate a final hidden state vector. This vector is then used to calculate the proximity between the target user and candidate items, and generate top-N recommendation for the user by ranking the proximity.

In summary, the main contributions of this paper are as follows:

- We propose the Deep Hybrid Knowledge Graph Embedding (DHKGE) method for top-N recommendation, which exploits a deep hybrid model to encode the path between users and items.
- We propose to use the attention mechanism to distinguish the importance of multiple semantic paths between a user-item pair, so that salient paths play a greater role in modeling user preferences.
- We evaluated our method on the MovieLens 100K and Yelp datasets. The experimental results show that our method overall outperforms RKGE and several typical recommendation methods in terms of Precision@N, MRR@N, and NDCG@N.

2 Related Work

2.1 Path-Based Recommendation Methods

Path-based methods make recommendations by building user-item graphs and exploiting connectivity patterns between the entities in the graph [7]. Traditional meta-path based methods rely heavily on handcrafted meta-paths. Deep neural network based methods can automatically mine the connectivity patterns between entities in the graph, thereby improving recommendation performance. For example, Hu et al. [9] proposed to leverage meta-path based context for top-N recommendation with a neural co-attention model. Sun et al. [14] proposed the RKGE approach that employs RNN to learn high-quality representations of both users and items, which are then used to generate better recommendations. Wang *et al.* [15] proposed the Knowledge-aware Path Recurrent Network (KPRN) which exploits KG to generate better recommendation, where the path embeddings in the KG are encoded with LSTM.

Existing path-based recommendation methods usually use only one type of neural network to encode path embeddings, while our proposed DHKGE exploits a deep hybrid model to encode path embeddings, which can generate a more comprehensive path representation for better recommendation.

2.2 Deep Neural Network-Based Recommendation

Deep neural networks have been widely used in recommender systems. The existing recommendation models can be divided into two categories: recommendation with neural building blocks and recommendation with deep hybrid models [4, 20].

In the first category, the recommendation models are divided into several subcategories [20] that exploit the deep learning models: CNN, recurrent neural network (RNN), and attentional model (AM), etc. For example, Kim *et al.* [10] proposed a context-aware recommendation model named convolutional matrix factorization (ConvMF) that integrates CNN into probabilistic matrix factorization.

Recently, researchers have proposed deep hybrid models, which can combine several neural building blocks to complement one another and form a more powerful recommendation model [20]. For instance, Lee *et al.* [12] proposed a deep learning recommender system that combines RNN and CNN to learn semantic representation of each utterance and build a sequence model for the dialog thread. To the best of our knowledge, existing deep hybrid models seldom use KGs for recommendation.

3 DHKGE: Deep Hybrid KG Embedding Method

In this section, we expatiate on our DHKGE method. After introducing concepts and notations, we first briefly explain its overall framework, then describe its main components, and finally describe model learning and recommendation generation.

Given a user set $\mathcal{U} = \{u_1, u_2, \ldots, u_m\}$ and an item set $\mathcal{V} = \{v_1, v_2, \ldots, v_n\}$ of the recommender system, we construct the users' implicit feedback matrix $\mathbf{R} \in \mathbb{R}^{m \times n}$, where each element is defined as follows: when user u_i interacted with item v_j set $r_{ij} = 1$ indicating that the user prefers the item, otherwise set $r_{ij} = 0$. Based on the matrix $\mathbf{R}$ and an external knowledge source (e.g., the IMDB dataset) that describes the items, we build a KG for recommendation, which contains the users, items, user's preference for the items, and the item descriptions extracted from the knowledge source, such as actors, directors and genres (as entities), as well as rating, categorizing, acting, and directing (as entity relations) in the domain of movie recommendation. We refer to all objects (e.g., users, items, actors, directors, and genres) except for various relations in the KG as *entities*. The definition [14] of the KG is given below.

Definition 1 (Knowledge Graph). KG is defined as a directed graph $\mathcal{G} = (\mathcal{E}, \mathcal{L})$, where $\mathcal{E} = \{e_1, e_2, \ldots, e_{|\mathcal{E}|}\}$ denotes the sets of entities and $\mathcal{L}$ the sets of links. An entity type mapping function $\phi : \mathcal{E} \to \mathcal{A}$ and a link type mapping function $\varphi : \mathcal{L} \to \mathcal{R}$

are defined for the graph. Each entity $e \in \mathcal{E}$ belongs to an entity type $\phi(e) \in \mathcal{A}$, and each link $l \in \mathcal{L}$ belongs to a link type (relation) $\varphi(l) \in \mathcal{R}$.

Based on the KG definition, we further define the connected semantic paths between entity pair (e_i, e_j) as $\mathcal{P}(e_i, e_j) = \{p_1, p_2, \ldots, p_s\}$ with s being the number of paths. A semantic path of length T in $\mathcal{P}$ is denoted as: $p = e_i \xrightarrow{r_1} e_1 \xrightarrow{r_2} \cdots \xrightarrow{r_T} e_j$.

Following the two semantic path mining strategies proposed in [14], DHKGE only considers user-item paths $\mathcal{P}(u_i, v_j)$, $u_i \in \mathcal{U}$, $v_j \in \mathcal{V}$ that connect user u_i with all her rated items v_j, and sets a length constraint for such paths, i.e., path length is T.

3.1   Overview

Our goal is to fully extract the information in the semantic path to model user preferences, which are then used to generate better recommendations. To achieve this goal, we propose the deep hybrid knowledge graph embedding (DHKGE) method.

The core ideas of DHKGE is as follows: Given a user and an item, DHKGE first automatically extracts all semantic paths between the user and the item from the KG according to the semantic path mining strategies. It then uses a deep hybrid model to obtain a final hidden vector for quantifying the relation (proximity) between the user and the item. Finally, it generates a top-N recommendation list for the user by sorting the proximity scores of the candidate items in descending order.

The overall framework of DHKGE is depicted in Fig. 1. As shown in the figure, DHKGE is composed of four key components: the embedding layer, CNN layer, LSTM layer, and attention layer, which are further described as follows:

- The embedding layer: This layer takes the semantic path of length T as input, learns $T + 1$ low-dimensional embedding vectors for $T + 1$ entities on the semantic path, and outputs these vectors as an embedding of the path.
- The CNN layer: This layer takes the path embedding as input, uses multiple filters to extract the local features of the path to form T local feature vectors, and outputs these vectors.
- The LSTM layer: This layer takes the ordered local feature vectors as input, encodes them to get a representation of the path, and outputs the path representation.
- The attention layer: This layer takes the representations of s paths as input, uses the attention mechanism to aggregate these path representations by weighting them to obtain a final hidden state vector, and outputs the vector.

3.2   Embedding Layer

Given a set of s semantic paths of length T between user u_i and item v_j, $\mathcal{P}(u_i, v_j) = \{p_1, p_2, \ldots, p_s\}$, where the start entity and end entity of each path in $\mathcal{P}$ are u_i and v_j, respectively. As shown in Fig. 1, $e_0 = u_i$ and $e_T = v_j$ in path p_1. The embedding layer maps each entity e_t in such a path into a d-dimensional vector $\mathbf{e}_t \in \mathbb{R}^d$, which captures the semantic meaning of the entity. The vectors of all entities in the path constitute an embedding $\mathbf{p}_1 = \{\mathbf{e}_0, \mathbf{e}_1, \ldots, \mathbf{e}_T\}$ of the path.

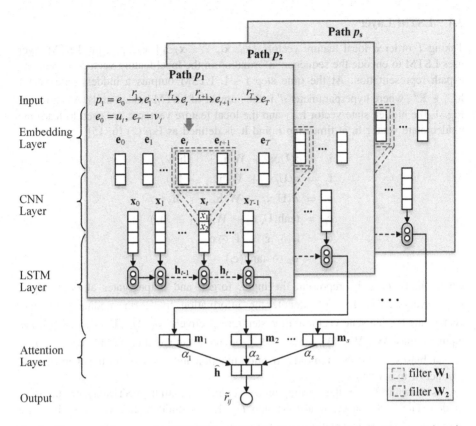

Fig. 1. The overall framework of DHKGE, which describes the case of a user-item pair (u_i, v_j).

3.3 CNN Layer

The CNN layer takes path embedding $\mathbf{p}_1$ as input, and then slides multiple filters with the same window size over the path embedding to extract local features of the path. Let $\mathbf{W}_1 \in \mathbb{R}^{2 \times d}$ be a filter with a window size of 2. As shown in Fig. 1, $\mathbf{W}_1$ is applied to two embeddings $\mathbf{e}_t$ and $\mathbf{e}_{t+1}$ of the adjacent entities to generate a local feature x_1, which is defined as Eq. (1) [5, 11].

$$x_1 = f(\mathbf{W}_1 \circ [\mathbf{e}_t, \mathbf{e}_{t+1}] + b_1) \tag{1}$$

where $\circ$ denotes the convolution operation, b_1 is the bias, and $f(\cdot)$ is the nonlinear activation function ReLU.

This way, k filters with the same window size, $\mathbf{W}_1, \mathbf{W}_2, \ldots$, are applied to the two entity embeddings $\mathbf{e}_t$ and $\mathbf{e}_{t+1}$ to obtain a local feature vector $\mathbf{x}_t = [x_1, x_2, \ldots, x_k]$, where k is a hyperparameter. The CNN layer slides k filters from entity embedding $\mathbf{e}_0$ to entity embedding $\mathbf{e}_{T-1}$ with stride 1, thus forming a sequence of local feature vectors $\{\mathbf{x}_0, \mathbf{x}_1, \ldots, \mathbf{x}_{T-1}\}$.

3.4   LSTM Layer

Taking T ordered local feature vectors $\{\mathbf{x}_0, \mathbf{x}_1, \ldots, \mathbf{x}_{T-1}\}$ as input, the LSTM layer uses LSTM to encode the sequence information in the local feature vectors to generate a path representation. At the time step $t-1$, LTSM outputs a hidden state vector $\mathbf{h}_{t-1} \in \mathbb{R}^{d'}$, where hyperparameter d' is the number of LSTM hidden units. As shown in Fig. 1, the hidden state vector $\mathbf{h}_{t-1}$ and the local feature vector $\mathbf{x}_t$ are used to learn the hidden state vector $\mathbf{h}_t$ at time step t, and $\mathbf{h}_t$ is defined as Eq. (2) [6, 15].

$$
\begin{aligned}
\mathbf{i}_t &= \sigma(\mathbf{U}_i \mathbf{x}_t + \mathbf{W}_i \mathbf{h}_{t-1} + \mathbf{b}_i) \\
\mathbf{f}_t &= \sigma(\mathbf{U}_f \mathbf{x}_t + \mathbf{W}_f \mathbf{h}_{t-1} + \mathbf{b}_f) \\
\mathbf{o}_t &= \sigma(\mathbf{U}_o \mathbf{x}_t + \mathbf{W}_o \mathbf{h}_{t-1} + \mathbf{b}_o) \\
\hat{\mathbf{c}}_t &= \tanh(\mathbf{U}_c \mathbf{x}_t + \mathbf{W}_c \mathbf{h}_{t-1} + \mathbf{b}_c) \\
\mathbf{c}_t &= \mathbf{i}_t \odot \hat{\mathbf{c}}_t + \mathbf{f}_t \odot \mathbf{c}_{t-1} \\
\mathbf{h}_t &= \mathbf{o}_t \odot \tanh(\mathbf{c}_t)
\end{aligned}
\tag{2}
$$

where, $\mathbf{i}_t$, $\mathbf{f}_t$, $\mathbf{o}_t \in \mathbb{R}^{d'}$ represent the input, forget, and output gates at time step t, respectively. $\hat{\mathbf{c}}_t$, $\mathbf{c}_t$, $\mathbf{h}_t \in \mathbb{R}^{d'}$ denote the information transform module, cell state vector, and hidden state vector at time step t, respectively. $\mathbf{U}_i$, $\mathbf{U}_f$, $\mathbf{U}_o$, $\mathbf{U}_c \in \mathbb{R}^{d' \times k}$ are input weights, $\mathbf{W}_i$, $\mathbf{W}_f$, $\mathbf{W}_o$, $\mathbf{W}_c \in \mathbb{R}^{d' \times d'}$ are recurrent weights, and $\mathbf{b}_i$, $\mathbf{b}_f$, $\mathbf{b}_o$, $\mathbf{b}_c \in \mathbb{R}^{d'}$ are biases. $\sigma(\cdot)$ is the sigmoid activation function and $\odot$ stands for the element-wise product of two vectors.

As shown in Fig. 1, the learning process continues until the LSTM layer obtains the hidden state vector at the final time step $T-1$. This hidden state vector is therefore output as a path representation, denoted $\mathbf{m}_1 \in \mathbb{R}^{d'}$.

3.5   Attention Layer

Once the path representations are obtained, the attention layer takes these path representations as input and uses the attention mechanism to generate a final hidden state vector and output it. The process of generating hidden state vectors is as follows: First, this layer learns an attention score $score(\mathbf{m}_i)$ for each path representation $\mathbf{m}_i$ in the path representation set $\{\mathbf{m}_1, \mathbf{m}_2, \ldots, \mathbf{m}_s\}$. Then these scores are normalized, and finally these path representations are aggregated by weighting them to obtain a final hidden state vector $\widehat{\mathbf{h}} \in \mathbb{R}^{d'}$, which characterizes the user preferences towards items. The above process is defined as Eq. (3) [17].

$$
\begin{aligned}
score(\mathbf{m}_i) &= \mathbf{w}_a^{\mathrm{T}} \tanh(\mathbf{W}_a \mathbf{m}_i) \\
\alpha_i &= \frac{\exp(score(\mathbf{m}_i))}{\sum\limits_{i=1}^{s} \exp(score(\mathbf{m}_i))} \\
\widehat{\mathbf{h}} &= \sum_{i=1}^{s} \alpha_i \mathbf{m}_i
\end{aligned}
\tag{3}
$$

where $\mathbf{w}_a \in \mathbb{R}^{d'}$ and $\mathbf{W}_a \in \mathbb{R}^{d' \times d'}$ are weights, and α_i the normalized attention score.

Finally, DHKGE uses a fully connected layer to quantify the proximity $\tilde{r}_{ij}$ of user u_i and item v_j, which is defined as Eq. (4) [14]:

$$\tilde{r}_{ij} = \sigma(\mathbf{W}_r\widehat{\mathbf{h}} + b_r) \tag{4}$$

where $\mathbf{W}_r \in \mathbb{R}^{1\times d'}$ and b_r are the weights and bias, respectively.

3.6 Method Learning and Recommendation Generation

Like RKGE [14], given the training data $\mathcal{D}_{\text{train}}$, which contains instances in the form of $(u_i, v_j, r_{ij}, \mathcal{P}(u_i, v_j))$, DHKGE also uses stochastic gradient descent (SGD) to minimize the loss function defined as Eq. (5) to learn all the parameters in DHKGE.

$$\mathcal{J} = \frac{1}{|\mathcal{D}_{\text{train}}|} \sum\nolimits_{r_{ij}\in\mathcal{D}_{\text{train}}} BCELoss(\tilde{r}_{ij}, r_{ij}) \tag{5}$$

where $BCELoss(\cdot)$ is the binary cross-entropy between the observed ratings and estimated ones.

The recommendation problem can be dealt with as a binary classification problem [8, 14]. When user u_i prefers item v_j, namely $r_{ij} = 1$, we expect the estimated proximity $\tilde{r}_{ij}$ to approach 1, otherwise it approaches 0. Once the learning process is completed, DHKGE can obtain all trained embeddings of the users and items.

Following [14, 18], during the testing process DHKGE can obtain the proximity scores between the target user and candidate items by calculating the inner products of the user embedding and the item embeddings. DHKGE finally generates top-N recommendation lists for the user by sorting the proximity scores in descending order.

4 Experimental Evaluation

4.1 Experimental Setup

Datasets. Our experiment used two datasets MovieLens 100K[1] and Yelp published on GitHub[2] by [14]. The former is a movie dataset containing user interaction with movies. Sun *et al.* [14] combined this dataset with the IMDB dataset[3] to add description information of movies, such as genre, actor, and director. Yelp contains user check-ins to local business, user reviews, and local business information, and no external information needs to be added to this dataset. The two datasets were used to build two KGs following **Definition** 1. The statistics of two datasets are shown in Table 1.

---

[1] The experiment of [14] used MovieLens 1M, but the pre-training vectors of users and items in MovieLens 1M were not published on GitHub, so we can only use MovieLens 100K.

[2] https://github.com/sunzhuntu/Recurrent-Knowledge-Graph-Embedding/tree/master/data.

[3] https://www.imdb.com/.

Table 1. Dataset and knowledge graph statistics

Datasets		MovieLens 100K	Yelp
User-item interaction	# Users	943	37,940
	# Items	1,675	11,516
	# Ratings	99,975	229,178
Knowledge graph	# Entities	7,744	50,028
	# Entity types	5	4
	# Links	112,321	272,057
	# Links types	7	5

Following [3, 14], we sorted the two datasets according to the feedback timestamp, and used the earlier 80% feedback as training data and the more recent 20% feedback as test data. For each user-item pair in the training set, we extracted all paths with a length of 3 and randomly selected five paths from them to train our model.

Evaluation Metrics. Three popular evaluation metrics [1], Precision at N (Prec@N), Mean Reciprocal Rank at N (MRR@N), and Normalized Discounted Cumulative Gain at N (NDCG@N), are adopted to evaluate the top-N recommendation methods in our experiment. We set N = {1, 5, 10, 20} for Prec@N, and N = {5, 10, 20} for MRR@N and NDCG@N.

Comparison Methods and Their Implementation. We compared our DHKGE with the following four recommendation methods:

- BPRMF [13]: It is a Bayesian personalized ranking method based on Matrix Factorization. We used the Cornac[4] framework to implement BPRMF.
- NCF [8]: It is a classic neural network-based recommendation method. It was also implemented by using the Cornac framework.
- CKE [19]: It is the recently proposed state-of-the-art KG embedding based recommendation method. This method directly used the Python code[5] provided in [2].
- RKGE [14]: It is a state-of-the-art recommendation method based on KG path. This method directly used the Python code published on GitHub[6] by the authors.

We used PyTorch to generate the code of DHKGE by modifying the recurrent network module and performance evaluation module in the RKGE code.

Hyperparameter Settings. For DHKGE, we used grid search to select both the dimension d of the entity embedding and the number k of convolution filters in {10, 20, 30, 40, 50, 100}, the number d' of LSTM hidden units in {16, 32, 64, 128}, and the learning rate λ of SGD in {0.001, 0.01, 0.1, 0.2}. The hyperparameters for

---

[4] https://github.com/PreferredAI/cornac.

[5] https://github.com/TaoMiner/joint-kg-recommender.

[6] https://github.com/sunzhuntu/Recurrent-Knowledge-Graph-Embedding.

DHKGE were set to $d = 10$, $k = 10$, $d' = 16$, $\lambda = 0.2$ on MovieLens 100K and $d = 20$, $k = 40$, $d' = 32$, $\lambda = 0.01$ on Yelp. For the four comparison methods, the hyperparameters were set as suggested by the original papers.

4.2 Experimental Results

Tables 2 and 3 show the results of top-N recommendation performed on the two datasets. In the tables, bold numbers indicate the best performance among all the methods; underlined numbers are the best performance among the four comparison methods; the numbers in the "Improve" column indicate the percentage (%) of performance improvement achieved by DHKGE relative to the best performance among the comparison methods. The same way as in [14], we also created two views for each dataset: "All Users" means that all users are considered in the test data, whereas "Cold Start" indicates that the test data only includes users with less than 5 ratings.

Observing these results, we can obtain the following findings:

1. The performance of both DHKGE and RKGE in terms of all metrics except for Prec@1 is significantly better than the other three methods. This indicates DHKGE and RKGE can make full use of the path information to model user's preference for items, thereby improving the recommendation performance.
2. DHKGE's performance is better than RKGE in all metrics (the performance in terms of Prec@1 on MovieLens 100K is the same). This indicates that deep hybrid

Table 2. Results of top-N recommendation on MovieLens 100K

Views	Metrics	BPRMF	NCF	CKE	RKGE	DHKGE	Improve (%)
All Users	Prec@1	0.0382	0.0509	0.1044	0.1469	**0.1548**	5.38
	Prec@5	0.0418	0.0492	0.0826	<u>0.1044</u>	**0.1103**	5.65
	Prec@10	0.0431	0.0467	0.0735	<u>0.0890</u>	**0.0962**	8.09
	Prec@20	0.0467	0.0481	0.0660	<u>0.0777</u>	**0.0822**	5.79
	MRR@5	0.1135	0.1247	0.2272	<u>0.2760</u>	**0.2883**	4.46
	MRR@10	0.1221	0.1388	0.2637	<u>0.3284</u>	**0.3416**	4.02
	MRR@20	0.1278	0.1416	0.3031	<u>0.3760</u>	**0.3876**	3.09
	NDCG@5	0.0412	0.0498	0.1976	<u>0.2552</u>	**0.2623**	2.78
	NDCG@10	0.0488	0.0541	0.2133	<u>0.2853</u>	**0.3006**	5.36
	NDCG@20	0.0705	0.0735	0.2597	<u>0.3256</u>	**0.3316**	1.84
Cold Start	Prec@1	0.0250	0.0500	0.0503	**0.0625**	**0.0625**	0.00
	Prec@5	0.0225	0.0275	0.0287	<u>0.0325</u>	**0.0350**	7.69
	Prec@10	0.0163	0.0213	0.0265	<u>0.0263</u>	**0.0288**	9.51
	Prec@20	0.0175	0.0194	0.0239	<u>0.0206</u>	**0.0213**	3.40
	MRR@5	0.0794	0.0872	0.0922	<u>0.1008</u>	**0.1056**	4.76
	MRR@10	0.0823	0.0953	0.1003	<u>0.1143</u>	**0.1206**	5.51
	MRR@20	0.0846	0.1067	0.1107	<u>0.1237</u>	**0.1289**	4.20
	NDCG@5	0.0278	0.0365	0.0744	<u>0.1029</u>	**0.1103**	7.19
	NDCG@10	0.0345	0.0459	0.0953	<u>0.1354</u>	**0.1404**	3.69
	NDCG@20	0.0530	0.0635	0.1184	<u>0.1604</u>	**0.1658**	3.37

Table 3. Results of top-N recommendation on Yelp

Views	Metrics	BPRMF	NCF	CKE	RKGE	DHKGE	Improve (%)
All Users	Prec@1	0.0038	0.0050	0.0076	0.0093	**0.0102**	9.68
	Prec@5	0.0041	0.0044	0.0053	0.0078	**0.0085**	8.97
	Prec@10	0.0039	0.0043	0.0050	0.0066	**0.0075**	13.64
	Prec@20	0.0036	0.0041	0.0047	0.0058	**0.0063**	8.62
	MRR@5	0.0134	0.0164	0.0171	0.0194	**0.0210**	8.25
	MRR@10	0.0162	0.0181	0.0195	0.0231	**0.0252**	9.09
	MRR@20	0.0185	0.0206	0.0226	0.0270	**0.0286**	5.93
	NDCG@5	0.0072	0.0084	0.0197	0.0243	**0.0253**	4.12
	NDCG@10	0.0103	0.0118	0.0213	0.0320	**0.0339**	5.94
	NDCG@20	0.0148	0.0172	0.0256	0.0418	**0.0432**	3.35
Cold Start	Prec@1	0.0031	0.0035	**0.0064**	0.0061	0.0062	−3.13
	Prec@5	0.0030	0.0034	0.0045	0.0048	**0.0050**	4.17
	Prec@10	0.0028	0.0031	0.0036	0.0040	**0.0043**	7.50
	Prec@20	0.0027	0.0030	0.0031	0.0032	**0.0035**	9.38
	MRR@5	0.0108	0.0114	0.0117	0.0121	**0.0127**	4.96
	MRR@10	0.0134	0.0142	0.0144	0.0143	**0.0150**	4.90
	MRR@20	0.0149	0.0161	0.0162	0.0159	**0.0168**	3.70
	NDCG@5	0.0069	0.0079	0.0134	0.0149	**0.0157**	5.37
	NDCG@10	0.0100	0.0114	0.0187	0.0200	**0.0212**	6.00
	NDCG@20	0.0151	0.0170	0.0212	0.0259	**0.0274**	5.79

models can encode semantic paths more efficiently than one type of neural network, because by extracting the local features of the path and encoding the sequence information in the path, DHKGE can generate a more comprehensive path representation for recommendation.

3. On most metrics, the performance improvement of DHKGE in the "All Users" view is higher than in the "Cold Start" view. This indicates that in the "Cold Start" view, the quality and quantity of the semantic paths extracted from the KG are limited, which affects the recommendation performance of DHKGE since this method relies on the user's historical interaction information to make recommendations.

Based on these findings, we can draw the conclusion that DHKGE's recommendation performance is generally better than RKGE and other comparison methods.

5 Conclusions

To overcome the weaknesses of existing KG path-based recommendation methods, in this paper we propose the DHKGE method for top-N recommendation. DHKGE exploits a deep hybrid model to encode the path between users and items, and uses the attention mechanism to distinguish the importance of multiple semantic paths between a user-item pair. Experiments on the MovieLens 100K and Yelp datasets show that

DHKGE overall outperforms RKGE and several typical recommendation methods in terms of Precision@N, MRR@N, and NDCG@N. In future work, we plan to improve our method by adding entity relations to path embeddings.

References

1. Aggarwal, C.C.: Evaluating recommender systems. Recommender Systems, pp. 225–254. Springer, Cham (2016). https://doi.org/10.1007/978-3-319-29659-3_7
2. Cao, Y., Wang, X., He, X., Hu, Z., Chua, T.: Unifying knowledge graph learning and recommendation: towards a better understanding of user preferences. In: Proceedings of the 28th International Conference on World Wide Web, WWW 2019, pp. 151–161. ACM (2019). https://doi.org/10.1145/3308558.3313705
3. Catherine, R., Cohen, W.W.: Personalized recommendations using knowledge graphs: a probabilistic logic programming approach. In: Proceedings of the 10th ACM Conference on Recommender Systems, RecSys 2016, pp. 325–332. ACM (2016). https://doi.org/10.1145/2959100.2959131
4. Da'u, A., Salim, N.: Recommendation system based on deep learning methods: a systematic review and new directions. Artif. Intell. Rev. **53**(4), 2709–2748 (2020). https://doi.org/10.1007/s10462-019-09744-1
5. Goodfellow, I.J., Bengio, Y., Courville, A.C.: Convolutional networks. In: Goodfellow, I.J., Bengio, Y., Courville, A.C. (eds.) Deep Learning, pp. 330–372. MIT Press (2016). http://www.deeplearningbook.org/contents/convnets.html
6. Goodfellow, I.J., Bengio, Y., Courville, A.C.: Sequence modeling: recurrent and recursive nets. In: Goodfellow, I.J., Bengio, Y., Courville, A.C. (eds.) Deep Learning, pp. 373–420. MIT Press (2016). http://www.deeplearningbook.org/contents/rnn.html
7. Guo, Q., Zhuang, F., Qin, C., et al.: A survey on knowledge graph-based recommender systems. CoRR abs/2003.00911 (2020). https://arxiv.org/abs/2003.00911
8. He, X., Liao, L., Zhang, H., Nie, L., Hu, X., Chua, T.: Neural collaborative filtering. In: Proceedings of the 26th International Conference on World Wide Web, WWW 2017, pp. 173–182. ACM (2017). https://doi.org/10.1145/3038912.3052569
9. Hu, B., Shi, C., Zhao, W.Z., Yu, P.S.: Leveraging meta-path based context for top-n recommendation with a neural co-attention model. In: Proceedings of the 24th ACM SIGKDD International Conference on Knowledge Discovery & Data Mining, KDD 2018, pp. 1531–1540. ACM (2018). https://doi.org/10.1145/3219819.3219965
10. Kim, D., Park, C., Oh, J., Lee, S., Yu, H.: Convolutional matrix factorization for document context-aware recommendation. In: Proceedings of the 10th ACM Conference on Recommender Systems, RecSys 2016, pp. 233–240. ACM (2016). https://doi.org/10.1145/2959100.2959165
11. Kim, Y.: Convolutional neural networks for sentence classification. In: Proceedings of the 2014 Conference on Empirical Methods in Natural Language Processing, EMNLP 2014, pp. 1746–1751. ACL (2014). https://doi.org/10.3115/v1/D14-1181
12. Lee, H., Ahn, Y., Lee, H., Ha, S., Lee, S.: Quote recommendation in dialogue using deep neural network. In: Proceedings of the 39th International ACM SIGIR conference on Research and Development in Information Retrieval, SIGIR 2016, pp. 957–960. ACM (2016). https://doi.org/10.1145/2911451.2914734

13. Rendle, S., Freudenthaler, C., Gantner, Z., Schmidt-Thieme, L.: BPR: Bayesian personalized ranking from implicit feedback. In: Proceedings of the Twenty-Fifth Conference on Uncertainty in Artificial Intelligence, UAI 2009, pp. 452–461. AUAI Press (2009). https:// dslpitt.org/uai/papers/09/p452-rendle.pdf
14. Sun, Z., Yang, J., Zhang, J., Bozzon, A., Huang, L., Xu, C.: Recurrent knowledge graph embedding for effective recommendation. In: Proceedings of the 12th ACM Conference on Recommender Systems, RecSys 2018, pp. 297–305. ACM (2018). https://doi.org/10.1145/3240323.3240361
15. Wang, X., Wang, D., Xu, C., He, X., Cao, Y., Chua, T.: Explainable reasoning over knowledge graphs for recommendation. In: Proceedings of the Thirty-Third AAAI Conference on Artificial Intelligence, AAAI 2019, pp. 5329–5336. AAAI Press (2019). https://doi.org/10.1609/aaai.v33i01.33015329
16. Xu, W., Xu, Z., Ye, L.: Computing user similarity by combining item ratings and background knowledge from linked open data. In: Meng, X., Li, R., Wang, K., Niu, B., Wang, X., Zhao, G. (eds.) WISA 2018. LNCS, vol. 11242, pp. 467–478. Springer, Cham (2018). https://doi.org/10.1007/978-3-030-02934-0_43
17. Yang, Z., Yang, D., Dyer, C., He, X., Smola, A.J., Hovy, E.H.: Hierarchical attention networks for document classification. In: Proceedings of the 2018 Conference of the North American Chapter of the Association for Computational Linguistics, HLT-NAACL 2016, pp. 1480–1489. ACL (2016). https://doi.org/10.18653/v1/N16-1174
18. Yu, X., Ren, X., Sun, Y., Sturt, B., Khandelwal, U., Gu, Q., Norick, B., Han, J.: Recommendation in heterogeneous information networks with implicit user feedback. In: Proceedings of the 7th ACM Conference on Recommender Systems, RecSys 2013, pp. 347–350. ACM (2013). https://doi.org/10.1145/2507157.2507230
19. Zhang, F., Yuan, N.J., Lian, D., Xie, X., Ma, W.: Collaborative knowledge base embedding for recommender systems. In: Proceedings of the 22nd ACM SIGKDD International Conference on Knowledge Discovery and Data Mining, KDD 2016, pp. 353–362. ACM (2016). https://doi.org/10.1145/2939672.2939673
20. Zhang, S., Yao, L., Sun, A., Tay, Y.: Deep learning based recommender system: a survey and new perspectives. ACM Comput. Surv. 52(1), 5:1–5:38 (2019). https://doi.org/10.1145/3285029

PS-LDA: A Course Item Model for Tutorial Personalized Recommendation

Yuefeng Du, Angzhi Liu, Xiaoguang Li$^{(\boxtimes)}$, and Baoyan Song

College of Information, Liaoning University, Shenyang 110036, China
{duyuefeng,liuangzhi,xgli,bysong}@lnu.edu.cn

Abstract. With the development of educational big data, personalized tutoring has become an important research direction to help people find interesting learning resources. However, due to limitation of learning resources, especially for the resource in unfamiliar subject areas, it may bring data sparseness of users' learning matrix. In this paper, we propose PS-LDA, a potential probability generation model for course item on learning preferences and subject area aware. By considering the mix of these two factors, our model provides personalized guidance for designated users. Moreover, we present a top-k method for online recommendation by matching the results from P-LDA and S-LDA. Finally, the experiments on two real-life datasets can verify the effectiveness and efficiency of our model.

Keywords: Personalized recommendation · Course item model · Online tutorial · Top-k recommendation

1 Introduction

With the widespread application in educational big data, varieties of online tutorial approaches have been proposed to acquire knowledge and skills, such as MOOC. Most of the existing learning systems have realized resource sharing, which helps users to study by resource categories. However, users may confuse their learning goals sometimes. In that case, it will lead to an inefficient guidance. Thus, personalized recommendations [1, 2] are required for capturing users' expectations.

Personal preferences and subject area are two factors of tutorial recommendations. Many researches [2–4] are existing for tutorial personalized recommendation. For example, the discussion on strategy behavior of teaching recommendation is based on the user's cognitive characteristics, cognitive style, learning motivation, personality structure characteristics, and personality type factor theories. Sarwar *et al.* [1] presented a method combining users' learning preferences and subject area aware. By establishing prediction model, LCARS [5] can analyze the relationship between personal preferences and hot topics. Unfortunately, it is difficult to make recommendations when users face to unfamiliar subject areas, even confusion. Hence, we focus on the perception of recommendation issues in different subject areas.

Note that, it brings a challenge to infer items from unfamiliar subject areas through using a user's historical learning data. CF (collaborative filtering) can make recommendation by tracing users' common interests. Usually, users only access a limited

© Springer Nature Switzerland AG 2020
G. Wang et al. (Eds.): WISA 2020, LNCS 12432, pp. 71–83, 2020.
https://doi.org/10.1007/978-3-030-60029-7_7

number of subject areas, this leads to data sparseness of users' learning preference matrix, even cold start to CF. In this case, it is not feasible to use only CF-based methods [6], especially when dealing with problems in unfamiliar subject areas, because query users often do not have sufficient activity history in their unfamiliar subject areas. To solve this problem, we propose a potential probability generation model PS-LDA, it consists of offline modeling and online recommendation. The offline model is designed to take into account the following two factors in a unified way at the same time. In fact, one has his own learning preference which can be obtained by trace his historical learning data. Besides, popular learning courses in various subject areas also attract one's interests. When users access a new subject area, especially unfamiliar subject areas, they are more likely to be interested in popular learning courses. Specifically, our model employs P-LDA to understand the user's learning preferences from the user's historical learning data. To pick the courses of subject areas aware, S-LDA utilizes subject-area-aware information from the subject areas. Next, given the query user u who visits the subject area s_u, the online recommendation of our model calculates the ranking score of each course item v within s_u by automatically combining u's learning preferences and s_u's popular courses. Thus, our model contributes to tutorial personalized recommendations both in one's own subject areas and unfamiliar subject areas.

The main contributions of our research are summarized as follows:

1) We propose a potential probability generation model PS-LDA. Specifically, P-LDA performs user topic modeling to obtain user learning preferences. S-LDA performs subject area topic modeling to obtain popular courses in the subject area. We also investigate the inference problem of our model.
2) We present a top-k method for personalized recommendations by matching the learning preferences and subject areas from the results of P-LDA and S-LDA.
3) We conducted experiments to evaluate the performance of our recommendation model on two real-life datasets. The results verified the effectiveness and efficiency of our model both in one's own subject areas and unfamiliar subject areas.

The rest of the paper is organized as follows: Sect. 2 reviews the related work. Section 3 details the model PS-LDA on learning preferences and subject area perception. Section 4 introduces the top-k method for online recommendation. The experimental results are reported in Sect. 5. The paper is summarized in Sect. 6.

2   Related Work

Recommender System. Collaborative filtering and content-based recommendation techniques are two widely applied methods for recommender systems. They can find relevant items according to the user's personal interests. Collaborative filtering [1, 6] automatically recommends related items to users by referencing item rating information from other similar users. The content-based recommendation [7] assumes that the descriptive characteristics of an item well reflect the user's preference for the item. Nevertheless, the data sparseness will affect CF, even cold start. It also brings limitation to content-based recommendations. Therefore, a great deal of researches [8] were proposed on the advantages of combining both these approaches. Our recommendations focus on incorporating popular courses in subject areas.

Personalized Generation Model. Many models [9] were presented for obtaining and analyzing users' preferences. Yu *et al.* [11] used the content sentiment analysis to improve the performance of recommendation algorithm based on CF. Based on the LOM (Learning Object Meta-data), Mei *et al.* [10] modeled user interests and educational resources for online course recommendation. Apaza *et al.* [9] used the LDA (Latent Dirichlet Allocation) model to extract the features of online courses. Chen *et al.* [6] used cluster analysis and multiple linear regression models to recommend students' interest courses from their behavioral information such as attendance. However, it is lack of studies on the interaction between personal preferences and unfamiliar subject areas.

Our recommendation model differs from the above in the following three aspects. 1) We abstract a preference from user's historical learning records to match unfamiliar subject areas. 2) We analyze the popular courses to obtain the hot topic. 3) We propose a course item model mixed with personal preferences and subject area aware.

3 Personalized Generation Model

In this section, we first introduce the key data structures and symbols used in this paper. Then we propose PS-LDA on learning preferences and subject area awareness for personalized recommendations.

3.1 Problem Definition

To facilitate the following demonstration, we have defined the key data structures and symbols used in this article. Table 1 lists the relevant symbols used in this article.

Definition 1 Course Item. Course item v refers to a specific course in an access subject area.

Definition 2 User Learning. The user learning is a triple (u, v, s_v), which indicates that the user u selects the course item v in the subject area s_v.

Definition 3 User Learning Record. For each user u in dataset D, we create a user learning record D_u, which is a set of quaternions associated with u. We denote users, course items, subject areas and labels as $(u, v, s_v, c_v) \in D$, where $u \in U$, $v \in V$, $s_v \in S$, $c_v \in C_v$. C_v represents the set of labels associated with the course item v. Note that, course items may contain multiple labels. For the learning record of the user activity, user u selects the course item v in s_v. Then we have a set of quaternions, which is $D_{uv} = \{(u, v, s_v, c_v): c_v \in C_v\}$. Obviously, $D_{uv} \subseteq D_u$.

Definition 4 Topic. A topic z in the course item set V is represented by the topic model ϕ_z, $\{P(v|\phi_z) : v \in V\}$ or $\{\phi_{zv} : v \in V\}$, which is the probability distribution of the geographic items. By analogy, the learning preference topic in the user set U is represented by the label c_v in the user's historical learning record, and is represented by the topic model ϕ'_z, $\{P(c|\phi'_z) : c \in C\}$ or $\{\phi'_{zc} : c \in C\}$, which is the probability distribution of the user's learning preferences. In summary, each topic z corresponds to two topic models in our work, namely ϕ_z and ϕ'_z.

Table 1. Definition of symbols.

Symbol	Description
N, V, M, C	The number of users, course items, subject areas, labels
U, V, S, C	The set of users, course items, subject areas, labels
V_s	The set of course items belong to subject areas s
s_v	The course item v of subject area s
c_v	The label describing course item v
K	The number of topics
D_u	The historical learning record of u
θ_u	The learning preferences of user u, expressed by a multinomial distribution over topics
θ'_s	The popular courses of subject Area s, expressed by a multinomial distribution over topics
ϕ_z	A multinomial distribution over course items specific to topic z
ϕ'_z	A multinomial distribution over labels specific to topic z
β, β'	Dirichlet priors to multinomial distributions ϕ_z, ϕ'_z
α, α'	Dirichlet priors to multinomial distributions θ_u, θ_s
λ_u	The mixing weight specific to user u
γ, γ'	Beta priors to generate λ_u

Definition 5 User Learning Preferences. The learning preference of user u is represented by θ_u, where θ_u is the probability distribution of the topic.

Definition 6 Popular Courses. Popular courses in subject area s are represented by θ'_s, the probability distribution of topics, which can mine popular courses in subject areas.

3.2   PS-LDA

The hybrid model considers the user's learning preferences and the influence of popular courses in a unified way. Given the querying user u and the visiting subject area s, the probability that user u chooses course item v when visiting the intersection of the subject area is sampled from the following model.

$$P(v|\theta_u, \theta'_{su}, \phi, \phi') = \lambda_u P(v|\theta_u, \phi, \phi') + (1 - \lambda_u)P(v|\theta'_{su}, \phi, \phi') \tag{1}$$

$P(v|\theta_u, \phi, \phi')$ is the probability of generating the curriculum item v based on learning preferences θ_u of u. And the process of generating $P(v|\theta_u, \phi, \phi')$ is denoted as P-LDA. $P(v|\theta'_{su}, \phi, \phi')$ is the probability of generating the curriculum item v according to popular courses θ'_s in the subject area s. And the process of generating $P(v|\theta'_{su}, \phi, \phi')$ is denoted as S-LDA. λ_u is the parameter mixed weight for controlling the selection.

In order to further alleviate the problem of data sparseness, PS-LDA combines the label information of user history learning records. We redefine Eq. 1 as follows:

$$P(v|\theta_u, \theta'_{su}, \phi, \phi') = \sum_{c \in C_v} P(v, c|\theta_u, \theta'_{su}, \phi, \phi') \tag{2}$$

$$P(v|\theta_u, \phi, \phi') = \sum_{c \in C_v} P(v, c|\theta_u, \phi, \phi') \tag{3}$$

$$P(v|\theta'_{su}, \phi, \phi') = \sum_{c \in C_v} P(v, c|\theta'_{su}, \phi, \phi') \tag{4}$$

Where C_v represents the set of labels associated with the course item v. In PS-LDA, users' learning interest θ_u and popular courses θ'_s are both modeled by polynomial distributions on potential topics. Each course item v is generated from a sample topic z. PS-LDA also parameterizes the distribution of labels associated with each topic z. So, z is responsible for generating course items and their labels at the same time.

$$P(v, c|\theta_u, \phi, \phi') = \sum_z P(v, c|z, \phi_z, \phi'_z)P(z|\theta_u) = \sum_z P(v|z, \phi_z)P(c|z, \phi'_z)P(z|\theta_u) \tag{5}$$

$$P(v, c|\theta'_{su}, \phi, \phi') = \sum_z P(v, c|z, \phi_z, \phi'_z)P(z|\theta'_{su}) = \sum_z P(v|z, \phi_z)P(c|z, \phi'_z)P(z|\theta'_{su})$$
$$\tag{6}$$

We assume that the course items and their labels are independent of the topic. $P(v, c|\theta_u, \phi, \phi')$ and $P(v, c|\theta'_{su}, \phi, \phi')$ are calculated according to formulas (5) and (6).

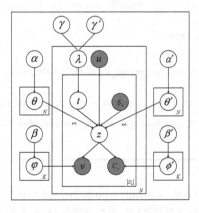

Fig. 1. Graphical representation of PS-LDA

By estimating the parameters of the PS-LDA model to obtain the topics of the course items and labels, this validates our prior knowledge that course items with many users. Otherwise, we cluster similar content into the same topic with high probability. Figure 1 illustrates the generation process with a graphical model. Algorithm 1 outlines the generation process, where Beta (.) is the Beta distribution. And γ, γ' are two of the parameters.

3.3 Model Inference

We use folded Gibbs sampling to obtain samples of hidden variable assignments, which helps to estimate unknown parameters $\{\theta, \theta', \phi, \phi', \lambda\}$ in PS-LDA. To simplify, we specify the hyperparameters α, α', β, β', γ, γ' with fixed values, $e.g.$, $\alpha = \alpha' = 50/K$, $\beta = \beta' = 0.01$, $\gamma = \gamma' = 0.5$. During the sampling process, we start with the joint probability of all user profiles in the dataset. Next, using the chain rule, we obtain the posterior probability of the sampled subject of each quadruplet (u, v, s_v, c_v). Specifically, we use a two-step Gibbs sampling procedure.

Algorithm 1. Probabilistic generative process in PS-LDA

Input: Topic z; User learning record D;
Output: Model parameters θ, θ', ϕ, ϕ' and λ;

1:　　**For** *each topic z* **do**
2:　　　　Draw $\phi_z \sim Dirichlet(\cdot \,|\beta)$;
3:　　　　Draw $\phi'_z \sim Dirichlet(\cdot \,|\beta')$;
4:　　**End**
5:　　**For** *each D_u in D* **do**
6:　　　　**For** *each record* $(u, v_{ui}, s_{ui}, c_{ui}) \in D_u$
7:　　　　　　Toss a coin t_{ui} according to $bernoulli(t_{ui}) \sim beta(\gamma, \gamma')$;
8:　　　　　　**If** $t_{ui} = 1$ **then**
9:　　　　　　　　Draw $\theta_u \sim Dirichlet(\cdot \,|\alpha)$;
10:　　　　　　　Draw a topic $z_{ui} \sim multi(\theta_u)$ according to the learning preference of user u;
11:　　　　　　**End**
12:　　　　　　**If** $t_{ui} = 0$ **then**
13:　　　　　　　　Draw $\theta'_{s_u} \sim Dirichlet(\cdot \,|\alpha')$;
14:　　　　　　　Draw a topic $z_{ui} \sim multi(\theta'_{s_u})$ according to the popular courses of s;
15:　　　　　　**End**
16:　　　　　　Draw a course item $v_{ui} \sim multi(\phi_{z_{ui}})$ from z_{ui} - specific course item distribution;
17:　　　　　　Draw a label $c_{ui} \sim multi(\phi'_{z_{ui}})$ from z_{ui} - specific label distribution;
18:　　　　**End**
19:　　**End**

Due to space constraints, we only show the derived Gibbs sampling formula, omitting the detailed derivation process. We sample t based on the posterior probability as show in Eq. 7 and 8:

$$P(t_{ui} = 1 | t_{\neg ui}, z, u, .) \propto \frac{n_{uz_{ui}}^{\neg ui} + \alpha_{z_{ui}}}{\sum_z \left(n_{uz}^{\neg ui} + \alpha_z \right)} \times \frac{n_{ut_1}^{\neg ui} + \gamma}{n_{ut_0}^{\neg ui} + n_{ut_1}^{\neg ui} + \gamma + \gamma'} \tag{7}$$

$$P(t_{ui} = 0 | t_{\neg ui}, z, u, .) \propto \frac{n_{s_{ui}z_{ui}}^{\neg ui} + \alpha'_{z_{ui}}}{\sum_z \left(n_{s_{ui}z}^{\neg ui} + \alpha'_z \right)} \times \frac{n_{ut_0}^{\neg ui} + \gamma'}{n_{ut_0}^{\neg ui} + n_{ut_1}^{\neg ui} + \gamma + \gamma'} \tag{8}$$

Where n_{ut_1} is the number of times when $t = 1$ in the user profile D_u. So is n_{ut_0} when $t = 0$. n_{uz} is the number of times when the topic z is sampled from a polynomial distribution specific to user. n_{sz} is the number of times when the topic z is sampled in the polynomial distribution of subject area s. The number $n^{\neg ui}$ with a superscript $\neg ui$ indicates that it does not include the number of current instances.

For $t_{ui} = 1$ and $t_{ui} = 0$, we sample the topic z according to the following posterior probability as show in Eq. 9 and 10:

$$P(z_{ui} | t_{ui} = 1, z_{\neg ui}, v, c, u, .) \propto \frac{n_{uz_{ui}}^{\neg ui} + \alpha_{z_{ui}}}{\sum_z \left(n_{uz}^{\neg ui} + \alpha_z \right)} \frac{n_{z_{ui}v_{ui}}^{\neg ui} + \beta_{v_{ui}}}{\sum_v \left(n_{z_{ui}v}^{\neg ui} + \beta_v \right)} \frac{n_{z_{ui}c_{ui}}^{\neg ui} + \beta'_{c_{ui}}}{\sum_c \left(n_{z_{ui}c}^{\neg ui} + \beta'_c \right)} \tag{9}$$

$$P(z_{ui} | t_{ui} = 0, z_{\neg ui}, v, c, u, .) \propto \frac{n_{s_{ui}z_{ui}}^{\neg ui} + \alpha'_{z_{ui}}}{\sum_z \left(n_{s_{ui}z}^{\neg ui} + \alpha'_z \right)} \frac{n_{z_{ui}v_{ui}}^{\neg ui} + \beta_{v_{ui}}}{\sum_v \left(n_{z_{ui}v}^{\neg ui} + \beta_v \right)} \frac{n_{z_{ui}c_{ui}}^{\neg ui} + \beta'_{c_{ui}}}{\sum_c \left(n_{z_{ui}c}^{\neg ui} + \beta'_c \right)} \tag{10}$$

Where n_{zv} is the number of times the topic z generates a course term v. n_{zc} is the number of times the label c is sampled from the topic z.

After a sufficient number of sampling iterations, we can estimate the parameters $\theta, \theta', \phi, \phi'$ and λ as shown in Eq. 11 to 15:

$$\hat{\theta}_{uz} = \frac{n_{uz} + \alpha_z}{\sum_{z'} \left(n_{uz'} + \alpha_{z'} \right)} \tag{11}$$

$$\hat{\theta}'_{sz} = \frac{n_{sz} + \alpha'_z}{\sum_{z'} \left(n_{sz'} + \alpha'_{z'} \right)} \tag{12}$$

$$\hat{\phi}_{zv} = \frac{n_{zv} + \beta_v}{\sum_{v'} \left(n_{zv'} + \beta_{v'} \right)} \tag{13}$$

$$\hat{\phi}'_{zc} = \frac{n_{zc} + \beta'_c}{\sum_{c'} \left(n_{zc'} + \beta'_{c'} \right)} \tag{14}$$

$$\hat{\lambda}_u = \frac{n_{ut_1} + \gamma}{n_{ut_1} + n_{ut_0} + \gamma + \gamma'} \tag{15}$$

4 Top-K Online Recommendation

In our recommendation, we denote a two-parameter pair (u, s_u) as query task with query user u and subject area s_u. The result of the query is a sequential list of course items, which matches the user's learning preferences. After we infer PS-LDA model parameters θ_u, θ'_s, ϕ_z, ϕ'_z, λ_u during the offline modeling phase, the online recommendation section calculates the ranking of each course item v in the query subject area s_u Scores.

$$S(u, s_u, v) = \sum_z F(s_u, v, z) W(u, s_u, z) \tag{16}$$

$S(u, s_u, v)$ is the ranking framework in Eq. 16, which separates offline process from online process for scoring calculation. Specifically, $F(s_u, v, z)$ represents the offline score part for the course item v with respect to the subject area s_u in the dimension z. $F(s_u, v, z)$ is independent to query users. The weight score $W(u, s_u, z)$ is calculated in the online part to find expected weight of the query task (u, s_u).

$$W(u, s_u, z) = \hat{\lambda}_u \hat{\theta}_{uz} + (1 - \hat{\lambda}_u) \hat{\theta}'_{s_u z} \tag{17}$$

$$F(s_u, v, z) = \begin{cases} \hat{\phi}_{zv} \sum_{c_v \in C_v} \hat{\phi}'_{zc_v} & v \in V_{s_u} \\ 0 & v \notin V_{s_u} \end{cases} \tag{18}$$

The main time-consuming components of $W(u, s_u, z)$ are implemented offline. The online calculation can combine the processes shown in Eq. 17. In the process of querying, the offline score $F(s_u, v, z)$ needs to be aggregated in the K dimension by a simple weighted sum function from Eqs. 17 and 18. $W(u, s_u, z)$ is composed of two components, which are used to simulate user learning preferences and popular courses. Each component is associated with a user motivation. $F(s_u, v, z)$ concerns about similarities between the project co-occurrence information and the project content to generate recommendations.

5 Experiments

In this section, we conduct several experiments to compare the recommendation quality of our model.

5.1 Data Setting

Data Sets. We employ the two real-life datasets to evaluate the performance of our model on the course recommendation task.

EdX^1. *EdX* is an online MOOC platform launched by Harvard and MIT. Users can learn the super-quality courses offered by these two famous schools on edX, covering different fields such as computer science, mathematics. EdX provides data on 290 Harvard and MIT online courses, 250 thousand certifications, 4.5 million participants, and 28 million participant hours since 2012.

$GCSE^2$. Google Custom Search Engine (GCSE) is designed to retrieve LinkedIn profiles with the keyword "coursera". Overall, the dataset consists of 15,744 coursera MOOC entries for 5,668 professionals from LinkedIn.

Comparison Methods. We compare our proposed PS-LDA with the following five recommendation methods.

User-Topic Model (UT) [12]: This model is similar to the classic author-topic model (AT model) which assumes that topics are generated according to user interests. The probabilistic formula of the user topic model is presented as follows, where θ_B is a background for smoothing. $P(v|u; \Psi) = \lambda_B P(v|\theta_B) + (1 - \lambda_B) \sum_z P(z|\theta_u)P(v|\phi_z)$.

Category-based k-Nearest Neighbors Algorithm (CKNN) [3]: CKNN projects a user's learning history into the category space and models user's learning preference using a weighted category hierarchy. When receiving a query, CKNN retrieves all the users and course items belong to the querying subject area. Then it applies a user-based CF method to predict the querying user rating of an unvisited course item. Note that the similarity between two users in CKNN is computed according to their weights in the category hierarchy, making CKNN a hybrid recommendation method.

Item-based k-Nearest Neighbors Algorithm (IKNN) [13]: This method utilizes the user's learning history to create a user-course item matrix. When receiving a query, IKNN retrieves all users to find k nearest neighbors by computing the Cosine similarity between two users' course item vectors. Finally, the course items in the user-specific querying subject area that have a relatively high ranking score will be recommended.

Learning Preference LDA (P-LDA): As a component of the proposed PS-LDA model, P-LDA means our method without exploiting the subject area information of course items. For online recommendation, the ranking score is computed by Eq. 16 with $F(s_u, v, z) = \hat{\phi}_{zv} \sum_{c_v \in C_v} \hat{\phi}'_{zc_v}$ and $W(u, s_u, z) = \hat{\theta}_{uz}$.

Subject Area Aware LDA (S-LDA): As another component of the PS-LDA model, S-LDA means our method without considering the content information of course items. For online recommendation, the ranking score is computed by Eq. 16 with $F(s_u, v, z) = \hat{\phi}_{zv}$ and $W(u, s_u, z) = \hat{\lambda}_u \hat{\theta}_{uz} + (1 - \hat{\lambda}_u)\hat{\theta}'_{s_u z}$.

---

[1] https://www.edx.org/.

[2] https://www.gcse.com/.

5.2   Evaluation Methods and Indicators

To make an overall evaluation of the recommendation effectiveness of our proposed PS-LDA, we first design the following two real settings: 1) querying subject areas are new areas to querying users; 2) querying subject areas are familiar to querying users. We divide a user's learning history into a test set and a training set. And we adopt two different dividing strategies with respect to the two settings. For the first setting, we select all course items visited by the user in an unfamiliar subject area as the test set. The rest of the user's learning history is used as the training set. For the second setting, we randomly select 20% of course items visited by the user in familiar subject area as the test set. The rest of personal learning history is used as the training set. We split the user learning history D_u into the training data set $D_{training}$ and the test set D_{test}. To evaluate the recommender models, we adopt the testing methodology and the measurement Recall $@k$ for each test case (u, v, s_v) in D_{test}.

1. We randomly select 1000 additional course in s_v and unrated by user u. We assume that most of them will not be of interest to user u.
2. We compute the ranking score for the test item v as well as the additional 1000 course items.
3. We form a ranked list by ordering all the 1001 course items according to their ranking scores. Let p denote the rank of the test item v within this list. The best result corresponds to the case where v precedes all the random items (*i.e.*, $p = 0$).
4. We form a top-k recommendation list by picking the top-k ranked items from the list. If $p < k$, we have a hit (*i.e.*, the test item v is recommended to the user). Otherwise, we have a miss. The probability of a hit increases with the increasing value of k. When $k = 1001$, we always have a hit.

The computation of Recall $@k$ proceeds as follows. We set hit $@k = 1$ for a single test case if the test course item v appears in the top-k results. If not, hit $@k$ will be set with 0. The overall Recall $@k$ are defined by averaging all test cases.

$$Recall@k = \frac{\#hit@k}{|D_{test}|} \tag{19}$$

Where #hit $@k$ denotes the number of hits in the test set, and $|D_{test}|$ are all test cases.

5.3   Experimental Results

Overall Performance. We first present the optimal performance with well-tuned parameters. And we also study the impact of model parameters. Figure 2 reports the performance of the recommendation algorithms on *EdX*. We show the performance where k is in the range from 1 to 20 since a greater value of k is usually ignored for a typical top-k recommendation task. It is apparent that the algorithms have significant performance disparity in terms of top-k recall. As shown in Fig. 2(a) where querying subject areas are new areas, the recall of PS-LDA is about 0.34 when $k = 10$ and 0.42

when $k = 20$ (*i.e.*, the model has a probability of 34% of placing an appealing event within the querying subject area in the top-10 and 42% of placing it in the top-20). Clearly, our proposed PS-LDA model outperforms other competitor recommendation methods. First, IKNN, CKNN and UT drop behind three other model-based methods, showing the advantage of using latent topic models to model users' preferences. Second, PS-LDA outperforms both P-LDA and S-LDA, showing the advantages of combining learning preferences and subject area in a unified manner.

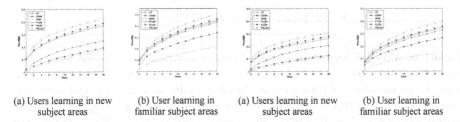

(a) Users learning in new subject areas

(b) User learning in familiar subject areas

(a) Users learning in new subject areas

(b) User learning in familiar subject areas

Fig. 2. Top-k performance on EdX   **Fig. 3.** Top-k performance on GCSE

In Fig. 2(b), we report the performance of all recommendation algorithms for the second setting where querying subject areas are familiar to querying users. We can see that the trend of comparison result is similar to that presented in Fig. 2(a). The main difference is that CKNN outperforms IKNN in Fig. 2(a) while IKNN exceeds CKNN significantly in Fig. 2(b). It shows that the CF-based method (*i.e.*, IKNN) better suits the setting if the user-item matrix is not very sparse. The hybrid method (*i.e.*, CKNN) is more capable of overcoming the difficulty of data sparseness, *e.g.*, the new subject area problem. Another observation is that UT almost performs as well as PS-LDA, and outperforms CKNN and IKNN in the familiar subject area setting, verifying the benefit brought with the subject area influence. However, UT is still less effective than PS-LDA under this setting. Furthermore, the performance of UT is poor in the new subject area setting, as shown in Fig. 2(a), which shows that exploiting subject area influence cannot alleviate the new subject area problem since there is no learning history of the querying user in the new subject area.

Figure 3 reports the performance of the recommendation algorithms on the GCSE dataset. We compare PS-LDA with UT, CKNN, IKNN, P-LDA and S-LDA. From the figure, we can see that the trend of comparison result is similar to that presented in Fig. 2, and PS-LDA performs best.

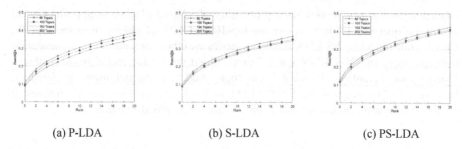

(a) P-LDA (b) S-LDA (c) PS-LDA

Fig. 4. Impact of the number of latent topics

Impact of Model Parameters. Tuning model parameters, such as the number of topics for all topic models, is critical to the performance of models. We therefore also study the impact of model parameters on *EdX* dataset. Because of space limitations, we only show the experimental results for the new subject area setting. As for the hyperparameters α, α', β, β', γ and γ', following the existing works [2], we empirically set fixed values (*i.e.*, $\alpha = \alpha' = 50/K$, $\beta = \beta' = 0.01$, $\gamma = \gamma' = 0.5$). We tried different setups and found that the estimated topic models are not sensitive to the hyperparameters. But the performances of the topic models are slightly sensitive to the number of topics. Thus, we tested the performance of P-LDA, S-LDA and PS-LDA models by varying the number of topics shown in Figs. 4(a) to 4(c). From the results, we observe 1) the Recall @k values of all latent topic-based recommender models slightly increase with the increasing number of topics. 2) The performance of latent topic-based recommender models does not change significantly when the number of topics is larger than 150. 3) P-LDA, S-LDA and PS-LDA perform better under any number of topics, and PS-LDA consistently performs best.

6 Conclusion

This paper proposed a personalized recommendation, PS-LDA, which can facilitate people's study not only in their familiar subject area but also in a new area where they have no learning history. By taking advantage of both the content and subject area information of course items, our system overcomes the data sparsity problem in the original user-item matrix. We evaluated our system using extensive experiments based on two real-life datasets. According to the experimental results, our approach significantly outperforms existing recommendation methods in effectiveness. The results also justify each component proposed in our system, such as taking learning preferences and subject area information into account.

Acknowledgement. This research was supported by the Joint Funds of the National Natural Science Foundation of China under Grant No. U1811261, the Project of Liaoning Provincial Public Opinion and Network Security Big Data System Engineering Laboratory.

References

1. Sarwar, B., Karypis, G., Konstan, J., Riedl, J.: Item-based collaborative filtering recommendation algorithms. In: Proceedings of the 10th International Conference on World Wide Web, pp. 285–295. ACM (2001). https://doi.org/10.1145/371920.372071
2. Du, M., Christensen, R., Zhang, W., Li, F.: Pcard: personalized restaurants recommendation from card payment transaction records. In: Proceedings of the 28th International Conference on World Wide Web, pp. 951–958. ACM (2019)
3. Bao, J., Zheng, Y., Mokbel, M.F.: Location-based and preference-aware recommendation using sparse geo-social networking data. In: Proceedings of the 20th International Conference on Advances in Geographic Information Systems, pp. 199–208. ACM (2012)
4. Fan, J., et al.: Octopus: an online topic-aware influence analysis system for social networks. In: Proceedings of the 34th International Conference on Data Engineering, pp. 1569–1572. IEEE (2018)
5. Yin, H., Sun, Y., Cui, B., Hu, Z., Chen, L.: LCARS: a location-content-aware recommender system. In: Proceedings of the 19th International Conference on Knowledge Discovery and Data Mining, pp. 221–229. ACM (2013)
6. Chen, W., Chu, J., Luan, J., Bai, H., Wang, Y., Chang, E.Y.: Collaborative filtering for orkut communities: discovery of user latent behavior. In: Proceedings of the 18th International Conference on World Wide Web, pp. 681–690. ACM (2009)
7. Hou, Y., Zhou, P., Wang, T., Yu, L., Hu, Y., Wu, D.: Context-aware online learning for course recommendation of MOOC big data. arXiv preprint arXiv:1610.03147 (2016)
8. Popescul, A., Ungar, L.H., Pennock, D.M., Lawrence, S.: Probabilistic models for unified collaborative and content-based recommendation in sparse-data environments. arXiv preprint arXiv:1301.2303 (2013)
9. Apaza, R.G., Cervantes, E.V., Quispe, L.C., Luna, J.O.: Online courses recommendation based on LDA. In: Proceedings of the 1st Symposium on Information Management and Big Data, pp. 42–48. CEUR (2014)
10. Mei, L., He, J., Liu, H., Du, X.: Latent path connected space model for recommendation. In: Shao, J., Yiu, M.L., Toyoda, M., Zhang, D., Wang, W., Cui, B. (eds.) APWeb-WAIM 2019. LNCS, vol. 11642, pp. 163–172. Springer, Cham (2019). https://doi.org/10.1007/978-3-030-26075-0_13
11. Yu, J., An, Y., Xu, T., Gao, J., Zhao, M., Yu, M.: Product recommendation method based on sentiment analysis. In: Meng, X., Li, R., Wang, K., Niu, B., Wang, X., Zhao, G. (eds.) WISA 2018. LNCS, vol. 11242, pp. 488–495. Springer, Cham (2018). https://doi.org/10.1007/978-3-030-02934-0_45
12. Takeuchi, Y., Sugimoto, M.: CityVoyager: an outdoor recommendation system based on user location history. In: Ma, J., Jin, H., Yang, L.T., Tsai, J.J.-P. (eds.) UIC 2006. LNCS, vol. 4159, pp. 625–636. Springer, Heidelberg (2006). https://doi.org/10.1007/11833529_64
13. Song, Y., Huang, J., Zhou, D., Zha, H., Giles, C.L.: IKNN: informative k-nearest neighbor pattern classification. In: Kok, J.N., Koronacki, J., Lopez de Mantaras, R., Matwin, S., Mladenič, D., Skowron, A. (eds.) PKDD 2007. LNCS (LNAI), vol. 4702, pp. 248–264. Springer, Heidelberg (2007). https://doi.org/10.1007/978-3-540-74976-9_25

An Explainable Recommendation Method Based on Multi-timeslice Graph Embedding

Huiying Wang, Yue Kou[✉], Derong Shen, and Tiezheng Nie

Northeastern University, Shenyang 110004, China
307326704@qq.com, {kouyue,shenderong,
nietiezheng}@cse.neu.edu.cn

Abstract. Deep neural networks (DNN) can be used to model users' behavior sequences and predict their interest based on the historical behavior. However, current DNN-based recommendation methods lack explainability, making them difficult to guarantee the credibility of the recommendation results. In this paper, a Multi-Timeslice Graph Embedding (MTGE) model is proposed. First, it can effectively obtain the embedded representations of user behavior (or items) on a single timeslice. Second, the dynamic evolution of user preferences can be analyzed through integrating the embedded representations on multi-timeslices. Then, an explainable recommendation algorithm based on MTGE is proposed, which can effectively improve the accuracy of recommendation and support the model-level explainability. The feasibility and effectiveness of the key technologies proposed in the paper are verified through experiments.

Keywords: Explainable recommendation · Multiple timeslices · Graph embedding · User behavior sequences

1 Introduction

Personalized recommendation systems have played an increasingly important role in people's lives. Among a large number of recommendation algorithms, the collaborative filtering algorithms has been widely used, but traditional collaborative filtering algorithms are less accurate. Deep neural networks (DNN) can be used to capture deep features of users and items to get more accurate representations. However, DNN is often seen as a black box with the disadvantages of unexplainable processing.

Let us consider the following motivating scenarios.

Example 1.1: A month ago, a user gave a high rating to a smartwatch, while a week earlier he was interested in a tennis racket. That is, the user's interests are not static. So, we need to describe the dynamic evolution of user preferences.

Example 1.2: Fig. 1 depicts DNN-based recommendation model and explainable recommendation model. DNN-based recommendation model does not support explainability, and it is difficult to convince users. On the contrary, the explainable recommendation model can give supporting arguments for the results. Some reasons such as "N users love this item" that can reflect the characteristics of the item. It can help people understand the model and improve its transparency.

G. Wang et al. (Eds.): WISA 2020, LNCS 12432, pp. 84–95, 2020.
https://doi.org/10.1007/978-3-030-60029-7_8

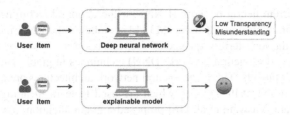

Fig. 1. Comparison of recommendation models

In this paper, we propose an explainable recommendation method based on multi-timeslice graph embedding model. Our main contributions are summarized as follows:

- A Multi-Timeslice Graph Embedding model (namely MTGE) is proposed. First, it can effectively obtain the embedded representations of user behavior (or items) on a single timeslice. Second, the dynamic evolution of user preferences can be analyzed through integrating these embedded representations on multiple timeslices.
- An explainable recommendation algorithm based on MTGE is proposed, which can effectively improve the accuracy of recommendation and support the model-level explainability. On the one hand, the ratings are predicted based on the user's latent vectors and item latent vectors. On the other hand, the explainability of the model is supported, which can effectively improve the transparency of the recommendation model.
- Compared with existing methods through experiments, the effectiveness of our method on real datasets is demonstrated.

The rest of this paper is organized as follows. Section 2 introduces the related work. Section 3 defines several important concepts used in the paper. Section 4 proposes our multi-timeslice graph embedding model. Section 5 proposes an explainable recommendation algorithm based on MTGE. Section 6 shows the experimental results. Section 7 concludes the paper and presents the future work.

2 Related Work

Many methods of recommendation have been proposed in recent years. First, we review the techniques for them. Then we analyze how our work differs from them.

The collaborative filtering recommendation algorithm [1–5, 7] has been widely researched and used in recent years. PMF [2] models the latent vectors of users and items through a Gaussian distribution. Ma Porteous et al. [5] added auxiliary information to Bayesian matrix factorization to improve the accuracy of the recommendation results.

Deep neural networks (DNN) [6, 8] have significant feature learning capabilities by learning deep network structures to represent information about users and items. NeuMF [6] presented a Neural Collaborative Filtering framework to learn latent features of users and items by multi-layer perceptron. Some works have used DNN modeling users' behavior sequences to predict users' preferences [12, 14, 15]. RRN

[12] used a recurrent neural network (RNN) for the temporal recommendation, it can capture the dynamic changes of users. M3 [14] combined the RNN model and attentional to model the long-term sequence of user behaviors.

More recently, deep neural network (DNN) techniques in graph data [13, 18] have made great developments. These deep neural network architectures are known as graph neural networks (GNNs) [9–11, 17], which are used to learn meaningful representations of graph data. Vaswani et al. [16] proposed a graph attention network to resolve the shortcomings of existing methods based on graph convolution. Fan et al. [17] proposed a graph neural network framework for rating predictions that models social graphs to learn user representations.

In previous work, traditional recommendation techniques have mainly considered the static preferences of users. Although some works used DNN to capture users' dynamic preferences, DNN are unexplainable, making the model difficult to understand. In this paper, we proposed the multi-timeslice graph embedding model to obtain the user dynamic preferences and MTGE-based explainable recommendation algorithm can support the model-level explainability.

3 Problem Definition

User-Item Graph. $\mathbf{R} \in \mathbf{R}^{n \times m}$ is user-item graph, also known as the user-item rating matrix, where n is the number of users and m is the number of items, r_{ij} is the rating score of user u_i for item v_j, which can be regarded as the opinion of u_i for v_j.

Latent Vector. The latent vectors of users and items can be regarded as their hidden features, which can predict the user's preference for items.

Temporal Recommendation. Given a user set U, an item set V, where the user-item preference sequences from time 1 to time T is: $\mathbf{R} = \left[\mathbf{R}^1, \mathbf{R}^2, \ldots, \mathbf{R}^T, \right]$. The goal of temporal recommendation is to predict the behavior of each user at time $T + 1$.

Rating Prediction. The rating prediction is based on the user-item rating matrix $\mathbf{R}$, $\hat{r}_{ij}$ represents the predicted rating of u_i for v_j, and we aim to predict the missing rating value in $\mathbf{R}$.

4 MTGE Model

In this section, we describe the model of MTGE. First, we introduce the overall framework of the model, then the single-timeslice graph embedding method is explained, finally, the integration method of multi-timeslice graph embedding is proposed.

4.1 Model Overview

The basic idea of our MTGE model is shown as Fig. 2. The model consists of three components: user embedding, item embedding and rating prediction.

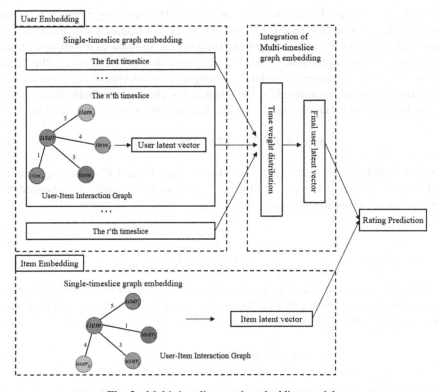

Fig. 2. Multi-timeslice graph embedding model

- User embedding. It is to obtain the user latent vectors. The features of users can be reflected in a set of items that users interact with. Considering the dynamic features of user preferences, the sequence of user behavior is first divided into T timeslices in time sequence.
- Item embedding. It is to obtain the item latent vectors. Similarly, item latent vector can be learned through a set of users that the item has interacted with. And since the item features are not easy to change in a short time, we regard them as static.
- Rating prediction. It is to get the user's predicted rating for the item. We integrate user and item modeling to predict ratings and learn model parameters.

4.2 Single-Timeslice Graph Embedding

The graph neural network updates the current node status by aggregating the information of neighboring nodes. Therefore, fusing neighborhood information from the user-item graph, and the user latent vector is learned through the set $N(i)$ of items that the user has interacted with. The latent vector $\mathbf{h}_i$ of user u_i can be expressed as:

$$\mathbf{h}_i = ReLU\left(\mathbf{W} \cdot \left\{\sum\nolimits_{k \in N(i) \cup \{i\}} \alpha_{ik}\mathbf{o}_{ik}\right\}\right) \tag{1}$$

where *ReLU* is the non-linear activation function and $\mathbf{W}$ is the weight matrix. $\mathbf{o}_{ik}$ is the opinion interaction vector between u_i and v_k, which is obtained by aggregating the embedding vector of v_k and the rating embedding vector. Since the user's rating of items can reflect the user's preference, encoding the interaction information into the latent vector can get a higher accuracy vector. To distinguish the influence of each item on user u_i, α_{ik} denotes the attention weight of the interaction with v_k in contribution to $u'_i s$ latent vector from the interaction history $N(i)$. And the attention network is defined as:

$$\alpha_{ik}^* = ReLU\left(\mathbf{W}_2 \cdot RELU\left(\mathbf{W}_1\left[\mathbf{e}_u^i; \mathbf{o}_{ik}\right]\right)\right) \tag{2}$$

where the inputs of the attention network is the embedded vector $\mathbf{e}_u^i$ of the target user u_i and opinion interaction vector $\mathbf{o}_{ik}$. Then the final attention scores are obtained by normalizing α_{ik}^* using the Softmax function:

$$\alpha_{ik} = \frac{\exp\left(\alpha_{ik}^*\right)}{\sum_{k \in N(i)} \exp\left(\alpha_{ik}^*\right)} \tag{3}$$

Likewise, we use a similar method to learn the latent vector of the items. For each item, information can be collected from a group of users (denoted as $D(j)$) who interacted with. The latent vector $\mathbf{z}_j$ for item v_j is:

$$\mathbf{z}_j = ReLU\left(\mathbf{W} \cdot \left\{\sum_{k \in D(j) \cup \{j\}} \alpha_{jt} \mathbf{s}_{jt}\right\}\right) \tag{4}$$

where $\mathbf{s}_{tj}$ is the opinion interaction vector of v_j with u_t and α_{jt} represents the attention weight to identify the importance of different users.

$$\alpha_{jt}^* = ReLU\left(\mathbf{W}_2 \cdot ReLU\left(\mathbf{W}_1[\mathbf{e}_v^j; \mathbf{s}_{jt}]\right)\right) \tag{5}$$

$$\alpha_{jt} = \frac{exp\left(\alpha_{jt}^*\right)}{\sum_{k \in D(j)} exp\left(\alpha_{jt}^*\right)} \tag{6}$$

4.3   Integration of Multi-timeslice Graph Embedding

User preferences for items are not always static, but can change over time. Therefore, user's preferences are summarized as a matrix sequence $\mathbf{R} = \left[\mathbf{R}^1, \mathbf{R}^2, \ldots, \mathbf{R}^T,\right]$. For a single timeslice, use the graph embedding method introduced in Sect. 4.2 to obtain the latent factor of user u_i on each timeslice: $\mathbf{h}_{i,1}, \mathbf{h}_{i,2}, \ldots, \mathbf{h}_{i,T}$.

However, even if a user's recent behavior is more likely to influence the current decision, the user's distant behavioral preferences can also influence current behavior to some extent. As shown in Fig. 3, we introduce a time decay function to artificially control the weight of each timeslice' influence on the current user's interest preferences:

$$\beta_t^* = e^{-\alpha t} \tag{7}$$

$$\beta_t = \frac{\exp(\beta_t^*)}{\sum_{t \in T} \exp(\beta_t^*)} \tag{8}$$

where β_t^* is the value of β at moment t and α is the "cooling factor", which takes a smaller value if wants to slow down the rate of decay of interest, otherwise, it takes a larger value. Then, the final time weight is normalized with a Softmax function.

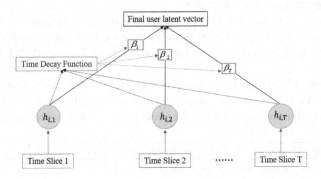

Fig. 3. Time weight distribution

For the weighted sum of the representations of user u_i in each timeslice, the final interest representation $\mathbf{H}_i$ of u_i can be obtained:

$$\mathbf{H}_i = \sum_{t=1}^{T} \beta_t \cdot \mathbf{h}_{i,t} \tag{9}$$

5   MTGE-Based Explainable Recommendation Algorithm

In this section, an explainable recommendation algorithm based on MTGE is proposed. We first introduce the algorithm, then the rating prediction and parameter learning section of the algorithm is represented, and finally analyze the explainability of the algorithm.

5.1   Algorithm Description

Training Stage. First, construct the MTGE model with the training set as input. Then initialize the model parameters, and calculate the predicted ratings of the model. Next, the model parameters are updated by gradient descent to minimize the loss function.

Testing Stage. First, construct the MTGE model with the testing set as input. Then the ratings were calculated using the trained model parameters as the final ratings predicted output.

MTGE-based explainable recommendation algorithm is showed in Algorithm 1, and the major steps are as follows.

Step 1 (line 1). Construct MTGE model based on rating matrix and the number of timeslices.

Step 2 (line 2–4). Divide the rating matrix **R** according to the number of timeslices T, initialize the user embedding vector matrix U, item embedding matrix V, score embedding matrix S and *loss* value.

Step 3 (line 5–13). Calculate user latent vector $\mathbf{H}_i$ and item latent vector $\mathbf{z}_j$.

Step 4 (line 14–15). Learning Stage. Calculate the rating error, update the model parameters by gradient descent until the stop condition is satisfied, and then stop the iteration.

Step 5 (line 17–19). Generate prediction results. For the users and items to be predicted, after obtaining the model parameters through training, the rating prediction is performed based on the user latent vector and the item latent vector.

Algorithm 1: MTGE-based explainable recommendation algorithm

Input:	Rating matrix **R**, Number of timeslices T
Output:	Predicted rating $\hat{\mathbf{R}}$

1.   Construct MTGE based on **R**, T
2. $\mathbf{R} = [\mathbf{R}^1, \mathbf{R}^2, \dots, \mathbf{R}^T,]$
3.   Initialize U, V, S, the neural network parameter
4.   Initialize *loss*
5.   For $u_i \in U$ do
6.   For each timeslice do
7. $\mathbf{h}_{i,t} = compu(\mathbf{R}^t, U, V, S)$
8.   End for
9. $\mathbf{H}_i = \sum_{t=1}^{T} \beta_t \cdot \mathbf{h}_{i,t}$
10.   End for
11.   For $v_j \in V$ do
12. $\mathbf{z}_j = compu(\mathbf{R}^t, U, V, S)$
13.   End for
14   Compute *loss*
15.   Update U, V, S, the neural network parameter
16.   IF meet the stop condition then
17. $\hat{r}_{ij} = MLP[\mathbf{H}_i; \mathbf{z}_j]$
18.   End IF
19.   Return $\hat{\mathbf{R}}$

5.2 Rating Prediction and Parameter Learning

Rating Prediction. The latent vectors of users and items can reflect their preferences and features. We can first connect $\mathbf{H}_i$ and $\mathbf{z}_j$, then feed it into multilayer perceptron (MLP) to obtain the rating prediction:

$$\hat{r}_{ij} = MLP\left[\mathbf{H}_i; \mathbf{z}_j\right] \tag{10}$$

Parameter Learning. To learn the model parameters for MTGE, we specify a loss function for optimization. The loss function is expressed as:

$$loss = \frac{1}{2|T|} \sum_{(u,i)\in T} \left(r_{ui} - \hat{r}_{ij}\right)^2 \tag{11}$$

5.3 Explainability Analysis

This paper consider that user preferences for items are change over time. Some researchers have used deep neural networks (DNN) to model dynamic user preferences, but the model is unexplainable, reducing the transparency of the recommended model.

Since the user's behavior can reflect his or her current preferences, we split the user behavior data based on timeslice and model the user behavior separately to obtain the user preference representation on each timeslice. Next, combining the user's interest decay phenomenon, and then weighing and integrating the user's preference representation in each timeslice, which makes the user's final feature vector have intuitive meaning. For example, a user's behavior a week ago usually has more influence on his current preferences than his behavior six months ago. Moreover, the model-level explainability has multiple significance for the recommendation system, which can improve the transparency and persuasiveness of the system.

6 Experiments

6.1 Dataset

We choose the Epinions and the Ciao dataset, where users can score the products. The statistics of the two datasets are shown in Table 1.

Table 1. Statistics of the experimental datasets

Dataset	Epinions	Ciao
Users	136	103
Items	7716	994
Ratings	12659	3016
Time windows	12	6

6.2 Evaluation Metrics

For the rating prediction task of the recommended model, Root Mean Square Error (RMSE) and Mean Absolute Error (MAE) are usually used to evaluate the prediction accuracy:

$$RMSE = \sqrt{\frac{1}{|T|} \sum_{(u,i) \in T} \left(r_{ui} - \hat{r}_{ui}\right)^2} \tag{12}$$

$$MAE = \frac{1}{|T|} \sum_{(u,i) \in T} \left(r_{ui} - \hat{r}_{ui}\right) \tag{13}$$

where r is the true rating data from the testing set, $\hat{r}$ is the predicted rating of the recommended system, and T is the testing set.

6.3 Performance Evaluation

Hyperparameter Settings. To obtain the optimal combination of parameters, we experimented with the main parameters.

Learning Rate. Figure 4(a) and 5(b) shows the changes in the RMSE and MAE at different learning rates. It can be seen that when the learning rate is 0.005, the RMSE and MAE are lowest on the two datasets. Therefore, we set the learning rate to 0.005.

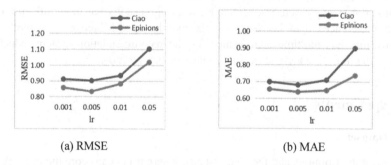

(a) RMSE   (b) MAE

Fig. 4. The effect of learning rate of MTGE algorithm

Embedded Vector Dimension. Figure 5(a) and 5(b) shows the changes in the RMSE and MAE at different embedding vector dimensions. It can be seen that when the embedding vector length is 64, the RMSE and MAE are lowest on the two datasets. Therefore, we set the embedding vector length to 64.

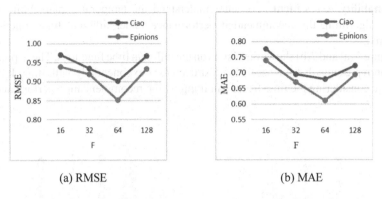

(a) RMSE	(b) MAE

Fig. 5. The effect of embedding vector length of MTGE algorithm

Baseline Algorithm. In order to evaluate the performance of the algorithm, we compare our algorithm with four baseline methods:

- PMF: A probability matrix factorization model that uses a rating matrix to model latent vectors of users and items.
- NeuMF: A matrix factorization model and neural network architecture, using a deeper neural network can provide better recommendation performance.
- RRN: It used a recurrent neural network (RNN) for the temporal recommendation, this model can capture the changes in users' dynamic preferences.
- GraphRec: It proposed a state-of-the-art graph neural network model, which can model graph data coherently for rating prediction.

Figure 6(a) and 6(b) shows the performance of methods. The PMF algorithm is a traditional matrix factorization method, while NeuMF is based on a neural network architecture. RRN takes into account the user's dynamic interests, and its performance has been significantly improved. The GraphRec algorithm good behavior implies the power of graph neural networks in node learning. The MTGE algorithm performs well in both RMSE and MAE, shows that using the graph neural network and considering the user's dynamic features can improve the recommendation performance.

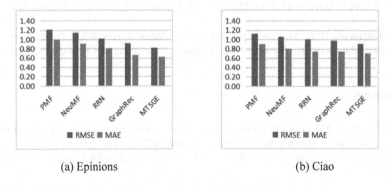

(a) Epinions	(b) Ciao

Fig. 6. Performance comparison between MTGE algorithm and other algorithms

Explainability Assessment. To better understand the temporal explainability of the MTGE algorithm, the recommended performance under different time functions is compared.

Figure 7(a) and 7(b) shows the results on the different time functions. The exponential function represents that the user's interest decreases exponentially with time. It can be seen that considering the time factor can improve the recommendation performance.

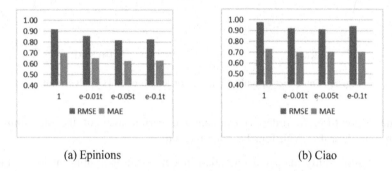

(a) Epinions   (b) Ciao

Fig. 7. Performance comparison of different time functions

For example, for the user preference representation within each timeslice, the closer the timeslice is to the current point in time, the greater the contribution of the timeslice to predicting the user's current interest, while the impact weight of the timeslice faraway is relatively reduced, i.e., the user's current interest is more dependent on his recent behaviors.

7   Conclusion

In this paper, a multi-timeslice graph embedding (MTGE) model is proposed, and the model considers the dynamics of users' interest preferences. Besides, this paper proposes an explainable recommendation algorithm based on MTGE, which has better results compared with existing methods. In the future, we will analyze the user interest drift on multiple timeslices, then improve the model based on user psychology.

Acknowledgment. This work is supported by the National Key R&D Program of China (2018YFB1003404) and the National Natural Science Foundation of China (61672142).

References

1. Wu, Y., Ester, M.: FLAME: a probabilistic model combining aspect based opinion mining and collaborative filtering. In: Proceedings of the ACM International Conference on Web Search and Data Mining, WSDM, pp. 199–208 (2015). https://doi.org/10.1145/2684822. 2685291

2. Salakhutdinov, R., Mnih, A.: Probabilistic matrix factorization. In: Proceedings of the International Conference on Neural Information Processing Systems, NIPS, pp. 1257–1264 (2007). https://doi.org/10.3233/ifs-141462

3. Ma, H., Yang, H., Lyu, M. R., King, I.: SoRec: social recommendation using probabilistic matrix factorization. In: Proceedings of the ACM Conference on Information and Knowledge Management, CIKM, pp. 931–940 (2008). https://doi.org/10.1145/1458082.1458205

4. Koren, Y.: Factorization meets the neighborhood: a multifaceted collaborative filtering model. In: Proceedings of the ACM SIGKDD International Conference on Knowledge Discovery and Data Mining, KDD, pp. 426–434 (2008). https://doi.org/10.1145/1401890.1401944

5. Porteous, I., Asuncion, A.U., Welling, M.: Bayesian matrix factorization with side information and Dirichlet process mixtures. In: Proceedings of the AAAI Conference on Artificial Intelligence, pp. 563–568 (2010). https://doi.org/10.5555/2898607.2898698

6. He, X., Liao, L., Zhang, H., Nie, L., Hu, X., Chua, T.: Neural collaborative filtering. In: Proceedings of the International Conference on World Wide Web, WWW, pp. 173–182 (2017). https://doi.org/10.1145/3038912.3052569

7. Xiong, X., Zhang, M., Zheng, J., Liu, Y.: Social network user recommendation method based on dynamic influence. In: Meng, X., Li, R., Wang, K., Niu, B., Wang, X., Zhao, G. (eds.) WISA 2018. LNCS, vol. 11242, pp. 455–466. Springer, Cham (2018). https://doi.org/10.1007/978-3-030-02934-0_42

8. Kim, D., Park, C., Oh, J., Lee, S., Yu, H.: Convolutional matrix factorization for document context-aware recommendation. In: Proceedings of the 10th ACM Conference on Recommender Systems, pp. 233–240 (2016). https://doi.org/10.1145/2959100.2959165

9. Ma, Y., Wang, S., Aggarwal, C.C., Yin, D., Tang, J.: Multi-dimensional graph convolutional networks. In: Proceedings of the 2019 SIAM International Conference on Data Mining, SIDM, pp. 657–665 (2019). https://doi.org/10.1137/1.9781611975673.74

10. Berg, R.V., Kipf, T., Welling, M.: Graph convolutional matrix completion. arXiv:1706.02263 (2017)

11. Hamilton, W.L., Ying, Z., Leskovec, J.: Inductive representation learning on large graphs. In: Proceedings of the Neural Information Processing Systems, NIPS, pp. 1024–1034 (2017)

12. Wu, C., Ahmed, A.A., Beutel, A., Smola, A., Jing, H.: Recurrent recommender networks. In: Proceedings of the 10th ACM International Conference on Web Search and Data Mining, WSDM, pp. 495–503 (2017). https://doi.org/10.1145/3018661.3018689

13. Kipf, T., Welling, M.: Semi-supervised classification with graph convolutional networks. In: Proceedings of the International Conference on Learning Representations, ICLR (2017)

14. Tang, J., et al.: Towards neural mixture recommender for long range dependent user sequences. In: Proceedings of the World Wide Web Conference, WWW, pp. 1782–1793 (2019). https://doi.org/10.1145/3308558.3313650

15. Kang, W., McAuley, J.J.: Self-attentive sequential recommendation. In: Proceedings of the International Conference on Data Mining, ICDM, pp. 197–206 (2018). https://doi.org/10.1109/icdm.2018.00035

16. Vaswani, A., Shazeer, N., Parmar, N., Uszkoreit, J., Jones, L., Gomez, A.N.: Attention is all you need. In: Proceedings of the Neural Information Processing Systems, NIPS, pp. 5998–6008 (2017)

17. Fan, W., et al.: Graph neural networks for social recommendation. In: Proceedings of the World Wide Web Conference, WWW, pp. 417–426 (2019). https://doi.org/10.1145/3308558.3313488

18. Xu, W., Xu, Z., Zhao, B.: A graph kernel based item similarity measure for top-n recommendation. In: Proceedings of International Conference on Web Information Systems and Applications, WISA, pp. 684–689 (2019). https://doi.org/10.1007/978-3-030-30952-7_69

Effective Knowledge-Aware Recommendation via Graph Convolutional Networks

Bo Zhao, Zhuoming Xu$^{(\boxtimes)}$, Yan Tang, Jian Li, Bei Liu,
and Haimei Tian

College of Computer and Information, Hohai University, Nanjing 210098, China
{bzhao,zmxu,tangyan,jli,liubei,hmtian}@hhu.edu.cn

Abstract. Most existing graph neural network (GNN)-based knowledge-aware recommendation models rely on handcrafted feature engineering and do not allow for end-to-end training. As a state-of-the-art end-to-end framework, the Knowledge-aware Graph Neural Networks with Label Smoothness Regularization (KGNN-LS) model can extend GNNs architecture to knowledge graphs to simultaneously capture semantic relations between entities as well as personalized user preferences for entities/items, thereby making effective recommendation. However, we believe that KGNN-LS still has two weaknesses: (1) In KGNN-LS, the weights of the edges in the graph are determined solely by user preferences for relations without considering user's (potential) personalized interests in entities/items. (2) The sum pooling adopted by KGNN-LS cannot effectively aggregate the most representative information of the neighborhood. In this paper, we propose the improved Knowledge-aware Graph Neural Networks with Label Smoothness Regularization (iKGNN-LS) model, which makes two improvements to KGNN-LS: (1) In iKGNN-LS, by introducing user-specific entity scoring functions, the edge weights are determined jointly by personalized user preferences for relations and for entities. (2) iKGNN-LS uses max pooling instead of sum pooling for neighborhood aggregation. Top-N recommendation experiments on three datasets show that iKGNN-LS outperforms KGNN-LS in terms of Precision@N, Recall@N, and F1-measure@N.

Keywords: Knowledge-aware recommendation · Knowledge graph · GCN · User-specific entity scoring function · Pooling aggregator

1 Introduction

Knowledge graphs (KGs) [5] have proven to be effective in enhancing recommendation performance by providing recommender systems with additional knowledge [2, 4, 7–10, 12, 13]. KG-based recommender systems exploit knowledge-aware recommendation models and apply KGs in three ways [3]: embedding-based methods, path-based methods, and unified methods. *Embedding-based methods* [2, 13] use KG embedding algorithms to translate KG elements into low-dimensional vector representations, which are further integrated into the recommendation models. *Path-based methods* [4, 10] leverage the informative connectivity patterns between the entities in the user-item KG for recommendation. *Unified methods* [7–9] leverage both the semantic

© Springer Nature Switzerland AG 2020
G. Wang et al. (Eds.): WISA 2020, LNCS 12432, pp. 96–107, 2020.
https://doi.org/10.1007/978-3-030-60029-7_9

representation and the connectivity information in the KG for recommendation. Unified methods generally adopt an architecture based on graph neural networks (GNNs) [11], such as graph convolutional networks (GCNs) [6].

However, most existing GNN-based knowledge-aware recommendation models rely on handcrafted feature engineering and do not allow for end-to-end training [8]. The Knowledge-aware Graph Neural Networks with Label Smoothness Regularization (KGNN-LS) model recently proposed by H. Wang *et al.* [7, 8] is a state-of-the-art end-to-end framework that can extend GNNs architecture to KGs to simultaneously capture semantic relations between entities as well as personalized user preferences for entities/items, thereby making effective recommendation.

Despite a state-of-the-art model, KGNN-LS still has two weaknesses: (1) In KGNN-LS, the weights of the edges in the user-specific weighted graph are determined solely by user preferences for relations without considering user's personalized interests in entities/items. This approach cannot distinguish the different contributions of different neighbor entities connected by the same relation to user interests. (2) The sum pooling adopted by KGNN-LS cannot effectively aggregate the most representative information of the neighborhood, resulting in less effective user-specific item embeddings.

To overcome the weaknesses of KGNN-LS, in this paper we propose the improved Knowledge-aware Graph Neural Networks with Label Smoothness Regularization (iKGNN-LS) model, which makes two improvements to KGNN-LS: (1) In iKGNN-LS, by introducing user-specific entity scoring functions, the user-specific edge weights are determined jointly by personalized user preferences for relations and for entities. (2) During feature propagation on the user-specific weighted graph, iKGNN-LS uses max pooling instead of sum pooling for neighborhood aggregation.

To verify the effectiveness of iKGNN-LS and its performance advantage over KGNN-LS, we conducted two comparative experiments of top-N recommendation. First, we used the MovieLens 20M dataset and KGNN-LS to study the influences of edge weights and pooling aggregators on the performance of top-N recommendation. The results show that the edge weights determined jointly by user preferences for relations and for entities, as well as the use of max pooling in neighborhood aggregation can improve recommendation performance. Second, we used three datasets, MovieLens 20M, Last.FM, and Yelp2018, to compare the top-N recommendation performance between KGNN-LS and iKGNN-LS. The results indicate that iKGNN-LS outperforms KGNN-LS in terms of Precision@N, Recall@N, and F1-measure@N.

In summary, the main contributions of this paper are as follows:

- We propose the knowledge-aware recommendation model iKGNN-LS that can calculate user-specific item embeddings more effectively. The model exploits user preferences for relations and for entities to jointly determine user-specific edge weights and uses max pooling instead of sum pooling to aggregate the most representative information of the neighborhood.
- Our comparative experiment on KGNN-LS indicates that both the improved edge weights and max pooling can improve recommendation performance.
- Our comparative experiment between KGNN-LS and iKGNN-LS demonstrates the recommendation performance advantage of iKGNN-LS over KGNN-LS.

2   Improved Knowledge-Aware Recommendation Model

In this section, we expatiate on our proposed iKGNN-LS model. We first give the overview of the model, and then describe three major steps of the model in detail.

In the following, we first introduce some concepts and notations, and then formulate the problem of knowledge-aware recommendation.

A recommender system (RS) has a set of m users $\mathcal{U} = \{u_1, u_2, \ldots, u_m\}$ and a set of n items $\mathcal{V} = \{v_1, v_2, \ldots, v_n\}$. According to users' implicit feedback, a user-item interaction matrix $\mathbf{Y} \in \mathbb{R}^{m \times n}$ can be defined for the system. The matrix's element $y_{uv} = 1$ indicates that user $u \in \mathcal{U}$ engages with item $v \in \mathcal{V}$, such as clicking, browsing, or purchasing; otherwise $y_{uv} = 0$.

A knowledge graph $\mathcal{G}(\mathcal{E}, \mathcal{R})$ has a set of entities $\mathcal{E}$ and a set of relations $\mathcal{R}$. The graph consists of entity-relation-entity triples (h, r, t), where $h \in \mathcal{E}, r \in \mathcal{R}$, and $t \in \mathcal{E}$ are the head, relation, and tail of a triple. In the recommendation setting, each item $v \in \mathcal{V}$ in the RS corresponds to an entity $e \in \mathcal{E}$ in the knowledge graph.

As formulated in [7, 8], given the interaction matrix $\mathbf{Y}$ and the knowledge graph $\mathcal{G}$, the goal of knowledge-aware recommendation is to predict whether user $u \in \mathcal{U}$ has potential interest in item $v \in \mathcal{V}$ with which the user has not engaged before. That is, the task is to learn a prediction function $\tilde{y}_{uv} = \mathcal{F}(u, v | \Theta, \mathbf{Y}, \mathcal{G})$, where $\tilde{y}_{uv}$ denotes the probability that user u will engage with item v, and Θ are model parameters of function $\mathcal{F}$.

2.1   Overview

The main goal of our proposed iKGNN-LS is to learn user-specific item embeddings more effectively. The overview of iKGNN-LS is depicted in Fig. 1, where the learning and recommendation process can be divided into the following steps:

- *Transforming the KG into a user-specific weighted graph:* The KG is transformed into a user-specific weighted graph, where a limited number of direct neighbors of each entity are sampled, and each relation (edge) between two sampled entities is given a user-specific weight to reflect the user's personalized interest.
- *Learning the prediction function:* iKGNN-LS takes the user-specific weighted graph as input, and uses knowledge-aware GCN to calculate user-specific item embeddings via feature propagation on the graph. Simultaneously, it performs label smoothness regularization on the edge weights via label propagation on the graph. Finally, iKGNN-LS uses the unified loss function to learn the prediction model.
- *Making top-N recommendation:* iKGNN-LS uses the learned prediction function to predict probabilities that the user will engage with candidate items, and then employs the probabilities to produce a top-N recommendation list for the user.

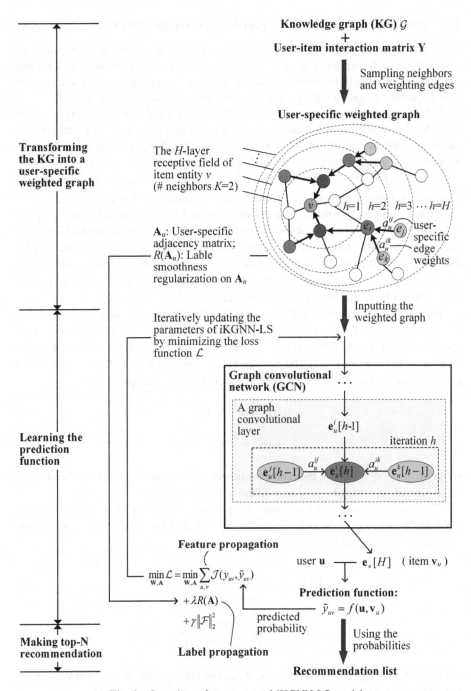

Fig. 1. Overview of our proposed iKGNN-LS model

2.2 Transforming the KG into a User-Specific Weighted Graph

Given a user in the RS, iKGNN-LS transforms the KG into a user-specific weighted graph [8]. In such a graph, at most K direct neighbors of each entity node are sampled (concerned), an edge (representing a relation) between each sampled entity pair is given a user-specific weight to reflect the user's personalized interest in the relation.

Let $N(e_i)$ represent a set of neighbor entities directly connected to entity $e_i \in \mathcal{E}$ in $\mathcal{G}(\mathcal{E}, \mathcal{R})$. In a real-world KG, the number of entities in $N(e_i)$ may vary greatly. Using the same approach as in KGNN-LS [7, 8], iKGNN-LS samples at most K direct neighbors of each entity to form the (single-layer) receptive field $S(e_i) \triangleq \{e_j | e_j \sim N(e_i)\}$, $|S(e_i)| = K$ of entity e_i. This receptive field is used to compute the user-specific neighborhood representation of e_i, denoted as $\mathbf{e}_u^{S(e_i)}$, where $u \in \mathcal{U}$ is a specific user. $\mathbf{e}_u^{S(e_i)}$ can capture structural proximity among entities in the KG. As stated in [7], the receptive field can be extended to multiple hops away (i.e., multiple layers) to model high-order structural proximity and capture users' potential interests. As in KGNN-LS, iKGNN-LS uses the h-layer receptive field to compute the h-order structural proximity, denoted as $\mathbf{e}_u^{S(e_i)}[h]$, $h = 1, 2, \ldots, H$, where H is the maximum depth of the receptive field. Figure 1 depicts the H-layer receptive field of item entity $v \in \mathcal{V} \subseteq \mathcal{E}$, where $K = 2$.

In each layer of receptive field, user-specific edge weights are calculated as follows. Given two entities $e_i \in \mathcal{E}$, $e_j \in S(e_i)$ and their relation r_{e_i,e_j}, iKGNN-LS uses user-specific relation scoring function $s_u(r_{e_i,e_j})$ defined as Eq. (1) [7, 8] to calculate the user-relation score between $u \in \mathcal{U}$ and r_{e_i,e_j}, and uses user-specific entity scoring function $t_u(e_j)$ defined as Eq. (2) to calculate the user-entity score between u and e_j.

$$s_u(r_{e_i,e_j}) = g(\mathbf{u}, \mathbf{r}_{e_i,e_j}) \tag{1}$$

$$t_u(e_j) = g(\mathbf{u}, \mathbf{e}_u^j) \tag{2}$$

where $\mathbf{u}$, $\mathbf{r}_{e_i,e_j}$, $\mathbf{e}_u^j \in \mathbb{R}^d$ are the representations of u, r_{e_i,e_j}, and e_j, respectively. d is the dimension of representations. g is a differentiable function (e.g., inner product).

The above two scores are normalized separately, that is, the user-relation score is normalized to $\tilde{s}_u(r_{e_i,e_j})$, as defined by Eq. (3) [7], and the user-entity score is normalized to $\tilde{t}_u(e_j)$, as defined by Eq. (4).

$$\tilde{s}_u(r_{e_i,e_j}) = \frac{\exp(s_u(r_{e_i,e_j}))}{\sum_{e \in S(e_i)} \exp(s_u(r_{e_i,e}))} \tag{3}$$

$$\tilde{t}_u(e_j) = \frac{\exp(t_u(e_j))}{\sum_{e \in S(e_i)} \exp(t_u(e))} \tag{4}$$

iKGNN-LS uses Eq. (5) to compute the weight of the edge (representing relation r_{e_i,e_j}) with respect to user u, which is referred to as user-specific edge weight.

$$a_u^{ij} = \tilde{s}_u(r_{e_i,e_j}) \cdot \tilde{t}_u(e_j) \tag{5}$$

The weight a_u^{ij} is used as an element to form a user-specific adjacency matrix $\mathbf{A}_u \in \mathbb{R}^{\mathcal{E} \times \mathcal{E}}$, which represents user-specific edge weights for the weighted graph.

2.3 Learning the Prediction Function

As mentioned earlier, our task is to learn a prediction function $\mathcal{F}(u, v | \Theta, \mathbf{Y}, \mathcal{G})$. The learning process includes three steps: feature propagation on the graph, label propagation on the graph, and model learning. Below we describe these three steps.

Feature Propagation on the Graph. The goal of feature propagation is to calculate user-specific item embeddings. The calculation process takes the user-specific weighted graph (including the H-layer receptive field and user-specific edge weights) as input, and employs GCN to compute the final H-order entity representation by aggregating and incorporating neighborhood information (i.e., structure information) in an iterative layer-by-layer manner. The total number of neighborhood aggregation iterations is H. In h-th iteration ($h = 1, 2, \ldots, H$), iKGNN-LS uses max-pooling instead of sum-pooling to aggregate all entities in $S(e_i)$ to form the $(h-1)$-order neighborhood representation of entity e_i, denoted as $\mathbf{e}_u^{S(e_i)}[h-1]$, which is defined by Eq. (6).

$$\mathbf{e}_u^{S(e_i)}[h-1] = \max(\{a_u^{ij} \times \mathbf{e}_u^j[h-1], \forall e_j \in S(e_i)\}) \tag{6}$$

where a_u^{ij} is the user u specific weight of the edge between e_i and e_j, and $\mathbf{e}_u^j[h-1]$ denotes the $(h-1)$-order representation of entity $e_j \in S(e_i)$.

Like KGNN-LS, iKGNN-LS then combines neighborhood representation $\mathbf{e}_u^{S(e_i)}[h-1]$ with the $(h-1)$-order representation of the entity itself, $\mathbf{e}_u^i[h-1]$, to form its h-order entity representation $\mathbf{e}_u^i[h]$, which is defined as Eq. (7) [7].

$$\mathbf{e}_u^i[h] = \sigma(\mathbf{W}_h \cdot (\mathbf{e}_u^i[h-1] + \mathbf{e}_u^{S(e_i)}[h-1]) + \mathbf{b}_h) \tag{7}$$

where $\mathbf{W}_h$ and $\mathbf{b}_h$ are transformation weight and bias, and σ is the nonlinear function such as *ReLU*.

For item entity $v \in \mathcal{V} \subseteq \mathcal{E}$, after H iterations, the final H-order entity representation $\mathbf{e}_u[H]$ (i.e., $\mathbf{v}_u[H]$) is the user-specific item embedding $\mathbf{v}_u$. We can thus input $\mathbf{v}_u$ and user representation $\mathbf{u}$ into a differentiable function (e.g., inner product) to predict the probability $\tilde{y}_{uv}$ that user u will engage with item v, as defined by Eq. (8) [7, 8].

$$\tilde{y}_{uv} = f(\mathbf{u}, \mathbf{v}_u) \tag{8}$$

Label Propagation on the Graph. The goal of label propagation is to assist the learning of entity representations and to help predict unobserved user-item interactions through label smoothness (LS) regularization on the edge weights [8]. As formulated in [8], let $l_u : \mathcal{E} \to \mathbb{R}$ denote a real-valued label function on $\mathcal{G}$, which is constrained to take a specific value $l_u(v) = y_{uv} \in \mathbf{Y}$ at node $v \in \mathcal{V} \subseteq \mathcal{E}$. The label smoothness assumption that adjacent entities in the graph are likely to have similar relevancy labels [8] leads to the following definition of energy function E.

$$E(l_u, \mathbf{A}_u) = \frac{1}{2} \sum_{e_i \in \mathcal{E}, e_j \in \mathcal{E}} a_u^{ij} (l_u(e_i) - l_u(e_j))^2 \tag{9}$$

where a_u^{ij} is the user specific edge weight. $l_u(e_i)$ and $l_u(e_j)$ are user relevancy scores of e_i and e_j, respectively. Like the approach in KGNN-LS, iKGNN-LS repeats the following two steps [8] to achieve the minimum-energy of function E, thereby predicting a user relevancy label/score for each unlabeled entity:

- Propagate labels for all entities: $l_u(\mathcal{E}) \leftarrow \mathbf{D}_u^{-1} \mathbf{A}_u l_u(\mathcal{E})$, where $l_u(\mathcal{E})$ is the vector of labels for all entities, $\mathbf{D}_u$ is a diagonal degree matrix with $D_u^{ij} = \sum_j a_u^{ij}$.
- Reset labels of all items to initial labels: $l_u(\mathcal{V}) \leftarrow \mathbf{Y}[u, \mathcal{V}]^\top$, where $l_u(\mathcal{V})$ is the vector of labels for all items and $\mathbf{Y}[u, \mathcal{V}] = [y_{uv_1}, y_{uv_2}, \ldots]$ are initial labels.

iKGNN-LS uses the same approach as KGNN-LS to perform label smoothness (LS) regularization on the edge weights. Specifically, as described in [8], a single item $v \in \mathcal{V} \subseteq \mathcal{E}$ is held out and it is treated as unlabeled; the label of v can then be predicted by using the rest of (labeled) items and (unlabeled) non-item entities. The LS regularization on the edge weights can thus be achieved via a learning procedure that uses the difference between the true relevancy label of v (i.e., y_{uv}) and the predicted label $\tilde{l}_u(v)$ as a supervised signal. The regularization is defined as Eq. (10) [8].

$$R(\mathbf{A}) = \sum_u R(\mathbf{A}_u) = \sum_u \sum_v \mathcal{J}(y_{uv}, \tilde{l}_u(v)) \tag{10}$$

where $\mathcal{J}$ is the cross-entropy loss function.

Model Learning via the Unified Loss Function. iKGNN-LS uses the same loss function as iKGNN-LS to learn the prediction model. The unified loss function, which combines knowledge-aware GCN and LS regularization, is defined as Eq. (11) [8].

$$\min_{\mathbf{W}, \mathbf{A}} \mathcal{L} = \min_{\mathbf{W}, \mathbf{A}} \sum_{u,v} \mathcal{J}(y_{uv}, \tilde{y}_{uv}) + \lambda R(\mathbf{A}) + \gamma \|\mathcal{F}\|_2^2 \tag{11}$$

where $\mathcal{J}$ is the cross-entropy loss function, λ and γ are balancing hyper-parameters, $R(\mathbf{A})$ is LS regularization on edge weights $\mathbf{A}$, $\|\mathcal{F}\|_2^2$ is the L2-regularizer. By minimizing the loss function, iKGNN-LS uses stochastic gradient descent (SGD) to simultaneously update model parameters: transformation matrix $\mathbf{W}$ and edge weights $\mathbf{A}$.

Note that in Eq. (11), the first term corresponds to feature propagation on the KG, whereas the second term $R(\mathbf{A})$ corresponds to label propagation on the KG. Once the trainable parameters are learned, the prediction function of iKGNN-LS is achieved.

2.4 Making Top-N Recommendation

For a specific user $u \in \mathcal{U}$ in the RS, iKGNN-LS can use the learned prediction function to compute a predicted probability that user u will engage with item $v \in \mathcal{V} \subseteq \mathcal{E}$ with which the user has not engaged before. As shown in Fig. 1, given user representation $\mathbf{u}$ and user-specific item embedding $\mathbf{v}_u$, prediction function $\tilde{y}_{uv} = f(\mathbf{u}, \mathbf{v}_u)$ (being inner product $< \mathbf{u}, \mathbf{v}_u >$ in our experiments) generates predicted probability $\tilde{y}_{uv}$. iKGNN-LS sorts the probabilities in descending order to produce a top-N recommendation list for the user.

3 Experiments

This section presents our two parts of top-N recommendation experiments: (1) we used KGNN-LS to study the influences of edge weights and pooling aggregators on top-N recommendation performance; (2) we used three datasets to compare the top-N recommendation performance of KGNN-LS and iKGNN-LS in order to show the performance advantages of iKGNN-LS over KGNN-LS. It is worth noting that the experiments in [8] have shown that KGNN-LS outperforms six state-of-the-art baselines. Therefore, our experiments do not need to compare iKGNN-LS with the baselines.

3.1 Experimental Setup

Datasets. Our experiments used the MovieLens 20M, Last.FM, and Yelp2018 datasets for movie, music, and local business recommendations. The first two datasets were published by H. Wang *et al.* [7, 8] on GitHub[1]. The authors used Microsoft Satori to construct the KGs for the MovieLens 20M and Last.FM datasets. The details on the datasets and the corresponding KGs can be found in [7, 8]. The Yelp2018 dataset, which is the 2018 edition of the Yelp challenge, was published by X. Wang *et al.* [9] on GitHub[2]. The authors extracted item knowledge from the local business information network (e.g., category, location, and attribute) to construct the KG. The details on the dataset and the corresponding KG can be found in [9]. Following [7, 8], for each dataset, the ratio of training set, validation set, and test set is 6:2:2. Table 1 shows the statistics of the datasets and the KGs.

---

[1] https://github.com/hwwang55/KGNN-LS/tree/master/data.

[2] https://github.com/xiangwang1223/knowledge_graph_attention_network/tree/master/Data.

Table 1. Statistics of the three datasets

	MovieLens 20M	Last.FM	Yelp2018
# users	138,159	1,872	45,919
# items	16,954	3,846	45,538
# interactions	13,501,622	42,346	1,185,068
# entities	102,569	9,366	90,961
# relations	32	60	42
# KG triples	499,474	15,518	1,853,704

Evaluation Metrics. Three popular evaluation metrics [1], Precision at N (P@N), Recall@N (R@N), and F1-measure@N (F1@N), are used to evaluate the top-N recommendation performance (N = 5 or 10).

Model Implementation. The Python code of KGNN-LS was obtained from the GitHub webpage[3]. The code of iKGNN-LS was generated by modifying the Python code of KGNN-LS, specifically, by adding the implementation of the user-specific entity scoring function and the improved edge weights, as well as replacing the sum pooling aggregator with the max pooling one.

Hyperparameter Setting. Like [7, 8], in both iKGNN-LS and KGNN-LS, we set σ as *ReLU* for non-last-layers and *tanh* for the last-layer. We used grid search to select the hyperparameters for the two models. More specifically, just as in [7, 8], we selected the number K of sampled neighbors for entities in $\{2, 4, 8, 16, 32\}$, the dimension d of hidden layers in $\{4, 8, 16, 32\}$, the number L of layers in $\{1, 2\}$, the label smoothness regularizer weight λ in $\{0.01, 0.1, 0.5, 1.0, 1.5\}$, the L2-regularizer weight γ in $\{10^{-9}, 10^{-8}, 5 \times 10^{-8}, 10^{-7}, 5 \times 10^{-7}, 10^{-6}, 5 \times 10^{-6}, 10^{-5}, 5 \times 10^{-5}, 10^{-4}\}$, and the learning rate η in $\{10^{-4}, 2 \times 10^{-4}, 5 \times 10^{-4}, 10^{-3}, 2 \times 10^{-3}, 5 \times 10^{-3}, 10^{-2}, 2 \times 10^{-2}\}$. The resulting optimal hyperparameter settings for the three datasets are shown in Table 2.

Table 2. Hyperparameter settings for the three datasets

	MovieLens 20M	Last.FM	Yelp2018
K	16	8	32
d	32	16	16
L	1	1	1
λ	1.0	0.1	1.0
γ	10^{-7}	10^{-4}	10^{-9}
η	2×10^{-2}	5×10^{-4}	5×10^{-3}

---

[3] https://github.com/hwwang55/KGNN-LS.

3.2 Experimental Results

As in [8], each experiment was repeated 5 times, and the average performance is reported here. For the influence of edge weights on recommendation performance, Table 3 shows the top-N recommendation results on MovieLens 20M, where the figures in columns KGNN-LS and KGNN-LS-entity mean the results of the original KGNN-LS and the KGNN-LS that adds the user-specific entity scoring function, respectively, and the "Improvement (%)" figures refer to the percentages of performance improvement of KGNN-LS-entity over KGNN-LS. The results suggest that the KGNN-LS that adds the user-specific entity scoring function outperforms KGNN-LS in terms of all the metrics. This indicates that the edge weights determined jointly by personalized user preferences for relations and for entities can improve recommendation performance.

For the influence of pooling aggregators on recommendation performance, Table 4 shows the top-N recommendation results on MovieLens 20M, where the figures in columns KGNN-LS and KGNN-LS-max mean the results of the original KGNN-LS and the KGNN-LS that uses max pooling instead of sum pooling for neighborhood aggregation, respectively, and the "Improvement (%)" figures refer to the percentages of performance improvement of KGNN-LS-max over KGNN-LS. The results suggest that the KGNN-LS that uses max pooling outperforms KGNN-LS in terms of all the metrics. This indicates that the max pooling is better than the sum pooling in aggregating neighborhood in the recommendation context.

For the comparative experiment between KGNN-LS and iKGNN-LS, Table 5 shows the top-N recommendation results on the MovieLens 20M, Last.FM, and Yelp2018 datasets, where the "Improvement (%)" figures refer to the percentages of performance improvement of iKGNN-LS over KGNN-LS. The results suggest that on the three datasets, iKGNN-LS outperforms KGNN-LS in terms of all the metrics. This indicates the performance advantage of iKGNN-LS over KGNN-LS.

Table 3. Top-N recommendation results on MovieLens 20M (influence of edge weights)

Metrics	KGNN-LS	KGNN-LS-entity	Improvement (%)
P@5	0.1260	**0.1280**	1.59
P@10	0.0940	**0.0980**	4.26
R@5	0.0989	**0.0996**	0.71
R@10	**0.1550**	**0.1550**	0.00
F1@5	0.1108	**0.1120**	1.08
F1@10	0.1173	**0.1201**	2.39

Table 4. Top-N recommendation results on MovieLens 20M (influence of pooling aggregators)

Metrics	KGNN-LS	KGNN-LS-max	Improvement (%)
P@5	0.1260	**0.1280**	1.59
P@10	0.0940	**0.0990**	5.32
R@5	0.0989	**0.0994**	0.51
R@10	0.1550	**0.1607**	3.68
F1@5	0.1108	**0.1118**	0.90
F1@10	0.1173	**0.1225**	4.43

Table 5. Top-N recommendation results on MovieLens 20M, Last.FM, and Yelp2018

Dataset	Metrics	KGNN-LS	iKGNN-LS	Improvement (%)
MovieLens 20M	P@5	0.1260	**0.1290**	2.38
	P@10	0.0940	**0.1020**	8.51
	R@5	0.0989	**0.1022**	3.34
	R@10	0.1550	**0.1559**	0.58
	F1@5	0.1108	**0.1141**	2.98
	F1@10	0.1173	**0.1232**	5.03
Last.FM	P@5	0.0300	**0.0320**	6.67
	P@10	0.0280	**0.0300**	7.14
	R@5	0.0589	**0.0649**	8.53
	R@10	0.1223	**0.1329**	8.67
	F1@5	0.0399	**0.0429**	7.52
	F1@10	0.0456	**0.0491**	7.68
Yelp2018	P@5	0.0100	**0.0110**	10.00
	P@10	**0.0070**	**0.0070**	0.00
	R@5	0.0122	**0.0128**	4.92
	R@10	0.0184	**0.0195**	5.98
	F1@5	0.0110	**0.0118**	7.27
	F1@10	0.0101	**0.0103**	1.98

4 Conclusions

To overcome the weaknesses of KGNN-LS and learn user-specific item embeddings more effectively, in this paper we propose the improved iKGNN-LS model, which exploits user preferences for relations and for entities to jointly determine user-specific edge weights and uses max pooling instead of sum pooling to aggregate the most representative information of the neighborhood. Our comparative experiments of top-N recommendation on three datasets demonstrate the performance advantage of iKGNN-LS over KGNN-LS. Our future work will focus on further enhancing the iKGNN-LS model by integrating knowledge about users from social networks into the model.

References

1. Aggarwal, C.C.: Evaluating recommender systems. Recommender Systems, pp. 225–254. Springer, Cham (2016). https://doi.org/10.1007/978-3-319-29659-3_7
2. Cao, Y., Wang, X., He, X., Hu, Z., Chua, T.: Unifying knowledge graph learning and recommendation: towards a better understanding of user preferences. In: Proceedings of the World Wide Web Conference, WWW 2019, pp. 151–161. ACM (2019). https://doi.org/10.1145/3308558.3313705
3. Guo, Q., Zhuang, F., Qin, C., et al.: A survey on knowledge graph-based recommender systems. CoRR abs/2003.00911 (2020). https://arxiv.org/abs/2003.00911
4. Ma, W., Zhang, M., Cao, Y., et al.: Jointly learning explainable rules for recommendation with knowledge graph. In: Proceedings of the World Wide Web Conference, WWW 2019, pp. 1210–1221. ACM (2019). https://doi.org/10.1609/aaai.v33i01.33015329
5. Nickel, M., Murphy, K., Tresp, V., Gabrilovich, E.: A review of relational machine learning for knowledge graphs. Proc. IEEE **104**(1), 11–33 (2016). https://doi.org/10.1109/JPROC.2015.2483592
6. Niepert, M., Ahmed, M., Kutzkov, K.: Learning convolutional neural networks for graphs. In: Proceedings of the 33nd International Conference on Machine Learning, ICML 2016, JMLR Workshop and Conference Proceedings, vol. 48, pp. 2014–2023. JMLR.org (2016). http://proceedings.mlr.press/v48/niepert16.html
7. Wang, H., Zhao, M., Xie, X., Li, W., Guo, M.: Knowledge graph convolutional networks for recommender systems. In: Proceedings of the World Wide Web Conference, WWW 2019 pp. 3307–3313. ACM (2019). https://doi.org/10.1145/3308558.3313417
8. Wang, H., Zhang, F., Zhang, M., et al.: Knowledge-aware graph neural networks with label smoothness regularization for recommender systems. In: Proceedings of the 25th ACM SIGKDD International Conference on Knowledge Discovery & Data Mining, KDD 2019, pp. 968–977. ACM (2019). https://doi.org/10.1145/3292500.3330836
9. Wang, X., He, X., Cao, Y., Liu, M., Chua, T.: KGAT: knowledge graph attention network for recommendation. In: Proceedings of the 25th ACM SIGKDD International Conference on Knowledge Discovery & Data Mining, KDD 2019, pp. 950–958. ACM (2019). https://doi.org/10.1145/3292500.3330989
10. Xian, Y., Fu, Z., Muthukrishnan, S., Melo, G., Zhang, Y.: Reinforcement knowledge graph reasoning for explainable recommendation. In: Proceedings of the 42nd International ACM SIGIR Conference on Research and Development in Information Retrieval, SIGIR 2019, pp. 285–294. ACM (2019). https://doi.org/10.1145/3331184.3331203
11. Xu, K., Hu, W., Leskovec, J., Jegelka, S.: How powerful are graph neural networks? In: Proceedings of the 7th International Conference on Learning Representations, ICLR 2019. OpenReview.net (2019). https://openreview.net/forum?id=ryGs6iA5Km
12. Xu, W., Xu, Z., Ye, L.: Computing user similarity by combining item ratings and background knowledge from linked open data. In: Meng, X., Li, R., Wang, K., Niu, B., Wang, X., Zhao, G. (eds.) WISA 2018. LNCS, vol. 11242, pp. 467–478. Springer, Cham (2018). https://doi.org/10.1007/978-3-030-02934-0_43
13. Ye, Y., Wang, X., Yao, J., et al.: Bayes EMbedding (BEM): refining representation by integrating knowledge graphs and behavior-specific networks. In: Proceedings of the 28th ACM International Conference on Information and Knowledge Management, CIKM 2019, pp. 679–688. ACM (2019). https://doi.org/10.1145/3357384.3358014

Geographical Information Enhanced POI Hierarchical Classification

Shaopeng Liu[1,2], Jifan Yu[1,2], Juanzi Li[1,2], and Lei Hou[1,2(✉)]

[1] Department of Computer Science and Technology, BNRist,
Tsinghua University, Beijing 100084, China
{lsp16,yujf18}@mails.tsinghua.edu.cn,
{lijuanzi,houlei}@tsinghua.edu.cn
[2] Knowledge Intelligence Research Center, Institute for Artificial Intelligence,
Tsinghua University, Beijing 100084, China

Abstract. Categories of Point of Interest (POI) facilitate location-based services from many aspects like location search and place recommendation [6]. However, POI categories are often incomplete and new POIs are increasing, this rises the problem of automatic POI classification. Current POI classification methods suffer from two problems: lack of textual information about POIs and not leveraging the hierarchical structure of the categories. In this paper, we propose an Ensemble POI Hierarchical Classification framework (EHC) consisting of three components: Textual and Geographic Feature Extraction, Hierarchical Classifier, and Soft Voting Ensemble Model. We conduct extensive experiments to demonstrate the effectiveness of our framework.

Keywords: Point of Interests · Hierarchical Classification

1 Introduction

Point of interests, or POIs, are specific point locations that someone may find useful or interesting places such as restaurants, parks etc. Given a Taxonomy of POI categories, POI classification is to assign the categories in the taxonomy for a POI. As shown in Fig. 1, we can easily recognize " 彤德顺海鲜火锅 (Tongdeshun Seafood Hotpot)" is a hotpot restaurant through its signboard, but it is difficult to recognize the categories of the places whose names are obscure such as "果海芳馨 (Fragrant Fruit Sea)". Because increasing amounts of data on POIs are becoming available online, automatic POI classification has becoming an important task and has a lot of applications in location based services, e.g.., street view navigation [4].

Since a POI itself has few information, usually the information of the name and location of a POI, many POI classification methods are proposed to leverage various auxiliary information to solve the problem. In [3], Krumm et al. extract a set of manually designed features based on individual demographics, the timing of visits and nearby businesses, etc. While other approaches exploit

G. Wang et al. (Eds.): WISA 2020, LNCS 12432, pp. 108–119, 2020.
https://doi.org/10.1007/978-3-030-60029-7_10

Fig. 1. Example of POI taxonomy

information from check-in activities, search queries, and image [1]. However, in many real applications, most of the information used in these methods cannot be obtained. Instead, we try to automatically classify a POI only based on its name (short text) and location (latitude and longitude coordinates), which is a critical challenge.

Second, most of the previous work only classified POIs into a coarse-grained and loosely organized category system [7,8]. In real applications, normally we can have a POI Taxonomy, which is a hierarchy of fine-grained categories. For example in Fig. 1, " 果品市场 (Fruit Market)" is a sub-category of " 综合市场 (Comprehensive Market)" which is further a sub-category of " 购物服务 (Shopping)". Therefore, how to classify POIs into categories in the taxonomy, i.e., POI hierarchical classification, has not been effectively explored. Due to the reasons above, it is far from sufficient to directly apply existing methods to the specific POI Hierarchical Classification Given only Name and Location.

In this paper, we propose an Ensemble POI Hierarchical Classification framework to address above problems. First, we utilize the information contained in the POI name and geographic location by constructing a set of effective features. Specifically, the name of a POI contains semantic information which can be used to infer its category. We constructed both word-level and character-level textual features using the names of the POIs with the help of pre-trained embeddings. In addition, according to the assumption that the POIs that are geographically close to each other usually have similar categories, we construct the geographic features for POIs based on the distribution of categories in their neighborhoods, to cope with the case where text information cannot work on.

Moreover, inspired by [5], we use a hierarchical classification algorithm, in which we train a local multi-class classifier on each intermediate node of taxonomy, to effectively use the hierarchical information in POI Taxonomy. By using a hierarchy of classifiers, the classification task is divided into several simple parts, each part has lower complexity and higher accuracy. We further ensemble the hierarchical classifiers trained on each feature mentioned above based on a soft-voting model for better performance and robustness. Our contributions are summarized as follows:

- We systematically investigate the problem of POI Hierarchical Classification given only the information of Name and Location for POIs.
- We propose a novel Ensemble POI Hierarchical Classification framework which can fully utilize the information in both word and character levels contained in the name, and the category information of geographical location of the POIs.
- To evaluate the effectiveness of our method, we construct four datasets consisting of millions of POIs and hundreds of POI categories using Gaode Map API. And extensive experiments are conducted to validate the effectiveness of our proposed framework.

2 Preliminaries

In this section, we first give some necessary definitions and then formulate the problem of POI hierarchical classification given only name and location.

Point of Interest (POI) is a specific point location that someone may find useful or interesting. An example is a point representing the location of the Summer Palace, or the location of Everest, the highest mountain on the Earth. Most consumers use the term when referring to hotels, restaurant, airport, markets or any other classes used in modern (automotive) navigation systems. We define a POI Instance i as (n, l), where n denotes the name of i, and l denotes the location of i, which is defined as the longitude and latitude coordinates.

POI Taxonomy is a tree structured hierarchy of POI Categories. We define POI Taxonomy as $\mathcal{T} = (\mathcal{C}, \mathcal{E})$, in which $\mathcal{C}$ is a set of POI Categories, and $\mathcal{E} = \{(c_j, c_k \in \mathcal{C}) | c_j \prec c_k\}$, where $\prec$ denotes the sub-category-of relationship. Fig. 1 shows a example of POI Taxonomy, we use rounded-rectangle to represent a category and the arrow to represent the sub-category-of relationship between two categories.

Example 1. Figure 1 shows a example of POI Taxonomy and POI instance. the POI Taxonomy is a three-level tree hierarchy consisting of categories such as "果品市场 (Fruit Market)", "综合市场 (Comprehensive Market)", and "购物服务 (Shopping)", etc. Specifically, "果品市场 (Fruit Market)" is a sub-category of "综合市场 (Comprehensive Market)" which is further a sub-category of "购物服务 (Shopping)". POI instance is a place which consists of name and location, e.g.., ("果海芳馨 (Fragrant Fruit Sea)", (116.852054, 40.368126)).

POI Hierarchical Classification is formally defined as follows. Given a POI instance i, POI Hierarchical Classification is to find a path of categories p in the POI taxonomy $\mathcal{T}$, where $p = (c^0, c^1, ..., c^l)$ is a category path from the ROOT to the leaf category in $\mathcal{T}$, and the super script of c denotes the level of the category.

More specifically, we divide all POI instances into $\mathcal{I}$ and $\hat{\mathcal{I}}$. $\mathcal{I}$ represents the POI Instances with category paths, that is $p \in \mathcal{P}$, and $p = (c^0, c^1, ..., c^l)$. $\hat{\mathcal{I}}$ does not have category paths. Our hierarchical classification task is to find the category path in the POI Taxonomy for each POI instance $\hat{i} \in \hat{\mathcal{I}}$ given $\mathcal{I}$ and $\mathcal{T}$.

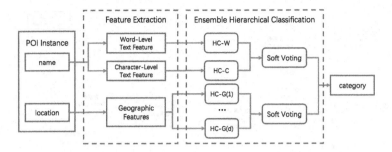

Fig. 2. Ensemble POI Hierarchical Classification framework

In real applications, such as online map services and location based social networks, categories of POI Instance are often incomplete, and many newly emerged POI instances may have no category information at all. Therefore, the POI hierarchical classification has become an important task in location based services. To appropriately solve this task, we need to address two crucial problems: (1) How to address the lack of information? (2) How to make use of the hierarchical information of POI categories?

3 Our Approach

In this section, we introduce our Ensemble POI Hierarchical Classification framework. Figure 2 illustrates the overview of our framework, which consists of two stage. In **Feature Extraction**, to address the lack of information, we construct word-level and character-level text features based on the name of POI, and construct geographic features based on the distribution of POI categories within geographic neighborhoods at different scales. In **Ensemble Hierarchical Classification**, to effectively use the hierarchical structure of POI Taxonomy, we train one hierarchical classifier based on each of the above features. HC-W and HC-C refers to the hierarchical classifiers trained on word-level and character-level textual features respectively, and HC-G(d) refers to the Hierarchical Classifier trained on Geographic feature at scale d. A Soft-Voting ensemble model is used to effectively integrate the output of all these hierarchical classifiers, to improve the performance and robustness of our framework. We describe the details about the construction of textual features and geographic features, as well as how to perform hierarchical classification and ensemble as follows.

3.1 Textual Features

Since the lack of textual information caused by the very short length of POI instance names, we construct word and character level textual features to obtain more comprehensive and fine-grained semantic information of POI instances with the help of external word (and character) embeddings pre-trained on massive Chinese corpus. Using both word and character-level textual features can mine

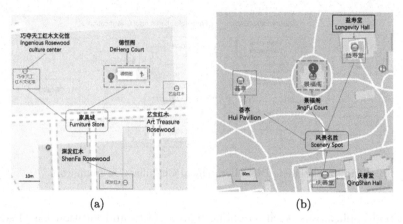

(a)   (b)

Fig. 3. Example of map. We use red dash box to denote target POI, blue solid box to denote its neighbors and rounded box to denote POI category. It is difficult to distinguish the categories of "德恒阁 " (Deheng Court) and "景福阁 (Jingfu Court)" through textual information. However, since "德恒阁 (Deheng Court)" is surrounded by several furniture stores, we can infer that it may by a furniture store, whereas "景福阁 (Jingfu Court)" may be a scenery spot according to its neighbors. (Color figure online)

textual information from different aspects: for POI names like "中国银行 (Bank of China)", we can infer its category from the word "银行 (Bank)", and for some POI names that are too abstract or artistic, character-level feature can provide more semantic details. For example, we cannot know the category of "果海芳馨 (Fragrant Fruit Sea)" explicitly only using word-level text information. However, the character "果 (fruit)" implies that this POI instance may be a fruit store.

Given a POI instance i, we cut its name n to a sequence of words and to a sequence of characters, denoted as $\mathbf{n}^w = (n_1^w, n_2^w, \cdots, n_{|\mathbf{n}^w|}^w)$ and $\mathbf{n}^c = (n_1^c, n_2^c, \cdots, n_{|\mathbf{n}^c|}^c)$, respectively, where $|\mathbf{n}^w|$ is the length of $\mathbf{n}^w$ and $|\mathbf{n}^c|$ is the length of $\mathbf{n}^c$. Then the word-level and character-level textual feature of i are defined as the average of the words or characters in its name, as follows.

$$t^w(i) = |\mathbf{n}^w|^{-1} \sum_{j=1}^{|\mathbf{n}^w|} emb^w(n_j^w), \tag{1}$$

$$t^c(i) = |\mathbf{n}^c|^{-1} \sum_{j=1}^{|\mathbf{n}^c|} emb^c(n_j^c), \tag{2}$$

where $emb^w(\cdot)$ and $emb^c(\cdot)$ are the embedding of a word and a character respectively.

3.2 Multi-scale Geographic Features

In some cases, we still cannot know the explicit categories of some POI instances, even after introducing character-level textual feature, such as "德恒阁 (Deheng Court)". When text information does not work, we need to introduce geographic information as supplement. Intuitively, the categories of a POI instance may be implied by its neighbors. For example, "德恒阁" in Fig. 3(a) and "景福阁 (Jingfu Court)" in Fig. 3(b) are very similar in text. It is very difficult to distinguish their categories among scenery spot, furniture store or others. However, in Fig. 3(a), there are several furniture stores around "德恒阁 (Deheng Court)", so it is more possible to be a furniture store. In contrast, "景福阁" is more likely to be a scenery spot, since it is surrounded by many scenery spots.

Moreover, the geographic neighborhoods at different scale may provide different aspects of geographic information. In some areas where the POIs are relatively dense, like Fig. 3(a), part of a building material market, the neighbors that are very close to the target POI can provide more accurate information. Whereas in some areas where the POIs are relatively sparse, like Fig. 3(b), part of Summer Palace, we need to consider more neighbors in a larger scale area to obtain more sufficient information.

According to these observations, we can propose the multi-scale geographic features based on the category distribution in the neighborhoods at different scale, using TF-IDF transformation. First we divide the map into grids with the same longitude and latitude span d, which controls the scale of the geographic neighborhood. Each grid is identified by gird index, a two-dimension real number coordinate. Given a POI instance i and parameter d, its grid index is defined as $g(i,d) = ([d^{-1}10^4 long(i)], [d^{-1}10^4 lat(i)])$, where $[\cdot]$ is the round function, $long(i), lat(i)$ are the longitude and latitude coordinates of the POI instance i respectively. And the d scale neighborhood $Nei(i,d)$ of i is defined as all POI instances which have same grid index as i, given the parameter d. Then the d scale geographic feature $g(i,d)$ of i is defined as a vector consisting of the tf-idf weights of all leaf categories in $Nei(i,d)$, as follows.

$$g(i,d) = [f(i,c_j,d)]_{j=1}^{|\mathcal{C}^l|}, c_j \in \mathcal{C}^l, \tag{3}$$

where $f(i,c,d) = tf(i,c,d) * (\log |\mathcal{C}^l| - \log df(c,d))$, $tf(i,c,d)$ indicates the term frequency of category c in $Nei(i,d)$, and $df(c,d)$ indicates the number of POI instances whose d scale geographic neighborhood contains category c.

3.3 Ensemble Hierarchical Classification

Hierarchical Classifier. In order to effectively utilize the hierarchical information of POI Taxonomy, we use the Hierarchical Classifier instead of Flat Classifier. In this approach, a base multi-class classifier is trained for each internal node (non-leaf node) of the POI Taxonomy to distinguish between its child nodes. Specifically, we trained one hierarchical classifier on each of the features constructed above. As shown in Fig. 2, HC-W refers to the Hierarchical Classifier trained on Word-level textual feature, HC-C refers to Hierarchical Classifier trained on Character-level textual feature, and HC-G(d) refers to the Hierarchical Classifier trained on Geographic feature at scale d.

Soft Voting Ensemble Model. To efficiently integrate the output of the hierarchical classifiers trained on different features, we use an ensemble model based

Table 1. The statistics of datasets. #POI means the count of POI instances, and # Category@n means the count of POI categories at level n, where n = 1, 2, 3.

Dataset	#POI	#Category@1	#Category@2	#Category@3
Beijing	345k	12	53	291
Shanghai	344k	12	55	293
Guangzhou	259k	12	54	264
Chongqing	260k	12	52	244

on soft voting, which takes the predicted probabilities of several classifiers as input, and then generates their weighted average as output, where the weights are the performance of those classifiers on validation dataset. Specifically, we first ensemble HC-W and HC-C as EHC-T which denotes the output probabilities based on text features, and then we ensemble HC-G(d) with different parameter d as EHC-G. Finally, we ensemble EHC-T and EHC-G as the output of our whole framework.

4   Experiments

4.1   Datasets

In order to validate the efficiency of our Ensemble POI Hierarchical Classification framework, we collect the POIs from Gaode Map[1], to construct four datasets. Specifically, we collect millions of POIs in Beijing, Shanghai, Guangzhou and Chongqing. And for each city, we have a POI Taxonomy which is a 3-level tree hierarchy of categories. We select 12 categories at level 1. Further, we drop the leaf-categories with less than 20 instances, since instances of these categories are too few to be trained and predicted effectively. Then, we randomly separate the dataset into three folds. One fold consisting of 80% of POIs is used as training data, one fold consisting of 10% of POIs is used as validation data and another fold consisting of 10% of POIs is used as testing data. The statistics of the datasets are illustrated in Table 1.

4.2   Experiment Settings

Basic Setting. We choose word embedding and character embedding pretrained on Baidu Encyclopedia corpus as the external embeddings we use to construct text features. And we use Multi-layer Perceptron (MLP) as the local classifier on each node in all hierarchical classifiers. Moreover, we use four metrics: Accuracy, Precision, Recall and F1-score to evaluate our framework. For the last three ones, we firstly compute the binary metric for each class(category), and then average the binary metrics across all classes in "macro" way.

---

[1] https://lbs.amap.com/api/webservice/guide/api/search/.

Table 2. Performance of different methods on datasets. EHC represents the Ensemble POI Hierarchical Classification framework using all features.

Dataset	Beijing				Shanghai			
Method	SVM	NB	MLP	EHC	SVM	NB	MLP	EHC
Accuracy	52.88%	9.05%	55.43%	**70.51%**	55.72%	8.36%	57.72%	**71.63%**
Precision	42.86%	25.30%	43.31%	**63.47%**	46.50%	27.01%	46.88%	**64.69%**
Recall	51.49%	20.63%	51.79%	**77.35%**	57.94%	26.48%	58.14%	**79.98%**
F1-score	43.22%	14.89%	44.93%	**67.62%**	48.76%	18.26%	49.57%	**69.11%**
Dataset	Guangzhou				Chongqing			
Method	SVM	NB	MLP	EHC	SVM	NB	MLP	EHC
Accuracy	56.58%	9.53%	59.56%	**72.58%**	56.58%	12.97%	59.39%	**75.00%**
Precision	51.78%	27.28%	52.18%	**65.15%**	50.67%	32.57%	51.99%	**66.96%**
Recall	59.64%	26.00%	59.37%	**80.53%**	59.15%	29.44%	60.42%	**82.66%**
F1-score	53.14%	16.67%	54.87%	**69.75%**	52.53%	21.76%	53.19%	**71.53%**

Baselines. In order to evaluate the overall effectiveness of our framework, we concatenate all the features constructed above (i.e., word and character level text features and geographic features) as input to train three representative baseline classifiers:

- Support Vector Machine with linear kernel (SVM) is a maximum margin linear classifier on feature space.
- Naive Bayes (NB) is a generative model based on applying Bayes' theorem with the "naive" assumption of conditional independence between every pair of features given the value of the class variable.
- Multi-layer Perceptron (MLP) is a feed-forward fully connected neural network model that can fit highly complex non-linear mappings.

Specifically, we choose "scikit-learn"[2] library to implement all the baseline methods with default parameters. We conduct a feature contribution analysis to prove all the proposed features are indispensable. Then we further discuss the influence of the scale parameter d of geographical features on performance. Finally, we give some interesting case studies and observations.

4.3   Overall Performance

Table 2 summarizes the comparison results of different methods on all datasets. Our method outperforms all baseline methods across all 4 datasets "EHC" refers to our method). For example, the F1-score of our method on Beijing outperforms SVM, NB, MLP by 24.40%, 52.73% and 22.69%, respectively. Further, we observe the performance in following aspects:

---

[2] https://scikit-learn.org/stable/.

For Different Methods. First, NB performs the worst across all 4 datasets. This seems to be caused by the fact that the independence assumption is not satisfied for our features. There is no independence between dimensions of pre-trained embeddings. And due to the correlation between different categories, geographic features cannot satisfy the independence assumption, either. Second, Since MLP can better fit nonlinear functions, it outperforms other baseline methods (SVM and NB) on all metrics.

Table 3. Contribution analysis of different features.

Dataset	Beijing			Shanghai		
Ignored feature(s)	W	C	G	W	C	G
Accuracy	−2.15%	−1.89%	−1.36%	−1.76%	−2.08%	−0.98%
Precision	−2.68%	−2.39%	−0.83%	−2.05%	−0.96%	−0.46%
Recall	−2.04%	−1.70%	−0.59%	−1.02%	−1.22%	−0.38%
F1-score	−2.62%	−2.28%	−0.91%	−1.75%	−1.17%	−0.48%
Dataset	Guangzhou			Chongqing		
Ignored feature(s)	W	C	G	W	C	G
Accuracy	−1.99%	−1.52%	−0.96%	−1.82%	−1.32%	−1.15%
Precision	−1.95%	−0.86%	−0.65%	−2.12%	−1.05%	−0.91%
Recall	−0.79%	−1.47%	−0.38%	−0.98%	−1.10%	−0.46%
F1-score	−1.69%	−1.13%	−0.62%	−1.75%	−1.03%	−0.81%

For Different Datasets. Our method steadily surpasses all baseline methods across different datasets. For example, F1-score of our method outperforms MLP which is the best baseline method by 22.69%, 19.54%, 14.88% and 18.35% on Beijing, Shanghai, Guangzhou, and Chongqing, respectively. Compared with Guangzhou and Chongqing, our method has more obvious advantages in Beijing and Shanghai, which reflects that the more complex the data set, the more effective our method is (As shown in Table 1, Beijing and Shanghai have more categories than Guangzhou and Chongqing).

4.4 Feature Contribution Analysis

In order to get an insight into the importance of each feature in our method, we perform a contribution analysis with different features. Here, we run our approach 3 times on the all 4 datasets. In each of the first 2 times, word-level (W) and character-level (C) textual features are removed respectively. In the rest one time, we remove all geographic features. Table 3 records the changes of Accuracy, Precision, Recall and F1-score for each setting. According to the decrements of F1-scores, we find that all the proposed features are indispensable in classification. word-level and character-level textual features are about equally important. Removing either of them decreases the F1-score in the range of 1%–3%. Besides, removing geographic features decreases the performance, which shows the necessity of using geographic information.

4.5 Parameter Analysis

During the construction of geographic features, the parameter d controls the scale of geographic neighborhood. As we increase d, more neighbors are taken into account, so that the geographic features contain richer information, however, some noise may also be introduced in some cases. Integrating geographic features at different scales can overcome the shortcomings of any single feature, thereby obtaining richer information and improving performance. Specifically, we use HC-G(d) to denote the hierarchical classifier trained on geographic feature at scale d, and EHC-G(d) to denote the ensemble from HC-G(1) to HC-G(d), In Fig. 4, we set parameter d from 1 to 10, and compare the average performance of HC-G(d) and EHC-G(d) across all the four datasets, respectively. As we increase d, the performance of HC-G(d) first increases and then decreases significantly, whereas EHC-G(d) continues to increase and the slope becomes smaller when d is bigger than 5. The results demonstrate that the ensemble framework lifts the performance of geographic features significantly.

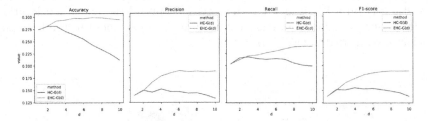

Fig. 4. Parameter analysis. HC-G(d) denotes the hierarchical classifier trained on geographic feature at scale d, and EHC-G(d) denotes the ensemble from HC-G(1) to HC-G(d), e.g., EHC-G(d) refers to the ensemble of HC-G(1), HC-G(2) and HC-G(3)

4.6 Case Study

In this section, we present two examples to intuitively illustrate the effectiveness of our framework and the indispensability of each component. First, as shown in Fig. 5(a), "果海芳馨 (Fragrant Fruit Sea)" is misclassified into "礼品饰品店 (Gift Store)" using only word-level textual information, which shows word-level text information is too coarse-grained to distinguish such abstract and artistic name. However, After adding character-level features, "果海芳馨 (Fragrant Fruit Sea)" can be correctly identified as a fruit market. Since the character "果 (fruit)" in its name appears very frequently among the fruit markets (e.g., >78% in Beijing). Second, only based on textual information from its name, we can easily mistake

"紫云阁 (Pavilion in Violet Cloud)" in Fig. 5(b) as a scenic spot. However, since it is surrounded by many antique and art shops (blue circles), we can easily correct its category as an "古玩字画店 (Antique and Art shop)". Moreover, we find that geographical features contribute much more to Chongqing and Beijing than the other two cities, which may be caused by the differences in regional characteristics, living habits and urban planning of different cities.

5 Related Works

Most of existing methods classify POIs into fine-grained categories, by constructing features from various auxiliary information, such as user behaviors. Ye et al. [7] designed a two-phase algorithm addressing the problem of semantic annotation. The first phase takes care of the feature extraction, whereas the second phase handles the semantic annotation. For each tag, the algorithm learns a binary SVM. However, their work only focuses on only three categories. Chon et al. [1] pushed the results to 7 categories through using image and audio data captured by mobile phones. Placer++ [3] identify place labels based on the timing of visits to certain places, nearby businesses, and simple demographics of the user, however they can only work on 15 high-level categories. While other approaches exploit the features of user check-in activities, search query [8]. Although the few works try to classify the POI into a hierarchy of categories [2], they still rely on a lot of auxiliary information beyond the name and location of POI. Thus, it is far from sufficient to directly apply these methods to the problem investigated in this work.

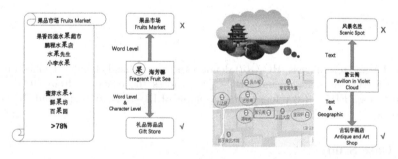

(a) character information.   (b) geographic information.

Fig. 5. Case study (Color figure online).

6 Conclusions and Future Work

We conduct a new investigation on POI hierarchical classification only given name and location. We precisely define the problem and propose an Ensemble

POI Hierarchical Classification framework, which can fully mine the information contained in the name and geographical location of the POI, while effectively using the hierarchical information of POI Taxonomy. Promising future directions would be to investigate how to expand the POI Taxonomy, as well as how to promote POI classification and POI taxonomy expansion by effectively utilizing the differences in culture, history, and customs between different regions.

Acknowledgments. This work is supported by the Key-Area Research and Development Program of Guangdong Province (2019B010153002), NSFC Key Projects (U1736204, 61533018), grants from Beijing Academy of Artificial Intelligence (BAAI2019 ZD0502) and Institute for Guo Qiang, Tsinghua University (2019GQB0003), and THUNUS NExT Co-Lab.

References

1. Chon, Y., Kim, Y., Cha, H.: Autonomous place naming system using opportunistic crowdsensing and knowledge from crowdsourcing. In: 2013 ACM/IEEE International Conference on Information Processing in Sensor Networks (IPSN), pp. 19–30. IEEE (2013). https://doi.org/10.1145/2461381.2461388
2. He, T., Yin, H., Chen, Z., Zhou, X., Sadiq, S., Luo, B.: A spatial-temporal topic model for the semantic annotation of POIs in LBSNs. ACM Trans. Intell. Syst. Technol. (TIST) **8**(1), 1–24 (2016). https://doi.org/10.1145/2905373
3. Krumm, J., Rouhana, D., Chang, M.W.: Placer++: semantic place labels beyond the visit. In: 2015 IEEE International Conference on Pervasive Computing and Communications (PerCom), pp. 11–19. IEEE (2015). https://doi.org/10.1109/PERCOM.2015.7146504
4. Liu, H., Li, T., Hu, R., Fu, Y., Gu, J., Xiong, H.: Joint representation learning for multi-modal transportation recommendation. In: Proceedings of the AAAI Conference on Artificial Intelligence, vol. 33, pp. 1036–1043 (2019). https://doi.org/10.1609/AAAI.V33I01.33011036
5. Silla, C.N., Freitas, A.A.: A survey of hierarchical classification across different application domains. Data Min. Knowl. Discov. **22**(1–2), 31–72 (2011). https://doi.org/10.1007/S10618-010-0175-9
6. Yang, R., Han, X., Zhang, X.: A multi-factor recommendation algorithm for POI recommendation. In: Meng, X., Li, R., Wang, K., Niu, B., Wang, X., Zhao, G. (eds.) WISA 2018. LNCS, vol. 11242, pp. 445–454. Springer, Cham (2018). https://doi.org/10.1007/978-3-030-02934-0_41
7. Ye, M., Shou, D., Lee, W.C., Yin, P., Janowicz, K.: On the semantic annotation of places in location-based social networks. In: Proceedings of the 17th ACM SIGKDD International Conference on Knowledge Discovery and Data Mining, pp. 520–528 (2011). https://doi.org/10.1145/2020408.2020491
8. Zhou, J., et al.: A collaborative learning framework to tag refinement for points of interest. In: Proceedings of the 25th ACM SIGKDD International Conference on Knowledge Discovery & Data Mining, pp. 1752–1761 (2019). https://doi.org/10.1145/3292500.3330698

GNE: Generic Heterogeneous Information Network Embedding

Chao Kong$^{(\boxtimes)}$, Baoxiang Chen, Shaoying Li, Yifan Chen, Jiahui Chen, and Liping Zhang

School of Computer and Information, Anhui Polytechnic University, Wuhu, China
kongchao@ahpu.edu.cn,
{bxchen1996,shyli1996,yfchen1999,jhchen2000,lpzhang1980}@yeah.net

Abstract. As an effective approach to solve graph mining problems, network embedding aims to learn low-dimensional latent representation of nodes in a network. We develop a representation learning method called GNE for generic heterogeneous information networks to learn the vertex representations for generic HINs. Greatly different from previous works, our model consists two components. First, GNE assigns the probability of each random walk step according to vertex centrality, weight of relations and structural similarity for neighbors on premise of performing a biased self-adaptive random walk generator. Second, to learn more desirable representations for generic HINs, we then design an advanced joint optimization framework by accounting for both the explicit (1st-order) relations and implicit (higher-order) relations.

Keywords: Network embedding · Heterogeneous information network · Link prediction · Recommendation

1 Introduction

With the popularization of deep learning method, a huge volume of complex network data has become available in a variety of domains. Network representation learning or network embedding is one of the core process in many complex network analysis tasks, e.g., link prediction. Early studies employ the adjacency matrix to store a graph, which can only represents the 1st-order relations. In addition, it has a very high dimension.

As an effective and efficient solution, network embedding converts the network data into a low dimensional space in which the network structural information and network properties are maximumly preserved [1,2]. But it is also a challenging task due to the following reasons: (1) it's extremely sparse for real-world network data since there may be quite a few of missing relations (i.e., high-order implicit relations); (2) it is arduous to assign the probability of random walk in a reasonable manner.

C. Kong and B. Chen—Contributed equally to this work.

© Springer Nature Switzerland AG 2020
G. Wang et al. (Eds.): WISA 2020, LNCS 12432, pp. 120–127, 2020.
https://doi.org/10.1007/978-3-030-60029-7_11

To address these difficulties, we propose a novel approach called GNE, to learn the vertex representations for generic HINs. For the provided approach, we would like to address two challenges highlighted earlier. It utilizes vertex centrality, weight of relations and structural similarity to model the implicit relations, then exploits an advanced biased self-adaptive random walk generator to generate corpus. Specifically, we formulate the task of network embedding for generic HINs as a joint optimization problem by accounting for both the explicit (1st-order) relations and implicit (higher-order) relations. In summary, our major contributions are as follows:

- We extend a biased self-adaptive random walk generator which can perform efficiently by vertex centrality, weight of relations and structural similarity for neighbors.
- We propose a novel method to reconstruct the network by considering both explicit and implicit relations for generic HINs.
- We illustrate the performance of our algorithm against comparable baselines on five real datasets. Experimental study results manifest that GNE outperforms baselines in link prediction and recommendation.

The remainder of paper is organized as follows. We shortly discuss the related work in Sect. 2. We describe the overview of our algorithm and delving into details of proposed generic network embedding method in Sect. 3 and report our empirical study in Sect. 4. Finally, we conclude this paper in Sect. 5.

2 Related Work

To date, substantial works have primarily focused on embedding homogeneous networks where vertices are of the same type. Due to the effectiveness and prevalence, word2vec [3] inspires many works [4–7] to exploit inner product to model the interaction between two instances. Perozzi et al. first introduce [4] word vectors to social networks, which can not only represent vertices but also topological relations (i.e., social connections). Following the pioneering work, these methods typically apply a two-step solution: first performing random walks on the network to obtain a "corpus" of vertices, and then employing word embedding methods to obtain the embeddings for vertices. To the best of our knowledge, few of existing works have paid special attention to embed generic HIN. Although there are some effective and prevalence methods, such as metapath2vec++ [8] for HINs, they do not address the problem generically. We try to propose a novel network embedding approach for generic HINs. However, it is not easy to obtain more desirable representations of HINs, because they overlook the implicit relations in networks. To tackle this problem, some advanced works have been proposed in recent years. Tang et al. [5] first attempt to optimize the objective function by employing the first proximity (i.e., explicit relations) and second proximity (i.e., implicit relations) simultaneously to obtain more desirable representations. Gao et al. [9] propose a novel optimization framework by accounting for both the explicit relations and implicit relations for bipartite network to learn the vertex representations. However, these two methods consider vertex centrality and weights of relations only to assign the probability of biased random walk, but ignoring the structural similarity. These are the focuses of this paper.

3 Network Embedding Approach

3.1 Overview of GNE

The proposed GNE approach consists of three components as follows:

Step 1: Biased self-adaptive random walk generator. We combine the Co-HITS algorithm [10] with euclidean distance to measure the weight of reconstructed network by Eq. (2) and propose a novel biased self-adaptive random walk generator.

Step 2: Modeling explicit relations. Edges exist between vertices of different types, providing an explicit relation on constructing the network. We model explicit relations by considering the local proximity between two connected vertices by Eq. (1).

Step 3: Modeling implicit relations. Calculating implicit relations between different types of vertices by Eq. (3) and Eq. (4).

3.2 Modeling Explicit and Implicit Relations

The joint probability between v_i^n and v_j^m is defined as: $P(i,j) = \frac{w_{ij}}{\sum_{e_{ij} \in E} w_{ij}}$, where w_{ij} represents the weight of two vertices v_i^n and v_j^m. It means the proportion of w_{ij} in the total weight associated with these two vertices. According to some previous works [3,4,9], the interaction between two vectors can be adequately represented by the inner product between them. We employ the sigmoid function to map the interaction value to the probability space as: $\hat{P}(i,j) = \frac{1}{1+exp((v_i^n)^\top v_j^m)}$. The empirical distribution and the reconstructed distribution can been calculated respectively. Through minimizing the KL-divergence of them, we obtain the embedding vectors $\boldsymbol{v}_i^n$ and $\boldsymbol{v}_j^m$:

$$minimize \quad O_1 = -\sum_{e_{ij} \in E} w_{ij} log \hat{P}(i,j). \tag{1}$$

Two vertices can be closely connected in the embedding space, which preserves the local proximity as desire by minimizing the O_1. According to the existing work [11], the introduction of implicit relation can reveal hidden relations between vertices more comprehensively because the 2nd-order proximity allows vertices to connect with more neighbors.

Random Walk with Weight Reconstruction. In our method, we introduce the comparison of the similarity between neighbors by combining the euclidean distance with Co-HITS. We can assume that the higher the similarity is, the greater the weight is. Then, we give the Eq. (2) as follows:

$$
\begin{aligned}
w_{ij}^{V_n} &= \sum_{k \in V_n} \frac{w_{ik} w_{jk}}{1 + \sqrt{(w_{ik} - \bar{w})^2 + (w_{jk} - \bar{w})^2}}; \\
w_{ij}^{V_m} &= \sum_{k \in V_m} \frac{w_{ki} w_{kj}}{1 + \sqrt{(w_{ki} - \bar{w})^2 + (w_{kj} - \bar{w})^2}},
\end{aligned}
\tag{2}
$$

where w_{ij} is the weight of e_{ij}, and $\bar{w}$ is the average of two weights. We can reconstruct weights within a set of vertices by utilizing above equation. First, all vertex sets will be reconstructed with weights, then the original explicit weight will be added to it. A new biased self-adaptive random walk generator is designed to generate the corpus by using the reconstructed weight as the walk probability.

Implicit Relation with Sampling. If two vertices often appear in the context of the same sequence, it means that they have similar embedding. Thus, we can use conditional probability for 2nd-order proximity. As shown in Eq. (3), the probability of v_i^n under the condition of v_j^m should be maximized:

$$
maximize \quad O_2 = \prod_{v_i^n \in S \wedge S \in C^{V_n}} \prod_{v_c^n \in T_S(v_i^n)} P(v_c^n | v_i^n),
\tag{3}
$$

where v_i^n is a vertex in sequence S, C^{V_n} is the corpus for vertex set V_n, and v_c^n is a vertex in $T_S(v_i^n)$ which denotes the context vertices of vertex v_i^n in sequence S. For a HIN with n vertex sets, there are n objective functions. We parameterize the conditional probability $P(v_c^n | v_i^n)$ using the inner product kernel with softmax for output:

$$
P(v_c^n | v_i^n) = \frac{exp((v_i^n)^\top \theta_c^n)}{\sum_{k=1}^{|V_n|} exp((u_i^n)^\top \theta_k^n)},
\tag{4}
$$

where θ represents the context vector of a vertex. It represents the possibility of observing v_c^n in the v_i^n so that vertices with the same context can be similar in the embedding space. Nevertheless, each evaluation of the softmax function needs to traverse all vertices, which is very time-costing. To reduce the learning complexity, we employ the idea of Locally Sensitive Hash (LSH) [12] to collect diverse and high-quality negative samples.

3.3 Joint Optimization

Suppose there are N vertex sets in a HIN. To embed a HIN by preserving both explicit and implicit relations simultaneously, we combine their objective functions to form a joint optimization framework: $maximize \quad L = -\alpha O_1 + \beta O_2 + ... + \gamma O_{N+1}$, where parameters α, β and γ are hyper-parameters to be specified to combine different components in the joint optimization framework.

Table 1. Descriptive statistics of training datasets.

Name	Diggvote[a]	Wikipedia[b]	VisualizeUs[c]	DBLP[d]	MovieLens[e]
Type	Undirected, unweighted		Undirected, weighted		
$\|V_1\|$	5,770	15,000	5,765	6,001	69,878
$\|V_2\|$	3,553	2,529	2,417	1,177	10,677
$\|E\|$	292,813	36,284	21,434	17,675	6,000,034

[a]http://konect.uni-koblenz.de/networks/digg-votes
[b]http://konect.uni-koblenz.de/networks/wikipedia_link_en
[c]http://konect.uni-koblenz.de/networks/pics_ti
[d]http://dblp.uni-trier.de/xml/
[e]http://grouplens.org/datasets/movielens/

To optimize the joint model, we utilize the Stochastic Gradient Ascent algorithm (SGA). $-\alpha O_1$ denotes the explicit relation which can be used to update v_i^n as: $v_i^n = v_i^n + \lambda\{\gamma w_{ij}[1-\sigma((v_i^n)^\top v_j^m)]\}$, where λ denotes the learning rate. We have already preserve the explicit relation and there is also implicit relation like βO_2 and γO_{N+1}. Center vertex v_i^n and its context v_c^n can update the embedding vector v_i^n as: $v_i^n = v_i^n + \lambda\{\sum_{z \in \{v_c^n\} \cup N_S^{ns}(v_i^n)} \beta[I(z, v_i^n) - \sigma((v_i^n)^\top \theta_z^n)]\theta_z^n\}$, where $I(z, v_i^n)$ is an indicator function. It can determine whether vertex z is in the context of v_i^n or not. The positive and negative instances in the context vector can be updated as: $\theta_z^n = \theta_z^n + \lambda\{\beta[I(z, v_i^n) - \sigma((v_i^n)^\top \theta_z^n)]v_i^n\}$.

4   Empirical Study

We use five real-world datasets to evaluate our proposed GNE method in link prediction and recommendation tasks. The descriptive statistics about the them are shown in Table 1. All datasets are randomly sampled 60% instances as the training set, evaluating performance on the remaining 40% of the testing set. We compare GNE with other state-of-the-art network embedding methods. They are **DeepWalk, Node2vec, LINE** and **BiNE** which are mentioned in related work.

4.1   Performance Comparison

Table 2 illustrates the performance of baselines and GNE: (1) DeepWalk may not be efficient enough, it does not consider implicit relations and performs unbiased random walks to generate the corpus of vertex sequence; (2) Node2vec introduces a different search strategy, the resulting sequence becomes more realistic, so the performance is improved; (3) LINE, BiNE and GNE all consider the explicit and implicit relations, but LINE ignores the joint optimization, which leads to the lower performance. Moreover, GNE outperforms BiNE and achieves the best performance on both datasets in both metrics. This improvement demonstrates the effectiveness and rationality of considering both neighbors' structure and weight on modeling the implicit relations.

Table 2. Link prediction and recommendation performance on five datasets.

Algorithm		DeepWalk	Node2vec	LINE	BiNE	GNE
VisualizeUs	F1@10	5.82%	6.73%	9.62%	13.63%	14.51%
	NDCG@10	8.83%	9.71%	13.76%	24.50%	24.74%
	MAP@10	4.28%	6.25%	7.81%	16.45%	16.77%
	MRR@10	12.12%	13.95%	14.99%	34.23%	34.52%
DBLP	F1@10	8.50%	8.54%	8.99%	11.37%	11.51%
	NDCG@10	24.14%	23.89%	14.41%	24.75%	26.34%
	MAP@10	19.71%	19.44%	17.13%	20.47%	20.62%
	MRR@10	31.53%	31.11%	20.56%	33.36%	33.18%
Movielens	F1@10	3.73%	4.16%	6.91%	9.14%	9.22%
	NDCG@10	3.21%	3.68%	6.50%	9.02%	9.17%
	MAP@10	0.90%	1.05%	1.74%	3.01%	3.23%
	MRR@10	15.40%	18.33%	38.12%	45.95%	46.02%
Diggvote	AUC-ROC	53.03%	84.99%	86.51%	87.42%	87.51%
	AUC-PR	54.95%	80.56%	82.18%	83.88%	84.33%
Wikipedia	AUC-ROC	79.23%	80.37%	94.38%	92.91%	94.69%
	AUC-PR	72.76%	81.22%	95.83%	94.45%	96.15%

Table 3. GNE with and without weight reconstruction.

Without weight reconstruction			With weight reconstruction	
Link prediction				
Dataset	AUC-ROC	AUC-PR	AUC-ROC	AUC-PR
Diggvote	86.28%	83.62%	87.51%	84.33%
Wikipedia	92.88%	**94. 32%**	94.69%	**96.15%**
Recommendation				
Dataset	F1@10	NDCG@10	F1@10	NDCG@10
VisualizeUS	12.79%	24.06%	14.51%	24.74%
DBLP	10.47%	**22.32%**	11.51%	**26.34%**
MovieLens	8.73%	8.28%	9.22%	9.17%

4.2 Utility of Weight Reconstruction

To demonstrate the effectiveness and rationality of weight reconstruction by integrating neighbors' structure and property of a vertex, we compare GNE with its variant that removes the weight reconstruction. We observe the largest improvements of GNE with weight reconstruction are 1.83% and 4.02% for link prediction and recommendation respectively from Table 3. It indicates that our proposed way of reconstructing weight plays a crucial role in the biased self-adaptive random walk to complement with common 2nd-order proximity.

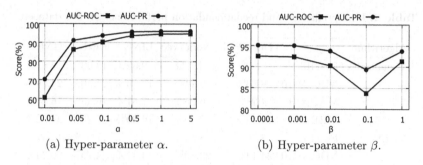

Fig. 1. Impact of hyper-parameters on link prediction.

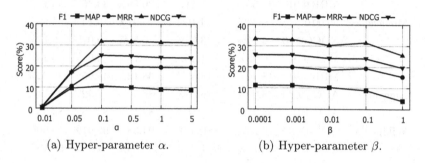

Fig. 2. Impact of hyper-parameters on recommendation.

4.3   Hyper-parameter Studies

Due to the page limitation, we only select two significant parameters α and β to study, where α is an optimization parameter for explicit relation, and β is for implicit relation. We perform link prediction and recommendation tasks on Diggvote and Wikipedia respectively in Fig. 1: (1) with the increase of α, the performance first increases and then keeps stable after certain values; (2) with the increase of β, the performance first increases and then decreases after certain values. When α is small, our joint optimization model will reduce the importance of the explicit relations for network embeddings and return the worse performance. However, when β is large, the joint optimization model may exaggerate the role of implicit relations. It indicates that both explicit and implicit relations are helpful for network embedding, but explicit relations are more important than implicit relations for the HINs (Fig. 2).

5   Conclusion

In this paper, we have studied the problem of network embedding in generic heterogeneous information networks. It is a challenging task due to the extremely sparse for real-world network data and the hardness to assign the probability of random walk in a reasonable manner. Experimental results indicate that GNE not only outperforms the comparable baselines but also obtains the promising performance. In our future work, we plan to extend our work to obtain more implicit relations through metapath or auxiliary information.

Acknowledgment. This work was supported by the National Natural Science Foundation of China Youth Fund under Grant No. 61902001 and Initial Scientific Research Fund of Introduced Talents in Anhui Polytechnic University under Grant No. 2017YQQ015.

References

1. Goyal, P., Ferrara, E.: Graph embedding techniques, applications, and performance: a survey. Knowl. Based Syst. **151**, 78–94 (2018)
2. Hamilton, W.L., Ying, R., Leskovec, J.: Representation learning on graphs: methods and applications. IEEE Data Eng. Bull. **40**(3), 52–74 (2017)
3. Mikolov, T., Chen, K., Corrado, G., Dean, J.: Efficient estimation of word representations in vector space. In: ICLR 2013, Scottsdale, Arizona, USA, May 2–4, 2013, Workshop Track Proceedings (2013)
4. Perozzi, B., Al-Rfou, R., Skiena, S.: Deepwalk: online learning of social representations. In: KDD 2014, New York, NY, USA, August 24–27, 2014, pp. 701–710 (2014)
5. Tang, J., Qu, M., Wang, M., et al.: LINE: large-scale information network embedding. In: WWW 2015, Florence, Italy, May 18–22, 2015, pp. 1067–1077 (2015)
6. Grover, A., Leskovec, J.: node2vec: scalable feature learning for networks. In: Proceedings of the 22nd ACM SIGKDD, San Francisco, CA, USA, August 13–17, 2016, pp. 855–864 (2016)
7. Chang, X., Shi, W., Zhang, F.: Signed network embedding based on noise contrastive estimation and deep learning. In: Ni, W., Wang, X., Song, W., Li, Y. (eds.) WISA 2019. LNCS, vol. 11817, pp. 40–46. Springer, Cham (2019). https://doi.org/10.1007/978-3-030-30952-7_5
8. Dong, Y., Chawla, N.V., Swami, A.: metapath2vec: scalable representation learning for heterogeneous networks. In: Proceedings of the 23rd ACM SIGKDD, Halifax, NS, Canada, August 13–17, 2017, pp. 135–144 (2017)
9. Gao, M., Chen, L., He, X., Zhou, A.: Bine: bipartite network embedding. In: SIGIR 2018, Ann Arbor, MI, USA, July 08–12, 2018, pp. 715–724 (2018)
10. Deng, H., Lyu, M.R., King, I.: A generalized co-hits algorithm and its application to bipartite graphs. In: Proceedings of the 15th ACM SIGKDD, Paris, France, June 28–July 1, 2009, pp. 239–248 (2009)
11. Yu, L., Zhang, C., Pei, S., et al.: WalkRanker: a unified pairwise ranking model with multiple relations for item recommendation. In: AAAI-18, New Orleans, Louisiana, USA, February 2–7, 2018, pp. 2596–2603 (2018)
12. Gionis, A., Indyk, P., Motwani, R.: Similarity search in high dimensions via hashing. In: VLDB 1999, September 7–10, 1999, Edinburgh, Scotland, UK, pp. 518–529 (1999)

Query Processing and Algorithm

SLPSO-Based X-Architecture Steiner Minimum Tree Construction

Xiaohua Chen[1], Ruping Zhou[1], Genggeng Liu[1]($\boxtimes$), and Xin Wang[2]

[1] College of Mathematics and Computer Science, Fuzhou University, Fuzhou, China
sherryhchen@163.com, 1647507716@qq.com, liugenggeng@fzu.edu.cn
[2] College of Intelligence and Computing, Tianjin University, Tianjin, China
wangx@tju.edu.cn

Abstract. The X-architecture Steiner Minimum Tree (XSMT) is the best connection model for multi-terminal nets in global routing algorithms under non-Manhattan structures, and it is an NP-hard problem. And the successful application of Particle Swarm Optimization (PSO) technique in this field also reflects its extraordinary optimization ability. Therefore, based on Social Learning Particle Swarm Optimization (SLPSO), this paper proposes an XSMT construction algorithm (called SLPSO-XSMT) that can effectively balance exploration and exploitation capabilities. In order to expand the learning range of particles, a novel SLPSO approach based on the learning mechanism of example pool is proposed, which is conductive to break through local extrema. Then the proposed mutation operator is integrated into the inertia component of SLPSO to enhance the exploration ability of the algorithm. At the same time, in order to maintain the exploitation ability, the proposed crossover operator is integrated into the individual cognition and social cognition of SLPSO. Experimental results show that compared with other Steiner tree construction algorithms, the proposed SLPSO-XSMT algorithm has better wirelength optimization capability and superior stability.

Keywords: Particle Swarm Optimization · Social learning · Steiner Minimum Tree · X-architecture · Wirelength optimization

1 Introduction

Steiner Minimum Tree (SMT) is the best connection model for multi-terminal nets in global routing of Very Large Scale Integration (VLSI). The SMT problem is to find a routing tree with the least cost to connect all given pins by introducing additional points (Steiner points). Therefore, SMT construction is one of the most important issues in VLSI routing.

Most current researches on routing algorithms are based on the Manhattan structure [6,7], which can only routing in horizontal and vertical directions. In

This work was supported in part by National Natural Science Foundation of China (No. 61877010, 11501114), Natural Science Foundation of Fujian Province, China (2019J01243).

order to make fuller use of routing resources, the scholars are gradually shifting their focus to non-Manhattan structures, thereby improving routing quality and chip performance.

Therefore, the construction of Steiner minimum tree based on non-Manhattan structure becomes a critical step in VLSI routing. In the early years, scholars use precise algorithms [2,14] to construct a non-Manhattan structure routing tree, which can obtain a shorter wirelength than the Manhattan structure, but the complexity is too high. So some heuristic algorithms [4,17,20] are proposed to solve larger-scale SMT problems. However, these traditional heuristics are prone to fall into local extrema. In recent years, evolutionary computing has developed rapidly in many fields, especially Swarm Intelligence (SI) technology [1,3,12,19]. Some routing algorithms [5,8,13] consider important optimization goals such as wirelength, obstacles, delay and bends based on Particle Swarm Optimization (PSO) technique. In [10], a hybrid transformation strategy is proposed to expand the search space based on self-adapting PSO. And an unified algorithm for constructing Rectilinear Steiner Minimum Tree (RSMT) and XSMT is proposed in [11], which can obtain multiple topologies of SMT to optimize the congestion in global routing. It can be seen that PSO technique is indeed a powerful tool to solve SMT problems.

Based on the analysis of the above related research work, this paper designs and implements an effective algorithm to solve the XSMT construction problem using Social Learning PSO (SLPSO), called SLPSO-XSMT. The contributions of this paper are as follows:

- A novel SLPSO approach based on the learning mechanism of example pool is proposed to enable particles to learn from different and better particles in each iteration, and enhance the diversity of population evolution.
- Mutation and crossover operators are integrated into the update formula of the particles to achieve the discretization of SLPSO, which can well balance the exploration and exploitation capabilities, thereby better solving the XSMT construction problem.

The rest of this paper is organized as follows. Section 2 presents the problem formulation. And the SLPSO method with example pool mechanism is introduced in Sect. 3. Section 4 describes the XSMT construction using SLPSO method in details. In order to verify the good performance of the proposed SLPSO-XSMT algorithm, the experimental comparisons are given in Sect. 5. Section 6 concludes this paper.

2   Problem Formulation

The XSMT problem can be described as follows: Given a set of pins $P = \{P_1, P_2, ..., P_3\}$, each pin is represented by a coordinate pair (x_i, y_i). Then connect all pins in P through some Steiner points to construct an XSMT, where the direction of routing path can be 45° and 135°, in addition to the traditional horizontal and vertical directions. Taking a routing net with 10 pins as an example, Table 1 shows the input information of the pins. The layout distribution of the given pins is shown in Fig. 1(a).

Table 1. The input information for the pins of a net

Pi	1	2	3	4	5	6	7	8	9	10
xi	35	19	24	24	38	10	15	4	20	7
yi	27	8	33	5	12	22	29	59	42	47

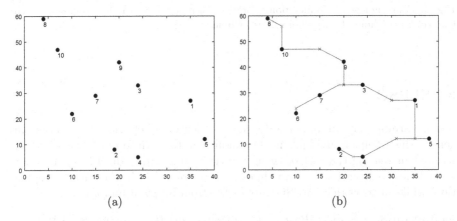

(a)   (b)

Fig. 1. Routing graph corresponding to Table 1: (a) the layout distribution of pins; (b) an X-architecture Steiner tree with the given pin set.

Definition 1. *Pseudo-Steiner point. In addition to original points formed by given pins, the final XMST can be constructed by introducing additional points called pseudo-Steiner points (PSP). In Fig. 2, the point S is PSP, and PSP contains the Steiner point.*

Definition 2. *Choice 0 (as shown in Fig. 2(b)). The Choice 0 of PSP corresponding to edge L is defined as leading rectilinear side first from A to PSP S, and then leading non-rectilinear side to B.*

Definition 3. *Choice 1 (as shown in Fig. 2(c)). The Choice 1 of PSP corresponding to edge L is defined as leading non-rectilinear side first from A to PSP S, and then leading rectilinear side to B.*

Definition 4. *Choice 2 (as shown in Fig. 2(d)). The Choice 2 of PSP corresponding to edge L is defined as leading vertical side first from A to PSP S, and then leading horizontal side to B.*

Definition 5. *Choice 3 (as shown in Fig. 2(e)). The Choice 3 of PSP corresponding to edge L is defined as leading horizontal side first from A to PSP S, and then leading vertical side to B.*

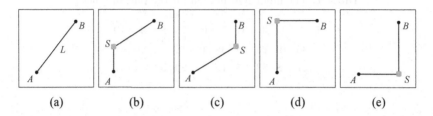

Fig. 2. Four choices of Steiner point for the given segment: (a) line segment L; (b) Choice 0; (c) Choice 1; (d) Choice 2; (e) Choice 3.

3   SLPSO

Social learning plays an important role in the learning behavior of swarm intelligence, which helps individuals in the population to learn from other individuals without increasing the cost of their own trials and errors. In SLPSO [18], each particle learns from better individuals (called *examples*) in the current population, while each particle in PSO only learns from its *pbest* and *gbest*.

Definition 6. *Example Pool. All particles in the swarm $S = \{X_i | 1 \leq i \leq M\}$ are arranged in ascending order according to the fitness: $S = \{X_1, ..., X_{i-1}, X_i, X_{i+1}, ..., X_M\}$, and then $EP = \{X_1, ..., X_{i-1}\}$ constitutes the example pool of particle X_i.*

Based on the example learning mechanism, the new formulas for updating particles are proposed as follows:

$$V_i^{t+1} = \omega \cdot V_i^t + c_1 \cdot r_1 \cdot (P_i^t - X_i^t) + c_2 \cdot r_2 \cdot (K_i^t - X_i^t) \tag{1}$$

$$X_i^{t+1} = V_i^{t+1} + X_i^t \tag{2}$$

where P_i is the personal historical best position of particle i, K_i is the historical best position of the Kth particle in the example pool, which is the social learning object for particle i. ω is the inertia weight. c_1 and c_2 are acceleration coefficients, which respectively adjust the step size of the particle flying to personal historical best position (P_i) and its social learning object (K_i). r_1 and r_2 are mutually independent random numbers uniformly distributed in the interval $(0, 1)$.

Figure 3 shows the example pool of a particle. For particle X_i, the particles with better fitness values than it including the global optimal solution X_G constitute its example pool. X_i randomly selects any particle in the example pool at

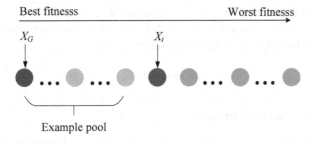

Fig. 3. Example pool of particle X_i

each iteration, and learns the historical experience of this particle to complete its own social learning process. This social learning mechanism allows particles to improve themselves through continuous learning from different excellent individuals during the evolution process, which is conducive to the diversified development of the population.

4 XSMT Construction Using SLPSO

4.1 Particle Encoding

The edge-vertex encoding strategy [11] is adopted in this paper, which is more suitable for evolutionary algorithms, especially PSO. For a net with n pins, the corresponding spanning tree has n-1 edges and one extra digit that is the fitness value of particle. Thus the length of a particle encoding is $3 \times (n - 1) + 1$.

For example, Fig. 1(b) shows an X-architecture routing tree ($n = 10$) corresponding to the layout distribution of pins given in Fig. 1(a), where the symbol 'x' represents PSP. And this routing tree can be expressed as the particle whose encoding is the following numeric string:

$$9\ 3\ \mathit{2}\ 3\ 7\ \mathit{0}\ 7\ 6\ \mathit{1}\ 3\ 1\ \mathit{1}\ 1\ 5\ \mathit{2}\ 9\ 10\ \mathit{1}\ 10\ 8\ \mathit{0}\ 5\ 4\ \mathit{0}\ 4\ 2\ \mathit{0}\ \mathbf{108.6686}$$

where the length of the particle is $3 \times (10 - 1) + 1 = 28$, the last bold number 108.6686 is the fitness of the particle and each italic number represents the choice of PSP for each edge. The first substring (9, 3, *2*) represents that Pin 9 and Pin 3 of the spanning tree in Fig. 1(a) are connected through *Choice 2*.

4.2 Fitness Function

The length of an X-architecture Steiner tree is the sum of the lengths of all the edge segments in the tree, which is calculated as follows:

$$L(T_x) = \sum_{e_i \in T_x} l(e_i) \tag{3}$$

where $l(e_i)$ represents the length of each segment e_i in the tree T_x.

The smaller the fitness value, the better the particle is represented. Thus the particle fitness function is designed as follows.

$$fitness = L(T_x) \tag{4}$$

4.3   Particle Update Formula

In order to better solve the XSMT problem, a new particle update method with mutation and crossover operators is proposed. The specific formula is as follows:

$$X_i^t = F_3(F_2(F_1(X_i^{t-1}, \omega), c_1), c_2) \tag{5}$$

where ω is the mutation probability, c_1 and c_2 are crossover probability. F_1 is the mutation operator, which corresponds to the *inertia component* of PSO. F_2 and F_3 are crossover operators, corresponding to the *individual cognition* and *social cognition*, respectively.

Inertia Component. The particle velocity of SLPSO-XSMT is updated through F_1, which is expressed as follows:

$$W_i^t = F_1(X_i^{t-1}, \omega) = \begin{cases} M(X_i^{t-1}), r_1 < \omega \\ X_i^{t-1}, \quad \text{otherwise} \end{cases} \tag{6}$$

where ω is the probability of mutation operation, and r_1 is a random number in [0,1].

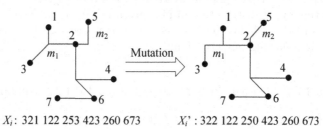

X_i: 321 122 253 423 260 673 X_i' : 322 122 250 423 260 673

Fig. 4. Mutation operator of SLPSO-XSMT

The proposed algorithm uses two-point mutation. If the generated random number $r_1 < \omega$, the algorithm will randomly replace the PSP choices of any two edges. Otherwise, keep the routing tree unchanged. Figure 4 gives a routing tree with 6 pins. It can be seen that after F_1, the PSP choices of m_1 and m_2 are replaced to *Choice 2* and *Choice 0*, respectively.

Individual Cognition. The SLPSO-XSMT algorithm uses F_2 to complete the *individual cognition* of particles, which is expressed as follows:

$$S_i^t = F_2(W_i^t, c_1) = \begin{cases} C_p(W_i^t), r_2 < c_1 \\ W_i^t, \quad \text{otherwise} \end{cases} \tag{7}$$

where c_1 represents the probability that the particle crosses with its personal historical optimum (X_i^P), and r_2 is a random number in $[0, 1)$.

Social Cognition. The SLPSO-XSMT algorithm uses F_3 to complete the *social cognition* of particles, which is expressed as follows:

$$X_i^t = F_3(S_i^t, c_2) = \begin{cases} C_p(S_i^t), r_3 < c_2 \\ S_i^t, \quad \text{otherwise} \end{cases} \tag{8}$$

where c_2 represents the probability that the particle crosses with the historical optimum of any particle X_k^P in the example pool, and r_3 is a random number in $[0, 1)$.

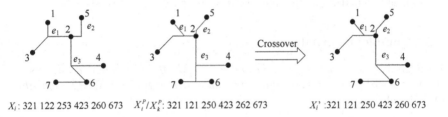

X_i: 321 122 253 423 260 673 X_i^P/X_k^P: 321 121 250 423 262 673 X_i' :321 121 250 423 260 673

Fig. 5. Crossover operator of SLPSO-XSMT

Figure 5 shows the crossover operation in *individual cognition* and *social cognition* of a particle. X_i is the particle to be crossed, and its learning object is X_i^P or X_k^P. The proposed algorithm firstly selects a continuous interval of the encoding, like the corresponding edges to be crossed e_1, e_2, and e_3. Then, replace the encoding on this interval of particle X_i with the encoding string of its learning object. After the crossover operation, the PSP choices of edges e_1, e_2, and e_3 in X_i are respectively changed from *Choice 2*, *Choice 3*, and *Choice 3* to *Choice 1*, *Choice 0*, and *Choice 3*, while the topology of the remaining edges remains unchanged.

Repeated iterative learning can gradually make particle X_i move closer to the global optimal position. Moreover, the acceleration coefficient c_1 is set to decrease linearly and c_2 is set to increase linearly, so that the algorithm has a higher probability to learn its own historical experience in the early iteration to enhance global search ability. While it has a higher probability to learn outstanding particles in the later iteration to enhance exploitation ability, so as to quickly converge to a position close to the global optimum.

4.4 Overall Procedure

Property 1. The proposed SLPSO-XSMT algorithm with example pool learning mechanism has a good balance between global exploration and local exploitation ability so as to effectively solve the XSMT problem.

The steps for SLPSO-XSMT can be summarized as follows.

Step 1. Initialize the population and PSO parameters, where the minimum spanning tree method is utilized to construct initial routing tree.

Step 2. Calculate the fitness value of each particle according to Eq. (4), and sort them in ascending order: $S = \{X_1, ..., X_{i-1}, X_i, X_{i+1}, ..., X_M\}$.

Step 3. Initialize *pbest* of each particle and its learning example pool $EP = \{X_1, ..., X_{i-1}\}$, and initialize *gbest*.

Step 4. Update the velocity and position of each particle according to Eqs. (5)–(8).

Step 5. Calculate the fitness value of each particle.

Step 6. Update *pbest* of each particle and its example pool *EP*, as well as *gbest*.

Step 7. If the termination condition is met (the set maximum number of iterations is reached), end the algorithm. Otherwise, return to step 4.

4.5 Complexity Analysis

Lemma 1. *Assuming the population size is M, the number of iterations is T, the number of pins is n, and then the complexity of SLPSO-XSMT algorithm is $O(MT \cdot n\log_2 n)$.*

Proof. The time complexity of mutation and crossover operations are both linear time $O(n)$. As for the calculation of fitness value, its complexity is mainly determined by the complexity of the sorting method $O(n\log_2 n)$. Since the example pool of each particle would change at the end of each iteration, the time for updating example pool is mainly spent on sorting, that is, its time complexity is also $O(n\log_2 n)$. Therefore, the complexity of the internal loop of the SLPSO-XSMT algorithm is $O(n\log_2 n)$. At the same time, the complexity of the external loop of the algorithm is mainly related to the size of the population and the number of iterations. Therefore, the complexity of proposed SLPSO-XSMT algorithm is $O(MT \cdot n\log_2 n)$.

5 Experiment Results

In order to verify the performance and effectiveness of the proposed algorithm in this paper, experiments are performed on the benchmark circuit suite [15]. The parameter settings in this paper are consistent with [8]. Considering the randomness of the PSO algorithm, the mean values in all experiments are obtained by independent run 20 times.

5.1   Validation of Social Learning Mechanism

In order to verify the effectiveness of proposed social learning mechanism based on example pool, this section applies PSO [8] and the proposed SLPSO method to seek the solution of XSMT, in which the social cognition of PSO is achieved through crossing with the global optimal solution (*gbest*). The experiments compare the wirelength optimization capabilities and stability of the two methods, as shown in Table 2. In all test cases, the SLPSO method can achieve shorter wirelength and lower standard deviation than the PSO method. On the three evaluation indicators (best wirelength, mean wirelength and standard deviation), the SLPSO method can achieve optimization rates of 0.171%, 0.289%, and 35.881%, respectively. The experimental data show that SLPSO method has better exploration and exploitation capability than PSO method.

Table 2. Comparison between PSO and the proposed SLPSO method

Test	Pins	Best wirelength			Mean wirelength			Standard deviation		
		PSO	SLPSO	Imp (%)	PSO	SLPSO	Imp (%)	PSO	SLPSO	Imp (%)
1	8	16918	16918	0.00	16918	16918	0.00	0.00	0.00	0.00
2	9	18041	18041	0.00	18041	18041	0.00	0.00	0.00	0.00
3	10	19696	19696	0.00	19696	19696	0.00	0.00	0.00	0.00
4	20	32193	32193	0.00	32217	32195	0.07	11.95	6.31	47.17
5	50	47960	47953	0.01	48103	47959	0.30	55.04	14.49	73.68
6	70	56357	56278	0.14	56536	56314	0.39	106.72	42.65	60.04
7	100	68650	68462	0.27	69047	68623	0.61	222.04	90.21	59.37
8	410	141520	140858	0.47	141908	141172	0.52	243.95	164.87	32.42
9	500	154365	153708	0.43	154760	153951	0.52	246.25	162.83	33.87
10	1000	220774	219928	0.38	221196	220132	0.48	235.45	112.40	52.26
Avg				0.17			0.29			35.88

5.2   Validation of SLPSO-Based XSMT Construction Algorithm

In order to verify the good performance of proposed SLPSO-XSMT algorithm, this section gives a comparison between SLPSO-XSMT and two SMT algorithms which are traditional RSMT (R) [9] and DDE-based XSMT (DDE) [16] algorithms. As shown in Table 3, ours performs well in wirelength optimization, and can reduce the average wirelength by 8.76% and 1.81%, respectively. It can be found from the comparison with DDE-based XSMT algorithm that our algorithm is more conducive to the construction of large-scale Steiner trees.

Additionally, SLPSO-XSMT algorithm has an overwhelming advantage in stability. It can be seen that ours is far superior to the two algorithms and can greatly reduce the standard deviation of the algorithm. Among them, the DDE-based algorithm has the worst stability, and ours can reduce the standard deviation by 97.39% on average.

Table 3. Comparison between SLPSO-XSMT and other SMT algorithms

Test	Mean wirelength					Standard deviation				
	Absolute values			Imp (%)		Absolute values			Imp (%)	
	R	DDE	Ours	O/R	O/D	R	DDE	Ours	O/R	O/D
1	17931	16911	16918	5.65	−0.04	0.00	33.65	0.00	–	100.00
2	20503	18039	18041	12.01	−0.01	0.00	41.47	0.00	–	100.00
3	21910	19469	19696	10.11	−1.16	0.00	132.73	0.00	–	100.00
4	35723	32342	32195	9.88	0.45	0.00	165.03	6.31	–	96.18
5	52509	48668	47959	8.66	1.46	54.67	827.25	14.49	73.50	98.25
6	60909	57255	56314	7.55	1.64	75.08	1134.68	42.65	43.19	96.24
7	74107	70686	68623	7.40	2.92	219.21	1802.17	90.21	58.85	94.99
8	154835	147115	141172	8.82	4.04	524.25	3615.35	164.87	68.55	95.44
9	167751	159672	153951	8.23	3.58	530.39	3687.87	162.83	69.30	95.58
10	242587	232359	220132	9.26	5.26	465.02	4088.80	112.40	75.83	97.25
Average				8.76	1.81				64.87	97.39

6   Conclusion

Aiming at the XSMT construction problem in VLSI routing, this paper proposes the SLPSO-based XSMT algorithm with the goal of optimizing the total wirelength. The algorithm adopts a novel social learning mechanism based on the example pool, so that particles can learn from different and better particles in each iteration, which expands searching range and helps to break through local extremes. At the same time, mutation and crossover operators are integrated into the update formula of particles to better solve the discrete XSMT problem.

The experimental results show that the proposed SLPSO-XSMT algorithm has obvious advantages in reducing wirelength and enhancing the stability of the algorithm, especially for large-scale Steiner trees. In future work, we will continue to improve this high-performance SLPSO to better solve various problems in the field of VLSI routing.

References

1. Chen, X., Liu, G., Xiong, N., Su, Y., Chen, G.: A survey of swarm intelligence techniques in VLSI routing problems. IEEE Access **8**, 26266–26292 (2020). https://doi.org/10.1109/ACCESS.2020.2971574
2. Coulston, C.S.: Constructing exact octagonal Steiner minimal trees. In: Proceedings of the 13th ACM Great Lakes symposium on VLSI, pp. 1–6 (2003). https://doi.org/10.1145/764808.764810
3. Guo, W., Liu, G., Chen, G., Peng, S.: A hybrid multi-objective PSO algorithm with local search strategy for VLSI partitioning. Front. Comput. Sci. China **8**(2), 203–216 (2014). https://doi.org/10.1007/S11704-014-3008-Y
4. Huang, X., Guo, W., Liu, G., Chen, G.: FH-OAOS: a fast four-step heuristic for obstacle-avoiding octilinear Steiner tree construction. ACM Trans. Des. Autom. Electron. Syst. **21**(3), 1–31 (2016). https://doi.org/10.1145/2856033

5. Huang, X., Liu, G., Guo, W., Niu, Y., Chen, G.: Obstacle-avoiding algorithm in x-architecture based on discrete particle swarm optimization for VLSI design. ACM Trans. Des. Autom. Electron. Syst. (TODAES) **20**(2), 1–28 (2015). https://doi.org/10.1145/2699862

6. Lin, K.W., Lin, Y.S., Li, Y.L., Lin, R.B.: A maze routing-based methodology with bounded exploration and path-assessed retracing for constrained multilayer obstacle-avoiding rectilinear steiner tree construction. ACM Trans. Des. Autom. Electron. Syst. (TODAES) **23**(4), 1–26 (2018). https://doi.org/10.1145/3177878

7. Lin, S.E.D., Kim, D.H.: Construction of all rectilinear Steiner minimum trees on the hanan grid. In: Proceedings of the 2018 International Symposium on Physical Design, pp. 18–25 (2018). https://doi.org/10.1145/3177540.3178240

8. Liu, G., Chen, G., Guo, W.: DPSO based octagonal Steiner tree algorithm for VLSI routing. In: 2012 IEEE Fifth International Conference on Advanced Computational Intelligence (ICACI), pp. 383–387. IEEE (2012). https://doi.org/10.1109/ICACI.2012.6463191

9. Liu, G., Chen, G., Guo, W., Chen, Z.: DPSO-based rectilinear Steiner minimal tree construction considering bend reduction. In: 2011 Seventh International Conference on Natural Computation, vol. 2, pp. 1161–1165. IEEE (2011). https://doi.org/10.1109/ICNC.2011.6022221

10. Liu, G., Chen, Z., Guo, W., Chen, G.: Self-adapting PSO algorithm with efficient hybrid transformation strategy for x-architecture Steiner minimal tree construction algorithm. Pattern Recogn. Artif. Intell. **31**(5), 398–408 (2018). https://doi.org/10.16451/j.cnki.issn1003-6059.201805002. (In Chinese)

11. Liu, G., Chen, Z., Zhuang, Z., Guo, W., Chen, G.: A unified algorithm based on HTS and self-adapting PSO for the construction of octagonal and rectilinear SMT. Soft Comput. **24**(6), 3943–3961 (2020). https://doi.org/10.1007/S00500-019-04165-2

12. Liu, G., Guo, W., Li, R., Niu, Y., Chen, G.: Xgrouter: high-quality global router in x-architecture with particle swarm optimization. Front. Comput. Sci. China **9**(4), 576–594 (2015). https://doi.org/10.1007/S11704-015-4017-1

13. Liu, G., Guo, W., Niu, Y., Chen, G., Huang, X.: A PSO-based timing-driven octilinear Steiner tree algorithm for VLSI routing considering bend reduction. Soft Comput. **19**(5), 1153–1169 (2015). https://doi.org/10.1007/S00500-014-1329-2

14. Thurber, A., Xue, G.: Computing hexagonal Steiner trees using PCX [for VLSI]. In: Proceedings of ICECS 1999, 6th IEEE International Conference on Electronics, Circuits and Systems, ICECS 1999 (Cat. No. 99EX357), vol. 1, pp. 381–384. IEEE (1999). https://doi.org/10.1109/ICECS.1999.812302

15. Warme, D., Winter, P., Zachariasen, M.: Geosteiner software for computing Steiner trees (2003). http://geosteiner.net

16. Wu, H., Xu, S., Zhuang, Z., Liu, G.: X-architecture Steiner minimal tree construction based on discrete differential evolution. In: Liu, Y., Wang, L., Zhao, L., Yu, Z. (eds.) ICNC-FSKD 2019. AISC, vol. 1074, pp. 433–442. Springer, Cham (2020). https://doi.org/10.1007/978-3-030-32456-8_47

17. Yan, J.T.: Timing-driven octilinear Steiner tree construction based on Steiner-point reassignment and path reconstruction. ACM Trans. Des. Autom. Electron. Syst. (TODAES) **13**(2), 1–18 (2008). https://doi.org/10.1145/1344418.1344422

18. Zhang, X., Wang, X., Kang, Q., Cheng, J.: Differential mutation and novel social learning particle swarm optimization algorithm. Inf. Sci. **480**, 109–129 (2019). https://doi.org/10.1016/J.INS.2018.12.030

19. Zhao, H., Xia, S., Zhao, J., Zhu, D., Yao, R., Niu, Q.: Pareto-based many-objective convolutional neural networks. In: Meng, X., Li, R., Wang, K., Niu, B., Wang, X., Zhao, G. (eds.) WISA 2018. LNCS, vol. 11242, pp. 3–14. Springer, Cham (2018). https://doi.org/10.1007/978-3-030-02934-0_1
20. Zhu, Q., Zhou, H., Jing, T., Hong, X.L., Yang, Y.: Spanning graph-based nonrectilinear Steiner tree algorithms. IEEE Trans. Comput.-Aided Des. Integr. Circuits Syst. **24**(7), 1066–1075 (2005). https://doi.org/10.1109/TCAD.2005.850862

Micro-nano Depth Information Recovery Method Based on TV Regularization

Mingxin Zhang[1(✉)], Qiuyu Wu[1,2], and Yongjun Liu[1]

[1] School of Computer Science and Engineering,
Changshu Institute of Technology, Changshu, China
mxzhang163@163.com
[2] School of Computer Science and Technology China,
University of Mining and Technology, Xuzhou, China

Abstract. In micro-nano computer vision, the defocus depth recovery method is used for micro-nano depth information recovery, and the Tikhonov regularization method is used to solve the problem that the objective function of the traditional defocus depth recovery method is ill-posed, resulting in low accuracy of depth information recovery. The TV regularization and L-curve method are introduced into the objective function of the traditional defocus depth recovery method. A depth information recovery algorithm based on TV regularization and L curve (L_TV algorithm) is proposed to improve the accuracy of depth information recovery. The algorithm introduces TV regularization in the objective function of the traditional defocus depth recovery method to avoid excessive punishment of the restored depth information and tends to be smooth. By introducing the L-curve method to select the appropriate regularization parameters, the depth of the recovery is avoided. The information is too smooth and retains more detail. The depth information recovery experiments of the standard 500 nm scale grid show that compared with the Tikhonov regularization method and the TSVD regularization method, the L_TV algorithm proposed in this paper can avoid excessive punishment of the recovered depth information, tend to be smooth, retain more details, and effectively improve the accuracy of deep information recovery.

Keywords: Micro-nano depth information recovery · Depth from defocus · TV regularization · L curve method

1 Introduction

In micro-nano computer vision, vision-based micro-nano scale 3D topography reconstruction is of great significance for a more comprehensive understanding of sample characteristics and evaluation of operational processes. The more commonly used 3D reconstruction methods [1] mainly include volume recovery methods, depth from sereo (DFS), depth from focus (DFF), and depth from defocus. Compared with other 3D reconstruction methods, the defocus depth recovery method has received extensive attention and in-depth research in recent years due to its advantages of fewer pictures, simple equipment, and convenient operation [2]. Defocus depth recovery was first proposed by Pentland [3]. This method uses the mapping relationship between the

© Springer Nature Switzerland AG 2020
G. Wang et al. (Eds.): WISA 2020, LNCS 12432, pp. 143–154, 2020.
https://doi.org/10.1007/978-3-030-60029-7_13

defocus degree feature of the two-dimensional image and the depth of the scene to inversely solve the three-dimensional depth information of the scene [4, 5].

In the process of defocus depth recovery of the scene, it is first necessary to obtain different degrees of defocus images of the scene, which leads to changes in the parameters of the camera. However, in micro-nano image observation, the observation space is very limited, and the camera with high magnification is used, so the imaging model of the camera will change with the change of camera parameters [6], limited to the conditional limitation in micro-nano observation. Wei Yangjie [7] proposed a new three-dimensional defocus shape recovery method based on parameter fixed monocular vision camera. This method uses the relative ambiguity [8] and the thermal radiation equation [9] to solve the depth information of the scene.

The traditional defocus depth recovery method uses the relative ambiguity and thermal radiation equation to establish the objective function of the defocus depth recovery method to solve the target object depth information, although the depth information of the target object can be obtained to some extent; however, the traditional defocus depth recovery The method objective function has ill-posedness and is often solved by the Tikhonov regularization method [10]. The Laplacian operator used in this method is easy to cause excessively penalized depth information and tends to be smooth due to its isotropic characteristics; The method determines the regularization parameter according to the empirical value, and sometimes the excessive regularization parameter often makes the restored depth information excessively smooth, and loses many details, resulting in low accuracy of depth information recovery.

Aiming at the problem that the objective function of the depth from defocus is not identifiable by the Tikhonov regularization method, and the accuracy of the depth information recovery is not high, the objective function of the traditional defocus depth recovery method is deeply studied, and the objective function of the traditional defocus depth recovery method is studied. The TV regularization and L-curve method are introduced, and a depth information recovery algorithm based on TV regularization and L-curve (L_TV algorithm) is proposed to improve the accuracy of depth information recovery. The algorithm introduces TV regularization into the objective function of the traditional depth from defocus. Avoid the depth information of the recovery to be over-punished and tend to be smooth; by introducing the L-curve method to select the appropriate regularization parameters, to avoid the parameter being too large, the depth information of the recovery is too smooth, and more details are retained.

The depth information recovery experiments of the standard 500 nm scale grid show that compared with the Tikhonov regularization method and the TSVD regularization method, the L_TV algorithm proposed in this paper can avoid excessive punishment of the recovered depth information, tend to be smooth, retain more details, and effectively improving the accuracy of deep information recovery.

2   Global Depth from Defocus

2.1   Defocus Imaging Model

In image processing, image blurring can usually be expressed as a convolution.

$$I_2(x, y) = I_1(x, y) * H(x, y) \tag{1}$$

where $I_1(x, y)$, $I_2(x, y)$ and $H(x, y)$ represent clear and blurred images and point spread functions, respectively. "$*$" represents a convolution. According to the principle of points spread, the point spread function can be approximated by a two-dimensional Gaussian function, and the fuzzy diffusion parameter therein σ indicates the degree of blurring of the image. Since the equation of heat radiation is isotropic, Eq. (1) can be expressed as

$$\begin{cases} \dot{z}(x, y, t) = c\nabla(x, y, t)c \in (0, \infty) \\ z(x, y, 0) = g(x, y) \end{cases} \tag{2}$$

Where "c" represents fuzz diffusion parameter, $\dot{z} = \partial z / \partial t$, "$\nabla$" represents laplacian operator.

$$\nabla z = \frac{\partial^2 z(x, y, t)}{\partial x^2} + \frac{\partial^2 z(x, y, t)}{\partial y^2} \tag{3}$$

$$\sigma^2 = 2tc \tag{4}$$

If the depth map is an ideal plane, then "c" is a constant, otherwise it can be expressed as

$$\begin{cases} \dot{z}(x, y, t) = \nabla \cdot (c(x, y)\nabla z(x, y, t)) \, t \in (0, \infty) \\ z(x, y, 0) = g(x, y) \end{cases} \tag{5}$$

Where "∇" 和 "$\nabla\cdot$" represent gradient and differential operators

$$\nabla = \begin{bmatrix} \frac{\partial}{\partial x} & \frac{\partial}{\partial y} \end{bmatrix}^T \nabla \cdot = \frac{\partial}{\partial x} + \frac{\partial}{\partial y} \tag{6}$$

From the above equation, we firstly need to know the clear image to solve the heat radiation equation, but this is a complicated process. Therefore, Favaro [8] proposed a concept of relatively ambiguity.

Suppose there are two different degrees of blurred images $I_1(x, y)$ and $I_2(x, y)$, the fuzzy diffusion coefficients are σ_1 and σ_2, $\sigma_1 < \sigma_2$, so $I_2(x, y)$ can be expressed as

$$\begin{aligned} I_2(x, y) &= \iint \frac{1}{2\pi\sigma_2^2} \exp\left(-\frac{(x-u)^2 - (y-v)^2}{2\sigma_2^2}\right) g(u, v) du dv \\ &= \iint \frac{1}{2\pi\Delta\sigma^2} \exp\left(-\frac{(x-u)^2 + (y-v)^2}{2\Delta\sigma^2}\right) I_1(u, v) du dv \end{aligned} \tag{7}$$

Where $\Delta\sigma^2 = \sigma_2^2 - \sigma_1^2$ represents relative ambiguity, Eq. (2) can be expressed as

$$\begin{cases} \dot{z}(x,y,t) = c\nabla z(x,y,t) \, c \in (0,\infty) \, t \in (0,\infty) \\ z(x,y,0) = I_1(x,y) \end{cases} \tag{8}$$

Equation (5) can be expressed as

$$\begin{cases} \dot{z}(x,y,t) = \nabla \cdot (c(x,y)\nabla z(x,y,t)) \, t \in (0,\infty) \\ z(x,y,0) = I_1(x,y) \end{cases} \tag{9}$$

And at the moment of Δt, $u(x,y,t) = I_2(x,y)$, Δt can be defined as

$$\Delta\sigma^2 = 2(t_2 - t_1)c \doteq 2\Delta tc \tag{10}$$

In addition

$$\Delta\sigma^2 = \lambda^2 \left(\eta_2^2 - \eta_1^2 \right) \tag{11}$$

Where $\eta_i (i = 1,2)$ represent fuzzy circle radius, λ represents the constant between the ambiguity and the radius of the fuzzy circle.

$$\eta = \frac{Dv}{2} \left| \frac{1}{f} - \frac{1}{v} - \frac{1}{s} \right| \tag{12}$$

Where v, f, s are the camera's image distance, focal length and object distance, respectively, and D is the convex lens radius.

2.2 The Objective Function of Depth from Defocus

Assume that the image $I_1(x,y)$ is a defocus image before depth variation, its depth information is $S_1(x,y)$, and the image $I_2(x,y)$ is a defocus image after the depth variation, its depth information is $S_2(x,y)$, s_0 is known as the ideal focal length. In this section, the depth information of the image is obtained by changing the object distance. The schematic diagram is shown in Fig. 1.

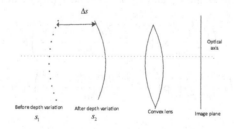

Fig. 1. Illustration of depth from defocus

First, establishing the heat radiation equation from the above conditions

$$\begin{cases} \dot{z}(x,y,t) = \nabla \cdot (c(x,y)\nabla z(x,y,t)) \, t \in (0,\infty) \\ z(x,y,0) = I_1(x,y) \\ z(x,y,\Delta t) = I_2(x,y) \end{cases} \tag{13}$$

Where relative ambiguity can be expressed as

$$\begin{aligned} \Delta\sigma^2(x,y) &= \lambda^2 \left(\eta_2^2(x,y) - \eta_1^2(x,y) \right) \\ &= \frac{\lambda^2 D^2 v^2}{4} \left[\left(\frac{1}{f} - \frac{1}{v} - \frac{1}{s_2(x,y)} \right)^2 - \left(\frac{1}{f} - \frac{1}{v} - \frac{1}{s_1(x,y)} \right)^2 \right] \\ &= \frac{\lambda^2 D^2 v^2}{4} \left[\left(\frac{1}{s_0} - \frac{1}{s_2(x,y)} \right)^2 - \left(\frac{1}{s_0} - \frac{1}{s_1(x,y)} \right)^2 \right] \end{aligned} \tag{14}$$

It defines in this paper

$$b = \frac{4\Delta\sigma^2}{\lambda^2 D^2 v^2} + \left(\frac{1}{s_0} - \frac{1}{s_1(x,y)} \right)^2 \tag{15}$$

Therefore, the final depth information is

$$s_2(x,y) = \frac{1}{\left((1/s_0 \pm \sqrt{b}) \right)} \tag{16}$$

In order to better solve the dynamic optimization problem of depth from focus, the following objective function is established to calculate the thermal radiation equation

$$\bar{s} = \arg\min \iint (z(x,y,\Delta t) - I_2(x,y))^2 dx dy \tag{17}$$

Since the above solution process may be ill-conditioned, a Tikhonov regularization term is added to the end of the objective function, expressed as

$$\bar{s} = \arg\min \iint (z(x,y,\Delta t) - I_2(x,y))^2 dx dy + \rho \|\nabla s_2(x,y)\|^2 + \rho l \|s_2(x,y)\|^2 \tag{18}$$

The third term of the above formula is about the smoothness constraint of the depth map, and the fourth is the guarantee of the bounded of the depth map, in fact the energy coefficient $\rho > 0$, $l > 0$ lost energy is expressed as

$$E(s) = \iint (z(x,y,\Delta t) - I_2(x,y))^2 dx dy + \rho \|\nabla s\|^2 + \rho l \|s\|^2 \tag{19}$$

Therefore, the scene depth information can be solved by solving the following optimization problem.

$$\tilde{s} = \arg\min E(s)$$
$$s.t(13)(16)$$

(20)

3 L_TV Algorithm

3.1   The TV Regularization Method

The objective function of the traditional defocus depth recovery method is ill-posed and often solved by the Tikhonov regularization method. The Laplacian operator used in this method is easy to cause excessive punishment of the recovered depth information due to its isotropic characteristics. Sometimes the excessive regularization para-meter often makes the restored depth information excessively smooth, and loses many details, resulting in low accuracy of depth information recovery.

At present, the regularization methods often used for ill-posed problems mainly include Tikhonov regularization, TSVD regularization and TV regularization [11]. The TV regularization method can avoid recovery by introducing the anisotropic diffusion equation of the partial differential equation. The depth information is over-punished, tends to be smooth, and retains more detail. Therefore, the TV regularization method is used to solve the ill-ness of the objective function of the traditional defocus depth recovery method, avoiding excessive punishment of the restored depth information, tending to smooth, and retain more details.

The TV regularization method mainly represents the sum of image gradient magnitudes [12], which can be expressed by the following expression

$$TV(u) = \sum_{i=1}^{N} \|D_i u\|_2 = \sum_{(i,j)} \sqrt{\left(\frac{\partial u}{\partial i}\right)^2 + \left(\frac{\partial u}{\partial j}\right)^2}$$

(21)

In the formula (21), u represents the original image, and D represents the difference operator, and by calculating the deviation of the pixels of each point, the total variation can be obtained.

For two-dimensional images

$$D_i u = \sqrt{D_i^x u^2 + D_i^y u^2}$$

(22)

Therefore, the objective function of the defocus depth recovery method based on the norm TV regularization can be expressed as follows

$$E(s) = \iint (z(x,y,\Delta t) - I_2(x,y))^2 dxdy + \sum_{i=1}^{N} \rho \|D_i s\| + \rho l \|s\| \qquad (23)$$

3.2 Select the Regularization Parameter by L-Curve Method

The objective function of the defocus depth recovery method based on the TV regularization method can avoid the excessive punishment of the restored depth information and tend to be smooth; however, the method determines the regularization parameter according to the empirical value, and the smaller regularization parameter can solve the original problem. A good approximate solution is given, but the influence of the error may make the solution unstable. Large regularization parameters may cause the obtained solution to be over-punished and deviate from the real solution. Therefore, the value of the regularization parameter is particularly important.

At present, regarding the ill-posed problem, the methods of regularization parameter selection include tolerance principle method, generalized cross-validation method and L-curve method [13]. The L-curve method is widely used for various discomforts because it can well determine the relative size of the residual and the real solution..

Lawson and Hanson [14] first proposed using the objective function and the corresponding constraint function to obtain a curve for the coordinates. The two sides of the maximum curvature of the curve are respectively composed of the objective function and the corresponding constraint function. In fact, the L-curve is composed of points with coordinates, and is obtained by taking logarithms to the horizontal and vertical coordinates, respectively, which correspond one-to-one with the regularization parameters.

According to $\left(\iint (z(x,y,\Delta t) - I_2(x,y))^2 dxdy, \|s_\rho\|\right)$, calculate the corresponding sequence $k(\rho)$ and draw the L-curve, find the maximum point of the graph, and determine the regularization parameter ρ^*. Equation (23) can be expressed as:

$$E(s) = \iint (z(x,y,\Delta t) - I_2(x,y))^2 dxdy + \sum_{i=1}^{N} \rho^* \|D_i s\| + \rho^* l \|s\| \qquad (24)$$

For Eq. (24), an iterative shrinkage threshold algorithm (ISTA) is used to solve.

4 The Flow Chart of the L_TV Algorithm

In summary, the Tikhonov regularization method is used to solve the problem that the objective function of the traditional defocus depth recovery method is ill-posed, and the accuracy of depth information recovery is not high. The objective function of the traditional defocus depth recovery method is deeply studied. The TV regularization method and the L-curve method are introduced into the objective function of the focal depth recovery method. A depth information recovery algorithm (L_TV) based on TV regularization and L-curve is proposed to effectively improve the accuracy of depth information recovery. The L_TV algorithm is implemented as follows:

Step 1: Give camera parameters focal length f, distance v, ideal object distance s_0, convex lens radius D, constant λ between ambiguity and fuzzy circle, two blurred images I_1, I_2, energy threshold ε, energy coefficient ρ_0, iterative step size β
Step 2: The initialization depth information is s_0;
Step 3: The Eq. (14) is calculated to obtain relative ambiguity $\Delta\sigma^2$;
Step 4: According to formula (21), the objective function expression of defocus depth recovery method based on TV regularization is established (23);
Step 5: According to the establishment of the function and constraints, calculate the corresponding sequence $k(\rho)$ and draw, find the maximum point of the graph, and determine the regularization parameters ρ^*;
Step 6: Bring the determined regularization parameters ρ^* into Eq. (24) and find the thermal diffusion conduction time Δt;
Step 7: According to the thermal diffusion conduction Δt and the thermal radiation Eq. (2), the solution of the diffusion equation $z(x, y, \Delta t)$ is obtained;
Step 8: Calculate the energy formula (19) using the solution obtained $z(x, y, \Delta t)$ in step 7. If the energy is less than the threshold ε, stop, and the depth information at this time is the desired; otherwise, use the iterative shrinkage threshold algorithm (ISTA) to solve the step size β;

$$\frac{\partial s}{\partial t} = -E'(s) \tag{25}$$

Step 9: Solve the formula (23), update the depth, and return to step two.

5 The Simulation

In order to verify the effectiveness of the proposed L_TV algorithm, the Tikhonov regularization method, the TSVD regularization method and the L_TV algorithm proposed in this chapter are used to perform depth information recovery experiments on the standard 500 nm scale grid. The HIROX-7700 microscope was used in the experiment, the magnification was 7000 times, the camera focal length f = 0.357 mm, the ideal object distance mm, the aperture size was 2, and the convex lens radius D = f/2.

In order to evaluate the performance of each algorithm more objectively, all algorithms will operate independently in the same environment. Experimental environment: Windows10 operating system, CPU Intel i5-4900 Murray dual-core, clocked at 3.3 GHz, 8G memory, MATLAB2012a.

The other parameters required in the experiment are set as follows: the constant between the ambiguity $\lambda = 3$ and the fuzzy circle $\varepsilon = 2$, the energy threshold, the energy coefficient $\rho_0 = 0.1$, and the iteration step size $\beta = 0.0001$.

The grid size is 120×110 pixels, 500 nm high and 1500 nm wide. The processing results for the standard 500 nm scale grid are shown below, where Fig. 2 is the two defocus images of the grid. The coordinate unit is a pixel, the pixel size is 115.36 nm * 115.36 nm, and the unit of height coordinate is mm (Figs. 3–5).

Fig. 2. Two defocus images of grid

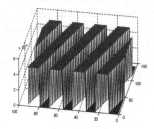

Fig. 3. Real surface map of grid

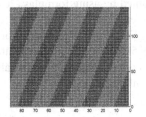

Fig. 4. Real depth map of grid

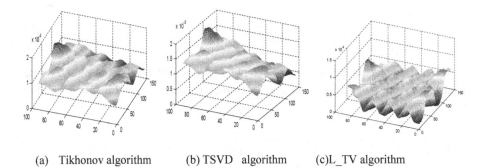

(a)   Tikhonov algorithm   (b) TSVD   algorithm   (c)L_TV algorithm

Fig. 5. The restored surface map of grid by three algorithms

In order to verify the accuracy of the L-TV algorithm proposed in this paper for the standard 500 nm scale raster depth information recovery, the angular part of the restored grid topography is partially intercepted. The angular detail of the corner is shown in Fig. 6:

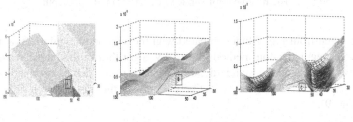

(a)The real surface map   (b) Tikhonov algorithm   (c) L_TV algorithm

Fig. 6. Local detail of the corner of the restored grid

It can be seen from Fig. 6 that for the local detail part of the lattice shape, the corner A of the real grid topography is a right angle. The angle C is smaller than the point B, the slope is steeper, and it is closer to the right angle, which is closer to the true grid depth information surface map.

In order to further verify the accuracy of the L-TV algorithm proposed in this paper for the standard 500 nm scale raster depth information recovery, the recovery grid depth is obtained by using formulas (26), (27) and (28) respectively. The relative error, mean square error and average value of the information were obtained by multiple experiments. Figure 7 is a relative error surface map using three algorithms to restore the standard 500 nm scale raster depth information; Fig. 8 is a graph of the mean square error convergence using three algorithms to restore the standard 500 nm scale raster depth information.

$$\phi = \tilde{s}/s - 1 \tag{26}$$

$$\psi = \sqrt{E\left[(\tilde{s}/s - 1)^2\right]} \tag{27}$$

$$E = \frac{1}{n}\sum_{k=1}^{n}\left|H_k - \tilde{H}_k\right| \tag{28}$$

Here, n represents the number of sampling points, which H_k represents the true height of the kth point of the grid, and $\tilde{H}_k$, represents the estimated height of the kth point.

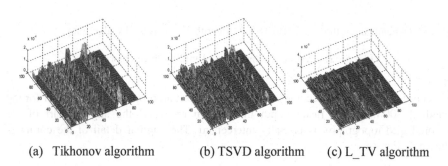

(a) Tikhonov algorithm   (b) TSVD algorithm   (c) L_TV algorithm

Fig. 7. The relative error surface map of restored grid depth

It can be seen from Fig. 7 that compared with the Tikhonov regularization method and the TVSD regularization method, the L_TV algorithm in this paper restores the relative error surface of the standard 500 nm scale raster depth information to be significantly smaller, which can effectively improve the raster depth information recovery (Fig. 9).

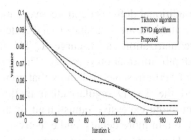

Fig. 8. Mean variance convergence curve of three algorithms

Finally, in order to better test the performance of the proposed algorithm, depth information recovery experiments were carried out for triangular probes using Tikhonov algorithm and L_TV algorithm respectively. Figure 10 shows the two defocus images of triangular probes.

Fig. 9. Two defocus images of triangular probes

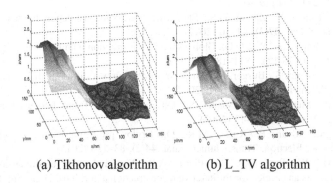

(a) Tikhonov algorithm   (b) L_TV algorithm

Fig. 10. The restored triangular proves by two algorithms

6 Conclusion

Based on the traditional defocus depth recovery method, this paper studies the accuracy of depth information recovery. Aiming at the problem that the objective function of the defocus depth recovery method is not identifiable by the Tikhonov regularization method, and the accuracy of the depth information recovery is not high, the objective function of the traditional defocus depth recovery method is deeply studied, and the objective function of the traditional defocus depth recovery method is studied. The TV regularization and L-curve method are introduced, and a depth information recovery algorithm based on TV regularization and L-curve (L_TV algorithm) is proposed to improve the accuracy of depth information recovery. The depth information recovery experiments of the standard 500 nm scale grid show that compared with the Tikhonov regularization method and the TSVD regularization method, the L_TV algorithm proposed in this paper can avoid excessive punishment of the recovered depth information, tend to be smooth, retain more details, and effectively improve the accuracy of deep information recovery. However, the average running time is relatively long, which is suitable for the occasions where the recovery accuracy is relatively high. The efficiency of depth information recovery will be studied below.

References

1. Li, C., Su, S., Matsushita, Y.: Bayesian depth-from-defocus with shading constraints. IEEE Trans. Image Process. **25**(2), 589–600 (2016)
2. Tao, M.W., Srinivasan, P.P.: Shape estimation from shaping, defocus, and correspondence using light-field angular coherence. IEEE Trans. Pattern Anal. Mach. Intell. **39**(3), 546–560 (2017)
3. Pentland, A.P.: A new sense for depth of field. IEEE Trans. Pattern Anal. Mach. Intell. **9**(4), 523–531 (1987)
4. Nayar, S.K., Watanabe, M., Noguchi, M.: Real time focus range sensor. IEEE Trans. Pattern Anal. Mach. Intell. **18**(12), 1186–1198 (1996)
5. Subbarao, M., Surya, G.: Depth from defocus: A spatial domain approach. Int. J. Comput. Vis. **13**(3), 271–294 (1994)
6. Yang, J., Tian, C.P., Zhong, G.S.: Stochastic optical reconstruction microscopy and its application. J. Opt. **37**(3), 44–56 (2017)
7. Wei, Y.J., Dong, Z.L., Wu, C.D.: Global shape reconstruction with fixed camera parameters. J. Image Graph. **15**(12), 1811–1817 (2010)
8. Favaro, P., Soatto, S., Burger, M.: Shape from defocus via diffusion. IEEE Trans. Pattern Anal. Mach. Intell. **30**(3), 518–531 (2008)
9. Favaro, P., Mennucci, A., Soatto, S.: Observing shape from defocused images. Int. J. Comput. Vis. **52**(1), 25–43 (2003)
10. Gazzola, S., Novati, P., Russo, M.R.: On Krylov projection methods and Tikhonov regularization. Electron. Trans. Numer. Anal. **44**(2), 83–123 (2015)
11. Zibetti, M.V.W., Pipa, D.R., De Pierro, A.R.: Fast and exact unidimensional L2–L1 optimization as an accelerator for iterative reconstruction algorithms. Digit. Sig. Proc. **48**(5), 2–4 (2015)
12. Miao, P., Qi, C., Fang, L.: Deep clipping noise mitigation using ISTA with the specified observations for LED-based DCO-OFDM system. IET Commun. **12**(20), 2582–2591 (2018)
13. Yahaghi, E., Mirzapour, M., Movafeghi, A.: FISTA algorithm for radiography images enhancement with background blurring removal. Res. Nondestr. Eval. **34**(5), 1–9 (2018)
14. Zhang, M., Wu, Q.: Optimization of depth from defocus based on iterative shrinkage thresholding algorithm. Web Information Systems and Applications **12**(5), 131–144 (2018)

An Approach for Progressive Set Similarity Join with GPU Accelerating

Lining Yu[✉], Tiezheng Nie, Derong Shen, and Yue Kou

College of Computer Science and Technology, Northeastern University,
Shenyang, China
2764717155@qq.com, {nietiezheng, shenderong,
kouyue}@cse.neu.edu.cn

Abstract. Set similarity join (SSJoin) is an important operation for searching similarity set pairs from the given database and play a core role in data integration, data cleaning, and data mining. In contrast to the traditional SSJoin methods, progressive SSJoin aims to resolve large datasets so that the efficiency of finding similarity pairs in the limited running time is improved. Progressive SSJoin can provide possible partial matching pairs of the dataset as early as possible in the processing. Moreover, recent research has shown that GPUs (Graphics Processing Units) can accelerate the similarity operation. This paper focuses on exploring progressive SSJoin algorithms and accelerating them with GPUs. We proposes two progressive SSJoin methods, PSSJM and PBM. PSSJM uses inverted index and PBM achieves its required functions by utilizing counting Bloom filter and prefix filtering techniques. In addition, we proposed a GPUs-based algorithm based on our proposed progressive method to accelerate the computation. Comprehensive experiments with real-world datasets show that our methods can generate better quality results than the traditional method under limited time and the method implementing on GPUs has high speedups over CPU-base method.

Keywords: Set similarity join · Progressive · Graphics processing units

1 Introduction

Similarity join [1] is an operation that identifies all similar sets pairs whose similarity meets the given threshold, in the given collection of sets. Set similarity join (SSJoin) is a kind of similarity join that executes on sets. When dealing with a very large amount of data, an SSJoin process is often so expensive because there are usually numerous record comparisons waiting for detected. Thus, tremendous working might need to be finished before we achieve the majority results. For resolving this challenge, we introduce the progressive method for SSJoin,

In this paper, we propose two novel, progressive SSJoin methods called progressive set similarity join method (PSSJM), which is based on building an inverted index, and progressive bloom filter method (PBM), which utilizes counting bloom filter techniques.

© Springer Nature Switzerland AG 2020
G. Wang et al. (Eds.): WISA 2020, LNCS 12432, pp. 155–167, 2020.
https://doi.org/10.1007/978-3-030-60029-7_14

2 Related Works

There have been many research on Set Similarity Join (SSJoin) [1–14]. The several state-of-the-art methods [4] implement a filter-verification framework. We focus on improving performance in filter phase, by introducing progressive methods, on the base of current methods. In the field of entity resolution and duplication detection, progressive methods are utilized to facilitate acquire as much results as possible under limited budget and we find many corresponding works [15–17]. As far as we known, the Sorted Neighborhood Method (SNM) [18] is often utilized by many progressive methods [15, 16]. However, when it processes some dataset which cannot be generated a key sorting where the closeness in ordering of keys is associated with the likelihood of sets similarity. SNM is difficult to achieve expected performance. In contrast, our methods are designed based on Jaccard Similarity and completely handle this case.

3 Preliminaries

3.1 Set Similarity Join

Given a collection of sets R, similarity threshold t and a similarity function $Sim(x, y)$ used to map two sets $x, y \in R$ into a value. Set Similarity Join can be utilized to identify all the set pairs,$\{x, y\}$, such that their similarities are no smaller than t, i.e., $Sim(x, y) \geq t$. The size of a set x (i.e., the number of tokens it contains) is denoted by $|x|$.

For example, the Jaccard similarity between sets "It is an apple tree" and "It is an apple" is $4/5 = 0.8$. If given similarity threshold $t = 0.6$, the pair consisted of the sets can be considered as a similar pair.

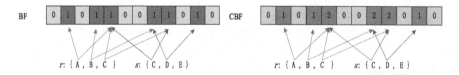

Fig. 1. Results of mapping two sets (r, s) using BF and CBF (both with $m = 12$, $k = 2$).

3.2 Bloom Filter and Counting Bloom Filter

A bloom filter (BF) [19] is a space-efficient probabilistic data structure that supports queries for whether an element is contained by a set [19]. It is a bit array of size m, where all bits are set as 0 initially. k different hash functions, $h_1, h_2, ..., h_k$, are utilized to map set element s to the bit array by setting the position $h_i(s)$, $i \in \{1, 2, ..., k\}$, to 1.

Different from BF, a counting bloom filter (CBFs) [19] provides a way to implement a delete operation. In a CBF, each position of its bit array is extended from being a single bit to being a multi-bit counter, denoted by $cbf[i]$, $i \in \{1, 2, ..., m\}$. Figure 1 shows an example of working of a BF and a CBF.

Fig. 2. Prefix filter.

3.3 Prefix Filtering

Prefix filtering [1], which is based on the pigeonhole principle and effectively decreases the number of candidate set pairs, is widely utilized in filtering phase of SSJoin. For example, considering sorted sets in Fig. 2, we shade prefix tokens (tokens included by prefix of each set respectively according to a given overlap threshold $t = 4$). The pair (r, z) and pair (s, z) can safely be pruned because there is no token match in the prefix. The pair (r, s) can be documented as a candidate pair because of a common token "C".

For each set x, we can define size bounds, which consist of two bounds: size upper bound and size lower bound [4], which respectively denoted by $lowerbound(x)$ and $upperbound(x)$. All sets whose size do not fall within the interval $[lowerbound(x), upperbound(x)]$ will be excluded, because they cannot match with x according to the given threshold. Besides, we define Max-prefix as x'smallest prefix needed for identifying $\forall y$ that the overlap, between x and y, meet the given overlap threshold. Max-prefix of x, $maxprefix(x)$, can be achieved by $|maxprefix(x)| = |x| - |lowerbound(x)| + 1$.

4 Our Approach

Our progressive methods use the blocking technique as the first procedure. That means the following operations are implemented on the base of blocking.

Blocking techniques can cluster similar sets into blocks according to some criteria called blocking keys [20]. Only sets that have been indexed into the same block, are considered to compare with each other. In this paper, we choose the prefix tokens as blocking keys. Figure 3 gives an example for blocking by keys.

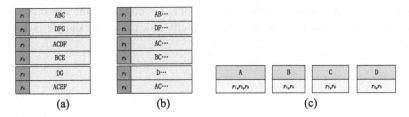

Fig. 3. (a) A set collection S which show id and tokens contained in each set. (b) Results of prefix processing. (c) Results of blocking by key.

Our core strategy is to generate a list of blocks sorted by scores which aim to evaluate proportion of possibly similar pairs, called candidate pairs, in every block. The set of candidate pairs can be obtained by filtering. We consider that a block with higher score have a higher priority to be processed in join phase according to its higher Proportion of Candidate Pairs (PCP). Obviously, what we focus on is how to get a reliable score to evaluate PCP of a block and ensure its processing priority. Figure 4 shows the overview of our strategy. In following sections, we propose two progressive methods on SSJoin.

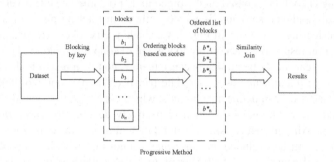

Fig. 4. Strategy overview of progressive method on SSJoin.

4.1 Pssjm

We now introduce the first proposed method, Progressive Set Similarity Join Method (PSSJM), using inverted index and prefix filtering techniques. In this method, we consider the proportion of sets which share at least one token in their prefix as the score for evaluating PCP of the block containing these sets. We classify the functionality of the method into two phases: indexing and score computing.

Indexing Phase. At first, we generate a count array for each block, which counts the frequency of occurrence for each prefix token contained in the block, and then build an inverted index for these prefix tokens. Max-prefix is utilized as size of the set prefix.

In practice, we implement a prefix index, *index*, as a two-dimensional array. *index* [i] points to an array that contains all ids of the sets involving t_i, a token whose id is marked as i. *count*[i] documents the frequency of t_i, i.e. number of the sets involving t_i. Algorithm 1 shows the process creating the prefix index and an example is given in Fig. 5.

Algorithm 1: Indexing phase for PSSJM

Input:(i) The collection of sets S, (ii) a Jaccard similarity threshold t
Output: *count, index*

1 *count* ← Build a array *count*;
2 *index* ← Build a data structure *index*;
3 initialize *count* with zeros;
4 count the occurrences of each token and store results in *count*;
5 **for** each set $r \in S$ **do**
6   **for** each token t in r's maxprefix **do**
7   add $r.id$ to *index*, according to t
8 **return** count, index

Algorithm 2: Score Computing

Input:(i) The collection of sets S, (ii) an prefix index of S: *index*,
           (iii) a Jaccard similarity threshold t
Output: *score* of S

1 $A \leftarrow \varnothing$;
2 $num \leftarrow 0$;
3 **for** each set $x \in S$ **do**
4   **for** each token w in x's prefix **do**
5   **for** each set $y \notin A$ in *index*$[w.id]$ **do**
6   **if** $x.\text{id} < y.\text{id}$ **then**
7   num ++; add y to A;
8      $A \leftarrow \varnothing$;
9 **return** $num / \binom{2}{|S|}$

Algorithm 3: PSSJM

Input: (i) The collection of sets S, (ii) a Jaccard similarity threshold t
Output: block list

1 BlockList ← $\varnothing$;
2 $B \leftarrow \varnothing$;
3 $B \leftarrow$ Blocking(S);
4 **foreach** $b_i \in B$ **do**
5   *count, index* ← indexing(b_i,t);
6   blockscore$_i$ ← ComputeScore ($S,index,t$);
7   BlockList add(getBlock (b_i,blockscore$_i$));
8 sortInDescreasingScore(*BlockList*);
9 **return** BlockList;

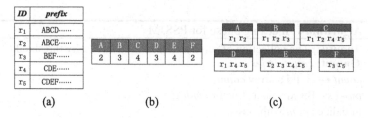

Fig. 5. (a) A set collection S which show id and tokens contain in prefix of each set. (b) Data structure *count* which is created about S. (c) Inverted index generated by S in indexing phase.

Score Computing. By analyzing frequency of tokens in *count*, we can obtain the number of sets which share at least one token in prefix by generated index. Algorithm 2 shows the process used to compute score and Algorithm 3 provides the complete process of PSSJM.

4.2   PBM

For PSSJM, building inverted index and computing scores would cause a high cost when processing a big dataset. Thus, in this section we present Progressive Bloom filter-based Set Similarity Join Method (PBM), based on the CBF, to solve the evaluation of likelihood for proportion of similar pairs in each block.

Before describing our algorithms, following theorems need to be outlined.

Theorem 1. Given two sets, x and y. If the two sets share at least one common token within their prefix, then there are at least k common positions between positions in array *cbf* set through mapping elements in prefix of x and y by k independent hash functions into the CBF.

Above theorem can simply explained by prefix filtering techniques, so we do not given corresponding proof here.

Theorem 2. Let sets $S = \{r_1, r_2, r_3,..., r_{|S|}\}$ be the given collection of sets, where $|S|$ denotes the size of S (i.e., the number of sets it contains), t denotes the given Jaccard similarity threshold and an integer array *cbf* of length l denotes the result of mapping all the tokens in Max-prefix of each set of S into a CBF (with k independent hash functions). Then, the PCP of S is restricted to the upper bound defined by Eq. 1.

$$Upperbound(PCP) = \min\left(\frac{\sum_{i}^{m}\binom{2}{cbf[i]}}{\binom{2}{|S|} \times k}, 1\right) \tag{1}$$

Proof. According the prefix filtering theories and definition of Max-prefix, we can consider that if two sets is similar, there is at least one common position, which CBF map the sets into, in array *cbf*. In other words, the pair which contain sets sharing none common position in *cbf* will be pruned.

Therefore, we define h as the number of the pairs meeting the given threshold. The PCP for S can be defined by Eq. 2.

$$PCP = h / \binom{2}{|S|} \qquad (2)$$

Considering $h \times k \leq \sum_{i}^{m} \binom{2}{|cbf[i]|}$, Eq. 2 can be directly transformed into the upper bound defined by Eq. 1.

Now, we take the Upper Bound of PCP (UBPCP) as the score which is used as our reference to create an ordering for blocks, a UBPCP correspond to possibly higher PCP. For example, with given two blocks, b_1 and b_2, if b_1's UBPCP is higher than b_2's, we consider that b_1 possibly has a higher PCP than b_2 and should have a high priority to be processed.

Algorithm 4 describes the initialization phase of PBM. Firstly, it creates its data structures (Lines 1–2). Then it operate blocking by key process which block all sets into blocks (Line 3) and for every set r_i in each block (Lines 4–5), it iterates over all tokens contained by r_i's prefix whose size is defined by Max-prefix of r_i. and map these tokens into an array cbf_i by using CBF(Lines 6–7). Based on built cbf_i, a score, UBPCP, can be generated for each block. Finally, all blocks are collected and sorted from the highest score to the lowest and a sorted list of blocks is returned.

However, for sets in each block, the upper bound given by PBM suffers by the performance of repeated comparisons because the sets with two or more common positions can increase the upper bound redundantly. As a result, in some cases, a block of lower PCP can generate a higher upper bound than another block with higher PCP. These can negatively impact the effect of PBM by sorting block list with an incorrect order.

4.3   Join Phase and Optimizing Strategy

In final join phase, we will find similar pairs in each block under an order according to block scores of blocks. However, there is a drawback that repeated comparisons will occur when two blocks contain more than one common pair. Therefore, we introduce the Least Common Block Index (LeCoBI) condition [17] to solve this problem.

Algorithm 4: PBM

Input:(i) The collection of sets D,
  (ii) a Jaccard similarity threshold t
Output: block list
1 BlockList $\leftarrow \varnothing$;
2 $B \leftarrow \varnothing$;
3 $B \leftarrow$ Blockbykey(D);
4 **foreach** $b_i \in B$ **do**
5   **foreach** $r_i \in b_i$ **do**
6   **foreach** tokens t in r_i's maxprefix **do**
7        $cbf_i =$ CBF(t);
8   blockscore$_i =$ ComputeScore(cbf $_i$);
9 *BlockList* add(getBlock (*b_i, blockscore$_i$*));
10 sortInDescreasingScore(*BlockList*);
11 return BlockList;

---

5 GPU-Based Similarity Join

Because of its expensive cost in creating of inverted index and counting of scores, we implement PSSJM in GPUs in order to achieve high speedups by taking advantage of the high parallelism of GPUs [6, 21].

In our GPUs' implementation, we use two main kernel functions, responsible for index phase and scoring phase. We iteratively process each block generated in blocking phase. As described in Algorithm 5, b_i denotes a block and t is the threshold. Before scoring blocks, we allocate and create two arrays, *count* and *index* (Lines 16–17), in the GPU memory and then create an inverted index for sets of each block. One GPU block is scheduled for processing one set, and each thread in this GPU block is responsible for a token in prefix, counting its frequency and updating *count* and *index* (Lines 2–6). After creation of the index for one block, we design a Scoring algorithm for GPU. We create and initial a single array in the GPU memory, *counting array*, for following score counting. With a similar processing scheme of allocating tasks of GPU blocks and threads, eventually, the IPR of the block is available by computing data stored in *counting array* and then the block scores also can be computed. When calculating scores of all blocks is finished, we can order these blocks in decreasing scores in order to produce a sorted list of blocks.

Algorithm 5: GPU-PSSJM Algorithm

---

1   **procedure** IndexingKernel(b_i, t, *count*, *index*)
2   **for** each set $x \in b_i$ **do** // executed by blocks
3   **for** each token w in x's maxprefix **do** //executed by threads
4   update *count* and *index*
5   **procedure** ScoringKernel (b_i, t, *count*, *index*, *blockscore$_i$*)
6   **for** each set $x \in S$ **do** // executed by blocks
7   **for** each token w in x's maxprefix **do** //executed by threads
8   *blockscore$_i$* ←ComputeScore (b_i, index, t);
9   **procedure** GPUHOST (S, t)
10   B ← BlockingbyKey(S);
11   BlockList← ∅;
12   **for** $b_i \in B$ **do**
13   *count* ← Build a array *count*;
14   *index* ← Build a data structure *index*;
15   initialize *count*;
16   IndexingKernel<<<Bl, T>>>(b_i, t, *count*, *index*);
17   blockscore$_i$ ← 0;
18   ScoringKernel<<<Bl, T>>>(b_i, t, *count*, *index*, *blockscore$_i$*);
19   BlockList.add(getBlock(b_i,blockscore$_i$));
20   **sortInDecreasingScore**(*BlockList*);
21   **return** BlockList;;

---

6   Experimental Results

6.1   Experiment Setup

Experiment Setup. All experiments have been performed on a machine with a six-core Intel Core i7-8700 CPU with 3.20 GHz, 16 GB of RAM and a Nvidia GeForce GTX 1060 with 6 GB of memory. Our experiments only cover the Jaccard similarity.

Data Sets. We employ 4 real-world datasets from different domains [5]. Their characteristics are reported in Table 1. KOSARAK_300 k: 300k records form KOSARAK. A set is the user-behavior recorded on Hungarian on-line news portal and a toke is a link. DBLP_100 k:100k articles from DBLP bibliography. A set is a publication; a token is an author contained in the authors list of the article. YOUTUBE: A set is a user; a token is a group membership of the user. BMS-POS: A set is a purchase in a shop; a token is a product category of the purchase.

Table 1. Datasets.

Dataset name	Number of rows	Different tokens
KOSARAK_300k	300k	41000
DBLP_100k	100k	161403
YOUTUBE	94238	30087
BMS-POS	320k	16570

6.2 CPU Only Experiments

We evaluate our methods PSNM and PBM on all four datasets and compare both with a conventional method as baseline to measure our methods' benefit, which is composed by blocking phase and an existing typical non-progressive SSJoin method.

We define metric to evaluate the effectiveness of all methods by recall, which measures the portion of detected similar pairs meeting given threshold. We assess the vertical axis with respect to recall progressiveness and consider the normalized number of executed comparisons as horizontal axis, $ec^* = ec/|D|$, where ec is the number of executed comparisons during the processing and $|D|$ is the number of sets included by given set collection D.

To facilitate the comparisons between our methods and baseline, we use the **area under the curve** (AUC) of the plots and for a method m, we define $AUC_m@ec^*$ as the value of AUC for a given ec^*. In addition, we introduce **normalized area under the**

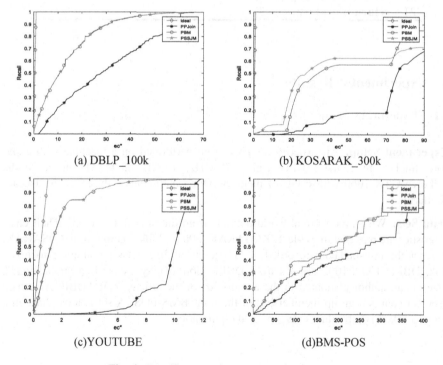

      (a) DBLP_100k   (b) KOSARAK_300k

      (c)YOUTUBE   (d)BMS-POS

Fig. 6. Recall progressiveness over the datasets

curve: a higher value meaning a better progressiveness [17]. We indicate the above metric with $AUC^*_m@\ ec^*$ and normalize it based on the ideal method: $AUC^*_m@ec* = \frac{AUC_m@ec*}{AUC_{ideal}@ec*}$.

The corresponding plots are shown in Fig. 6. It can be seen that for the four datasets, our methods can outperform the baseline and achieve a high recall level with less comparison whereas the baseline need more work to reach the same value. PPJ method is not a progressive approach and the obvious gap between our approach and baseline demonstrates the importance of a good strategy for ordering blocks and selecting which block to load next. We set three values for every datasets: {20, 40, 60} for DBLP_100k; {30, 60, 90} for KOSARAK_300k; {3, 6, 9} for YOUTUBE; {100, 200, 300} for BMS-POS.

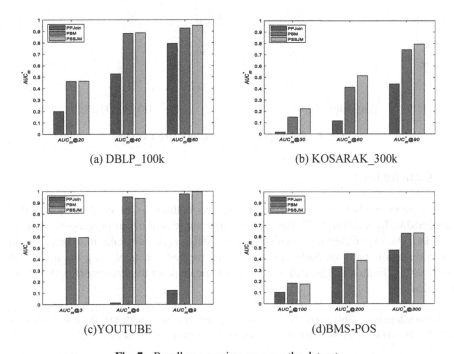

(a) DBLP_100k   (b) KOSARAK_300k

(c)YOUTUBE   (d)BMS-POS

Fig. 7. Recall progressiveness over the datasets

Figure 7 shows the **normalized area under the curve** for these different values of ec^* in four datasets. We can see that our methods are good performers in most comparisons and the gap between them with performance in progressiveness is not obvious in many cases. Overall, these results clearly show that our progressive methods can produce most of the similar pair much earlier than the baseline, traditional non-progressive set similarity join.

6.3   GPU-Based Experiments

We tested our method on two datasets DBLP_100k and YOUTUBE. PSSJM implemented on CPU is selected as our baseline. Figure 8 shows the execution times as we
vary the threshold from 0.5 to 0.9 with 0.1 increments. We can observe that the best
speedups which up to 20× have been obtained when processing dataset YOUTUBE.
Obviously, our algorithm achieved considerable speedups over another one which is
not accelerated by GPU.

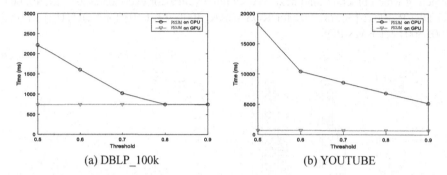

(a) DBLP_100k   (b) YOUTUBE

Fig. 8. Recall progressiveness over the datasets

7   Conclusions

In this paper, we have proposed two progressive methods to set similarity join, PBM
and PSSJM. In addition, we design and give the implement of our progressive strategy
on GPU. Our experimental evaluation with real datasets demonstrates that the proposed
methods significantly outperform conventional methods on improving the quality of
result with a limited budget and that expected speedups of the implement on GPU are
revealed.

Acknowledgments. This work is supported by the Nation Key R&D Program of China
(2018YFB1003404), the National Nature Science Foundation of China (61672142, U1811261).

References

1. Chaudhuri, S., Ganti, V., Kaushik, R..: A primitive operator for similarity joins in data
   cleaning. In: Proceedings of the ICDE, pp. 5–16 (2006)
2. Xiao, C., Wang, W., Lin, X., Yu, J.X., Wang, G.: Efficient similarity joins for near-duplicate
   detection. TODS **36**(3), 15 (2011)
3. Arasu, A., Ganti, V., Kaushik, R.: Efficient exact set-similarity joins. In: Proceedings of the
   VLDB, pp. 918–929 (2006)
4. Mann, W., Augsten, N.: PEL: Position-enhanced length filter for set similarity joins. In:
   Proceedings of the GvD (Foundations of Databases), pp. 89–94 (2014)

5. MannMann, W., Augsten, N., Bouros, P.: An empirical evaluation of set similarity join techniques. Proc. VLDB End. **9**, 636–647 (2016)
6. Zhou, J., et al.: A generic inverted index framework for similarity search on the GPU. In: 2018 IEEE 34th International Conference on Data Engineering (ICDE). IEEE (2018)
7. Sandes, E.F.O., Teodoro, G., Melo, A.C.M.A.: Bitmap filter: Speeding up exact set similarity joins with bitwise operations. (2017)
8. Li, C., et al.: A GPU Accelerated Update Efficient Index for kNN queries in road networks. In: 2018 IEEE 34th International Conference on Data Engineering (ICDE). IEEE Computer Society (2018)
9. Kruliš, M., Osipyan, H., Marchand-Maillet, S.: Optimizing sorting and Top-k selection steps in permutation based indexing on GPUs. In: Morzy, T., Valduriez, P., Bellatreche, L. (eds.) ADBIS 2015. CCIS, vol. 539, pp. 305–317. Springer, Cham (2015). https://doi.org/10.1007/978-3-319-23201-0_33
10. Wang, Y., et al.: FLASH: Randomized algorithms accelerated over CPU-GPU for ultra-high dimensional similarity search (2017)
11. Gowanlock, M., Casanova, H.: Distance threshold similarity searches: Efficient trajectory indexing on the GPU. IEEE Trans. Parallel Distrib. Syst. **27**(9), 2533–2545 (2016)
12. Xiao, C., et al.: Top-k set similarity joins. In: Proceedings of the 25th International Conference on Data Engineering, ICDE 2009, March 29 2009–April 2 2009, Shanghai, China. IEEE Computer Society (2009)
13. Vernica, R., Carey, M.J., Li, C.: Efficient parallel set-similarity joins using mapreduce. In: Proceedings of the ACM SIGMOD International Conference on Management of Data, SIGMOD 2010, Indianapolis, Indiana, USA, June 6–10, 2010. ACM (2010)
14. Ma, Y., Zhang, R., Zhang, Y.: Similarity histogram estimation based top-k similarity join algorithm on high-dimensional data. In: Ni, W., Wang, X., Song, W., Li, Y. (eds.) WISA 2019. LNCS, vol. 11817, pp. 589–600. Springer, Cham (2019). https://doi.org/10.1007/978-3-030-30952-7_60
15. Papenbrock, T., Heise, A., Naumann, F.: Progressive duplicate detection. IEEE Trans. Knowl. Data Eng. **27**(5), 1316–1329 (2015)
16. Whang, S.E., Marmaros, D., Garcia-Molina, H.: Pay-as-you-go entity resolution. IEEE TKDE **25**(5), 1111–1124 (2013)
17. Giovanni, S., George, P., Themis, P., et al.: Schema-agnostic progressive entity resolution. IEEE Trans. Knowl. Data Eng. **31**(6), 1208–1221 (2018)
18. Hernández, M.A., Stolfo, S.J.: The merge/purge problem for large databases. In: SIGMOD, pp. 127–138 (1995)
19. Bloom, B.: Space/time tradeoffs in hash coding with allowable errors. Commun. ACM **13** (7), 422–426 (1970)
20. Christen, P.: A survey of indexing techniques for scalable set linkage and deduplication. IEEE TKDE **24**(9), 1537–1555 (2012)
21. Nvidia. Nvidia CUDA Programming Guide 8.0.Nvidia (2017)

Prediction Method of Code Review Time Based on Hidden Markov Model

Weifeng Zhang, Zhen Pan, and Ziyuan Wang[✉]

School of Computer, Nanjing University of Posts and Telecommunication,
Nanjing 210093, China
wangziyuan@njupt.edu.cn

Abstract. Pull-Request (PR) is the primary method for developers to contribute code in GitHub. Code review can effectively ensure the quality of the code to be merged. The time of code review will affect the development progress of the entire project. Some researchers predict the duration of the review based on the initial attributes of PR as input attributes for building their prediction model. However, these methods ignore the temporal nature of these activities. In this paper, we propose a new method that uses the hidden Markov model (HMM) and the time series of developer activities. By considering the chronological sequence of developer activities, critical activities in PR are extracted to form a key activity sequence, by which HMM is used. To classify sequences, we collected the historical data of 5 projects from GitHub and conducted experiments. The results show that this method can effectively identify and predict PR's duration to be reviewed at an early stage.

Keywords: Pull request · Hidden markov model · Code review

1 Introduction

The workload estimation model has been studied for a long time in software engineering research. The workload estimation model can help organizations and individuals plan and track their software projects' progress and individual tasks to help plan delivery milestones better. With the advent of development environments like GitHub [1–3], Pull Requests (PR) became the norm for development. PR is a request from the developer to merge the modified content of the local branch into the main branch and complete all testing and code review activities before integrating it into the code branch. Bug fixes, new features, security fixes, and content changes can all be part of PR. Previous work by Microsoft [4] also focused on functional and project-level workload estimates, rather than the level of individual PR changes. After that, they estimated the workload of a single PR and used several metrics from the defect prediction literature, such as code loss, reviewer information, and ownership information to build their PR duration prediction model [5].

Some researchers have begun to study the reviewers in the code review, and they have studied various methods to recommend the reviewers. Yu et al. [6] combined information retrieval with social network analysis to make full use of the textual semantics of PR and the social relationship of developers. Measure the semantic

© Springer Nature Switzerland AG 2020
G. Wang et al. (Eds.): WISA 2020, LNCS 12432, pp. 168–175, 2020.
https://doi.org/10.1007/978-3-030-60029-7_15

similarity between the new PR and the historical PR, and predict the developer's expertise score based on the number of comments he submitted. Yu et al. [7] investigated the impact of review and code reviewer allocation issues on review time. A code reviewer recommendation method based on file location is proposed. We take advantage of the similarity of previously reviewed file paths to recommend appropriate code reviewers. Files located in similar file paths will be managed and reviewed by similarly experienced code reviewers. Mohammad et al. [8] considered not only the relevant cross-project work history (for example, external library experience) and the developer's experience in specific expertise related to the pull request. Potential code reviewer. Yang et al. [9] reproduced a popular and effective IR-based code reviewer recommendation algorithm, generating technical terms for each PR in the project based on the title and description, by deleting pre-defined stop words before this process To get the technical focus of each PR. All technical terms can form a corpus. When the test PR arrives, the cosine similarity is used to calculate the relationship between each PR and then rank it. Yang et al. [10] developed a two-tier reviewer recommendation model to recommend PR reviewers in the GitHub project from a technical and management perspective. For the first layer, it is recommended that suitable developers view the target PR according to the hybrid recommendation method. After obtaining the recommendation results from the first layer, the second layer specifies whether the target developer participates in the review process technically or administratively. Motahareh et al. [11] proposed a method, cHRev, to automatically recommend the most suitable reviewers (quantity of review reviews and their recency) on the historical contributions shown in their previous reviews.

We considered the life cycle of PR. For example, the developer created a PR and gave it to the reviewer to review it, raise the issue, discuss the problem, fix the problem, and continuously integrate detection. The sequence of events continues until the PR is merged or closed. We built a predictive model using the Hidden Markov Model, taking into account the time characteristics of various activities in the PR, to estimate the total time required to complete the PR, help the code review process, and improve developer productivity. It can also help each developer plan their work better and identify and avoid delays that may ultimately affect the entire project. Through experiments, our model prediction accuracy is about 70%, F-measure also reached about 75%, and the highest reached 82%. Compared with the latest method, the indicators of our method are slightly higher. In this article, we make the following two contributions:

1) Propose critical activities in the PR life cycle and retrieve them from PR history.
2) The method of using HMM to predict PR duration.

2 Motivation

In this section, we will discuss the motivation for this research. In GitHub, Pull-Requests (PR) is a request for developers to merge the content modified by the local branch into the main branch. It is a development specification. All testing and code review activities must be completed before integrating them into the code branch. PR includes bug fixes, new features, security fixes, and content changes. The speed of PR's progress largely determines the progress of the entire project. There are a large number of developers in such an open community. Anyone can participate in the development of the project. If the project owner wants to maintain the project's efficient progress, this is a challenge for the project owner and the developer. For example, we investigated five project warehouses. We considered that PR should be completed as soon as possible, and it will be defined as very slow after more than ten days, as shown in Table 1. We found that many projects have the phenomenon that the PR duration is too long. PRs exceeding ten days account for 10% to 30% of the total number of PRs. They perform differently in different projects, and the longest can reach more than 100 days.

Table 1. Number of PRs over 10 days in different projects.

Project	Pull-Requests	240 h
Vue	1279	292
React	7568	1642
React-Native	7667	2500
Node	8423	826
Bootstrap	6999	1573

3 Our Approach

Our method can be summarized in three steps, as shown in Fig. 1.

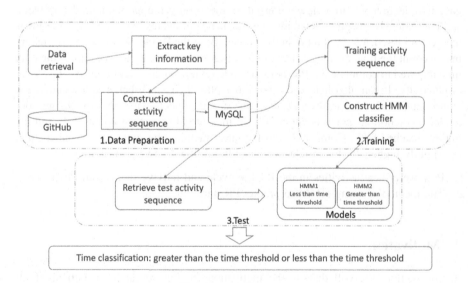

Fig. 1. Overall process

First, we extract the time series of developer activities from the data set. Second, we use the active time series of closed PR to train the HMM. We train two HMMs by dividing the closed PR into two categories: (a) PR that does not exceed the time threshold; (b) PR that exceeds the time threshold. Set the median PR duration in the training set as the time threshold. Then train an HMM on each category. We extract the first few time series of the latest PR and pass it to the trained HMM. The trained HMM will distinguish whether each PR will exceed the time threshold.

3.1 Train

Throughout the life cycle of Pull-Request, the code quality of the submitted code is hidden from us, but its state can be inferred by a series of observations of activities in PR, as shown in Fig. 2.

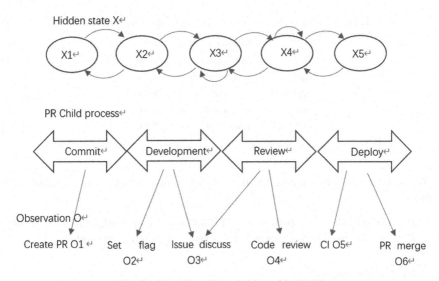

Fig. 2. Modeling PR activities with HMM

In Fig. 2, we have defined five hidden states of code quality. The code is submitted for review, the code has potential bugs, the potential bugs have been fixed, the code has not passed the merge request, and the code has passed the merge request. These states are hidden. The only visible is a series of PR process activities, such as submitting PR, setting tags, issue discussion, and code review. These observation points can occur in each hidden state, leading to a transition from one state to another.

4 Evaluation

In this section, we evaluate the effectiveness of this method through two different experiments. In the first experiment, we focused on demonstrating the predictive power of this method. In the second experiment, we focused on comparing our existing methods and evaluating our methods by comparison.

4.1 Experimental Data and Research Questions

We selected five sizeable open source projects from GitHub as the target data of the experiment. Whether these projects are in popularity, user usage, or in the watch, star, and fork projects, the ranking is very high and has very active. The core data of these projects is shown in Table 2.

Table 2. Five open source projects downloaded from Github.

Project	Language	Star	Fork
Vue	Javascript	157k	23.6k
React	Javascript	143k	27.5k
React-Native	Javascript	84.7k	18.9k
Node	Javascript	67.6k	16.1k
Bootstrap	Javascript	139k	68.1k

We will design experiments for the following two problems and evaluate our method.

RQ1: How accurate is the prediction of PR duration using a hidden Markov model?
RQ2: How effective is it compared with existing prediction methods?

4.2 Evaluation Index

To evaluate our proposed method, we divide the historical PR data set into two parts: (a) historical PR for training; (b) historical PR for testing. We used approximately 90% of closed historical PRs in the data set and 10% of closed historical PRs in the test set. We used a standard evaluation technique called 10-fold cross-validation. The duration of the historical PR in the test set allows us to evaluate the HMM's performance.

We first measure the classifier's performance by using a confusion matrix and then determining the accuracy of the classifier, recall rate, F measure, and accuracy. The confusion matrix stores correct and incorrect predictions made by the classifier. For example, if the HMM classifies the PR as taking a long time (slow), and it does take a long time (slow), the classification is real positive (TP). If an error is classified as slow, but in fact, it is not very slow, then the classification is a false alarm (FP). If the PR is classified as requiring a short time (fast) and is actually in the slow classification, the classification is a false negative (FN). If it is classified as fast and belongs to the fast category, it is a true negative (TN).

We use Precision, Recall, F-measure, and Accuracy to evaluate the performance of the HMM model.

4.3   Experiment

In this experiment, we first obtained the median days of all historical PRs in the data set. Secondly, we divided the data set into 10, one of which was the test set, and the remaining 9 were the training set repeated ten times. We train two HMMs in the training set, one of which uses PR with duration less than the median number of days for training, and the other uses PR with duration longer than the median number of days for training.

After training HMM, we use the corresponding test set to simulate such a scenario: two, three, four, and five active unclosed PRs in the project warehouse, and use HMM to predict its closure (slow or Fast) time.

We compare our model with the existing Maddila's model. It provides a method that uses a gradient boost model (GB) based on a series of initial properties of PR to provide an estimate of the workload of a single developer PR and predict the duration of the PR. The reasons why we chose this work to compare with our work are as follows. First, there is currently little work done to predict PR duration. Secondly, this work is up-to-date, which allows us to test their methods on the dataset and further compare the results of our algorithm with theirs. Third, most of the attributes used in their work are available on GitHub.

Since the dataset used in Maddila et al.'s work is not available, we chose to test their method on our dataset. First, we use the multiple features proposed in the method of Maddila et al. And extract them using the Github API. In their method, the regression model of the gradient lifting algorithm was used to predict the PR duration. To compare with our experiment, we used the gradient lifting algorithm (GB) classification model in the process of reducing the experiment and Taking the median of all PR durations as the threshold.

We select the prediction results when there are only the first two active sequences from our experimental results. The attributes in Maddila et al.'s method are some of the initial attributes of PR, and the action sequence is 2 Time, closer to the initial properties of PR. To more intuitively reflect the effects of different models, we use box plots to make further presentations.

As can be seen from Figs. 3 to 4, in terms of accuracy, there is an excessively high outlier, mainly because the prediction model performs differently in different projects. There are five projects in our experiment, which is also a reason for outliers. The worst performance of HMM in React projects is 70%, and the best performance in Bootstrap is 88%, while the gradient boost model has the worst performance in React-Native, only 57%, in Bootstrap. The best performance is 70%. In terms of recall, the two models are similar. In two projects Vue and Node, the results of the gradient lifting model are slightly better than HMM. In terms of accuracy and F-measure, HMM is slightly better than the gradient boost model. The reason for this difference may be that the data set used in Maddila et al.'s experiment comes from Microsoft's internal development community. Compared with Github, the developer's behavior and project management may be different.

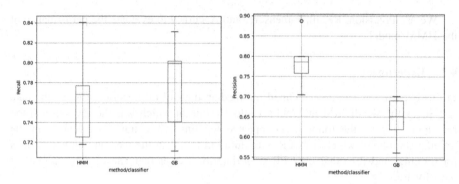

Fig. 3. Recall and Precision

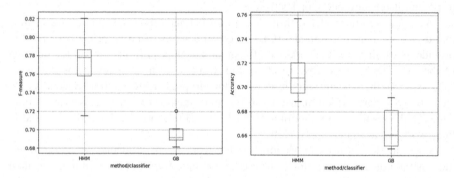

Fig. 4. F-measure Accuracy

5 Conclusion

In this article, we introduced the use of HMM to predict the time of PR completion. Our method takes into account the chronological sequence of activities in the PR life cycle. We provide a framework for extracting time series of developer activities from PR historical data, and we also train HMM into two categories: (a) PRs with durations greater than the threshold, and (b) PRs with durations less than the threshold.

Through experiments, the prediction accuracy of the model is about 70%, and the F-measure also reaches about 75%, and the highest reaches 82%. Compared with the latest method, the indicators of our method are slightly higher. Therefore, our method can predict the duration of PR at an early stage, and it is more helpful for developers and reviewers to prioritize their work.

In this experiment, the main language of our data set is JavaScript, so we consider using the model in other language projects in our future works. In addition, we are ready to consider the content of the discussions between developers to further increase the accuracy of our predictions.

References

1. Liu, Z., Xia, X., Treude, C., Lo, D., Li, S.: Automatic generation of pull request descriptions. In: 34th IEEE/ACM International Conference on Automated Software Engineering (ASE), pp. 176–188. IEEE (2019)
2. Mirhosseini, S., Parnin, C.: Can automated pull requests encourage software developers to upgrade out-of-date dependencies? In: 2017 32nd IEEE/ACM International Conference on Automated Software Engineering (ASE), pp. 84–94. IEEE (2017)
3. Ram, A., Sawant, A.A., Castelluccio, M., Bacchelli, A.: What makes a code change easier to review: an empirical investigation on code change reviewability. In: Proceedings of the 2018 26th ACM Joint Meeting on European Software Engineering Conference and Symposium on the Foundations of Software Engineering, pp. 201–212. IEEE (2018)
4. Layman, L., Nagappan, N., Guckenheimer, S., Beehler, J., Begel, A.: Mining software effort data: preliminary analysis of visual studio team system data. In: Proceedings of the 2008 International Working Conference on Mining Software Repositories, pp. 43–46. IEEE (2008)
5. Liu, X., et al.: Mining core contributors in open-source projects. In: Ni, W., Wang, X., Song, W., Li, Y. (eds.) WISA 2019. LNCS, vol. 11817, pp. 690–703. Springer, Cham (2019). https://doi.org/10.1007/978-3-030-30952-7_70
6. Yu, Y., Wang, H., Yin, G., Wang, T.: Reviewer recommendation for pull-requests in GitHub: What can we learn from code review and bug assignment? Information and Software Technology, pp. 204–218. IEEE (2016)
7. Yu, Y., Wang, H., Yin, G., Ling, C.X.: Who should review this pull-request: Reviewer recommendation to expedite crowd collaboration. In: 2014 21st Asia-Pacific Software Engineering Conference, pp. 335–342. IEEE (2014)
8. Rahman, M.M., Roy, C.K., Collins, J.A.: Correct: code reviewer recommendation in github based on cross-project and technology experience. In: Proceedings of the 38th International Conference on Software Engineering Companion, pp. 222–231. IEEE (2016)
9. Yang, C., Zhang, X., Zeng, L., Fan, Q., Yin, G., Wang, H.: An empirical study of reviewer recommendation in pull-based development model. In: Proceedings of the 9th Asia-Pacific Symposium on Internetware, pp. 1–6. IEEE (2017)
10. Zhang, X., et al.: DevRec: a developer recommendation system for open source repositories. In: Botterweck, G., Werner, C. (eds.) ICSR 2017. LNCS, vol. 10221, pp. 3–11. Springer, Cham (2017). https://doi.org/10.1007/978-3-319-56856-0_1
11. Zanjani, M.B., Kagdi, H., Bird, C.: Automatically recommending peer reviewers in modern code review. IEEE Transactions on Software Engineering 42(6), 530–543 (2015). IEEE

Evaluation and Prediction of COVID-19 Based on Time-Varying SIR Model

Liu Yanzhe and Liu Bingxiang[(✉)]

Jingdezhen Ceramic Institute, Jingdezhen 333403, China
lbx1966@163.com

Abstract. This article based on the law of the COVID-19 epidemic situation until May 6 improves the SIR model. Through the reverse solution of the parameters in the model, the parameters are modeled and predicted, and the parameters that change with time are obtained. Compared with setting fixed parameters, the accuracy of the model is greatly improved. Using the improved model to analyze the COVID-19 epidemic situation and study the virus spreading trend in different countries. The results show that the improved model is basically reliable in predicting the development trend of the COVID-19 epidemic; the epidemic in Italy will basically end in July; the development trend in Britain and America is similar, and the inflection point is expected to appear in mid-June. The results of the study confirm the effectiveness of measures such as reducing the movement of people and providing medical assistance to the epidemic-stricken areas, and provide reference for subsequent epidemic prevention and control.

Keywords: COVID-19 · SIR model · Prediction

1 Introduction

Since the outbreak of COVID-19, the number of infected people worldwide has reached more than 3 million, and almost all countries have suffered huge losses. Many countries have taken measures to shut down various organizations to reduce population contact to prevent the spread of the virus, but these measures have a huge impact on the economy. Therefore, it is necessary to evaluate the development of the epidemic and provide a reference for policy formulation.

Most of the current studies are based on the SIR model to assess and predict the development of the epidemic [1–3], but in these studies, the most critical parameters of the SIR model are set estimates mostly. Since the spread of the epidemic will change with policy changes and the number of patients, this treatment can cause large errors. In this paper, based on the epidemic development rules, the equations are fitted to the parameters, and the SIR model is established using the parameters that change over time, finally, the model is verified, and this model is used to make predictions and analysis of the epidemic development.

© Springer Nature Switzerland AG 2020
G. Wang et al. (Eds.): WISA 2020, LNCS 12432, pp. 176–183, 2020.
https://doi.org/10.1007/978-3-030-60029-7_16

2 Model Theory

The SIR model is a classic infectious disease dynamic model, this model is established by Kermack and McKendrick in 1927 [4]. Based on the SIR model and the characteristics of the COVID-19 epidemic, this article makes the following basic assumptions:

1) Since the change rate of epidemic situation changes over time is much more significant than that of births and deaths over time, and most countries have adopted strict immigration control measures, the total population change in a country is very small, so it is assumed that the total population in a warehouse keeps is a constant.
2) Infection rate coefficient = average number of patients in daily contact × probability of infection of susceptible persons after contact with patients, the average number of patients in contact is closely related to the government's prevention and control measures, people's awareness of isolation, etc. The average contact rate is high at the beginning of the outbreak. After a period of development, the average contact rate will gradually decrease after the outbreak is paid attention to. Therefore, the infection rate coefficient should be set as a variable parameter that changes with time.
3) The removal rate is related to the cure rate and mortality rate, however, because the number of dead patients accounts for a small proportion of the total number of patients, the removal rate and cure rate have a greater correlation. In the early stage of the outbreak, due to insufficient knowledge of the virus, the cure rate of patients is low. After accumulating a large amount of treatment experience, the patient's cure rate will rise and the removal rate will also rise. Therefore, the removal rate coefficient should also be set to change with time.
4) The data is in units of days and does not consider continuous changes.

Based on the above assumptions, this paper constructs the following balance equation:

$$\begin{cases} \frac{dS}{dt} = -S \times \frac{I}{N} \times \beta \\ \frac{dI}{dt} = S \times \frac{I}{N} \times \beta - \gamma \times I \\ \frac{dR}{dt} = \gamma \times I \end{cases} \tag{1}$$

$$\beta = a \times \ln(b \times t) + c \tag{2}$$

$$\gamma = k \times t + m \tag{3}$$

Among them, S, I, R denote susceptible persons, patients, and removed persons. β indicates the probability that a susceptible group will be infected after being exposed to infected crowd. γ is the coefficient of removal rate, indicating the probability of the patient being removed (dead or cured). At the same time, according to the above formula, the solving formula of basic reproduction number ($R0$) can be derived:

$$RO = \frac{\beta}{N} \times \frac{1}{\gamma} \times S_0 \tag{4}$$

In this paper, based on the existing data, the parameter values in the model are solved in reverse. According to the obtained parameter data, a parameter model is constructed and trained. The parameter model obtained is used to predict the time-varying parameter data, then the time-varying parameter is used to construct a time-varying SIR model. When solving parameters in reverse, it can be solved directly, or optimization algorithms such as ant colony algorithm and genetic algorithm can be used to find the optimal parameter value of the model. After a comparative experiment, the results obtained in multiple ways are the same. Machine learning methods can be used as alternatives for situations where direct solutions are not possible.

This article attempts to use four methods of linear regression, polynomial fitting, exponential smoothing, and LSTM [5, 6] to establish the model of parameters β and γ. Since the parameter changes are more dramatic in the early stage of the epidemic and tend to be gentle in the later stage, the LSTM algorithm will easily cause the gradient to disappear. For the same reason, the use of exponential smoothing will also cause large errors. According to the law of epidemic development, the logarithmic function is finally used to establish the model of parameter β, and the univariate linear regression is used to establish the model of parameter γ.

3   Prediction Experiment

This article obtained COVID-19 data from four countries including China, Italy, Britain and America, including the number of patients, the cumulative number of deaths, and the cumulative number of cures, and calculate the number of existing patients in each country. Use the improved model to process data, verify the accuracy of the model, and predict the development of the epidemic.

3.1   Model Verification

Since the epidemic in China has basically ended, and the epidemic in Italy has also passed its inflection point, the accuracy of the model can be verified by predicting the development of the epidemic in China and Italy.

China's data began on February 5, 2020, with 21 training data, 70 prediction data, and 70 verification data. The fitting equation of β obtained by training is: y = −0.069 11675658517227 * ln (5.086007728270503 * x) + 0.3247851719047432, and the fitting equation of γ is: y = 0.0016630688923199239 * x + 0.014222881482399054; The Italy's data began on March 11, 2020, with 25 training data, 80 prediction data, and 30 verification data. The fitting equation of β obtained by training is: y = −0.059 88265296840833 * ln (5.095313307692628 * x) + 0.35327576998120064, and the fitting equation of γ is: y = $1.1499257648340639e^{-10}$ * x + 0.03115033914997176. The parameter fitting curves is shown in Fig. 1:

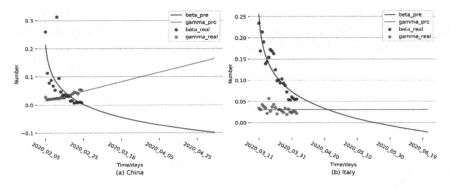

Fig. 1. Parameter fitting curves of China and Italy

The predicted parameters are used to build the model, and the resulting epidemic development curve is shown in Fig. 2. It can be seen from the fitting degree of the curve in the figure that the model has made a very good prediction on the development of the epidemic in China and Italy, and accurately predicts the development trend of the epidemic, the inflection point time and the number of infections. The results show that the Italian epidemic will basically end in July 2020. Data from China and Italy prove that the model has reliable effects.

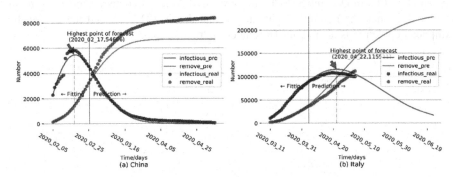

Fig. 2. Parameter fitting curves of China and Italy

3.2   Prediction

Apply the model to Britain and America to predict the development trend and inflection point of the epidemic in both countries.

Britain's data began on April 4, 2020, with 31 training data, 200 prediction data. The fitting equation of β obtained by training is: y = −0.026034231995124547 * ln (5.045657485354332 * x) + 0.16468626328436686, and the fitting equation of γ is: y = 1.009298676638269e^{-10} * x + 0.009474665494227158; American data began on April 5, 2020, with 30 training data, 200 prediction data. The fitting equation of β

obtained by training is: y = −0.0217276978288313 * ln (5.036525868374025 * x) + 0.13948441755802096, and the fitting equation of γ is: y = 1.3701824743687221e^{-08} * x + 0.012148954330530802. The parameter fitting image is shown in Fig. 3:

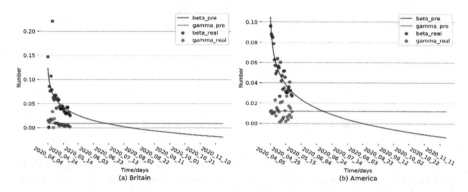

Fig. 3. Parameter fitting curves of Britain and America

Using the predicted parameters to build a model, the epidemic development curve is shown in Fig. 4. The model predicts that the inflection point of Britain epidemic will be June 19, when the number of patients on that day is 257673, and the end of Britain epidemic will be in December 2020. The predicted America epidemic inflection point is June 13, when the number of patients on that day is 1312227, and the end of the American epidemic is also about December 2020.

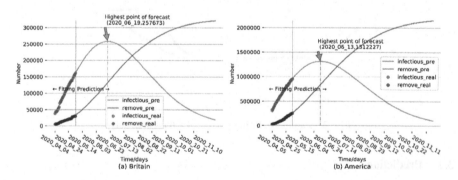

Fig. 4. The epidemic development curve of Britain and America

3.3 Analysis

The four countries have similar fitting equations for the parameter β, and the fit is very good. It shows that the infection rate coefficient of the epidemic has a fixed development law, and at the same time, it will cause some differences due to different national policies. The equations in China and Italy are similar, and the equations in Britain and

America are similar. It is linked to that China and Italy have adopted stricter prevention and control measures, proving that the prevention and control measures adopted by the government on the epidemic have a good impact on the development of the epidemic.

However, except for the model on China that has a good fitting effect on the parameter γ, none of the other three models can accurately obtain the fitting curve of γ. After the outbreak, China quickly mobilized national resources to support severely affected areas, so the cure rate has been significantly improved. However, the other three countries did not receive timely assistance after the nationwide epidemic broke out, so there is no obvious trend in the remove rate. At the same time, the fitting equation of γ shows that even without assistance, Italy's remove rate is still higher than that of Britain and America, proving that Italy's response measures have played a positive role.

According to the calculation formula of $R0$, the change curve of $R0$ of four countries can be obtained. In order to make the comparison results more intuitive, the forecast days of Britain and America are reduced to 100 days, and the training days are unchanged (ensure the fitting equation remains unchanged), and the drawn curve is shown in Fig. 5.

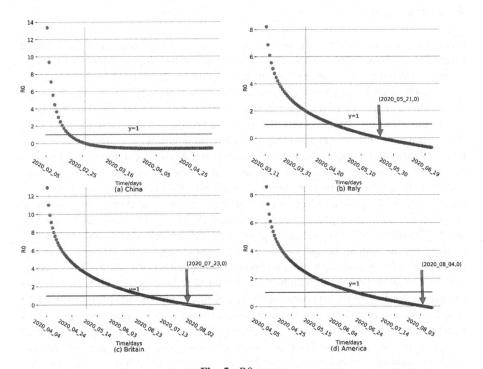

Fig. 5. R0 curve

It can be seen from the figure that China's R0 decreases the fastest, and it takes less than 20 days to reduce the R0 from 8 to 0. Italy's rate of decrease is also very fast, it is estimated that it takes about 40 days to reduce the R0 from 8 to 1. And Britain and America are estimated to take 70 days to reduce the R0 value from 8 to 1.

When the R0 is less than 1, the epidemic will no longer continue to spread, and it is not suitable for carrying out production activities before that. If judged according to the condition of R0 < 1, Italy, Britain and America can resume work on April 22, June 19 and June 13 respectively. However, when China started to resume work at the end of February, the R0 value was already less than 0. If judged according to this condition, Italy, Britain and America can resume work on May 21, July 23 and August 4.

In addition, a large number of studies on the R0 of the epidemic, it shows that the R0 of COVID-19 is between 2–8 [7–10].According to the predicted R0 curve analysis, if it is assumed that the initial stage with less human intervention is 20 days, the R0 in this stage is mainly distributed between 3 and 8, so the study believes that the R0 should not be less than 3, or even may possible be higher. The study also provides a reference for the estimation of R0 from the side.

4   Conclusion

Based on the SIR model, this paper establishes a fitting equation for the model parameters, estimates the parameters, and predicts the development trend of the epidemic according to the estimated parameters. Based on the data of COVID-19 diagnosis, death, and cure cases in 4 countries including China, Italy, Britain and America, the improved model is used to simulate and predict the data from the four countries. The results show that the improved SIR model can predict the epidemic trend reliably; the government's prevention and control measures can reduce the epidemic's infection rate coefficient and reduce the epidemic's spread rate; Assistance to the medical system in the outbreak area helps to increase the removal rate of patients; the average removal rate of Britain and America are similar, and both are significantly lower than the removal rate of Italy; The inflection point of the epidemic in Britain is on June 19, at this time the number of patients is 257673, and the end of the outbreak is approximately December 2020; the inflection point of the epidemic in America is on June 13, at this time, the number of patients is 1312227, and the end of the outbreak is also at December 2020; According to the conditions for resumption of work in China, people in Italy, Britain and America can return to work on May 21, July 23 and August 4 at the earliest; According to the R0 curve simulated in this study, the R0 of the COVID-19 epidemic should be at least 3 or more. Research results confirm that measures such as reducing crowd travel, closing out the severely affected areas, and providing medical assistance to the severely affected areas can effectively reduce the speed of the outbreak. In addition, the epidemics in Britain and America are still developing rapidly, and isolation measures should continue to be implemented.

References

1. Zhou, T., Liu, Q., Yang, Z., et al.: Preliminary prediction of the basic reproduction number of the Wuhan novel coronavirus 2019-nCoV. J. Evid. Based Med. **13**(1), 3–7 (2020). https://doi.org/10.1111/jebm.12376
2. Qianqian, S., Han, Z., Liqun, F., et al.: Study on assessing early epidemiological parameters of coronavirus disease epidemic in China. Chin. J. Epid. **41**(04), 461–465 (2020). https://doi.org/10.3760/cma.j.cn112338-20200205-00069
3. D'Arienzo, M., Coniglio, A.: Assessment of the SARS-CoV-2 basic reproduction number, R0, based on the early phase of COVID-19 outbreak in Italy. Biosaf. Health (2020). https://doi.org/10.1016/j.bsheal.2020.03.004
4. Barlow, N.S., Weinstein, S.J.: Accurate closed-form solution of the SIR epidemic model. Phys. D Nonlinear Phenom. **408**, 132540 (2020). https://doi.org/10.1016/j.physd.2020.132540
5. Guo, C., Guo, W., Chen, C.-H., Wang, X., Liu, G.: The air quality prediction based on a convolutional LSTM network. In: Ni, W., Wang, X., Song, W., Li, Y. (eds.) WISA 2019. LNCS, vol. 11817, pp. 98–109. Springer, Cham (2019). https://doi.org/10.1007/978-3-030-30952-7_12
6. Zheng, H., Shi, D.: Using a LSTM-RNN based deep learning framework for ICU mortality prediction. In: Meng, X., et al. (eds.) WISA 2018. LNCS, vol. 11242, pp. 60–67. Springer, Cham (2018). https://doi.org/10.1007/978-3-030-02934-0_6
7. Zhuang, Z., Zhao, S., Lin, Q., et al.: Preliminary estimating the reproduction number of the coronavirus disease (COVID-19) outbreak in Republic of Korea and Italy by 5 March 2020. Int. J. Infect. Dis. (2020). https://doi.org/10.1016/j.ijid.2020.04.044
8. Zheng, Z., Wu, K., Yao, Z., et al.: The prediction for development of COVID-19 in global major epidemic areas through empirical trends in China by utilizing state transition matrix model (3/10/2020). Available at SSRN: https://ssrn.com/abstract=3552835 or https://doi.org/10.2139/ssrn.3552835
9. Tang, B., Bragazzi, N.L., Li, Q., et al.: An updated estimation of the risk of transmission of the novel coronavirus (2019-nCov). Infect. Dis. Model. **5**, 248–255 (2020). https://doi.org/10.1016/j.idm.2020.02.001
10. Li, Q., Guan, X., Wu, P., et al.: Early transmission dynamics in Wuhan, China, of novel coronavirus–infected pneumonia. New Engl. J. Med. (2020). https://doi.org/10.1056/NEJMoa2001316

Quality Control Method for Peer Assessment System Based on Multi-dimensional Information

Peng Li[1], Zhuoran Yin[2], and Fengyun Li[2(✉)]

[1] School of Intelligent Engineering, Shenyang City University,
Shenyang 110112, China
[2] School of Computer Science and Technology, Northeastern University,
Shenyang 110819, China
lifengyun@mail.neu.edu.cn

Abstract. In recent years, online learning has received widespread attention, and has played an important role in recent years. As a new application field of crowdsourcing system, peer assessment can solve student's performance evaluation problem in massive online courses. Traditional crowdsourcing quality control algorithm has not been able to use effective information to the evaluation of workers. Aiming at this problem, a quality control algorithm based on multi-dimensional information is proposed. The user's behavior, comment text information and other useful elements are combined together. The feature vectors of reliability are extracted from a variety of information based on the frame of log-linear model. And then, the gradient descent algorithm model is used to study the optimal parameters. The experimental results show that when comparing with the traditional Expectation Maximization algorithm, our multi-dimensional quality control algorithm has better performance in the accuracy and mean square error.

Keywords: Peer assessment · Quality control · Crowdsourcing · Multi-dimensional information fusion

1 Introduction

As a new application of crowdsourcing system, peer assessment can solve the problem of student assignment evaluation in massive online courses [1]. Peer assessment is an important part of mutual exchange and learning between learners. Therefore, the effectiveness of peer assessment is an important factor affecting student satisfaction. In massive online courses, peer reviews can quickly provide learners with learning feedback, but there are also some comings. In the process of mutual review, there are individual reviewers who are not serious or even make malicious assessments. This leads to problems such as low quality of peer assessment and high error rate of total score calculation [2]. The existence of these problems has seriously affected the application and development of peer assessment technology in MOOC [3–5].

The quality control methods used in existing crowdsourcing applications can be divided into two categories [6]. One is the gold standard data method [7], that is, the

© Springer Nature Switzerland AG 2020
G. Wang et al. (Eds.): WISA 2020, LNCS 12432, pp. 184–193, 2020.
https://doi.org/10.1007/978-3-030-60029-7_17

supervised method. The other is the unsupervised method, including: majority voting [8], expectation maximization algorithm [9, 10]. In the peer assessment system, the system is different from the behavior of users on the traditional crowdsourcing platform [11]. It is difficult to take into account the above mentioned multi-dimensional information in accordance with traditional quality control methods [12]. Therefore, there is still a certain distance between the assessment results and the actual needs.

In order to improve the quality of peer assessment, a quality control algorithm based on multi-dimensional information was proposed. Different from traditional quality control methods those are based on unsupervised learning, the users' behavior information and text analysis information are applied so as to measure worker's credibility. The validity of this method has been verified, and it has a significant effect on identifying the credibility of workers and improving the quality of peer assessment results.

2 Model Definition

Assuming that the known student set $w = \{w_1, w_2 \ldots w_n\}$, and the job submitted by each student w_i is denoted as J_i. For each job J_i, there is a set E_i, E_i is a true subset of the set W. A total of $|E_i|$ students evaluate work J_i. $|E_i| < n$ and w_i does not belong to E_i, that is, student w_i cannot give his job a score. The evaluation result is denoted by s_{ij}, which indicates the score that the student w_j evaluated for the assignment J_i of the student w_i.

The goal of this article is to provide an accurate score S_i^* for each student's assignment based on the scores of existing evaluation scores and other diverse information. Based on this goal, one of the simplest ideas is to calculate S_i^* uing the average of the algorithms. According to the above definition, the formula is as follows.

$$S_i^* = \frac{\sum_J S_{ij}}{|E_i|} \tag{1}$$

Obviously, the arithmetic mean actually means that for every evaluator w_j, they have the same scoring weights, all of which are $\frac{1}{|E_i|}$. However, in the actual work process, some workers evaluate seriously, and some workers evaluate sloppy. If they are treated equally, only one malicious evaluation will easily affect the overall score. In order to distinguish this difference, the ideal solution is to establish a corresponding mathematical model to determine the trueness value P_{ij} of a student w_j's evaluation score for task J_i. The formula for the final calculation of the correct score S_i^* or student work is as follows.

$$S_i^* = \sum P_{ij} \times S_{ij}, \sum P_{ij} = 1 \tag{2}$$

The focus at this time is shifted to how to solve P_{ij} through modeling.

In the actual environment, in addition to the scoring information of workers, there is much information that can be used, such as the text information of task comments, or workers' behavioral information analyzed from the worker's operation logs, etc. This can help to judge the truth of workers' scores from more angles from more aspects.

3 Multi-dimensional Information Based Quality Control Algorithm

3.1 Training Model

In order to be able to fuse the above-mentioned multi-dimensional information, this paper uses the log-linear model [12, 13] to mathematically model the authenticity of workers' scores. The formula of log-linear model is as follows.

$$P_{ij} = \frac{e^{(\lambda \times f(l,j))}}{Z} \tag{3}$$

Among them, Z is a normalization term, let P_{ij}'s value range between [0, 1], and satisfy the condition of $\sum P_{ij} = 1$. $f(l, j)$ is an feature vector of dimension d, indicating that there are d features that describe the authenticity of worker w_j to task J_i. λ is a feature weight vector, and the dimension is also d, which indicates the degree of importance of each feature. An evaluation indicator of the degree of truth is represented by the inner product of the feature vector $f(l, j)$ and the weight vector λ.

It can be known from the above formula that since e^x s a monotonically increasing function, P_{ij} is proportional to $f(l, j) \times \lambda$ That is, the larger the inner product, the higher the realism. By substituting (3) into (2), the final model formula for determining task J_i' s correct score S_i^*:

$$S_i^* = \frac{\sum e^{(\lambda \times f(l,j))} \times S_{ij}}{Z} \tag{4}$$

Where λ is a parameter that needs to be solved during the training phase.

Since the estimated true score is a real number, the problem is a regression problem. Here, the loss is defined using the minimum mean square error loss function that is commonly used when dealing with regression problems. The formula is:

$$L\left(S_i^*, Y_i\right) = \frac{1}{2}\left(Y_i - S_i^*\right)^2 \tag{5}$$

The goal of this paper is to minimize the loss function mentioned above by optimizing parameter λ. In order to solve the optimal solution here, Stochastic Gradient Descent (SGD) algorithm is used to solve the optimal solution. The basic principle of the stochastic gradient descent algorithm is to follow the direction of the gradient, that is, the fastest falling/rising direction. Its formula is:

$$\lambda_{i+1} = \lambda_i - l_r \times G(L) \tag{6}$$

l_r is the learning rate, which indicates the magnitude of change when each parameter is updated. If the learning rate is too small, the learning speed is slow and the convergence time is long. If the learning rate is too large, the optimal solution may not be found, and unable to get optimal parameters. In (6), G(L) represents the gradient of the objective function for λ.

According to the chain rule of derivation, the partial derivative of L for $λ_i$ is calculated as follows:

$$\frac{dL}{d\lambda_i} = \frac{dL}{dS_i^*} \times \frac{dS_i^*}{d\lambda_i} \tag{7}$$

$$\frac{dL}{dS_i^*} = S_i^* - Y_i \tag{8}$$

$$\frac{dL}{d\lambda_i} = S_i^* \times \sum_j f(l,j) \tag{9}$$

That is, the partial derivative of L for $λ_i$ is calculated as follows:

$$\frac{dL}{d\lambda_i} = \left(S_i^* - Y_i\right) \times \left(S_i^* \times \sum_j f(l,j)\right) \tag{10}$$

To sum up, as long as the task with the correct answer that has been reviewed by the teacher is used as training data, according to (7), the gradient descent algorithm can be used to solve the optimal parameter.

3.2   Overall Description of the Algorithm

The algorithm mainly includes the following major stages: assessment task execution stage, feature extraction stage, weight calculation stage, and final score acquisition stage.

(1) Assessment Task Execution Stage
During this phase, teachers put forward the detailed requirements for the task on the peer assessment system, and give the corresponding necessary information such as the standard answer used as a reference. The system assigns the tasks to the students according to the pre-set allocation algorithm. Then, the students perform the assessment tasks in the peer assessment. In the process of worker executing the task, the peer assessment system will automatically record the relevant information. The information including workers' answers, the log information during the task execution (such as time for finishing the task), worker's comment text, and other relevant information.

(2) Feature Extraction Phase
When the worker has completed corresponding assignment, the system will extract the features needed from the recorded information according to the algorithm. In general, the features are divided into three categories, as shown in Table 1, including numerical calculation features, text analysis features, and user behavior features.

Table 1. Characteristic description

Feature category	Feature number
Numerical calculation	1
Text analysis	4
User behavior analysis	2

- Numerical calculation feature: it is the workers' credibility estimated based on the expectation maximization algorithm. Since this value is completely inferred from the worker's answer task information, it is added to the log-linear model.
- Text analysis features: Those features are derived from students' comments on the evaluated work. Natural language processing technology can also provide certain information for judging students' seriousness. The features are as follows:

Comment length feature, extracting the length information of the comments made by workers.

Comment similarity feature, using edit distance to measure the similarity between different comments made by workers for feature extraction.

Comment and content relevance feature: The fast-text method is used for text representation, and the word vector is obtained through the skip-gram algorithm to vectorize the text. The cosine similarity of the vector is used to measure the correlation between the comment and the content, and the feature is extracted.

Comment confusion feature, using a feedforward neural network language model to measure the fluency of comments made by workers and extract features. The idea behind the characteristics of comment confusion is that the more fluent the comment, the more likely it is to be written seriously. In actual situations, in order to complete the task quickly, there are students who randomly input in the comments to meet the word count requirements, but the text input in this way is often without clear semantics and does not help crowdsourced tasks. The main reasons for the unsuccessful comments are they contain typos, malicious meaningless information, etc., as shown in Table 2.

Table 2. The real dataset

Id	Review_id	Tno	Comment
20160256	20160045	19	Haha, I don't know
20160257	20160045	19	Expression is good, consistent with the theme
20160258	20160045	19	**********
20160259	20160045	19	**********
20160260	20160045	19	**********
20160261	20160045	19	**********
20160262	20160045	19	**********

Perplexity is an evaluation index used in the field of natural language to measure the performance of a language model. The role of a language model is that describe the text content in terms of fluency of the text. The process is as follows: given an arbitrary text S, it is composed of m words, denoted as w_1, w_2, ...w_m. The language model is defined as P(s), which gives the probability of text appears, and the perplexed degree pp formula of the text S is:

$$PP(S) = e^{\left(\frac{1}{m}*\sum_i \log\left(P\left(w_i/w_j\right)\right)\right)} \tag{11}$$

The neural probabilistic language model (NPLM) proposed by Bengio in 2003 is used to calculate P(wi | wj < i) [14]. Compared to the traditional count-based n-gram statistical language model, the neural probabilistic language model (NPLM) introduces a distributed representation of words, named word vectors [15]. After converting the original one-hot discrete word-based representation into continuous and real number vector, the problem of data sparse is greatly alleviated. At the same time, due to introduced word vectors, the neural probabilistic language model can distinguish more synonym than the traditional statistical language model methods, so the model has a stronger ability to express.

- User behavior characteristics mainly include: global average time-consuming characteristics of answering questions, by comparing the average time-consuming of workers with those of other workers, so as to measure the credibility of workers' results and extract features. The time-consuming feature of task answering, by comparing the time consumed by different workers on the same task, measures the credibility of the worker's results and extracts the features.

(3) Weight Calculation Stage
The log-linear model is used to mathematically model the worker's scoring reality, and the seven features of different types obtained in the feature extraction stage are merged. The stochastic gradient descent algorithm is used to optimize the loss function so as to calculate the credibility of each worker, and use it as the weight of the worker's score.

(4) Final Score Acquisition Stage
The algorithm performs a linear weighting operation based on the different weights obtained by the algorithm for the credibility of different workers in the weight calculation stage, and the actual scores made by the workers in the task completion stage to obtain the final score of the task.

4   Experiments and Results

4.1   Experimental Data

The experimental data comes from the peer assessment system that has been developed by our team. The system is firstly implemented by PHP and R language, and then we improved it with Python language. The system is applied to the computer courses taught by teacher of the research group. Corresponding curriculum design and accumulated a large number of data sets are available for research needs. The platform has been put into operation for five years. It is mainly used in non-computer professional courses. The total number of users has exceeded 10,000, and more than 400,000 mutual evaluation data have been collected.

The experimental data of this article has selected a total of 326 students to participate, each student have submitted his own assignments, each assignment contains 10 multiple-choice questions, each choice question contains 4 options, the total number of tasks is 3260, each to answering 10 multiple-choice questions and write comment of the evaluated work. Each student's submitted job also contains the score manually scored by the teacher. Details of real data show in Table 3.

Table 3. The real dataset

Number of workers	Number of tasks	Number of options	Number of reviews	Average length of review's word	Manual scoring <= 60	Manual scoring [70, 80]	Manual scoring [80, 90]	Manual scoring [90, 100]
326	3260	4	318	11.2	6.8%	34.83%	42.47%	15.9%

4.2 Assessment Criteria

In order to evaluate the effectiveness of different methods, this paper uses the accuracy of predictive labels and the Least Mean Square (LME) to predict workers' credibility, and then evaluate the performance of experimental results using simulated data. Using predicted scores' Least Mean Square to evaluate the performance of the experimental results and real data. Here, the accuracy of the forecasting tag accuracy is as follows:

$$Accuracy = \frac{\sum_{i \in T} sign(t_i, \hat{t}_i)}{|T|}. \tag{12}$$

Among them, t_i represents the correct answer of the task i, $\hat{t}_i$ presents the correct answer of the task predicted by the model, T is the set of all tasks, |T| represents the total number of tasks, sign is the step function, specifically:

$$sign(x, y) = \begin{cases} 1 & \text{if } x = y \\ 0 & \text{else} \end{cases} \tag{13}$$

The Least Mean Square of the prediction credibility is as follows:

$$LME = \frac{\sum_{i \in W} (d_i - \hat{d}_i)^2}{|W|} \tag{14}$$

d_i represents the correct credibility of the worker i, $\hat{d}_i$ represents the worker confidence predicted by the model, W is the set of all workers, and |W| represents the number of all workers. Obviously, the more accurate the predictions used, the higher the accuracy and the lower the LME.

4.3   Experimental Results

This section discusses that how the user behavioral features and text analysis features contribute to the final model prediction based on multi-dimensional information quality control algorithm. By comparing with traditional features based on numerical calculations as baselines, we will observe the effects of different types of multi-information on the performance evaluation system's forecasting performance. In order to make the experimental results more stable, this experiment uses a 10-fold crossover test method, which randomly cuts out 1/10 of the training data as a test set, records the results of each experiment, and repeats 10 times. In ten experiments, the average value is the final result. Among them, the detailed results of the 10 experiments are shown in Table 4. The average experimental results of different dimensions of information are shown in Table 5, and the performance curves are shown in Fig. 1. EM is short of Expectation Maximization algorithm [9], UB stands for user behavior characteristics, and TA stands for text analysis feature.

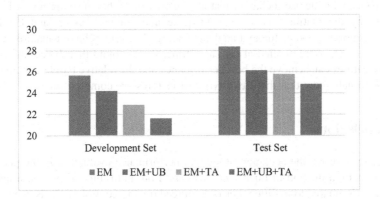

Fig. 1. Comparison between multi-dimensional and expectation-maximization

Table 4. Cross-validation experiment in different dimension

Number	Em	Em + ub	Em + ta	Em + ub + ta
1	25.452	24.079	22.819	21.732
2	25.139	24.138	22.642	21.943
3	25.477	24.281	22.918	21.799
4	25.824	24.174	22.619	21.352
5	25.391	23.869	22.774	21.375
6	26.004	24.592	23.182	21.489
7	26.182	24.771	23.489	21.871
8	25.743	24.357	22.347	21.613
9	25.812	24.281	23.192	21.652
10	25.436	23.248	22.948	21.264

Table 5. Average result of experiment in different dimension

System	Development Set LME	Test Set LME
Expectation to maximize	25.646	28.375
Expectation to maximize + User behavior characteristics	24.179	26.104
Expectation to maximize + Text analysis features	22.893	25.772
Expectation to maximize + User behavior characteristics + Text analysis features	21.609	24.828

From the experimental results, it can be seen that the performance of the model prediction is improved wherever it is adding user behavior features or text analysis features, and the performance is better than the baseline-based numerical calculation features based on the expectation maximization.

In comparison, the text analysis feature has a greater impact on the prediction performance than the user behavior feature. The reason is that the current user behavior obtained by the platform is not yet adequate and comprehensive, and the feature richness is not as good as the text analysis feature. The user behavior features and text analysis features were applied at the same time. It was found that the combination of the two features can achieve the best performance and the relative baseline increased by 14.43%, which show that these two types of features are complementary.

5 Conclusions

In order to improve the accuracy of student performance evaluation in the peer evaluation system, a method of quality control based on multi-dimensional information is proposed. Compared to traditional crowd-sourced quality control methods, we focus on how to effectively use users' behavior information, text analysis information to comprehensively measure and evaluate the credibility of individual workers through more diversified information, so as to improve the accuracy of the final results of predict evaluation.

In order to verify the effectiveness of the application of multi-dimensional information methods, the 10-fold cross validation was used to verify the effectiveness of the algorithm in real data sets. From the experimental results, we can see that whether it is adding user behavior information or text analysis information on the basis of the expectation maximization algorithm, it can improve the predictive ability of the final model. Since the above two types of information are complementary, when the two types of information are used at the same time, the model performance is optimal. Compared with the baseline system, the model's prediction accuracy is improved by 14.43%, which verifies the effectiveness of the method in this paper.

Acknowledgements. This work is supported by the Fundamental Research Funds for the Central Universities under Grant N181604015, and the National Natural Science Foundation of China under Grant No. 61602106.

References

1. Kulkarni, C., et al.: Peer and self-assessment in massive online classes. Des. Think. Res. **20** (6), 1–31 (2013)
2. Zhang, Z., et al.: Crowdsourcing quality control strategy and evaluation algorithm. Chin. J. Comput. **8**, 1636–1649 (2013)
3. Shah, N.B., Bradley, J.K., et al.: A case for ordinal peer-evaluation in MOOCs. In: NIPS Workshop on Data Driven Education, pp. 1–8 (2013)
4. Daniel, J.: Making sense of MOOCs: musings in a maze of myth, paradox and possibility. J. Interact. Media Educ. **2012**(3), 3–12 (2012)
5. Zelinski, M., Hicks, N.M., et al.: Instructor outcomes of teaching a STEM MOOC. In: IEEE Frontiers in Education Conference 2017, Indianapolis, IN, USA, vol. 1, pp. 1–7 (2017)
6. Dong, K., et al.: The evolution path and future prospect of research on crowdsourcing in China and abroad. Sci. Technol. Prog. Policy **8**, 154–160 (2016)
7. Zheng, Z., et al.: Crowdsourcing quality evaluation algorithm based on sliding task window. Mini-micro Syst. **9**, 2125–2129 (2017)
8. Felsenthal, D.S., et al.: The treaty of nice and qualified majority voting. Soc. Choice Welf. **18**(3), 431–464 (2001)
9. Moon, T.K.: The expectation-maximization algorithm. IEEE Sig. Process. Mag. **13**(6), 47–60 (2000)
10. Zhang, Z., et al.: Research on quality control of peer grading in MOOCs. Softw. Eng. **2**, 7–10 (2018)
11. Sun, J., Yang, X., Wang, B.: Crowdsourced indoor localization for diverse devices with RSSI sequences. In: Ni, W., Wang, X., Song, W., Li, Y. (eds.) WISA 2019. LNCS, vol. 11817, pp. 614–625. Springer, Cham (2019). https://doi.org/10.1007/978-3-030-30952-7_62
12. Ding, Y., et al.: Research on crowdsourcing quality control algorithm based on social platform. Softw. Guide **12**, 90–93 (2017)
13. Wang, B., et al.: Application of log-linear model in part-of-speech tagging. Comput. Sci. **35** (5), 163–166 (2008)
14. Christensen, R.: Log-linear models and logistic regression, pp. 267–268. Springer, New York (2006)
15. Bengio, Y., et al.: A neural probabilistic language model. J. Mach. Learn. Res. **3**(6), 1137–1155 (2003)
16. Kesorn, K., et al.: An enhanced bag-of-visual word vector space model to represent visual content in athletics images. IEEE Trans. Multimed. **14**(1), 211–222 (2012)

Fast Dynamic Density Outlier Detection Algorithm for Power Quality Disturbance Data

Siyu Liu[✉] and Jun Fang

North China University of Technology, Beijing 100144, China
julyand131@163.com, fangjun@ncut.edu.cn

Abstract. The existing anomaly detection methods have the problems of low accuracy and slow identification speed for outliers detection of high-dimensional power quality disturbance data with time sequence characteristics and large fluctuations. In order to solve the above problems, a fast dynamic density anomaly detection method for power quality disturbance data is proposed. The data set is divided into different time slices according to time, and only the changed data in different time slices are dynamically clustered, so that the data on the next time slice can get accurate clustering results with less time cost. The experimental results show that the proposed method not only ensures the accuracy of anomaly detection results, but also improves the time efficiency.

Keywords: Outlier detection · Power quality disturbance data · Dynamic algorithm

1 Introduction

Abnormal power quality disturbance events will bring adverse effects to users, and even cause incalculable economic losses. In order to reduce these negative effects, power supply companies need to take corresponding measures to deal with the abnormal power quality disturbance. Therefore, accurate and rapid identification of abnormal power quality disturbances is of great significance to improve power quality.

There are many common methods for data anomaly detection, such as anomaly detection based on statistics, outlier detection based on classification, outlier detection based on nearest neighbor model and outlier detection based on clustering. Among them, the statistical based anomaly detection method has poor effect on the processing of high-dimensional power quality disturbance data with high volatility. And the method based on classification is not suitable for power quality disturbance data without enough correct class labels. For the power quality disturbance data of many neighbors, it is sometimes difficult to define the distance between the data, so it is not a good choice to use the anomaly detection method based on the nearest neighbor model.

© Springer Nature Switzerland AG 2020
G. Wang et al. (Eds.): WISA 2020, LNCS 12432, pp. 194–201, 2020.
https://doi.org/10.1007/978-3-030-60029-7_18

Clustering-based anomaly detection method is a good choice for power quality disturbance data. Most of the detection indexes of power quality disturbance data are dynamic and change with time, but at present, most of the existing anomaly detection algorithms based on clustering use static clustering method to process time series data, the anomaly detection of dynamic and time-varying power quality disturbance data has poor adaptability. In this paper, a fast dynamic density anomaly detection method for power quality disturbance data is proposed based on the improvement of the existing clustering algorithm. This method is mainly based on the ordering points to identify the clustering structure (OPTICS) algorithm [4], the data set is divided into different time slices. On the basis of the previous time slice clustering results, only the change data is calculated, so that the accurate clustering results can be obtained with less time cost when clustering the data on the later time slice, at the same time, the accuracy of anomaly detection is guaranteed and the time efficiency of the algorithm is improved.

2 Related Work

At present, there are many outlier detection methods, such as outlier detection based on statistics, outlier detection based on classification, outlier detection based on nearest neighbor model, outlier detection based on clustering, etc.

In the aspect of outlier detection based on statistics, journal articles [12] realized anomaly detection by calculating the fraud coefficient of user electricity consumption from the data of power supply end and user end. Journal articles [9] used isolated forest algorithm to detect anomaly. Reference [6] proposed a novel online anomaly detection algorithm using incremental tensor decomposition However, this anomaly detection method can not get good results in the face of large volatility data.

In terms of outlier detection methods based on classification, reference [1] proposed a data modeling method based on neural network to identify the abnormal behavior of boilers in thermal power plants. Journal articles [3] proposed an anomaly detection method based on support vector machines that integrates information. However, this kind of anomaly detection method depends on whether there are enough training samples with correct category labels. This method is not suitable for power quality disturbance data with large amount of data, many data features and not enough correct class labels.

In the aspect of outlier detection based on nearest neighbor model, an adaptive outlier detection cloud computing scheme based on local outlier factor is proposed in [7]. Reference [5], outlier detection was performed by calculating the outlier factors of each object in the subset. The disadvantage of this method is that the nearest neighbor search needs some time complexity, and it is difficult to select a suitable distance function for complex data.

The anomaly detection method based on clustering includes the detection of static data and dynamic and time-varying data. For static data, the k-means algorithm was used for anomaly detection in [11]. For dynamic and time-varying data, journal articles [10] proposed a clustering based anomaly detection method for large-scale time series data. Journal articles [8] combined with sliding window and k-means clustering algorithm, proposed a clustering algorithm based on sliding window unequal time series of short time series distance, which could solve the problem of unequal time series clustering, but the optimal number of clusters is difficult to determine.

Reference [4], an improved algorithm of DBSCAN, OPTICS algorithm, was proposed. OPTICS algorithm is not sensitive to initial parameters and can cluster non-spherical data. From the perspective of anomaly detection, it has a strong ability to find outliers. However, for the power quality disturbance data, the OPTICS algorithm can not detect the abnormal data according to its time sequence feature, so this paper improves it and proposes a fast dynamic density abnormal value detection algorithm for power quality disturbance data.

3   Detection Method of Abnormal Data

A static clustering process on a time slice can be described as follows: Data set $Dataset = \{x_1, x_2, \ldots, x_n\}$ consists of n points, and each point can be represented by a d-dimensional vector $x_i = \{x_{i1}, x_{i2}, \ldots, x_{id}\}$. Given the data set, the algorithm finds a set of cluster class set $C = \{C_1, C_2, \ldots, C_k\}$ making the structure of clusters as different as possible, and the structure of clusters as similar as The center of mass of each cluster is determined by C_{E_j} represents, $j = 1, 2, \ldots, k$. Different from that, the clustering of time series data is a continuous and dynamic process, in which the data and cluster changes of two adjacent times can be expressed as follows: When t_i-time, set $Dataset^i$ medium n_i elements make up k_i classes. And in t_{i+1}-time, the number of elements in the $Dataset^{i+1}$ becomes n_{i+1} and form k_{i+1} classes.

The traditional clustering algorithms are mostly used to analyze the static data set on a time slice, but the power quality disturbance data has a continuous dynamic evolution process on the time series. The dynamic density clustering algorithm is to calculate the change data only locally on the basis of the result of the previous time slice clustering, and get the same result as the complete clustering of the data on the later time slice, at the same time, the amount of computation dropped significantly [2]. Based on this idea, a fast dynamic density detection method for power quality disturbance data is proposed.

The fast dynamic density (FDD) anomaly detection algorithm for power quality disturbance data includes two kinds of cluster structure local adjustment operations: elements remove (ER) and elements add (EA). The algorithm stores the changed data according to its type in 2 queues ER and EA, and then calls 2 processes for batch processing, reducing the number of calls and high computation efficiency. First the data at t_i-time is processed by OPTICS algorithm, and then the data of t_{i+1}-time whose attribute labels have changed is detected by the algorithm proposed.

Figure 1 is the initial state diagram of the elements and Fig. 2 shows the elements state diagram after change. By comparing Fig. 1 with Fig. 2, it can be found that element A is the vanishing element, element J is the new element, forming the data set at t_{i+1}-time.The proposed algorithm is used to cluster the changed data set. The specific element attributes and the process of category change are shown in Table 1, in which different numbers in the categories represent different categories. In the attribute description, 1 represents the core point, 0 represents the boundary point, −1 represents the exception point, Nan represents the disappearance of the element, and New represents the addition of the element.

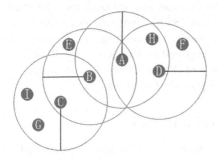
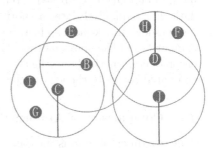

Fig. 1. Initial elements state.   **Fig. 2.** Elements state after change.

Table 1. Elements attribute label table

Element	A	B	C	D	E	F	G	H	I	J
t_i-time	1	1	1	1	1	0	0	0	0	
After change	Nan	0	1	1	−1	0	0	0	0	New
t_{i+1}-time		0	1	1	−1	0	0	0	0	0

It can be seen from Fig. 1 that at t_i-time all elements are the same class at any time. After the change of elements, elements D, F, J, H become Fig. 2, and elements E become abnormal point category. A large category is divided into three small categories. As can be seen from Table 1, at t_i-time, elements A, B, C and D are the core points. After element A is operated by ER, element B changes from the core point to the boundary point, and element E changes from the boundary point to the abnormal point. After the EA operation of element J, the attributes and categories of other elements have not changed, indicating that the addition of element J has no impact on the structure within the cluster. Finally, the result of t_{i+1}-time clustering shown in Fig. 2 is formed.

4   Experimental Analysis

The experiment is mainly divided into two parts: the first part is the comparative analysis of anomaly detection results; the second part is the comparative analysis of algorithm time efficiency.

The proposed anomaly detection method is essentially based on clustering. The anomaly detection method based on statistics, classification and nearest neighbor model does not belong to the same system as the method in this paper, so this paper does not make a comparison. k-means based anomaly detection method and DBSCAN based anomaly detection method are both used for static data. For sequential data, a short time series (STS) algorithm combining sliding window and k-means clustering algorithm is proposed in [8]. Through the comparison of the above methods and the proposed methods, the recall and precision of each method and the time efficiency of the algorithm are calculated and compared to verify the effectiveness of the proposed method.

4.1   Comparison of Abnormal Detection Results

The experimental data is the data of a monitoring point on June 7, 2019, including voltage deviation, effective value of voltage, total harmonic distortion rate of voltage, effective value of fundamental voltage, active power of fundamental wave, positive sequence voltage, negative sequence voltage, etc. The data collection time is from 0:00 a.m. to 23:59 p.m. every minute. There are 1440 min in a day, and 1440 pieces of data can be taken for each index every day. There are 1293 normal data and 147 abnormal data in total harmonic distortion rate data, 1332 normal data and 108 abnormal data in effective value data. Taking total harmonic distortion rate of voltage and effective value of voltage data as examples, the method mentioned above and the method in this paper are used to detect the abnormal data. Among them, Fig. 3 shows the detection results of four anomaly detection methods for voltage total harmonic distortion rate and voltage effective value respectively, the left side is the detection results of voltage total harmonic distortion rate, the right side is the detection results of voltage effective value; Table 2 shows the results of four anomaly detection algorithms, and Table 3 shows the recall and precision of four detection methods for testing data.

The k-means algorithm and the DBSCAN based anomaly detection method do not take the timing of the data into account. The data that should be judged as abnormal is also judged as normal. The data that should be judged as normal is judged as abnormal, and the detection result is poor. However, compared with the method proposed in this paper, STS algorithm proposed in [8] has poor detection results due to the influence of k-value selection. The method proposed in this paper fully considers the dynamic and time-varying trend of time series data, and is not affected by the selection of k-value. It can well adapt to the change trend of data, and the detection result is better, both the recall rate and the precision rate are higher. It has great advantages in the detection of abnormal values of power quality disturbance data with time series characteristics.

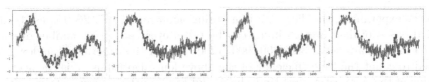

(a) k-means abnormal voltage total har- (b) DBSCAN abnormal voltage total harmonic distortion rate & effective value   monic distortion rate & effective value

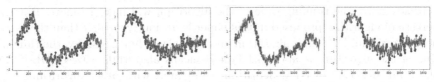

(c) STS abnormal voltage total harmonic (d) FDD abnormal voltage total harmonic
distortion rate & effective value   distortion rate & effective value

Fig. 3. Four abnormal detection methods are used to detect voltage deviation and voltage effective value respectively.

Table 2. Abnormal detection results

Algorithm name		K-means	DBSCAN	STS	FDD
Voltage total harmonic distortion rate	TP	21	60	71	120
	FP	24	9	84	0
	TN	1269	1277	1210	1393
	FN	126	94	75	27
Voltage effective value	TP	16	29	40	87
	FP	25	4	63	4
	TN	1307	1331	1269	1328
	FN	92	76	68	21

Table 3. Comparison of recall rate and precision rate of anomaly detection

Algorithm name		K-means	DBSCAN	STS	FDD
Voltage total harmonic distortion rate	P	46.67%	86.96%	45.81%	100%
	R	14.29%	40.82%	48.30%	81.63%
Voltage effective value	P	39.02%	96.97%	38.83%	87.88%
	R	14.81%	29.63%	37.04%	80.55%

4.2 Algorithm Time Efficiency Comparison

In the experiment, the data of a monitoring point on June 7, 2019 is used, and 1440 pieces of data of each index can be taken every day. The number of data used in the data set is 300, 600, 900, 1200, 140. In the experiment, index T is used to count the time efficiency of the above four algorithms. T is the average value of the data set on five time slices, and Fig. 4 is the experimental result.

It can be seen from Fig. 4 that the running time of the four algorithms increases with the increase of data, and when the total number of elements increases more, the running time of the algorithm will increase more. But the increase of algorithm proposed in this paper is obviously smaller than that of k-mean algorithm, DBSCAN algorithm and STS algorithm. Secondly, the running time of the algorithm is about 51% of K-means algorithm, 70% of DBSCAN algorithm and 72% of STS algorithm. Compared with other algorithms, the algorithm proposed in this paper can reduce the computational cost of clustering process by dynamic continuous local clustering adjustment, especially when the number of elements in the data set is more than 600, the time efficiency is improved more obviously. In comparison, the algorithm proposed in this paper takes less time for anomaly detection, which is very suitable for anomaly detection of power quality disturbance data.

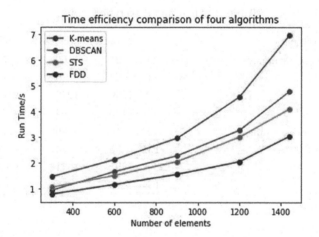

Fig. 4. Time efficiency comparison of anomaly detection.

5   Conclusion

In this paper, on the basis of the OPTICS algorithm, the time factor is introduced to cluster the data set according to the time slice. On the basis of the previous time slice clustering results, through the calculation of some changed data and the local adjustment of the cluster structure, the data on the later time slice can get accurate clustering results at a small time cost, reduce the calculation and marketing of the clustering process, and improve the time efficiency of the algorithm. The experimental results show that this method is more efficient and accurate to detect the power quality disturbance data, and it is suitable for anomaly detection of large fluctuation data in time series.

References

1. Banjanovicmehmedovic, L., Hajdarevic, A., Kantardzic, M., Mehmedovic, F., Dzananovic, I.: Neural network-based data-driven modelling of anomaly detection in thermal power plant. Automatika **58**(1), 69–79 (2017)
2. Chen, H., Ji, M., Guo, Z., Xia, Y.: A dynamic density clustering algorithm for time series data. Control Theor. Appl. **8** (2019)
3. Chen, Y., Qian, J., Saligrama, V.: A new one-class SVM for anomaly detection, pp. 3567–3571 (2013)
4. Duan, M., Chaolin, T.: Realization of clustering algorithm based on density. J. Jishou Univ. (Nat. Sci. Edn.) **34**(1), 26–27 (2013)
5. Fu, P., Hu, X.: Biased-sampling of density-based local outlier detection algorithm, pp. 1246–1253 (2016)
6. Gao, M., et al.: Online anomaly detection via incremental tensor decomposition. In: Ni, W., Wang, X., Song, W., Li, Y. (eds.) WISA 2019. LNCS, vol. 11817, pp. 3–14. Springer, Cham (2019). https://doi.org/10.1007/978-3-030-30952-7_1
7. Huang, T., et al.: An LOF-based adaptive anomaly detection scheme for cloud computing, pp. 206–211 (2013)
8. Liu, Q., Wang, K., Rao, W.: Non-equal time series clustering algorithm with sliding window STS distance. J. Front. Comput. Sci. Technol. **9**(011), 1301–1313 (2015)
9. Susto, G.A., Beghi, A., Mcloone, S.: Anomaly detection through on-line isolation forest: an application to plasma etching, pp. 89–94 (2017)
10. Truong, C.D., Anh, D.T.: An efficient method for motif and anomaly detection in time series based on clustering. Int. J. Bus. Intell. Data Min. **10**(4), 356–377 (2015)
11. Yin, C., Zhang, S.: Parallel implementing improved k-means applied for image retrieval and anomaly detection. Multimed. Tools Appl. **76**(16), 16911–16927 (2016). https://doi.org/10.1007/s11042-016-3638-1
12. Yip, S., Tan, W., Tan, C., Gan, M., Wong, K.: An anomaly detection framework for identifying energy theft and defective meters in smart grids. Int. J. Electr. Power Energy Syst. **101**, 189–203 (2018)

Word Embedding-Based Reformulation for Long Queries in Information Search

Wei Yan, Yarong Wang, Chunlan Huang, and Shengli Wu[✉]

School of Computer Science, Jiangsu University, Zhenjiang 212013, China
swu@ujs.edu.cn

Abstract. It has been found that very often long queries are more challenging than short queries for information search engines to obtain good results. In this paper, we present a word embedding-based approach. First short queries or concepts are extracted from the original query. Then with the help of a trained word embedding model, all of the query elements go through a series of reformulation operations including deletion, substitution, and addition of terms so as to obtain more profitable query representations. Finally all the reformulated elements are linearly combined with the original query. Experiments are conducted on three TREC collections, and the experimental results show that the proposed method is able to improve retrieval performance on average and especially effective for long queries. Compared with several state-of-the-art baseline methods, the proposed method is very good.

Keywords: Information search · Word embedding · Long queries · Query rewrites

1 Introduction

Although most of the queries submitted to Web search engines are short, long queries are also frequent [16]. Very often Web search engines do much worse with long queries than with short queries. Therefore, it is worthwhile to investigate why this happens and what can be done to improve retrieval performance of long queries.

Long or verbose queries are usually composed of multiple concepts, which may be useful for describing a user's information need more clearly. However, more information does not always mean a good thing to a search engine. It is challenging for a search engine to identify the significance of different concepts of a long query. For example, Goggle has a limit for the number of words that can be used in a query. It was 10 for some time and it allows up to 32 words now[1]. Long queries are also likely to bring more noise. Figure 1 shows an example of a long query, which is one of the queries (Query 453) used in TREC (Text Retrieval Conference, an annual event of information retrieval evaluation, held by the national institute of standards and technology, USA, its website is located at https://trec.nist.gov/).

---

[1] https://searchengineland.com/20-googles-limits-may-not-know-exist-281387, retrieved on 21 January, 2020.

G. Wang et al. (Eds.): WISA 2020, LNCS 12432, pp. 202–214, 2020.
https://doi.org/10.1007/978-3-030-60029-7_19

```
<num> Number: 453
<title> hunger
<desc> Description: Find documents that discuss organizations/groups that are aiding
in the eradication of the worldwide hunger problem.
```

Fig. 1. An example of a verbose query

This query has a title and a description part. As other queries in TREC, its title is short and can be treated as a short query; while its description is long and can be treated as a long query. Its title only has one term "hunger" and it is easy for a search engine to deal with. In its description part, words like "find", "documents" are obviously irrelevant to the query. Both "organizations" and "groups" may be relevant to the query moderately, but they are synonymous and interchangeable. "Worldwide hunger problem" is a key concept in this query and "worldwide" is a modifier of the "hunger problem". So it is a complex task for information search engines to decide the impact of each term or phrase in a long query like this accurately.

According to many previous research works [1–6], some operations on long queries may lead to better retrieval performance. The main operations on long queries mainly include query term weighting, query segmentation, query reduction, query expansion, and query reformulation. One common ground for these operations is they need to estimate the relative importance of multiple elements (terms or concepts) in the long query, but how to represent the original long query is different for each of the operations. For query term weighting and query segmentation, all the words in the original query remain; while some words in the original query are deleted and added for query reduction and query expansion, respectively. Query reformulation may include a combination of some of the above-mentioned operations.

In the last couple of years, the distributed representation of text, also known as word embedding, has attracted a lot of attention in the information retrieval community. This technique can be used to calculate syntactic or semantic similarity between different terms or phrases. Although word embedding has been used in many applications and in some information search tasks [8, 9, 20–22] as well, to the best of our knowledge, it has not been used to deal with the long query issue.

A long query may include multiple elements (terms or phrases) and each of them can be represented as a vector. Here "elements" are similar to "concepts" in [1]. They can be a single term, a phrase, or a free combination of terms. By using the distributed representation of texts, we are able to find similar new words for a given word or phrase by searching the vector space. Thus queries can be rewritten in a variety of ways by reformulating the elements involved. For those newly generated elements, we may compare its similarity with the whole original query to decide its goodness. Experiments are conducted on three TREC collections, and the results show that our method is effective for all the queries on average, and especially for long queries.

Next section is about related work. The proposed approach and empirical results are presented in the next two sections, respectively. Section 5 concludes the paper.

2 Related Work

In the last couple of years, word embedding has been used in some information retrieval tasks. Compared with the one-hot representation, it can build a continuous vector space with much lower dimensions, where words are represented as embedding vectors by training a large corpus to obtain their proximity information in the context. Based on the hypothesis that semantically or syntactically similar words often share similar contexts, word embedding aims to accurately predict related words for a given word or context as well as catch semantic similarities between two different pieces of text by calculating their geometric distance (i.e. cosine similarity or Euclidean's distance).

Proposed by Mikolov et al. in [7], word2vec is a group of related models that are used to produce word vectors. These models are shallow, two-layer neural networks that are trained to reconstruct linguistic contexts of words. Two types of model architectures, continuous bag of words (CBOW) and skip-gram, were utilized to produce a distributed representation of words. It can be used to predict adjacent words or the surrounding window of context words according to a given word. Other types of word embedding such as GloVe[2] and ConceptNet Numberbatch[3] (referred to as CNN later in this paper) are also available.

Word embedding has been applied to a number of IR tasks, including query expansion, query classification, short text similarity, and document model estimation. Zheng and Callan proposed a supervised embedding-based method to re-weight terms within some existing IR models such as BM25 [8], while Zamani and Bruce Corft presented some embedding-based query language models for ad hoc retrieval [9]. For sponsored search, [11] built three embedding-based models with query contexts and contents to generate query rewrites in response to the initial query. Word embedding-based query expansion for ad-hoc retrieval was investigated in [10, 18], while word embedding-based query expansion for Arabic retrieval was investigated in [21]. Both word embedding-based query expansion and entity embedding-based query expansion were investigated for ad hoc retrieval in [17]. Word embedding-based question classification and retrieval is investigated in [22]. It is found that word embedding is useful for these retrieval tasks.

In this paper, we propose a word-embedding based method for the reformulation of long queries. Noun phrases and other elements are extracted from the original query and then three types of operations are performed on all of them repeatedly for a number of rounds. At each step, those concepts are evaluated and some good ones are kept for next round and finally all chosen query elements along with the original query are combined to form the final query. Note in [10, 17, 18], only query expansion was investigated; while in this work we investigate three types of operations (expansion, deletion, and substitution) together. To our knowledge, this approach is novel and has not been investigated before.

---

[2] https://nlp.stanford.edu/projects/glove/.

[3] http://blog.conceptnet.io/posts/2016/conceptnet-numberbatch-a-new-name-for-the-best-word-embeddings-you-can-download/.

3   The Proposed Method

In this section, we present our method WE2R (Word Embedding-based Element Rewrites) for long queries. First of all, noun phrases and other types of semantic elements are extracted from the original query as initial elements. Then all of them go through a process of reformulation. There are three types of operations for each of them: deleting one term, adding one term, and using a new term to replace an existing term [12, 13, 19]. All of them are evaluated according to a given criteria and some top-ones are kept. The above process is repeated for several times. Thus a reformulation tree is generated for each of the elements. The deeper a tree we generate, the more words we search for the element. The rationale behind this process is to try to find more suitable terms for the query, especially for those extracted key elements (concepts).

In the above reformulation process, a pre-trained word-embedding model is used. Therefore, every term is represented as a vector. By a simple average of all the terms, we are also able to represent a phrase as a vector which is the centroid of all the term vectors. Then the syntactic and semantic similarity between two query elements or between a query element and the original query can be measured by calculating their distance in the vector space.

More specifically, given an element $r = w_1 \ldots w_{i-1} w_i w_{i+1} \ldots w_k$ $(1 \leq i \leq k, 2 \leq k)$, new elements are generated by three basic operations: deleting one term (related to query deduction), substitution of one term, and adding one term (related to query expansion). For term deletion, there are k options [12]. When term w_i is removed, we have $r_d = w_1 \ldots w_{i-1} w_{i+1} \ldots w_k$ $(1 \leq i \leq k)$. Similarly, for term substitution, we may replace w_i by a new term w_{new} in r ([13]): $r_s = w_1 \ldots w_{i-1} w_{new} w_{i+1} \ldots w_k$ $(1 \leq i \leq k)$. For term addition, we may add a new term w_{new} to r: $r_e = w_1 \ldots w_{i-1} w_i w_{i+1} \ldots w_k w_{new}$ $(1 \leq i \leq k)$. We use $C_d = \{r_d\}$, $C_s = \{r_s\}$ and $C_e = \{r_e\}$ to represent the set of element rewrites generated by the three basic operations, respectively. In both term substitution and addition, we need to find some suitable terms as candidates. Pre-trained word vector space models such as word2vec can be used to help find them. For term substitution, terms that are very close to w_i are chosen as candidates; while for term addition, terms that are very close to the centroid of r are chosen as candidates.

To measure the suitability of a reformulated query element r, we consider the similarity of the reformulated element to its parent and also to the original query. The similarity score of r is defined as

$$SimilarityScore(r) = sim(r, r_p) * log\left(\frac{sim(r, Q)}{sim(r_p, Q)}\right) \qquad (1)$$

In Eq. 1, r_p is the parent of r, or r is re-formulated from r_p. Q is the query. v_r, v_{r_p} and v_q are the vector representations of r, r_p, and Q. Cosine similarity is used for measuring the similarity of any two vectors, or $sim(r, v) = \frac{v_r \bullet v_v}{|v_r||v_v|}$. If $sim(r, Q) > sim(r_p, Q)$, then $SimilarityScore(r) > 0$; otherwise, $SimilarityScore(r) \leq 0$. In one round, all newly generated element rewrites in sets C_d, C_s, and C_e are put together and scored with Eq. 1. A number of top-ranked elements with the highest scores are kept for the next round.

After a few rounds, we end up with a group of elements for the final query. Then we need to define suitable weights for those elements involved. First let us introduce the concept of specificity of an element inside a group of elements as

$$Specificity(r, C_g) = \frac{1}{|C_g|} \sum_{r' \in C_g} \frac{\#(t_r)}{\#(t_r) + \#(t_{r'}) + \#(t_{r,r'})} \qquad (2)$$

where C_g is a group of elements and r is a given element in C_g. $\#(t_r)$ denotes the number of terms in r but not in r', and $\#(t_{r'})$ denotes the number of terms in r' but not in r, while $\#(t_{r,r'})$ denotes the number of terms which appear in both r and r'. C_g is the number of elements in C_g.

With a group of elements C_g, we may define their weights by considering both similarity and specificity. For element r, its weight is defined as

$$Weight(r) = Sim(r, Q) * Specificity(r, C_g) \qquad (3)$$

Finally, from all the elements we choose n of them with the highest weights (denoted by C_n), and normalize them by the following equation

$$N\_weights(r|Q) = \frac{Weight(r)}{\sum_{r \in C_n} Weight(r)} \qquad (4)$$

Algorithm 1. Query reformulation based on word embedding

1 Extract noun phrases and other semantic elements and put them into set C
2 For each element $c \in C$:
3   delete each of the terms in c to generate a new element
4   replace each of the terms in c with a new term for k times
5   add one new term to c for m times
6   score each of the new elements with Eq. 1
7   update C by replacing all the old elements with t top-ranked new elements
8 Repeat Lines 2-7 until the stop condition is satisfied
9 Score all the elements in C with weights calculated by Eq. 3
10 Normalize weights of top-n elements by Eq. 4 and return these elements with weights

---

Table 1. Statistics of three TREC collections and corresponding query information

Collections	No. of documents	Topics	No. of short queries (length <= 6)	No. of long queries (length > 6)
WT10	1,692,096	451–550	60	40
GOV2	25,205,179	701–850	78	71
ClueWeb09B	50,220,423	1–150	125	23

Algorithm 1 shows the process of query reformulation. It starts with some noun phrases and other semantic elements extracted from the original query as initial elements, and set C is used to contain all the updated candidates. There are three parameters: k, m, and n. In line 4, for each term in C, we seek top-k most similar terms in word2vec and generate k new element rewrites. In line 5, we seek top-m similar terms in word2vec and generate m new ones accordingly. With several rounds of search for more suitable elements, we obtain a group of elements with their associated weights.

4 Experiments

Experiments were conducted on three TREC collections: WT10G, GOV2 and Clue-Web09B (a subset of ClueWeb09 with data in Category B). Additionally, for Clue-Web09B, the Waterloo Fusion spam[4] scores were used for removing spam documents with the percentile-score of less than 70. We used all the topics related to each collection and treated the description of each topic as a verbose query. After removing stop words, those topics with more than 6 words are regarded as long queries; otherwise, they are regarded as short queries. Information about the three collections and queries is given in Table 1. Note that one query in GOV2 and two queries in Clue-Web09B have been removed because they do not have relevant documents.

We used Indri[5] as the retrieval software to index documents and search results for queries. A language model is adopted where the Dirichlet prior is 1500 and the normalized mutual information threshold is set as $\tau = 0.0001$. A standard list of English stop words is maintained for query processing. Additionally, in line with the treatment carried out in [14], we remove some extra stop words, such as "find", "information", and "documents", for queries before the retrieval process. Three trained word embedding models including weod2vec, GloVe, and ConceptNet Numberbatch (CNN) were used for the proposed method WE^2R. For evaluation, we use three metrics to measure retrieval effectiveness: mean average precision (MAP), precision at 5 document level (P@5), and precision at 10 document level (P@10). Statistical significance in terms of retrieval performance is judged by the two-tailed t-test.

Given a certain query, we extracted initial query elements by MontyLingua[6], which is a natural language procession tool. In the process of generating more query rewrites, we used the pre-trained word2vec [7] model conducted on the Google news dataset. It contains 300-dimensional vectors for 3 million words and phrases. The element rewrites are stored in set C, and at any stage, we keep a maximal number of 10 elements in it ($t = 10$). For term substitution and expansion, new words are obtained from word2verc. Here, we set $k = 3$ new words for term addition and $m = 2$ new words for term substitution. n is set to 5, which means that we choose five best elements to be

---

[4] http://plg.uwaterloo.ca/ ~ gvcormac/clueweb09spam.

[5] http://www.lemurproject.org.

[6] https://pypi.org/project/MontyLingua.

in the final query. The number of terms in each element is limited to up to 6, and an element with 6 or more words will not be expanded anymore.

4.1 Performance Evaluation of the Proposed Method

In this section, we present the performance of our method WE2R with some baseline methods. As baseline methods, we use the title part and description part of the TREC topics as queries for retrieval. In line with KeyConcept queries in [1], we also take a similar approach. Apart from the description part, some noun phrases and other semantic elements extracted from it and the top two are used as a part of the final query. Note that the key concept queries are different from their counterparts in [1]: key concepts are extracted in different ways and the weighting schemes used are also different. As for WE2R, we use Eq. 3 and Eq. 4 to calculate weights for the elements selected.

All of the above query models are represented by the Indri query language with #combine and #weight operations [15]. Figure 2 shows an example, which is the reformatted query for Topic 486. The initial query (description with stop words removed) and a group of selected extracted elements are combined with weights of 0.8 and 0.2 respectively. On the one hand, a weight of 0.8 for the original query stabilizes the performance; on the other hand, a weight of 0.2 for extracted and reformulated elements may lead to better performance in many cases. Note we did not try to optimize the weights so as to obtain better results in our experiments.

KeyConcept:
#weight(0.8 #combine(Eldorado El Dorado Casino reportedly located Reno address)
         0.2 #weight(0.5029 #combine(El Dorado)
                      0.4971 #combine(Reno)))
WE^{2}R:
#weight(0.8 #combine(Eldorado E1 Dorado Casino reportedly located Reno address)
         0.2 #weight(0.2171#combine(E1 Dorado)
                      0.2146 #combine(Reno)
                      0.1937 #combine(Eldorado Mesquite)
                      0.1924 #combine(Casino)
                      0.1822 #combine(Reno Las Vegas)))

Fig. 2. A query example of KeyConcept and WE^{2}R

In order to generate the final query, WE2R may go through multiple rounds of reformulation. After some trials, we find that in many cases the first iteration leads to good results, and more round of iteration is not helpful. Another option is the number of elements chosen in the final query. We tried six different options from one to six. It is found that five elements can achieve the best results. Therefore, we first report results of WE2R with such a setting in Tables 2, 3, and 4. More details about different options will be discussed later.

In Table 2, we can make some observations in general. Comparing Title with Desc, the former is better than the latter in two out of three data sets. Comparing Desc with KeyConcept, we find that KeyConcept is better in most cases. Query reformulation with three word embedding models works better than both Desc and KeyConcept in all the cases. Especially, the word2vec model is the best. The other two models can make modest improvement over the KeyConcept approach.

We also compare the performance of our approach with that of a number of baseline approaches including QL, SDM, KC, CRF-QL, and RTree. These baseline methods represent the state-of-the-art technology for long query processing. From Table 3, we can see that WE2R(word2vec) is the best in most cases.

In the following discussion we focus on WE2R (word2vec). It would be interesting to see the effect of query term reformation on short and long queries separately. Because each query is different from the others, it does not make much sense to compare the average performance of these two groups for any particular method such as WE2R. Instead, we look at the improvement rates of WE2R over Desc. We believe such a treatment makes them comparable. Figure 3 and 4 show the improvement rates of WE2R over Desc for these two groups of queries. We find that WE2R can make more improvement over Desc for long queries than for short queries.

4.2   Query Element Reformulation with Different Number of Iterations

In Sect. 4.1, we present the results of WE2R, which is from the first-round query element reformulation. In fact, multiple rounds can be carried out for that and the results are shown in Table 4. As in Sect. 4.1, top five element rewrites are chosen. From Table 4, we can see that in all but one case WE2R achieves the best performance in the first round. The only exception is: it performs best in P@10 on WT10G in the second round of reformulation. It indicates that with more rounds of iteration, it is possible that more irrelevant or not qualified element rewrites may be included or more relevant terms may be removed, which will hurt the performance of WE2R. Therefore, we conclude that just one round of reformulation is enough. In a sense, this is good news because more rounds mean lower efficiency. It is preferable to achieve good performance quickly.

4.3   Alternative Similarity Score and Weight Functions

In this section, we go a step further to see how different alternatives can be used to calculate similarity scores for a query element in Eq. 1 and weights for a query element in Eq. 3. Therefore, the experimental procedure is similar to the one in Sect. 4.1 and we choose the top five element rewrites from the first iteration. The only difference is one of those two functions is changed and all the others stay the same.

Equation 3 can be rewritten as $Weight(r) = Sim * Spec$. We may try other options such as $Sim^2 * Spec$ and $Sim * Spec^2$. Equation 1 is $SimilarityScore(r) = sim(r, r_p) * log\left(\frac{sim(r,Q)}{sim(r_p,Q)}\right)$, we may use $sim(r, r_p)$, $sim(r, Q)$, $sim(r, r_p) * \frac{sim(r,Q)}{sim(r_p,Q)}$, and $\frac{sim(r,Q)}{sim(r_p,Q)}$ to replace the right side of Eq. 1. Tables 5 and 6 show the results of them.

From Table 5, we can see that both similarity (Sim) and specificity (Spec) have positive impact for weighing query element rewrites. This can be confirmed by comparing the result of Desc with that of Sim and Spec. The combination of them is shown in three cases. On average and in most cases, Sim*Spec is more effective than all other alternatives. Overall, the results show that the weight function used in Eq. 3 is more effective than other alternatives.

From Table 6, we can see that sim(r,rp) * log(sim(r,Q)/sim(rp,Q)) is the most effective similarity score function on average. Therefore, we conclude that the selection function we used in Eq. 1 is a good choice for selecting query element rewrites.

Table 2. Retrieval performance for all the queries in three collections (best performance is shown in bold and improvement rate is shown over *Desc*)

Collections	Query models	MAP	P@5	P@10
WT10G	Title	0.1925	0.3367	0.2653
	Desc	0.1811	0.3720	0.3250
	KeyConcept	0.1985 (+09.61%)	0.3900 (+04.84%)	0.3360 (+03.38%)
	WE^2R (word2vec)	**0.2238** (+23.58%)	**0.4120** (+10.75%)	**0.3530** (+08.62%)
	WE^2R (GloVe)	0.2187 (+20.76%)	0.4040 (+08.60%)	0.3520 (+08.31%)
	WE^2R (CNN)	0.2050 (+13.20%)	0.4000 (+07.53%)	0.3450 (+06.15%)
GOV2	*Title*	**0.2957**	0.5651	0.5369
	Desc	0.2296	0.5072	0.4980
	KeyConcept	0.2688 (+17.08%)	0.5409 (+06.64%)	0.5148 (+03.37%)
	WE^2R (word2vec)	0.2718 (+18.38%)	**0.5732** (+13.01%)	**0.5463** (+09.70%)
	WE^2R (GloVe)	0.2602 (+13.33%)	0.5463 (+07.71%)	0.5208 (+04.57%)
	WE^2R (CNN)	0.2647 (+15.29%)	0.5517 (+08.77%)	0.5302 (+06.47%)
ClueWeb09B	*Title*	**0.1091**	0.3095	0.3007
	Desc	0.0794	0.2770	0.2743
	KeyConcept	0.0809 (+01.89%)	0.2784 (+00.51%)	0.2730 (−00.10%)
	WE^2R (word2vec)	0.0861 (+08.44%)	**0.3108** (+12.20%)	**0.3095** (+12.83%)
	WE^2R (GloVe)	0.0811 (+02.14%)	0.2946 (+06.35%)	0.2919 (+06.42%)
	WE^2R (CNN)	0.0845 (+06.83%)	0.2824 (+01.95%)	0.2784 (+01.49%)

Table 3. Performance comparison with a group of state-of-the-art methods (See Table 4 in [19] for performance of all baseline methods. Non-stemmed index was used for all the methods.)

Collections	Query models	MAP	P@10
WT10G	*QL*	0.1643	0.2897
	SDM	0.1676	0.3165
	KC	0.1746	0.3082
	CRF-QL	0.1681	0.2979
	CRF-SDM	0.1825	0.3113
	RTree	0.1944	0.3402
	WE²R (word2vec)	**0.2238**	**0.3530**
GOV2	*QL*	0.2246	0.4913
	SDM	0.2398	0.5101
	KC	02488	0.5087
	CRF-QL	0.2336	0.5081
	CRF-SDM	0.2482	0.5336
	RTree	0.2670	0.5396
	WE²R (word2vec)	**0.2718**	**0.5463**
ClueWeb09B	*QL*	0.1096	0.2163
	SDM	0.1152	0.2276
	CRF-QL	0.1100	0.2184
	CRF-SDM	0.1153	0.2194
	RTree	**0.1294**	0.2643
	WE²R (word2vec)	0.0861	**0.3095**

Table 4. Performance of WE^2R with multiple rounds of rewrites (best performer shown in bold)

Collections	Round	MAP	P@5	P@10
WT10G	0	0.1908	0.3840	0.3240
	1	**0.2238**	**0.4120**	0.3530
	2	0.2142	0.4080	**0.3550**
GOV2	0	0.2459	0.5345	0.5083
	1	**0.2721**	**0.5758**	**0.5463**
	2	0.2587	0.5436	0.5235
ClueWeb09B	0	0.0732	0.3527	0.3196
	1	**0.0997**	**0.3708**	**0.3291**
	2	0.0956	0.3651	0.3195

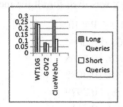   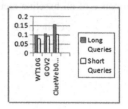

Fig. 3. Improvement rates of WE^2R over *Desc* (long queries vs short queries, MAP)

Fig. 4. Improvement rates of WE^2R over *Desc* (long queries vs short queries, P@10)

Table 5. Performance of WE^2R with alternative weighting functions

Collections	Weight function	MAP	P@5	P@10
WT10G	*Desc*	0.1811	0.3720	0.3250
	Sim	0.2062	0.3800	0.3300
	Spec	0.2165	0.4100	0.3460
	*Sim*Spec*	**0.2346**	0.4241	**0.3758**
GOV2	*Desc*	0.2296	0.5072	0.4980
	Sim	0.2647	0.5717	0.5455
	Spec	0.2745	0.5758	0.5423
	*Sim*Spec*	**0.2890**	**0.5878**	**0.5576**
ClueWeb09B	*Desc*	0.0669	0.323	0.2885
	Sim	0.0851	0.3665	0.3358
	Spec	0.0907	0.3595	0.3281
	*Sim*Spec*	**0.1026**	**0.3824**	**0.3377**

Table 6. Performance of WE^2R with alternative Selection Score functions

Collections	SelectionScore function	MAP	P@5	P@10
WT10G	$sim(r,r_p)$	0.2286	0.4198	0.3581
	$sim(r,Q)$	0.2263	0.4226	0.3593
	$sim(r,Q)/sim(r_p,Q)$	0.2366	0.4123	0.3647
	$sim(r,r_p)*sim(r,Q)/sim(r_p,Q)$	**0.2390**	0.4198	0.3490
	$sim(r,r_p)*log(sim(r,Q)/sim(r_p,Q))$	0.2346	**0.4241**	**0.3758**
GOV2	$sim(r,r_p)$	0.2808	0.5477	0.5462
	$sim(r,Q)$	0.2861	0.5857	0.5402
	$sim(r,Q)/sim(r_p,Q)$	0.2749	0.5784	0.5349
	$sim(r,r_p)*sim(r,Q)/sim(r_p,Q)$	0.2708	0.5463	0.5375
	$sim(r,r_p)*log(sim(r,Q)/sim(r_p,Q))$	**0.2890**	**0.5878**	**0.5576**
ClueWeb09B	$sim(r,r_p)$	0.0774	0.3724	0.3203
	$sim(r,Q)$	0.0757	0.3676	0.3142
	$sim(r,Q)/sim(r_p,Q)$	0.0837	0.3611	0.32
	$sim(r,r_p)*sim(r,Q)/sim(r_p,Q)$	0.0776	0.3743	0.3289
	$sim(r,r_p)*log(sim(r,Q)/sim(r_p,Q))$	**0.1026**	**0.3824**	**0.3377**

5   Conclusions and Future Work

In this paper, we have proposed a word embedding-based query reformulation method in supporting of long queries. The process of query reformulation includes three basic operations: deletion, substitution, and expansion. Word embedding was mainly used to measure the semantic similarity between query elements and/or the initial long query. Therefore, some good candidates can be chosen for term deletion, replacement or expansion. Compared with those baseline methods involved, the proposed method obtains considerably better retrieval performance. It demonstrates that our approach is effective and promising.

In this piece of work, we used word2vec, GloVe, and ConceptNet Numberbatch, each of which was trained by using a very different corpus from the ones we use for our experiments. If the same or similar corpora are used for both training and testing, then better results can be expected. Another direction is to combine these different word-embeddings by some fusion methods [23] to improve search performance. These remain to be our future work.

References

1. Bendersky, M., Bruce Croft, W.: Discovering key concepts in long queries. In: Proceedings of SIGIR 2008, pp. 491–498 (2008). https://doi.org/10.1145/1390334.1390419
2. Huston, S., Bruce Croft, W.: Evaluating verbose query processing techniques. In: Proceedings of SIGIR 2010, pp. 291–298 (2010). https://doi.org/10.1145/1835449.1835499
3. Park, J.H., Bruce Croft, W.: Query term ranking based on dependency parsing of long queries. In: Proceedings of SIGIR 2010, pp. 829–830 (2010). https://doi.org/10.1145/1835449.1835637
4. Xue, X., Bruce Croft, W.: Modeling subset distributions for long queries. In: Proceedings of SIGIR 2011, pp. 1133–1134 (2011). https://doi.org/10.1145/2009916.2010085
5. Maxwell, K.T., Bruce Croft, W.: Compact query term selection using topically related text. In: Proceedings of SIGIR 2013, pp. 583–592 (2013). https://doi.org/10.1145/2484028.2484096
6. Yang, B., Parikh, N., Singh, G.: A study of query term deletion using large-scale e-commerce search logs. In: Proceedings of ECowlIR 2014, pp. 235–246 (2014). https://doi.org/10.1007/978-3-319-06028-6_20
7. Mikolov, T., Chen, K., Corrado, G., Dean, J.: Efficient estimation of word representations in vector space. In: Proceedings of Workshop at ICLR (2013)
8. Zheng, G., Callam, J.: Learning to reweight terms with distributed representations. In Proceedings of SIGIR 2015, pp. 575–584. https://doi.org/10.1145/2766462.2767700
9. Zamani, H., Bruce Corft, W.: Embedding-based query language models. In: Proceedings of ICTIR 2016, pp. 147–156 (2016). https://doi.org/10.1145/2970398.2970405
10. Fernández-Reyes, F., Valadez, J., Montes-y-Gómez. M: A prospect-guided global query expansion strategy using word embeddings. Inf. Process. Manage. 54(1), 1–13 (2018). https://doi.org/10.1016/j.ipm.2017.09.001
11. Grbovic, M., Djuric, N., Radosavljevic, V., Silvestri, F., Bhamidipati, N.: Context- and content-aware embeddings for Query Rewriting in Sponsored Search. In: Proceedings of SIGIR 2015, pp. 383–392 (2015). https://doi.org/10.1145/2766462.2767709

12. Bendersky, M., Metzler, D., Bruce Croft, W.: Learning concept importance using a weighted dependence model. In: Proceedings of WSDM 2010, pp. 31–40 (2010). https://doi.org/10.1145/1718487.1718492
13. Xue, X., Huston, S., Bruce Croft, W.: Improving long queries using subset distribution. In: Proceedings of CIKM 2010, pp. 1059–1068 (2010). https://doi.org/10.1145/1871437.1871572
14. Xue, X., Tao, Y., Jiang, D., Li, H.: Automatically mining question reformulation patterns from search log data. In: Proceedings of ACL (2), pp. 187–192 (2012). https://doi.org/10.5555/2390665.2390712
15. Zamani, H., Bruce Croft, W.: Relevance-based word embedding. In: Proceedings of SIGIR 2017, pp. 505–514 (2017). https://doi.org/10.1145/3077136.3080831
16. Gupta, M., Bendersky, M.: Information retrieval with verbose queries. Found. Trends Inf. Retriev. 9(3–4), 91–208 (2015). https://doi.org/10.1561/9781680830453
17. Bagheri, E., Ensan, F., Al-Obeidat, F.: Neural word and entity embeddings for ad hoc retrieval. Inf. Process. Manage. 54(4), 657–673 (2018). https://doi.org/10.1016/j.ipm.2018.04.007
18. Amer, N O., Mulhem, P., Gery, M.: Toward word embedding for personalized information retrieval. CoRR abs/1606.06991 (2016)
19. Xue, X., Bruce Croft, W.: Generating reformulation trees for complex queries. In: Proceedings of SIGIR 2012, pp. 525–534 (2012). https://doi.org/10.1145/2348283.2348355
20. Liu, X., Nie, J.-Y., Sordoni, A.: Constraining word embeddings by prior knowledge – application to medical information retrieval. In: Ma, S., et al. (eds.) AIRS 2016. LNCS, vol. 9994, pp. 155–167. Springer, Cham (2016). https://doi.org/10.1007/978-3-319-48051-0_12
21. El Mahdaouy, A., El Alaoui Ouatik, S., Gaussier, E.: Word-embedding-based pseudo-relevance feedback for Arabic information retrieval. J. Inf. Sci. 45(4) (2019). https://doi.org/10.1177/0165551518792210
22. Bae, K., Ko, Y.: Efficient question classification and retrieval using category information and word embedding on cQA services. J. Intell. Inf. Syst. 53(1), 27–49 (2019). https://doi.org/10.1007/s10844-019-00556-x
23. Zhang, Z., Xu, C., Wu, S.: Evaluation of Score standardization methods for web search in support of results diversification. In: Proceedings of WISA 2018, pp. 182–190 (2018). https://doi.org/10.1007/978-3-030-02934-0_17

DSQA: A Domain Specific QA System for Smart Health Based on Knowledge Graph

Ming Sheng[1], Anqi Li[2(✉)], Yuelin Bu[3], Jing Dong[4], Yong Zhang[1], Xin Li[5], Chao Li[1], and Chunxiao Xing[1]

[1] BNRist, DCST, RIIT, Tsinghua University, Beijing 100084, China
{shengming,zhangyong05,li-chao,xingcx}@tsinghua.edu.cn
[2] Beihang University, Beijing 100191, China
anqili99@hotmail.com
[3] Beijing University of Post and Telecommunications, Beijing 100876, China
graceyuelin.bu@gmail.com
[4] University of Queensland, Brisbane, QLD 4072, Australia
j.dong1@uq.net.au
[5] Beijing Tsinghua Changgung Hospital, School of Clinical Medicine,
Tsinghua University, Beijing 102218, China
Horsebackdancing@sina.com

Abstract. In recent years, medical Question Answering (QA) systems have gained a lot of attentions, since they can not only help professionals make quick decisions, but also give ordinary people advice when they seek for helpful information. However, the interpretability and accuracy problems have plagued QA systems for a long time. In this paper, we propose a QA system, DSQA, based on knowledge graph for answering domain-specific medical questions. The data we use are from Electronic Medical Records (EMRs) that are complex and heterogeneous with varied qualities. We introduce doctor-in-the-loop mechanism to design the knowledge graph, which improves the interpretability and the accuracy of our QA system. For every natural-language question, we generate the answer with a hybrid of the traditional deep learning model and the reasoning on the knowledge graph. Finally, the answer is returned to the user in the format of a subgraph with corresponding natural-language sentences. Our system shows high interpretability and gives excellent performance according to the experiment results.

Keywords: QA System · Causality knowledge graph · Doctor-in-the-loop · EMR

1 Introduction

With the widespread use of online healthcare systems, the role of medical QA systems is becoming increasingly important. QA system is a type of system in which a user asks a question using natural language, and the system provides a concise and correct answer [1]. The goal of QA is to automatically return

© Springer Nature Switzerland AG 2020
G. Wang et al. (Eds.): WISA 2020, LNCS 12432, pp. 215–222, 2020.
https://doi.org/10.1007/978-3-030-60029-7_20

answers according to corresponding questions [17]. To this end, it requires to understand the semantics of web pages that are widely explored in studies related to text search [16] and understanding. QA system not only saves people time to retrieve information, but also provides people with a more accurate and flexible interface [9].

Among QA's application domains, medical QA is a special kind of QA that is involved with medical or clinical knowledge [5]. There are a few challenges faced by medical QA systems: how to build the QA system with high quality dataset to improve the accuracy of the system; and how to integrate doctors' prior knowledge to improve the interpretability of the system.

In recent years, knowledge graph comprising vast amounts of facts [10] has been made extensive use in question answering. A knowledge graph is a graph with concepts as nodes and their relations as edges [6,14]. Using knowledge graphs, distributed medical data sources can be aggregated into one meaningful data source [2].

In this work, we mainly consider the following aspects to deal with the challenges faced by QA systems.

1. We choose EMRs as our dataset due to their high quality compared to other datasets and their domain-specific features.
2. To improve our system's interpretability, we introduce doctor-in-the-loop mechanism to build the knowledge graphs.
3. For the answers, query subgraphs are returned to the user with natural-language answers to demonstrate relationships between concepts.

This paper is organized as follows: Sect. 2 discusses the related work. Section 3 introduces the framework of the QA system we built. Section 4 discusses the use case. Section 5 concludes our work and proposes some future directions.

2   Related Work

In Table 1 , we list some existing QA systems and compare their attributes in turn. For the data sources, some QA health systems collect medical data from Internet [15]. Compared with that, using EMRs which can be traced and used to support clinical decision-making is much more reliable [11]. The other general QA systems like CSQA [13] tend to use large scale public datasets.

As we can see from the table, traditional QA systems like HHH [4] and DIK-QA [8] usually provide text information as final output, which makes it a little bit difficult for users to understand the relationships of entity nodes contain in results. In order to make the query output easier for users to comprehend, DSQA we form is capable of providing the shape of causality knowledge graph in the output interface. Moreover, DSQA is medical domain-specific. Compared with general QA systems like CSQA [13], it provides more accurate and professional answers for users.

While there is a lack of prior knowledge in most of the existing QA systems, DSQA applies doctor-in-the-loop mechanism. With the help of doctors, the answers provided prove to be more reasonable. This significantly improves DSQA's interpretability.

Table 1. QA Systems.

Name	Data source	Medical domain specific	Application of prior knowledge	Answer representation
QAnalysis [12]	EMR	✓	✗	Tables and charts
KnowHealth [7]	Web data, online encyclopedia, literature	✓	✓	Answers and visual results to display relationships between the data
HHH [4]	Medical data collected from Internet	✓	✗	Natural-language answers
DIK-QA [8]	Chinese EMR	✓	✗	Natural-language answers
CSQA [13]	Wikidata	✗	✗	Natural-language answers
Data+oracle approach QA [19]	QALD dataset	✗	✓	Natural-language answers
DSQA	EMR	✓	✓	Natural-language answers and knowledge graphs

3 DSQA System

In this paper we propose DSQA, a domain-specific medical QA system based on knowledge graphs.

3.1 Causality Knowledge Graph

The causality knowledge graph is a knowledge graph with edges indicates causal relationships related to the QA process. The causal relationships are extracted and redefined from the relationships of the knowledge graph with doctors' prior knowledge. The causality knowledge graph contains information about causes and consequences of abundant examples.

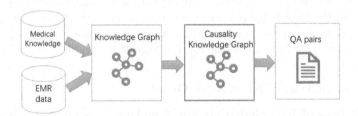

Fig. 1. The dataflow of causality knowledge graph.

The dataflow of the causality knowledge graph is shown in Fig. 1. The causality knowledge graph is built with causality-related data from the knowledge graph. Then the QA pairs are generated from the causality knowledge graph to train the machine learning models for the retrieval. The causality knowledge graph efficiently helps with the answer generation process.

3.2 Framework of DSQA

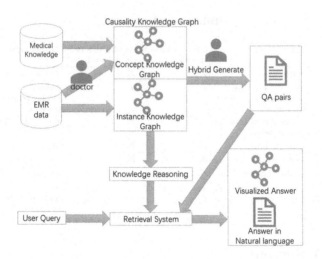

Fig. 2. The framework of DSQA.

Figure 2 shows the overview of the DSQA. We integrate the doctors' prior knowledge and apply some medical knowledge into the construction process of our concept knowledge graph. Then the instance data in EMRs are converted to the instance knowledge graph. The causality knowledge graph is generated based on the knowledge graph with doctors' prior knowledge to help redefine the casual relationships. It displays the causal relationships between concepts, and the QA pairs dataset can be generated from it according to different types of problems to train different models.

For the question process, we use a hybrid of the traditional machine learning method and the knowledge reasoning process based on the knowledge graph to generate answers. For the answer format, the user will not only get the answer in natural-language but also in the form of knowledge graph.

3.3 Construction of Knowledge Graph

Construction of Knowledge Graph. The knowledge graph consists of two parts: the concept knowledge graph and the instance knowledge graph.

The concept knowledge graph displays the schema of our data. The nodes of the concept knowledge graph are classes, and the property edges display relations between the classes. To improve our system's completeness, we also applied other medical knowledge like UMLS. We first invite doctors to give rules to convert the classes into triples. The experts have abundant prior knowledge about the relationships of the concepts in the certain medical domain, so they can well determine the structure of the concept graph.

The instance knowledge graph is the instantiation of the concept knowledge graph with data. Nodes of the instance knowledge graph are specific entities like drugs or diseases, and the edges indicates relationships between the entities.

Construction of Causality Knowledge Graph. The causality knowledge graph is built based on the knowledge graph. Doctors redefine the causal relationships based on the knowledge graph we have built, and conduct the construction of the causality knowledge graph with the causal relationships they defined and the related concepts. The causality knowledge graph demonstrates causal relationships clearly.

3.4 QA Pairs Generation

We generate QA pairs from the causality knowledge graph we built. Doctors first determine several question types that have different features and result types, for example, the hospitalization expenses and the risk of readmission. We should use different functions to deal with different kind of problems, thus we need different datasets to train them. Doctors will select data of corresponding features from the causality knowledge graph according to the specific question type. They also determine the hop. Then the QA pairs can be generated automatically from the causality knowledge graph. The QA pair dataset can be used to train corresponding machine learning models that serve questions in certain aspect.

3.5 Answer Retrieval

We use a hybrid method to generate the answer. Specifically, we first train a model with the QA pairs generated from the causality knowledge graph. For every natural language question, the system will try both the traditional deep learning model and the query method based on the causality graph we build, and provide two answers for the user to choose.

For the traditional method, we apply both support vector machine (SVM) and the sequence-to-sequence deep learning model. Specifically, we use SVM model to do the classification for the discrete or continuous variables, while using the sequence-to-sequence model to deal with the texts.

Another method is to conduct a structured query based on the knowledge graph. Figure 3 shows the query process. We intend to translate natural language questions submitted by users into SPARQL query statements. First, the system will execute word segmentation using Jieba word segmentation tool. We also added medical dictionaries to the tool to help improve its reliability. On the other hand, the system converts the questions into vectors, input to the CNN model to classify questions, thus find the best-fit predefined search pattern for the query. Then it will execute the SPARQL query and return the results to the users.

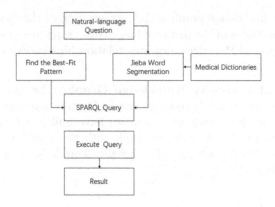

Fig. 3. The detailed workflow of the query process.

3.6   QA Presentation

The answer presentation format we design is a subgraph with corresponding natural-language sentences as shown in Fig. 4. The knowledge graph shown is a subgraph of the knowledge graph we built, which shows the reasoning process behind the answer and thus makes the answer reasonable. Moreover, if the user clicks on one node, the website will return to another subgraph connected the node and give the introduction about the concept node. This will help the user know more information about the concepts related to his or her question.

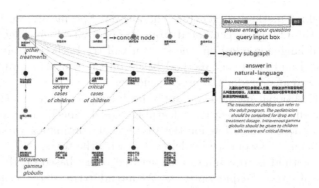

Fig. 4. The presentation of answers.

4   Use Case

According to the data and guidelines provided by the cooperative hospital, we build a QA system for the Novel Coronavirus Pneumonia (NCP) domain based on the methods described in Fig. 5. We use EMRs as our data source. When users enter some questions about NCP, the QA system will also generate some related questions for users to choose. Users can receive answers about their questions related to NCP. They can also check the causality knowledge graph on the website, which contains information about diagnosing and treating methods. By looking through the information we give on the QA system, users are sure to benefit a lot especially on guarding against the NCP.

Fig. 5. Demonstration of a use case.

5   Conclusion and Future Works

In this paper, we propose a domain-specific QA system, DSQA, based on knowledge graph for answering medical questions. There are still some questions remaining to be explored. Reducing more labor cost and constructing knowledge graphs much more rapidly by doctors to help diagnosis are sure to be promising topics. In addition, the unstructured data like clinical notes in EMRs also can be used to enrich the knowledge graph by automatically annotation. We also consider extending the methodology to deal with more data types in different applications such as streaming [3] and spatial [18].

Acknowledgments. This work was supported by NSFC (91646202), National Key R&D Program of China (2018YFB1404401, 2018YFB1402701).

References

1. Alqifari, R.: Question answering systems approaches and challenges. In: Student Research Workshop, pp. 69–75 (2019). https://doi.org/10.26615/issn.2603-2821.2019_011
2. Ansong, S., Eteffa, K.F., Li, C., Sheng, M., Zhang, Y., Xing, C.: How to empower disease diagnosis in a medical education system using knowledge graph. In: Ni, W., Wang, X., Song, W., Li, Y. (eds.) WISA 2019. LNCS, vol. 11817, pp. 518–523. Springer, Cham (2019). https://doi.org/10.1007/978-3-030-30952-7_52
3. Ao, X., Shi, H., Wang, J., Zuo, L., Li, H., He, Q.: Large-scale frequent episode mining from complex event sequences with hierarchies. ACM TIST **10**(4), 36:1–36:26 (2019). https://doi.org/10.1145/3326163

4. Bao, Q., Ni, L., Liu, J.: HHH: an online medical chatbot system based on knowledge graph and hierarchical bi-directional attention. In: Proceedings of the Australasian Computer Science Week Multiconference, pp. 1–10 (2020). https://doi.org/10.1145/3373017.3373049

5. He, J., Fu, M., Tu, M.: Applying deep matching networks to chinese medical question answering: a study and a dataset. BMC Med. Inform. Decis. Mak. **19**(2), 52 (2019). https://doi.org/10.1186/s12911-019-0761-8

6. Huang, X., Zhang, J., Li, D., Li, P.: Knowledge graph embedding based question answering. In: WSDM, pp. 105–113 (2019). https://doi.org/10.1145/3289600.3290956

7. Li, C., Hang, S., Chu, D., Zheng, H., Hu, X.: Knowhealth: a knowledge graph based question-answer platform for elderly people. In: EAI (2019). https://doi.org/10.4108/eai.29-6-2019.2282866

8. Li, M., Huang, M., Zhang, Y., Feng, W.: A DIK-based question-answering architecture with multi-sources data for medical self-service (S). In: Perkusich, A. (ed.) SEKE, pp. 1–10 (2019). https://doi.org/10.18293/SEKE2019-112

9. Liu, H., Hu, Q., Zhang, Y., Xing, C., Sheng, M.: A knowledge-based health question answering system. In: Chen, H., Zeng, D.D., Karahanna, E., Bardhan, I. (eds.) ICSH 2017. LNCS, vol. 10347, pp. 286–291. Springer, Cham (2017). https://doi.org/10.1007/978-3-319-67964-8_29

10. Lukovnikov, D., Fischer, A., Lehmann, J., Auer, S.: Neural network-based question answering over knowledge graphs on word and character level. In: WWW, pp. 1211–1220 (2017). https://doi.org/10.1145/3038912.3052675

11. Pampari, A., Raghavan, P., Liang, J., Peng, J.: emrQA: a large corpus for question answering on electronic medical records. arXiv preprint arXiv:1809.00732 (2018). https://doi.org/10.18653/v1/d18-1258

12. Ruan, T., Huang, Y., Liu, X., Xia, Y., Gao, J.: Qanalysis: a question-answer driven analytic tool on knowledge graphs for leveraging electronic medical records for clinical research. BMC Med. Inform. Decis. Mak. **19**(1), 82 (2019). https://doi.org/10.1186/s12911-019-0798-8

13. Saha, A., Pahuja, V., Khapra, M.M., Sankaranarayanan, K., Chandar, S.: Complex sequential question answering: towards learning to converse over linked question answer pairs with a knowledge graph. In: AAAI (2018)

14. Sheng, M., et al.: DEKGB: an extensible framework for health knowledge graph. In: Chen, H., Zeng, D., Yan, X., Xing, C. (eds.) ICSH 2019. LNCS, vol. 11924, pp. 27–38. Springer, Cham (2019). https://doi.org/10.1007/978-3-030-34482-5_3

15. Tian, Y., Ma, W., Xia, F., Song, Y.: Chimed: a Chinese medical corpus for question answering. In: BioNLP Workshop and Shared Task, pp. 250–260 (2019). https://doi.org/10.18653/v1/w19-5027

16. Wu, J., Zhang, Y., Wang, J., Lin, C., Fu, Y., Xing, C.: Scalable metric similarity join using mapreduce. In: ICDE, pp. 1662–1665 (2019). https://doi.org/10.1109/ICDE.2019.00167

17. Zhang, M., Tian, G., Zhang, Y.: A home service-oriented question answering system with high accuracy and stability. IEEE Access **7**, 22988–22999 (2019). https://doi.org/10.1109/ACCESS.2019.2894438

18. Zhao, K., et al.: Discovering subsequence patterns for next POI recommendation. In: IJCAI, pp. 3216–3222 (2020). https://doi.org/10.24963/ijcai.2020/445

19. Zheng, W., Cheng, H., Zou, L., Yu, J.X., Zhao, K.: Natural language question/answering: let users talk with the knowledge graph. In: CIKM, pp. 217–226 (2017). https://doi.org/10.1145/3132847.3132977

Natural Language Processing

A Text Representation Model Based on Convolutional Neural Network and Variational Auto Encoder

Canyang Guo[1], Lin Xie[1], Genggeng Liu[1(✉)], and Xin Wang[2]

[1] College of Mathematics and Computer Science, Fuzhou University, Fuzhou, China
Canyangguo@163.com, xlin540@163.com, liugenggeng@fzu.edu.cn
[2] College of Intelligence and Computing, Tianjin University, Tianjin, China
wangx@tju.edu.cn

Abstract. In the era of big data, the internet produces vast amounts of data every day, among which text data occupies the main position. It is difficult for manual processing to deal with the increasing growth rate of text data. As basis of most natural language processing (NLP) tasks, text representation aims to transform text into a vector that can be processed by computer without losing the original important semantic information. It has become an important research direction in the field of NLP that effectively organize, manage and quickly use the complex text information to extract useful semantics from it. Therefore, a text feature representation model based on convolutional neural network (CNN) and variational auto encoder (VAE) is proposed to extract the text features and apply the obtained text feature representation to text classification scene. CNN is used to extract local features and VAE makes the extracted features more consistent with Gaussian distribution. The proposed method has best performance compared with w2v-avg and CNN-AE in k-nearest neighbor (KNN), random forest (RF) and support vector machine (SVM) classification algorithms.

Keywords: Natural language processing · Variational auto encoder · Convolutional neural network · Text representation

1 Introduction

In recent year, neural network algorithm is widely used in NLP with the development of computer processing power. The common tasks of NLP [2] using neural network model contain automatically generating abstract, part-of-speech tagging and named entities recognition, Chinese word segmentation, emotion analysis, etc [10,16]. Text modeling for different granularity (including words, sentence, paragraph and document) are built in a progressive way, which have

This work was supported in part by National Natural Science Foundation of China (No. 61877010, 11501114), Natural Science Foundation of Fujian Province, China (2019J01243).

© Springer Nature Switzerland AG 2020
G. Wang et al. (Eds.): WISA 2020, LNCS 12432, pp. 225–235, 2020.
https://doi.org/10.1007/978-3-030-60029-7_21

some connections in the local scope and differences in the overall model [14]. The text representation of word granularity is the basis of text representation [13,19]. The text representation method of sentence granularity is obtained by combining word vectors according to certain rules. As for the text representations of paragraph and document, they can be obtained by the method of sentence granularity or cascade method. The traditional method of word vector representation is one-hot. The dimension of word vector representation is the same as the number of all different words in the corpus. The vector position corresponding to words is set to 1 and the rest is set to 0. For a corpus composed of n different words, each word vector dimension after unique coding is n dimension. The dimension of the word vector generated by this method results in the high dimension and sparsity for the word vector is directly proportional to the size of the corpus, which makes it difficult to obtain semantic information. Moreover, it is unable to calculate the correlation between different words with similar meanings by coding in the way. Considering the problem of word vector representation, a word embedding (i.e. distributed representation) method based on neural network language model (NNLM) is proposed. The method employs the probability of predicting the next word by the current word to develop neural network. This model is mainly for the prediction task of words and the word vector is the byproduct of the model. However, how to apply neural network to train word vector has become research hotspot. Mikolov et al. proposed word vector training model word to vector (word2vector), as the term suggestions, the main function of this model is to transform text into word vector through neural network [15]. Ji et al. proposed another word vector training model wordrank which calculate the similarity of words and other different standards to get word vector representation and is clever at representation of similar words [8]. Considering the complexity of the information interaction platform in the internet, Dhingra et al. proposes the tweet2vec model, which is a distributed representation model based on character combination in the complex social media environment [3]. Hill et al. considered that there is an optimal representation method for different application tasks and proposed a sentence level text representation method to learning from unlabeled text data [6]. It can not only optimize the training time, but also improve the portability by employing unsupervised method. Le et al. proposed two levels of text representation models including sentence and document, both of which are based on the word2vec model [11]. Kalchbrenner et al. proposed a text representation model based on dynamic convolution neural network (DCNN) which adopts dynamic pooling technology and is expert in emotion recognition [9]. Considering the advantages of convolution neural network (CNN) in extracting local features, Hu et al. proposed employing CNN to extract semantic information between words in sentences and achieved excellent results in sentence matching tasks [7]. On this basis, Yin et al. proposed Bi-CNN-MI text representation model which can extract four different granularity text representations from sentences and realize the interaction of these four different granularity features through the CNN so as to adapt to the synonymous sentence detection task [17]. Zhang et al. proposed a text representation model based on dependency sensitive CNN at the sentence and document level, which applies sensitive information and feature extracting on inputting word vectors [18]. This paper proposes a text feature representation model based on CNN and variational auto encoder (VAE), which is a feature representation method from word vector to text vector. In this model, CNN is used to extract the features of text vector as text representation to get the semantics between words. In addition, VAE is introduced to make the text feature space more consistent with Gaussian distribution. Four evaluation indexes, including accuracy, recall, precision and F-score, are used to evaluate the performance of the proposed model. Experimental results show that the model has better performance than w2v-avg and CNN-AE in k-nearest neighbor (KNN), random forest (RF) and support vector machine (SVM) classification.

The remains of this paper are organized as follows. Section 2 and Sect. 3 introduce the related algorithm and proposed model. Section 4 illustrates a case study of open dataset Cnews. Section 5 shows the conclusion and future research directions.

2   Related Algorithms

2.1   Word2vec

Word2vec algorithm proposed by Mikolov aims to obtain the text word vector model by training a neural network to get the weight matrix of the network. Word2vec model includes CBOW and Skip-gram training models and they only contain simple neural network structures including input layer, projection layer and output layer. CBOW and skip-gram realize the network based on different conditions. CBOW predicts the probability of the central word through the context word while skip-gram model predicts the probability of the context word through the central word [15].

The input of CBOW is the context $Context\,(w_t)$ of the central word w_t in the sliding window and the object is to predict the central word w_t. The output of CBOW model is a softmax classifier function, which is usually implemented by using negative sampling method. The central word w_t in the sliding window is judged as a positive sample and other words are judged as a negative sample. In the output layer, there is a negative sample set $NEG(w_t)$ for the central word w_t so that any word u in the corpus meets the Eq. (1).

$$R^{w_t}(u) = \{ \begin{matrix} 1, & u = w_t \\ 0, & u \neq w_t \end{matrix} \tag{1}$$

where $R^{w_t}(u)$ determines whether u is a positive sample. For the central word and its context $\langle w_t, Context(w_t)\rangle$, CBOW model needs to maximize the objective function $g(w_t)$ shown in Eq. (2).

$$g(w_t) = \prod_{u \in (\{w_t\} \cup NEG(w_t))} p(u|Context(w_t)) \tag{2}$$

The formal of $p(u|Context(w_t))$ is shown in Eq. (3).

$$p(u|Context(w_t)) = \begin{cases} \sigma(X_{w_t}^T \theta^{w_t}), & R^{w_t}(u)=1 \\ 1-\sigma(X_{w_t}^T \theta^{w_t}), & R^{w_t}(u)=0 \end{cases} \quad (3)$$

where X_{w_t} stands the sum of the word vectors of the words in the context $Context(w_t)$. The objective function can be converted to Eq. (4).

$$g(w_t) = \sigma(X_{w_t}^T \theta^{w_t}) \prod_{u \in NEG(w_t)} [1 - \sigma(X_{w_t}^T \theta^u)] \quad (4)$$

where $\sigma(X_{w_t}^T \theta^{w_t})$ represents the prediction probability of getting central word w_t via CBOW when the context is $Context(w_t)$. $\sigma(X_{w_t}^T \theta^u)$ is the prediction probability of getting central word u when the context is $Context(w_t)$. In order to maximize $g(w_t)$, it is necessary to maximize $\sigma(X_{w_t}^T \theta^{w_t})$ to increase the probability of positive samples. At the same time, minimize all the $\sigma(X_{w_t}^T \theta^u)$ to reduce the probability of negative samples.

Skip-Gram Based Method. In contrast to CBOW model, Skip-gram firstly determines the central word w_i and then predicts the context $Context(w_t)$ in its sliding window through the central word. The output of skip-gram model corresponds to a positive sample $u \in Context(w_t)$. On the contrary, $NEG(u)$ represents negative sample set which don't belong to the positive sample $Context(w_t)$.

Therefore, for the data $\langle w_t, Context(w_t) \rangle$ in the sliding window meet the objective function shown in Eq. (5).

$$g(w_t) = \prod_{u \in Context(w_t)} \prod_{x \in \{u\} \cup NEG(u)} p(x|w_t) \quad (5)$$

Similar to CBOW method, the relation $R^u(x)$ is introduced to determine whether x is a positive sample. When it is a positive sample, $x = u$ and $R^u(x) = 1$. When it is a negative sample, $x = NEG(u)$ and $R^u(x) = 0$. The conditional probability is shown in Eq. (6).

$$p(x|w_t) = [\sigma(v(w_t)^T v_x)]^{R^u(x)} \cdot [1 - \sigma(v(w_t)^T v_x)]^{1-R^u(x)} \quad (6)$$

where $\sigma(v(w_t)^T v_x)$ represents the prediction probability of getting $context(w_t)$ when the input of central word is w_t. $v(w_t)$ is the relationship between input layer and hidden layer. v_x is the output of network. For maximizing $g(w_t)$, it is necessary to maximize $\sigma(v(w_t)^T v_x)$ when positive sampling and minimize the $\sigma(v(w_t)^T v_x)$ when negative sampling.

2.2   Convolution Neural Network

CNN is a deep artificial neural network including convolution calculation. It has excellent performance in the field of computer vision and NLP with its powerful feature extraction ability [4,12]. In the field of NLP, CNN employed convolution kernel to convolute text matrix of different length and the vectors through convolution kernels are calculated by pooling layer to extract features for text classification [1].

Convolution Layer. Convolution layer can extract local features by convolution kernel and get the final output by activation function. For $k-th$ convolution kernel, the convolution process is shown in Eq. (7).

$$c_i = f(w_k * x + b_k) \tag{7}$$

where w_k and b_k are the weight matrix and bias of $k - th$ convolution kernel. x stands for the input matrix. f represents the activation function. For a sentence of length n, its feature vector is shown in Eq. (8).

$$c = [c_1, c_2, c_3...c_n] \tag{8}$$

Pooling Layer and Full Connection Layer. In the pooling layer of the model, the maximum pooling technology is used to extract the feature value C, which contains the highest semantic information in the local features of the convoluted window, as is shown in Eq. (9).

$$C = \max(c) \tag{9}$$

The complexity of the parameters in the convolutional neural network can be effectively reduced by using max pooling. For a window with k convolution kernels, the feature vectors obtained are shown in Eq. (10).

$$\hat{C} = [\hat{C}_1, \hat{C}_2, \hat{C}_3...\hat{C}_n] \tag{10}$$

Next, the activation function is used to predict the labels of sentences as is shown in Eq. (11).

$$y = f(\hat{C}) \tag{11}$$

where y is the predicted label (1 stands for positive label and 0 stands for negative label). The loss function can be expressed as Eq. (12).

$$loss = \frac{1}{2n} \sum_{i=1}^{n} \|y_i - \hat{y}_i\|^2 \tag{12}$$

where y_i is the actual label. The loss function can be used to iterate and learn the parameters of the network.

3   Text Feature Representation Model Based on CNN-VAE

This paper combine the network framework of the VAE with the CNN to extract the text feature representation from the word vector. The traditional AE employs full connection layer, which is replaced by convolutional neural network for CNN can learn local features efficiently. Moreover, VAE makes the vector feature space conform to the function of Gaussian distribution, which makes the final text feature representation richer in semantic information. The structure of CNN-VAE is shown in Fig. 1.

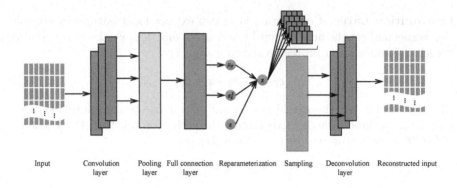

Fig. 1. The structures of CNN-VAE model

The CNN is used to realize the network structure of VAE, because CNN can learn better local features from the input matrix. In this part, CNN network is used to build the VAE network framework, so that CNN combines the VAE feature extraction and the function of making the vector feature space conform to the Gaussian distribution in its own text feature extraction, making the final text feature representation richer in semantic information.

Similar to AE, the proposed method includes decoding and encoding. The decoding process can be regarded as a general CNN which can achieve the purpose of feature extraction via convolution and pooling. A matrix x can be obtained by splicing the word vectors corresponding to the words appearing in an article is putted into the model. Random number z corresponding to the corresponding Gaussian distribution is generated according to the mean value μ and variance σ of the output of the convolution encoder. Suppose that there is a set of functions $p_\theta(x|z)$ for generating x from z, each of which is uniquely determined by θ. The goal of the VAE is to maximize $p_\theta(x)$ by optimizing θ to make the generated data $\hat{x}$ similar to original data x as possible and the formula is shown Eq. (13).

$$p_\theta(x) = \int p_\theta(x|z) \, p_\theta(z) \, dz \tag{13}$$

In order to obtain $p_\theta(z)$, the encoder network $p_\theta(z|x)$ is introduced. An approximate posterior $q_\varphi(z|x)$ obeying Gaussian distribution is employed to take the place of $p_\theta(z|x)$ considering that the latter is difficult to obtain by calculation. Kullback Leibler (KL) divergence is applied to measure the similarity between two distributions [5] and the formula is shown in Eq. (14).

$$\begin{aligned} &D_{KL}(q_\varphi(z|x) \parallel p_\theta(z|x)) \\ &= \sum q_\varphi(z|x) \left[\log q_\varphi(z|x) - \log p_\theta(x|z) - \log p_\theta(z)\right] + \log p_\theta(x) \end{aligned} \tag{14}$$

Therefore, the formal of $\log p_\theta(x)$ can be expressed as Eq. (15).

$$\begin{aligned} \log p_\theta(x) &= D_{KL}(q_\varphi(z|x) \parallel p_\theta(z|x)) \\ &- \sum q_\varphi(z|x) \left[\log q_\varphi(z|x) - \log p_\theta(x|z) - \log p_\theta(z)\right] \end{aligned} \tag{15}$$

Considering the KL divergence is not negative, the loss function can be expressed as Eq. (16).

$$L(\theta, \varphi; x) = -\sum q_\varphi(z|x) [\log q_\varphi(z|x) - \log p_\theta(x|z) - \log p_\theta(z)]$$
$$= -D_{KL}(q_\varphi(z|x) \parallel p_\theta(z|x)) + \sum q_\varphi(z|x)[\log p_\theta(x|z)] \quad (16)$$

where $L(\theta, \varphi; x)$ is loss function. $-D_{KL}(q_\varphi(z|x) \parallel p_\theta(z|x))$ is regularizer and $\sum q_\varphi(z|x)[\log p_\theta(x|z)]$ is reconstruction error. Considering $p_\theta(z)$ obeys the Gaussian distribution $N(0; I)$ and $q_\varphi(z|x)$ obeys the Gaussian distribution $n(\mu; \sigma^2)$ so that the regularizer can be expressed as Eq. (17).

$$-D_{KL}(q_\varphi(z|x \parallel p_\theta(z)) = \frac{1}{2}\sum (1 + \log(\sigma^2 - \mu^2 - \sigma^2)) \quad (17)$$

where j is the dimension of z. Monte Carlo evaluation is used to solve the reconstruction error shown as Eq. (18).

$$\sum q_\varphi(z|x)[\log p_\theta(x|z)] = \log p_\theta(x|z) \quad (18)$$

The technique of reparameterization is employed considering that is not derivable, whose formal is shown in Eq. 10. In this way, the original derivation of z can be converted into the derivation of μ and σ shown as Eq. (19).

$$z = \mu + \varepsilon \cdot \sigma \quad (19)$$

In this way, the original derivation of z can be converted into the derivation of μ and σ. $\log p_\theta(x|z)$ can be expressed as Eq. (20).

$$\log p_\theta(x|z) = -\sum \left(\left(\frac{1}{2} \left\| \frac{x - \mu}{\sigma} \right\| \right) + \log\left(\sqrt{x\pi}\sigma \right) \right) \quad (20)$$

In conclusion, the loss function can be expressed as Eq. (21).

$$L(\theta, \varphi; x) = \frac{1}{2}\sum (1 + \log(\sigma^2 - \mu^2 - \sigma^2)) - \sum \left(\left(\frac{1}{2} \left\| \frac{x - \mu}{\sigma} \right\| \right) + \log\left(\sqrt{x\pi}\sigma \right) \right)$$
$$(21)$$

The training process of the network is divided into the following steps. (1) The data of training set is used to find the optimal parameters of the model when the loss function is minimized, (2) when the network loss converges, the verification set is inputted to the encoder network obtained by training set to get the corresponding text representation vector. The corresponding classification accuracy is obtained by inputting the text representation vector into the classification model and (3) the learning rate is adjusted many times to obtain the optimal parameters of the model. After several adjustments, the optimal model parameters are shown as Table 1.

Table 1. Parameter settings

Parameter	Setting
Dimension of word vector	128
Number of iterations	30
Dropout	0.5
Size of hidden layer z	128
Learning rate	0.001
Padding (Maximum words per article)	100

This paper adopts the open data set Cnews which is generated by filtering the historical data from RSS subscription channel of Sina News from 2005 to 2011. The dataset contains 10 categories of news, namely sports, entertainment, home furnishing, real estate, education, fashion, current affairs, games, technology and finance.

Each row in the Cnews dataset represents an article and the beginning of each row corresponds to the news category of the article. Therefore, this paper extracts the tag information (i.e. news category) in the dataset and the dataset is divided into tag and corpus. Moreover, The Jieba segmentation is applied to process the article and the stoppage word is employed to filter out some words to improve the performance and computational efficiency of the subsequent experiments. Next, this paper makes word frequency statistics for the words appearing in the corpus and rank them according to the word frequency from large to small. Finally, the final data set can be obtained by eliminating the words that appear less frequently than the first 10000.

4   Experimental Results and Analysis

4.1   Evaluating Indicator

This paper employs four evaluation indicators including accuracy, recall, precision and F1-score in order to verify the performance of the proposed model. The true positive (TP) indicates that the prediction category is positive and the actual category is positive. The false positive (FP) indicates that the prediction category is positive and the actual category is negative. The false negative (FN) indicates that the prediction category is negative and the real category is positive. The true negative (TN) indicates that the prediction category is negative and the actual category is negative. The formulas of four indictors are shown in Eqs. (22), (23), (24) and (25) respectively.

$$Accuracy = \frac{TP + TN}{TP + FP + FN + TN} \tag{22}$$

$$Recall = \frac{TP}{TP + FN} \tag{23}$$

$$Precision = \frac{TP}{TP + FP} \tag{24}$$

$$F1 - score = 2 \times \frac{precision \times recall}{precison + recall} \tag{25}$$

4.2 Experimental Results

This paper compares the proposed model with some common unsupervised text representation methods that convert word vectors into document vectors. The control models are shown as follows.

w2v-avg: This model averages all word vectors of each document to get the final representation. This experiment sets the word vector dimension of training to 128 and get the text representation vector dimension to 128.

CNN-AE: An AE model based on CNN, which can get the text representation via using word vectors obtained by CNN as the input of encoder.

The experimental results are shown in Table 2, 3 and 4. In Tables 2, 3 and 4, the performance of CNN-AE is better than that of w2v-avg. Therefore, CNN is conducive to extracting the local semantic features of the components between words. In addition, when the text representation dimension is 128, the proposed model is superior to CNN-AE and w2v-avg in KNN, RF and SVM classifiers. That is to say, if AE is improved to VAE, the text feature space can better fit the Gaussian distribution, which makes the semantic information conform to

Table 2. The experimental results under KNN classification algorithm

	Accuracy	Recall	Precision	F1-Score
w2v-avg	87.32 ± 7.40	77.02 ± 7.31	80.82 ± 7.61	77.28 ± 7.38
CNN-AE	92.94 ± 5.06	86.44 ± 10.69	89.56 ± 11.12	86.66 ± 10.70
CNN-VAE	$\mathbf{94.87 \pm 4.61}$	$\mathbf{91.81 \pm 7.46}$	$\mathbf{91.87 \pm 9.12}$	$\mathbf{90.87 \pm 8.08}$

Table 3. The experimental results under RF classification algorithm

	Accuracy	Recall	Precision	F1-Score
w2v-avg	85.48 ± 0.93	66.23 ± 2.27	74.84 ± 4.25	68.27 ± 2.73
CNN-AE	92.43 ± 0.66	80.26 ± 1.42	92.64 ± 3.18	83.97 ± 1.85
CNN-VAE	$\mathbf{94.98 \pm 0.65}$	$\mathbf{92.54 \pm 1.67}$	$\mathbf{95.29 \pm 1.00}$	$\mathbf{93.37 \pm 1.22}$

Table 4. The experimental results under SVM classification algorithm

	Accuracy	Recall	Precision	F1-Score
w2v-avg	86.51 ± 3.18	87.61 ± 3.08	88.68 ± 2.83	87.58 ± 3.27
CNN-AE	90.00 ± 1.64	90.94 ± 1.48	90.10 ± 2.98	90.64 ± 1.55
CNN-VAE	$\mathbf{93.30 \pm 2.63}$	$\mathbf{93.90 \pm 2.08}$	$\mathbf{94.29 \pm 1.90}$	$\mathbf{93.83 \pm 1.99}$

the real distribution. In conclusion, the VAE text feature representation model based on CNN is effective and feasible.

5   Conclusion

Text feature representation plays a significant role in the field of NLP as the first step of machine recognition of natural language, which extracts the features of the text to express semantic information contained in the text. With the improvement of hardware performance, deep learning and neural network development, text representation has a leap forward development. At the same time, the increasing number of Internet users produces a large number of unstructured text data. How to employ NLP technology to analyze and process huge text data has become the current research hotspot. Therefore, this paper proposes a text feature representation model based on CNN-VAE, which is a feature representation method from word vector to text vector. On the one hand, the model applies CNN to extract the features of the text vector as the text representation, so as to better extract the semantics between words. On the other hand, the model uses VAE instead of AE to make the text feature space better fit the Gaussian distribution, so that the semantic information is more consistent with the real distribution. In order to verify the performance of the model, the proposed model is compared with w2v-avg and CNN-AE. Experimental results show that the proposed model has better performance in KNN, RF and SVM classification. There are four levels of language structure including document, paragraph, sentence and word in NLP. This paper only involves two different levels of text vector representation and attention should be paid to different levels of language structure in the future.

References

1. Ceylan, A.M., Ayta, V.: Concolutional auto encoders for sentence representation generation. Turkish J. Electrical Eng. Comput. Sci. **1135**(28) (2020). https://doi.org/10.3906/elk-1907-13
2. Collobert, R., Weston, J., Bottou, L., Karlen, M., Kavukcuoglu, K., Kuksa, P.: Natural language processing (almost) from scratch. J. Mach. Learn. Res. **12**(1), 2493–2537 (2011). https://doi.org/10.1016/j.chemolab.2011.03.009
3. Dhingra, B., Zhou, Z., Fitzpatrick, D., Muehl, M., Cohen, W.W.: Tweet2vec: Character-based distributed representations for social media. In: Proceedings of the 54th Annual Meeting of the Association for Computational Linguistics (2016)
4. Guo, C., Guo, W., Chen, C.H., Wang, X., Liu, G.: The air quality prediction based on a convolutional LSTM network. In: Web Information Systems and Applications (2019). https://doi.org/10.1007/978-3-030-30952-7_12
5. Hershey, J.R., Olsen, P.A.: Approximating the Kullback Leibler divergence between Gaussian mixture models. In: IEEE International Conference on Acoustics (2007). https://doi.org/10.1109/ICASSP.2007.366913

6. Hill, F., Cho, K., Korhonen, A.: Learning distributed representations of sentences from unlabelled data. In: Proceedings of the 2016 Conference of the North American Chapter of the Association for Computational Linguistics: Human Language Technologies (2016). https://doi.org/10.18653/v1/N16-1162

7. Hu, B., Lu, Z., Li, H., Chen, Q.: Convolutional neural network architectures for matching natural language sentences. In: Proceedings of Advances in Neural Information Processing Systems 27: Annual Conference on Neural Information Processing Systems (2015)

8. Ji, S., Yun, H., Yanardag, P., Matsushima, S., Vishwanathan, S.V.N.: Wordrank: Learning word embeddings via robust ranking. In: Computer Science (2015)

9. Kalchbrenner, N., Grefenstette, E., Blunsom, P.: A convolutional neural networkfor modelling sentences. Eprint Arxiv 1 (2014).https://doi.org/10.3115/v1/P14-1062

10. Kofler, C., Larson, M., Hanjalic, A.: User intent in multimedia search: A survey of the state of the art and future challenges. ACM Comput. Surv. 49(2), 1–37 (2016). https://doi.org/10.1145/2954930

11. Le, Q.V., Mikolov, T.: Distributed representations of sentences and documents. In: Proceedings of the International Conference on International Conference on Machine Learning (2014)

12. Liang, H., Sun, X., Sun, Y., Gao, Y.: Text feature extraction based on deep learning: A review. EURASIP J. Wirel. Commun. Netw. 2017(1), 1–12 (2017). https://doi.org/10.1186/s13638-017-0993-1

13. Liu, G.Z.: Semantic vector space model: Implementation and evaluation. J. Assoc. Inf. Sci. Technol. 48(5), 395–417 (2010). https://doi.org/10.1002/(SICI)1097-4571(199705)48:53.0.CO;2-Q

14. Liu, N., et. al.: Text representation: From vector to tensor. In: IEEE International Conference on Data Mining (2005). https://doi.org/10.1109/ICDM.2005.144

15. Mikolov, T.: Distributed representations of words and phrases and their compositionality. Adv. Neural Inf. Process. Syst. 26, 3111–3119 (2013)

16. Weeds, J., Weir, D.: Co-occurrence retrieval: A flexible framework for lexical distributional similarity. Comput. Linguist. 31(4), 439–475 (2005). https://doi.org/10.1162/089120105775299122

17. Yin, W., Schutze, H.: Convolutional neural network for paraphrase identification. In: Proceedings of Conference of the North American Chapter of the Association for Computational Linguistics: Human Language Technologies, pp. 901–911 (2015)

18. Zhang, R., Lee, H., Radev, D.: Dependency sensitive convolutional neural networks for modeling sentences and documents. In: Proceedings of the 2016 Conference of the North American Chapter of the Association for Computational Linguistics: Human Language Technologies (2016). https://doi.org/10.18653/v1/N16-1177

19. Zhang, Y., Jin, R., Zhou, Z.H.: Understanding bag-of-words model: A statistical framework. Int. J. Mach. Learn. Cybernet. 1(1–4), 43–52 (2010). https://doi.org/10.1007/s13042-010-0001-0

Cross-Language Generative Automatic Summarization Based on Attention Mechanism

Feiyang Yang, Rongyi Cui, Zhiwei Yi, and Yahui Zhao[✉]

Department of Computer Science and Technology, Yanbian University,
977 Gongyuan Road, Yanji 133002, China
903873610@qq.com

Abstract. Generative automatic summarization is a basic problem in natural language processing. We propose a cross-language generative automatic summarization model. Unlike the traditional methods that have to go through machine translation, our model can directly generate a text summary of another language from a text body in one language. We use the RNNLM(Recurrent Neural Network based Language Model) structure to pre-train word vectors to obtain semantic information in different languages. We combined the Soft Attention mechanism in the Seq2Seq model, using Chinese, Korean and English to build a parallel corpus to train the model, thereby, cross-language automatic summarization can be achieved without the help of machine translation technology. Experiments show that the improvement of our proposed model on ROUGE-1, ROUGE-2 and ROUGE-L indicators reached 6%, 2.46%, and 5.13%, respectively. The experimental effect is friendly.

Keywords: Automatic summarization · Abstractive · Word vector · Cross-language

1 Introduction

A text abstract usually refers to a text that covers the main information of the original text, which mainly generated by summarizing and condensing the information from the original document. The text abstract not only can save readers a lot of reading time, but also plays the role of information compression for the length is greatly reduced compared with the original document. For the first time in 1958, *Luhn* proposed using computers to automatically generate abstracts of texts [12]. Compared with manual abstract writing, automatic abstract technology can greatly improve the efficiency of abstract writing, so it has attracted wide attention in natural language processing [6]. At present, automatic generation of abstracts can be divided into two categories, extraction and production, according to the way of generation. Abstract extraction is generally to extract the sentence containing the main information from the original text to produce abstract text, so it is also called sentence extraction.

© Springer Nature Switzerland AG 2020
G. Wang et al. (Eds.): WISA 2020, LNCS 12432, pp. 236–247, 2020.
https://doi.org/10.1007/978-3-030-60029-7_22

This kind of method usually uses TextRank [13] and LexRank [4] algorithms to extract summary sentences. And both algorithms are iteratively calculated on topology graph based on PageRank [7] algorithm. The method extracts complete sentences from original text by segment, so a large amount of redundant information is hard to be eliminated, and the generated abstract is generally lack of coherence and usually has poor readability. Generative abstract is characterized by the fact that the sentences in the abstract are not from the original text. They are new sentences that are re-refined, mainly based on the deep learning method of the Seq2Seq(sequence-to-sequence) model [5]. Generative abstract has attracted more attention due to the abstract generated has the advantages of short length, low redundancy and strong generality of sentences [14,16,17,19].

The traditional method of automatically generating abstracts across languages is mainly based on machine language translation technology, which translates the source language into the target language, and then generates the abstract for the text of the target language; or first generates the abstract for the source language text, and then translates it into the target language [21]. *Bahdanau et al.* proposed the simultaneous translation and alignment using the attention mechanism method, which effectively improved the quality of translation and confirmed its great improvement effect in sequence learning tasks [3]. With parallel corpus, using Seq2seq model combined with attention mechanism, and using RNNLM pre-trained word vectors, a cross-language automatic summarization method that does not require a machine translation process is proposed.

2   Combined RNNLM Generative Summary

Mikolov et al. introduced recurrent neural networks into the establishment of language model [11], which can store historical information in hidden states. A structure diagram of the generated RNNLM (Recurrent Neural Network based Language Model), RNNLM is shown in Fig. 1.

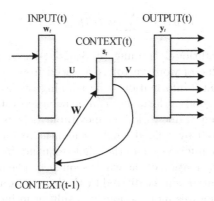

Fig. 1. RNNLM structure

$$s(t) = f(U \cdot w(t) + W \cdot s(t-1)) \tag{1}$$

$$y(t) = g(V \cdot s(t)) \tag{2}$$

$$f(x) = \frac{1}{1 + e^x} \tag{3}$$

$$g(z_m) = \frac{e^{z_m}}{\sum_k e^{z_k}} \tag{4}$$

where f is the sigmoid activation function and g is the softmax function for converting the vector $y(t)$ of the output layer to the corresponding probability distribution. U and W represent the input layer to the hidden layer weight matrix, V the hidden layer to the output layer weight matrix. The $w(t)$ here represents a specific word for a sequence in the corpus $w(1), ..., w(t-1), w(t)$, and the corresponding target words need to be predicted by RNNLM model. A maximum likelihood function L required for this model is:

$$L = P(w_t | w_1, ..., w_{t-1}) \tag{5}$$

The sequence model proposed in this paper consists of Bi-LSTM and LSTM two cyclic neural networks: Bi-LSTM [1] is responsible for encoding the input sequence, called Encoder, and LSTM [2] is responsible for decoding the target sequence, called Decoder. The efficiency of the Seq2Seq model can be improved in the framework of "encoder-decoder" by adding Attention model [18] to the coding end to carry out a reasonable weighted transformation of the input sequence data.

The input sequence is recorded as x, target sequence as y, T represents the output time series size, $y1 : t-1$ represents the output corresponding to the first $t-1$ time points. The automatic summary problem is regarded as a conditional language model, and the purpose of the model is to maximize the probability of generating the whole target summary sentence under the condition of original input text.

$$P(y|x) = \sum_{t=1}^{T} log P(y_t | y_{1:t-1}, x) \tag{6}$$

Based on the traditional training of Seq2Seq-based generative summary model, the initial value of word vector is replaced from random value to value which carrying semantic information. The training of the model will converge faster, and relatively good performance can be achieved with less training times. However, the Seq2Seq framework has inevitable defects. The encoding and decoding are connected by a fixed length semantic vector C, and the encoder wants to compress the information of the whole sequence into the C, so it can not fully represent the information of the whole sequence. The information carried in the previous input content will be diluted by the later information, the longer the input sequence deteriorates more seriously, resulting in incomplete information coverage and reduced accuracy when decoding. Hence, the attention mechanism is of great significance in sequence learning tasks. The generative summary model in this paper combines the attention mechanism based on RNNLM. The detailed structure of the model is shown in Fig. 2.

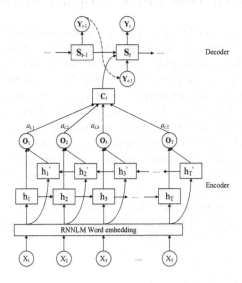

Fig. 2. Generative summary model based on RNNLM and attention mechanism

The first layer is the input layer, the second is the word embedding layer, the third is the hidden layer, and the fourth is the output layer. An intermediate vector Ct is a semantic vector containing attention information from the input. The initial value of the word embedding layer is composed of RNNLM pre-trained word vectors, and the Encoder part uses the semantic vectors of the text using the Attention mechanism. All the work flow of the model is as follows: the word embedding layer first converts the input word sequence into the corresponding vector sequence to the encoder Bi-LSTM, which receives the vector sequence passed by the word embedding layer and generates the output and gives it to the next layer of the model. Attention mechanism receives the output of the encoder, calculates the attention size of each step input, and then weighted summation to obtain the semantic vector Ct. of the whole text. The Attention mechanism receives the output of the encoder, calculates the attention size of each step input, and then obtains the semantic vector Ct. of the whole text by weighting sums. Decoder part uses LSTM to output the term items predicted at the current moment according to the $Y(t-1)$ of the output with the hidden state $S(t-1)$ and the current intermediate vector.

During the training stage, the input and output give the word sequence of the text of the model and the word sequence of the abstract, respectively. The model uses the Dropout mechanism [20] and the gradient clipping mechanism [15] to prevent overfitting and gradient explosion phenomena. Compared with the traditional model of random initialization word vector, the proposed model will have better performance under the same training times.

3   Translanguage Generative Abstract Method

A summary sequence can be generated for a text sequence when the Seq2Seq model is used for a text abstract task. The mapping relationship between one language and another can be well captured by the model. Therefore, the model can generate a summary sequence of another language for the text sequence of one language, and realize automatic summary across languages without using machine translation technology. In Seq2Seq-based machine translation, the training data consists of the original text and the translation pairs, which requires a large number of original text-translation pairs to train Seq2Seq models. In the cross-language automatic summary method proposed in this paper, the training data not only needs text and summary to be composed in pairs, but also requires them to come from different languages which have the same meaning. Therefore, we need to use parallel documents to build text (one language) - abstract (another language) pairs for automated abstracts across languages. The parallel documents are shown in Fig. 3.

DOC	Text	Summary
A	转座子(transposon)是基因组上的一段一定长度的DNA序列,能在自身编码的转座酶作用下,以"剪切-粘贴"的方式在基因组中进行高效转座,基于转座子的插入诱变突略,在包括鱼类在内的脊椎动物重要性状主控基因的捕选、功能解释以及基因组学研究方面有着重要的研究前景……	鱼类活性DNA转座子的发掘与应用概况
B	Transposon is a DNA sequence with a certain length in the genome. DNA transposons are transposable elements which can move within their host genomes by changing their insertional positions in a process called transposition. The transposition can efficiently mediate by its transposase on "cut-paste" approach. The strategy based on transposon insertional mutagenesis will be an extremely powerful tool for trapping the target genes that control important trait in fish…..	Overview of fish DNA transposon discovery and into application
C	전이 인자(transposon)란 일정한 길이를 갖춘 게놈의 DNA 서열로서, 자아 인코딩 전이효소의 작용을 통하여,오려두기-붙여넣기;의 방식으로 게놈에서 효과적인 전이를 진행한다,전이 인자에 기반한 삽입 돌연변이 유발계획은, 어류를 포함한 척추동물의 중요 성질인 마스터 유전자 선별, 기능 해석, 게놈학 연구에 전망을 제시하였다……	어류 활성 DNA 전이 인자의 발굴과 응용

Fig. 3. Sample parallel documents

Sample	Text	Summary
English-Chinese	Transposon is a DNA sequence with a certain length in the genome. DNA transposons are transposable elements which can move within their host genomes by changing their insertional positions in a process called transposition. The transposition can efficiently mediate by its transposase on " cut-paste " approach. The strategy based on transposon insertional mutagenesis will be an extremely powerful tool for trapping the target genes that control important trait in fish species…..	鱼类活性DNA转座子的发掘与应用概况
Korean-Chinese	전이 인자(transposon)란 일정한 길이를 갖춘 게놈의 DNA 서열로서, 자아 인코딩 전이효소의 작용을 통하여,오려두기-붙여넣기;의 방식으로 게놈에서 효과적인 전이를 진행한다,전이 인자에 기반한 삽입 돌연변이 유발계획은, 어류를 포함한 척추동물의 중요 성질인 마스터 유전자 선별, 기능 해석, 게놈학 연구에 전망을 제시하였다……	鱼类活性DNA转座子的发掘与应用概况

Fig. 4. Sample training data across language abstracts

The documents A, document B and document C in Fig. 3 are a set of parallel documents in different languages, but the meaning of the body part and the abstract part is consistent. When training the cross-language automatic abstract model of English-Chinese, we should use the text part of the document B and the summary part of the document A to form the text-abstract pair. A sample of training data across language automated summaries is shown in Fig. 4.

The proposed cross-language automatic summary processing flow is described as follows:First, the training corpus of cross-language automatic summary is constructed using parallel documents. The Sample training data is shown in Figs. 3 and 4. Secondly, the corpus of input language is preprocessed at the beginning, and the dictionary is constructed for the corpus of output language. Then, the LSTM-based Seq2Seq model is trained using the constructed cross-language automatic abstract corpus. Finally, a large number of cross-language body (source language)-abstract (target language) pairs are constructed through parallel documents, which are used to train the Seq2Seq model of automatic summary across languages, so that the model has the ability to generate target language summary for source language text. Through training, the model not only has the ability of abstract, but also establishes the mapping relationship between the source language text and the target language summary text. It completes the abstract task and also has the function of translation

4 Experimental Results and Analysis

4.1 Experimental Setup

The corpus used in this experiment come from a project commissioned by Yanbian Science and Technology Information Service Center, China and South Korea Science and Technology Information processing Comprehensive platform. In the early stage of the project, more than 30000 parallel scientific and technological literature abstracts and titles of China, Britain and Korea have been collected in three fields: ocean, aviation and biology. The preprocessing of experimental corpus mainly includes the following steps:

(1) Chinese corpus: We first remove non-native words (unknowns and symbols), numerals and English. Then, we found that the deactivation words had a great effect on the fluency and semantic representation of the statements, so the deactivation words were not removed. Finally, the word segmentation tool used in this paper is the stutter word segmentation kit, and the preprocessed Chinese language corpus is the pure Chinese character corpus.

(2) English corpus: First, the English text is word segmentation, and then all the uppercase letters in the generated string are converted to lowercase letters. Because Snow and snow are the same meaning, so they can not be considered two words. Finally, the words and some special symbols of non-pure letters are removed, so that the corpus is all English text. A NLTK word segmentation tool is used to do this task.

(3) Korean language corpus: We remove the foreign language words, symbols and numbers from the corpus, and retain the Korean words after using the Korean language participle to the Korean language corpus. A konlpy kit is used for word segmentation in this experiment.

After removing the blank text and some missing text in the corpus, it is divided into training set, validation set and test set according to the ratio of 8:1:1, as shown in Table 1.

Table 1. Corpus statistics for experiments

Language	Statistical data		
	Train	Verification	Test
Chinese	24489	3061	3062
English	24489	3061	3062
Korean	24489	3061	3062

4.2 Model Training

Our experiments were conducted on the Chinese data set. First, we conducted experiments on the batch size and epoch of the Seq2Seq model to determine the optimal choice of parameters. The experiments are carried out on the Chinese dataset. The epoch is set to 20 and iterated 20 times on the Chinese training set with different batch size, as well as keeping the rest parameters consistent. After the model is trained, we do experiments on Chinese test set and use ROUGE to evaluate the model performance, which is widely used in summary tasks and is a recognized evaluation standard [8]. This method was proposed by *ISI Chin-Yew Lin* in 2004. It is divided into five evaluation indexes, namely ROUGE-N, ROUGE-L,ROUGE-W,ROUGE-S,ROUGE-SU, each index should be calculated separately for P (precision, accuracy), R (recall) and F (Table 2).

Table 2. Results of different evaluation index for different batch size

Batch size	ROUGE-1, ROUGE-2, ROUGE-LF value in evaluation index (%)			
	ROUGE-1	ROUGE -2	ROUGE-L	Time
1	–	–	–	10.4 h
16	15.85	2.28	14.85	4996.69 s
32	16.82	2.77	15.71	3883.33 s
64	**16.91**	**2.85**	**15.77**	3363.08 s
128	16.3	2.62	15.34	3225.42 s
256	14.29	2.25	13.58	3260.15 s
512	9.66	1.45	9.34	3235.47 s
1024	9.78	0.86	9.66	3098.40 s
2048	Memory error			

According to the data in the table, this paper sets the value of batch size to 64. The experiment is carried out with different epoch under the condition that the other parameters are the same, and the experiment is carried out on the training set and the verification set. In Fig. 5, since each step of the Seq2Seq model at the output end is a multi-classification task, the training results of the model can be measured by accuracy. The horizontal axis represents the number of iterations on the training set epoch, the vertical axis represents the accuracy and the value of the cross entropy of the loss (loss function). The curves with square and triangle represent the accuracy of the model on the training set, the test set; and the curves with round and cross represent the loss values of the model on the training set, the test set, according to the order from bottom to top. When the epoch value is too large, the model is easy to overfit, and the epoch value is too small, the model will be in the underfitting state. It can be seen that when epoch = 15, the model has the highest accuracy and the smallest loss on the verification set. At this point, the model reaches the best state, so the value epoch in the experiment in this paper is set to 15.

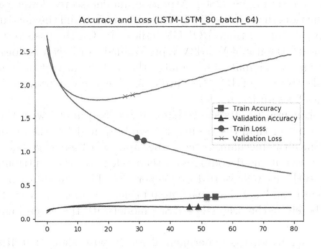

Fig. 5. Results of different epoch experiments

4.3   Analysis of Experimental Results

Compared to the baseline - traditional Seq2Seq, this paper sets up different methods for experiments. The experimental results of the model on the test set are shown in Table 3.

Table 3. Automatically summarise experimental results for across languages

Model	ROUGE-1, ROUGE-2, ROUGE-LF value in evaluation index (%)		
	ROUGE-1	ROUGE-2	ROUGE-L
Seq2Seq	23.88	4.71	20.34
Seq2Seq+Attention	26.57	6.71	20.57
word2vec(fine-tuning)+Seq2Seq+Attention	24.58	5.22	18.98
RNNLM(fine-tuning)+Seq2Seq+Attention	**32.57**	**9.17**	**25.7**

In Table 3, Seq2Seq is the baseline model and the initial values of word vectors are all random values. Seq2Seq Attention, the Attention mechanism is introduced in the Seq2Seq model, and the initial value of the word vector is also the random value. w2v (fine-tuning) refers to the word vector is no longer a randomly initialized value, but a word vector obtained after training on the training set with Word2Vec [10], and the word vector continues to adjust the parameters as the abstract model is trained. RNNLM (fine-tuning) refers to the initial value of a word vector RNNLM trained on a training set. To be fair, both Word2Vec and RNNLM iterate five times on the training set when training word vectors [9].

We can see from the experimental results that the Seq2Seq Attention model has higher indicators on ROUGE-1,ROUGE-2 and ROUGE-L than the Seq2Seq model. The fine-tuning Seq2Seq Attention model scores lower on all three ROUGE indicators than the Seq2Seq Attention model, and the fine-tuning model scores the highest on all three ROUGE indicators. On the basis of the Seq2Seq Attention model, the use of Word2Vec pre-trained word vectors does not have a lifting effect on the original model, makes the performance of the model decline a little, while the use of RNNLM pre-trained word vectors has a significant improvement effect on the model.

The experimental results show that the Seq2Seq model with the Attention mechanism is better than the traditional Seq2Seq model in the automatic summary task. Since the mechanism scans the input word sequence once when each word is generated at the decoding end, then calculates the contribution of each input word to the current word to be generated. The trained model learns the input pertinently and associates each item in the output sequence with the input sequence. Therefore, the original Seq2Seq model with the Attention mechanism works well.

On the other hand, the experimental results also show that RNNLM pre-trained word vector can promote the model, and is better than Word2Vec pre-trained word vector. By using the word vector obtained from upstream task training to help the summary model in the downstream automatic summary task, this is actually a process of transfer learning. In transfer learning, it is best to have common parts or similarities between upstream task and downstream task, so that the knowledge learned by upstream task can promote downstream task. Since the Seq2Seq model is actually a language model when generating abstracts at the output, and the RNNLM model is a language model based on recurrent neural networks, the word vectors learned by the upstream RNNLM can promote the downstream abstract model. In Word2Vec, there is no similarity between CBOW and Skip-gram models in the process of using context to predict the current word and using current word to predict the context of the word. The abstract and reference abstract generated by the proposed cross-linguistic automatic summary method in this paper are semantically close, so it shows that this method is effective (Fig. 6).

Source and target language	original	Reference summary	Generate summary
ch-ch	利用AFLP(amplified fragment length polymorphism)技术筛选中华绒螯蟹性别相关标记,共使用192对引物组合检测中华绒螯蟹雄、雌和雌雄混合3个基因组池的多态性,共扩增出5376条多态性条带,平均每对引物组合扩增出28条多态性条带,获得88条雌雄差异条带.利用雌雄各10个个体再次进行AFLP验证,共62条差异条带具有性别差异……	利用AFLP方法筛选中华绒螯蟹性别相关标记	中华绒螯蟹性别标记
en-en	high frequency motion is always ignored in dynamic positioning operation and relevant paper pays attention to the effect of first order response on dynamic positioning accuracy via a deep water semi submersible drilling numerical model is calculated in both time domain simulations considering low frequency part only and considering that combined with first order motion then the statistic motion results with and without first order response are analyzed……	influence of the motion of deepwater platform on dynamic positioning precision	study on the effect of dynamic
kr-kr	아미노산 분석 방법 은 유리 아미노산 폴리펩티 드 또는 단백질 을 함유 한 샘플 중 의 아미노산 함유량 을 측정 하는 방법 이다 지난 여 년 동안 아미노산 분석 에 주로 전통 적 인 후컬럼 유도체 화 이온교환 크로마토그래피 와 전 컬럼 유도체 화……	질량 분석 기 술 이 아미노산 분석 방법 에서 의 이용	아미노산 의 아미노 산 및 질량 분석

(a)

Source and target language	original	Reference summary	Generate summary
en-ch	One EST sequence with high homology with ferritin gene of other species was found from the cDNA library of Sinonovacula constricta and then the complete expression sequence was obtained by PCR.The cDNA of this gene was 1106 bp, which consists of a 128 bp 5untranslated region (UTR), a 669 bp open reading frame (ORF) and a 309 bp 3 UTR……	缢蛏铁蛋白基因的分子特性及其表达分析	基因的克隆及表达分析
kr-ch	생식선 자극 호르몬 방출 호르몬 의 주요 작용 은 뇌하수체 성 선 자극 호르몬 의 방출 을 자극 하여 어류 의 성장 호르몬 방출 을 추진 하는 것 이다 황체 형성 호르몬 방출 호르몬 유사 체 는 포유류 의 유사 체 이다 본 논문 에서는 가 나 일 ….	LHRH-A 对 尼罗罗非鱼生长及生长轴相关基因表达的影响	外源对不同生长和基因的影响

(b)

Fig. 6. Examples of automatic summarization across languages

5 Conclusion and Future Work

This paper proposes a cross-linguistic automatic abstract method. The word vector is pre-trained based on the RNNLM model and then the attention mechanism is introduced based on the Seq2Seq framework. In the coding and decoding end, the text-abstract is trained by using the short text of the scientific and technological literature on Chinese, Korean and English languages, so that the text of one language can be the abstract to another directly. The experimental results show that the proposed method achieves 6%, 2.46% and 5.13% improvement in ROUGE-1,ROUGE-2 and ROUGE-L indicators on the cross-language automatic summary task under the same training times. In the case of no use of machine translation, it is also good to complete the cross-language automatic summary task. Since text and abstract are a unique one-to-one correspondence in the text abstract task. Training this unique correspondence requires large-scale training corpus, so the next step of this paper needs to collect more text-summary pairs to expand the training data. At the same time, because the corpus used in this experiment is short text, the effect of the system to generate long text abstracts needs to be verified, so in the next step, it is necessary to collect long text corpus to improve the ability of model processing long text.

Acknowledgements. This work was supported by the National Language Commission Scientific Research Project (YB135-76); Yanbian University Foreign Language and Literature First-Class Subject Construction Project (18YLPY13).

References

1. Graves, A., Schmidhuber, J.: Frame wise phoneme classification with bidirectional LSTM and other neural network architectures. Neural Netw. **18**(5–6), 602–610 (2005). https://doi.org/10.1016/j.neunet.2005.06.042
2. Vaswani, A., et al.: Attention is all you need. In: Advances in Neural Information Processing Systems, pp. 5999–6009 (2017)
3. Bahdanau, D., Cho, K.H., Bengio, Y.: Neural machine translation by jointly learning to align and translate. In: Computer Science (2014)
4. Erkan, G., Radev, D.R.: Lexrank: Graph-based lexical centrality as salience in text summarization. J. Artif. Intell. Res. **22**(1), 457–479 (2004). https://doi.org/10.1613/jair.1523
5. Sutskever, I., Vinyals, O., Le, Q.V.: Sequence to sequence learning with neural network advances. In: Neural Information Processing Systems (2014)
6. Kai, L., Hongling, W.: Research on coherence of automatic abstracts based on text rhetoric structure. J. Chin. Inf. **33**(1), 77–84 (2019)
7. Page, L., et al.: The pagerank citation ranking: Bringing order to the web. Stanf. Digital Libraries Work. Paper **9**(1), 1–14 (1998)
8. Lin, C.Y.: Rouge: A package for automatic evaluation of summaries. In: Proceedings of Workshop on Text Summarization Branches Out, Text Summarization Branches Out, pp. 74–81. Association for Computational Linguistics, Barcelona, Spain (2004)
9. Meng, X., Cui, R., Zhao, Y., et al.: Multilingual short text classification based on LDA and BiLSTM-CNN neural network. Web Inf. Syst. Appl. **2019**, 319–323 (2019)
10. Mikolov, T., et al.: Efficient estimation of word representations in vector space. In: Computer Science (2013)
11. Mikolov, T., Karafit, M., et al.: Recurrent neural network based language model. In: Proceedings of INTERSPEECH 2010, 11th Annual Conference of the International Speech Communication Association, pp. 26–30. Makuhari, Chiba, Japan (2010). https://doi.org/10.1109/EIDWT.2013.25
12. Luhn, H.P.: The automatic creation of literature abstracts. IBM J. Res. Dev. **2**(2), 159–165 (1958). https://doi.org/10.1147/rd.22.0159
13. Mihalcea, R., Tarau, P.: Textrank: bringing order into texts. In: Proceedings of the 2004 Conference on Empirical Methods in Natural Language Processing, pp. 404–411. EMNLP (2004)

14. Qingyu, Z., Nan, Y., Furu, W., et al.: Selective encoding for abstractive sentence summarization. In: Proceedings of the 55th Annual Meeting of the Association for Computational Linguistics, pp. 1095–1104 (2016). https://doi.org/10.18653/v1/P17-1101

15. Pascanu, R., Mikolov, T., Bengio, Y.: On the difficulty of training recurrent neural networks. In: International Conference on Machine Learning, pp. 1310–1318 (2013). https://doi.org/10.1007/s12088-011-0245-8

16. Chopra, S., Auli, M., Rush, A.M.: Abstractive sentence summarization with attentive recurrent neural networks. In: Proceedings of NAACL-HLT, pp. 93–98. NAACL, San Diego (2016). https://doi.org/10.18653/v1/N16-1012

17. Ma, S., et al.: A hierarchical end-to-end model for jointly improving text summarization and sentiment classification. In: IJCAI 2018: International Joint Conference on Artificial Intelligence, IJCAI (2018). https://doi.org/10.24963/ijcai.2018/591

18. Senin, P., Malinchik, S.: SAX-VSM: Interpretable time series classification using SAX and vector space model. In: 2013 IEEE 13th International Conference on Data Mining (ICDM), pp. 1175–1180. IEEE (2013). https://doi.org/10.1109/ICDM.2013.52

19. Shuai, W.: Tp-as: A two-stage automatic summary method for long texts. J. Chin. Inf. 32(6), 71–79 (2018)

20. Srivastava, N., Hinton, G., Krizhevsky, A., et al.: Dropout: A simple way to prevent neural networks from overfitting. J. Mach. Learn. Res. 15(1), 1929–1958 (2014)

21. Wan., X.J.: Bilingual information for cross-language document summarization. In: Proceedings of the 49th Annual Meeting of the Association for Computational Linguistics ACL, Portland, USA, pp. 1546–1555. ACL (2011)

Text Keyword Extraction Based on Multi-dimensional Features

Yu Jin, Rong Chen, and Lizhen Xu[✉]

School of Computer Science and Engineering, Southeast University,
Nanjing 211189, China
seu_yjin@163.com, {220181702,lzxu}@seu.edu.cn

Abstract. Keyword extraction is a fundamental task of text mining, so extracting high-quality keywords is of great significance. Typical keyword extraction algorithms usually rely on the statistical features, but lack of the semantic information. At the same time, the supervised keyword extraction algorithms rely too much on sample labeling. Therefore, in this paper, an unsupervised keyword extraction algorithm based on multi-dimensional features called MDFKE is proposed, which combines statistical features, external knowledge-based features and semantic features. MDFKE mainly studies the semantic information of candidate keywords. LDA model is used to obtain text topic, and Word2vec word embedding is used to generate word vectors. Based on these, the similarity between candidate keyword and text topic is quantified as semantic feature. Nine specific features are extracted from five aspects: term frequency, length, position, external knowledge base, and semantics. Finally, this paper clusters on feature vectors to obtain the final keyword set. The experiment turns out that, compared with traditional keyword extraction algorithms based on statistical features, MDFKE can significantly improve extraction performance, and can also make up for the shortage of supervised learning overly relying on labels.

Keywords: Keyword extraction · Multi-dimensional features · Text topic · Semantic feature

1 Introduction

Keyword extraction is widely used in many applications, such as document retrieval, text summary [1], and text classification [2], and has always been a research hotspot in NLP. Researchers have proposed many supervised and unsupervised KE methods.

The supervised method treats keyword extraction as a binary classification problem, and uses the learned classification model to determine whether the candidate keyword is a keyword. Training datasets are not easy to obtain and the labeling cost is very high. As an unsupervised learning method, clustering does not require labeling in advance. The traditional unsupervised keyword extraction algorithm including TF-IDF and TextRank ranks the importance of each candidate keyword. TF-IDF method ranks importance according to term frequency and inverse document frequency while TextRank is based on word co-occurrence [3]. On the basis of traditional algorithms, research in recent years has gradually introduced more dimensional features to describe

G. Wang et al. (Eds.): WISA 2020, LNCS 12432, pp. 248–259, 2020.
https://doi.org/10.1007/978-3-030-60029-7_23

the information of words as much as possible. In [4], the authors pay attention to complex network structure characteristics of word graphs to compensate for the limitations of TextRank refering to local co-occurrence relationship among text words. Literature[5] uses a combination of TF-IDF and TextRank algorithms to extract keywords from the text by constructing a word graph model, calculating word frequency and inverse document frequency and considering the weight of title position. In addition, the YAKE [6] algorithm proposed in 2016 comprehensively considers term frequency, length, position and other statistical information, but the evaluation algorithm relies on expert experience. At the same time, the algorithm lacks the consideration of semantic information, which is the common shortcoming of most research including traditional keyword extraction algorithms. Therefore, this article focuses on combining the semantic features of the text topic relevance of words with traditional statistical features to construct a new keyword extraction algorithm.

In this paper, an unsupervised keyword extraction algorithm based on multidimensional features (MDFKE) is studied. Nine features are selected from term frequency, length, position, external knowledge base, and semantics. Based on LDA and Word2vec, the relationship between candidate keyword and text topic is quantified as semantic feature. Finally, K-Means is used for unsupervised keyword extraction.

This paper is organized as follows. In Sect. 2, some related work is discussed. In Sect. 3, MDFKE algorithm is introduced. In Sect. 4, the experimental setup and results are described. Finally, in Sect. 5, the final conclusion is presented.

2   Related Work

2.1   Topic Model

Topic model [7] is a clustering model that mines the implicit semantic structure of the corpus in an unsupervised learning manner. The topic model is composed of three dimensions: document, topic and word. Each word of a text is obtained through a process of "selecting a certain topic with a certain probability and selecting a certain word from this topic with a certain probability".

At present, mature topic models include Probabilistic Latent Semantic Analysis (PLSA or PLSI) [8] and Latent Dirichlet Allocation (LDA) [9]. Both are improvements to Latent Semantic Indexing (LSI). LSI maps documents from a high-dimensional vector space to a low-dimensional latent semantic space, but it cannot provide a clear semantic interpretation. PLSA introduces probability and enhances the matching relationship between potential topics and vocabulary documents. Based on PLSA, LDA introduces a polynomial conjugate prior distribution Dirichlet to enrich the parameters.

In the field of NLP, more and more studies tend to use the LDA topic model to enrich the text representation. Literature[10] proposes a semantic Web service discovery method based on LDA clustering, which can supplement the semantic information of the text.

This paper obtains text topic by LDA, and uses the similarity between candidate keyword and text topic as the semantic feature. The topic words of the text are used as the initial clustering centers for iterative convergence to obtain the final keyword set.

2.2 Feature-Driven Keyword Extraction

Feature extraction is one key step in the keyword extraction process. The purpose is to select some features that distinguish keywords and non-keywords as well as possible.

Frank [11] for the first time regarded the keyword extraction as a binary classification problem, using TFIDF and relative position as features, and proposed a KEA algorithm based on the Naive Bayes model. The algorithm proposed by Aquino [12] involves 20 attributes such as the term frequency, the first occurrence position, the last occurrence position, and whether it is a proper noun. In [13] it is proved that the words usually used as Wikipedia entries are more likely to be keywords. Literature [14] uses sequence pattern mining algorithms to generate candidate keywords, and then uses different classification algorithms to extract keywords. However, as a supervised learning, binary classification depends on the result of labeling and the labeling process is subjective.

Keywords should be as consistent as possible with the topic of the text, so this paper proposes semantic features to quantify the relevance of words to the topic of the text. A clustering algorithm combining statistical features, linguistic features, and semantic features is used to generate the final keyword set.

3   Text Keyword Extraction Based on Multi-dimensional Features

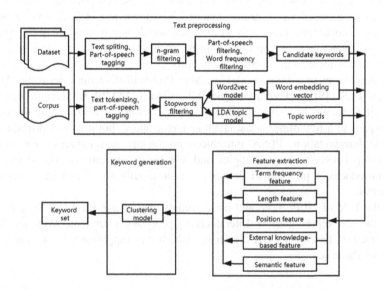

Fig. 1. Flow chart of MDFKE

Some existing KE algorithms only use statistical features for binary classification learning, ignoring the correlation between keyword and topic, and relying heavily on sample labeling. This paper combines statistical features, external knowledge-based features and semantic features, and proposes an unsupervised keyword extraction method (MDFKE). The method is composed of three parts: text preprocessing, feature extraction and cluster-based keyword extraction. Figure 1 shows the overall process.

3.1 Text Preprocessing

The first goal of text preprocessing is to obtain candidate keywords which may be composed of multiple words. The process of generating candidate keywords is as follows:

(1) Use stopwords and symbols as separators to split text and mark the part-of- speech of each word.
(2) Use n-gram filtering to obtain all n-gram word sequences (n $\leq$ 3).
(3) Use part-of-speech filtering to preserve noun phrases and adjective phrases.
(4) Use word-frequency filtering to keep phrases with higher frequency.

Another purpose of text preprocessing is to obtain text topic. LDA model assumes that each document is composed of multiple topics and each topic is a probability distribution on the set of words. The probability graph model of LDA is shown in Fig. 2.

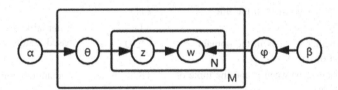

Fig. 2. LDA model diagram

The joint probability in LDA model is as (1):

$$P(\theta, z, w | \alpha, \beta) = P(\theta|\alpha) \prod_{n=1}^{N} P(z_n|\theta)P(w|z_n, \beta) \tag{1}$$

The document-topic distribution θ and topic-word distribution φ are both polynomial distributions. α is the Dirichlet prior parameter of θ, and β is the Dirichlet prior parameter of φ. EM algorithm and Gibbs sampling are two algorithms to solve the LDA model. EM algorithm is used to estimate the maximum likelihood of parameters in a probability model that depends on unobservable implicit variables, but the entire process is complicated and slow. At present, a more concise and fast Gibbs sampling is widely used in practice. The number of topics needs to be specified in advance.

The document-topic and topic-word distribution are obtained after training. This paper defines the concept of the topic importance $P\_T(w_j|d)$ to characterize the probability that the word w_j is the topic word of document d. The calculation formula is as (2):

$$P\_T(w_j|d) = \sum_{i=1}^{T} P(w_j|t_i) * P(t_i|d) \tag{2}$$

Here, $P(t_i|d)$ is the probability that document d belongs to topic t_i, and $P(w_j|t_i)$ is the probability that word w_j belongs to topic t_i. The k words with the largest topic importance value are selected as topic word set of the text recorded as $V = \{v_1, v_2, \ldots v_k\}$. In addition, the importance set matching V is recorded as $P = \{p_1, p_2, \ldots p_k\}$.

3.2 Feature Extraction

Table 1. Feature type and description

Num	Definition	Abbreviation	Type
1	Term frequency and inverse document frequency	TFIDF	Term-frequency feature
2	Document maximum phrase index	DMPI	
3	Word length	WL	Length feature
4	Average sentence length	ASL	
5	Position span	SPAN	Position feature
6	Whether it exists in the title	EIT	
7	Keyword degree	KD	External knowledge-based feature
8	Whether it exists in the domain dictionary	EIDD	
9	Distance between it and the topic of article	TS	Semantic feature

Keywords should reflect the topic of the text and the importance of the word. Statistical features show the position and frequency of terms in the text, and external knowledge-base features show the importance of the word out of the text. In addition, semantic feature is proposed to describe the similarity between candidate keyword and text topic. 9 specific features are introduced to describe the importance of keywords (Table 1).

Term-Frequency Feature
Term frequency refers to the frequency of words or phrases appearing in a given document. It is generally believed that the higher the term frequency, the higher its importance. TFIDF combines term frequency and inverse document frequency to measure the importance of candidate keywords. Document Maximum Phrase Index (DMPI) [15] is extended by term frequency and characterizes the overlap of candidate keywords.

$$\text{DMPI(w, d)} = 1 - \max_{s \in \sup(w,d)} \frac{f(s,d)}{f(w,d)} \tag{3}$$

$\sup(w, d)$ is the set of phrases in the document d that contains the word w. $f(w, d)$ is the number of times the word w appears in the document d.

Length Feature
The length feature refers to the length of the candidate keyword itself and the sentence in which it is located. Word length (WL) refers to the number of words contained in the candidate keywords. Because the length of the keyword is usually less than or equal to 3, WL has a good distinction. Average sentence length (ASL) refers to the average number of words of all sentences containing candidate keywords. It is generally considered that the longer the sentence where the candidate keyword is located, the greater the likelihood that it becomes a keyword.

Position Feature
Keywords often appear in certain important positions, such as the beginning of a document and the beginning of a paragraph. In this paper, SPAN and EIT are used as position features. SPAN considers the difference between the first and last occurrence of word in a text. EIT indicates whether it appears in the title.

$$\text{SPAN}(w, d) = \text{LP(w, d)} - \text{FP(w, d)} \tag{4}$$

FP (w, d) is the relative position of the first occurrence, and LP (w, d) is the relative position of the last occurrence. The candidate keywords that appear in the title are usually more likely to be keywords. If the candidate keyword appears in the title as a whole, EIT is 1, the partial occurrence is 0.5, and the non-appearance is 0.

External Knowledge-Based Feature
This article uses external knowledge-based features to characterize the importance of candidate keywords. Keyword degree (KD) [16] is the frequency of candidate keywords as keywords in a given corpus. The candidate keywords that can be queried in the domain dictionary are more important [17]. The open Chinese thesaurus launched by Tsinghua University contains dictionaries in IT, finance and other fields. If the candidate keyword appears in the domain dictionary as a whole, EIDD is 1, the partial occurrence is 0.5, and the non-appearance is 0.

Semantic Feature
Semantic feature is one key point of this paper. In this paper, LDA is used to obtain text topic. Word2vec is used to train a distributed representation of the words to use a low-dimensional and dense vector to represent the words, so that the word similarity involving contextual semantic information can be calculated. The distance between candidate keyword and topic is used as semantic feature. The smaller the distance, the greater the similarity and the closer the candidate keyword are to the text topic.

The candidate keywords are n-gram word sequences, n $\leq$ 3. The distance between word sequence and topic is the average of the distances between each word and topic in the sequence. The calculation formula of the semantic feature $TS(w, V)$ is as (5):

$$TS(w, V) = \frac{\sum_{i=1}^{n} Dis(w_i, V)}{n} \tag{5}$$

$$Dis(w_i, V) = \frac{\sum_{j=1}^{k} p_j * dis(w_i, v_j)}{\sum_{j=1}^{k} p_j} \tag{6}$$

$Dis(w_i, V)$ is the distance between the i-th word in the candidate keyword and text topic. v_j is the j-th component of topic word set V, that is, the j-th topic word. The cosine distance between the word vector of w_i and v_j based on Word2vec is $dis(w_i, v_j)$. p_j is the topic importance of v_j as weight or probability.

3.3   Cluster-Based Keyword Extraction

Supervised keyword extraction needs to label samples in advance, and the classification performance depends heavily on labeling accuracy. The clustering is unsupervised, so this paper uses the classic K-Means clustering algorithm which uses the k topic words as initial clustering centers, and calculates the distance between each candidate keyword and each clustering center based on multi-dimensional features. Based on the initial clustering center, the Euclidean distance between each candidate keyword and each clustering center is calculated, and then each candidate keyword is assigned to the nearest cluster to obtain k clusters $\{S_1, S_2, \ldots, S_k\}$.

$$dis(v_i, c_j) = \sqrt{(v_{i,1} - c_{j,1})^2 + \ldots + (v_{i,d} - c_{j,d})^2} = \sqrt{\sum_{t=1}^{d} (v_{i,t} - c_{j,t})^2} \tag{7}$$

$v_{i,t}$ represents the t-th attribute of the i-th feature vector v_i; $c_{j,t}$ represents the t-th attribute of the cluster center of the j-th cluster c_j. m is the number of all candidate keywords in the text, and k is the number of cluster centers. Here, $i \in [1, m], j \in [1, k]$. The goal of clustering is to get the division that can minimize the distance between each cluster center and the objects in this cluster. The objective function is as (8):

$$J = argmin \sum_{j=1}^{k} \sum_{i=1}^{m} dis(v_i, c_j) \tag{8}$$

If the objective function does not meet the convergence criterion, the members in the cluster are updated repeatedly until the final clustering partition is reached. The final clustering centers correspond to the final keywords of the text.

4 Experiments

4.1 Datasets and Evaluation Indicators

Taking into account the reality that vocabulary are huge, this paper selects the computer field as a smaller research area to verify the results and performance of the algorithm. This article collects the latest academic papers in the computer field, and selects the title, abstract and keywords as the experimental data. The experimental data set contains 1,250 academic articles and 6,149 keywords provided by the author.

In order to evaluate the performance of MDFKE, the dataset contains the keywords marked in advance. This paper calculates the precision rate P, recall rate R and F1 value by comparing the keywords obtained by MDFKE with those provided keywords. P is "the number of correct predictions in the data predicted as keywords". R is "the number of correct predictions in the data that are truly keywords" (Table 2).

Table 2. Parameters required for evaluation indicators

Actual value	Predictive value	
	Non-keyword	Keyword
Non-keyword	TN	FP
Keyword	FN	TP

$$p = \frac{TP}{TP + FP} \tag{9}$$

$$R = \frac{TP}{TP + FN} \tag{10}$$

$$F1 = \frac{2 \times P \times R}{P + R} \tag{11}$$

F1 score is the reconciled average of precision and recall. Only when both Precision and Recall are very high, the F1 score will be high. If one of them is very low, the harmonic average will be pulled close to that very low number.

4.2 Experiment Settings

The LTP tool of Harbin Institute of Technology is used to segment the Chinese text and tag the part-of-speech of each word in text preprocessing. First of all, this paper splits text and marks part-of-speech of each word, then uses n-gram model to obtain n-gram word sequences, and performs part-of-speech filtering and term frequency filtering. However, the term frequency is not easy to display here, so Table 3 shows the candidate keywords initially obtained after part-of-speech filtering of two sample sentences.

Table 3. Examples of candidate keywords after part-of-speech filtering

Raw text	Candidate keywords		
	n=1	n=2	n=3
将这些特征作为神经网络的输入,再经过卷积层进一步提炼类别表达能力更强的高层次文本特征,从而提高模型分类的准确率。	特征 神经 网络 卷积层 类别 能力 强 高层次 文本 特征 模型 准确率	神经网络 经过卷积层 卷积层进一步 提炼类别 类别表达 表达能力 能力更 更强 高层次 层次文本 文本特征 提高模型 模型分类	再经过卷积层 经过卷积层进一步 卷积层进一步提炼 进一步提炼类别 提炼类别表达 类别表达能力 表达能力更 能力更强 高层次文本 层次文本特征 提炼提高模型分类
图像分类是计算机视觉领域内一个重要研究问题,在自然场景理解和工业检测等图像分析任务中具有广泛的应用。	图像 分类 计算机 视觉 领域 重要 问题 自然 场景 工业 图像 任务 广泛	图像分类 计算机视觉 视觉领域 领域内 一个重要 重要研究 研究问题 自然场景 场景理解 工业检测 图像分析 分析任务 任务中 具有广泛	计算机视觉领域 视觉领域内 领域内一个 内一个重要 一个重要研究 重要研究问题 自然场景理解 图像分析任务 分析任务中 任务中具有 中具有广泛

The word embedding model used in this paper is the skip-gram model of Word2vec, which uses the default parameters for training. The corpus used for training Word2vec is full text of the academic journal papers collected in this paper. The training of LDA and the training of the classification and clustering model respectively use the topic model toolkit genism and the machine learning module sklearn based on python.

4.3   Experiment Results

Through feature extraction, this paper extracts 9 specific features in five aspects: term frequency, length, position, external knowledge base, and semantics. K-Means clustering uses k topic words acquired based on LDA as the initial clustering centers to obtain the final keyword set.

Table 4. Comparison of results for different k

k	Precision	Recall	F1
3	0.39	0.65	0.49
4	0.42	0.63	0.50
5	0.45	0.61	0.52
6	0.48	0.54	0.51
7	0.51	0.47	0.49
8	0.52	0.45	0.48

In order to verify the performance of MDFKE, this paper conducts a comparative test for different number of keywords k in the range of 3 to 8. Precision is "the number of correct predictions in the data predicted as keywords" and recall is "the number of correct predictions in the data that are truly keywords". Therefore, Table 4 and Fig. 3 show that as k increases, recall continues to decrease and precision continues to increase. However, the F1 score, as the reconciled average of precision and recall, can better reflect the performance of the algorithm. F1 score increases first and then decreases, and the method performs best when k is 5.

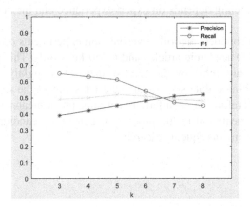

Fig. 3. Comparison of results for different k

MDFKE combines statistical features with external knowledge-based features and semantic features. Table 5 shows that MDFKE is significantly better than classic keyword extraction algorithms, such as TFIDF, TextRank, and YAKE.

Table 5. Comparison of different keyword extraction algorithms

Num	Algorithm	Precision	Recall	F1
1	TFIDF	0.33	0.43	0.37
2	TextRank	0.28	0.35	0.31
3	YAKE	0.39	0.48	0.43
4	MDFKE(NO-TS)	0.42	0.50	0.46
5	MDFKE	0.45	0.61	0.52
6	MDFKE(SVM)	0.47	0.63	0.54
7	MDFKE(NB)	0.47	0.65	0.55
8	MDFKE(DT)	0.45	0.64	0.53

Row 4 and 5 in Table 5 are the comparative experiments after removing and retaining the semantic feature in MDFKE. The experiment proves that after adding semantic features, keyword extraction performs better and F1 value is increased by 0.06. The importance of semantic feature means that the words that are closer to the text topic are more likely to become keywords.

In order to conduct a comparative experiment, this paper conducted a supervised keyword extraction experiment based on multi-dimensional features and classification algorithms (SVM, NB, DT). Comparing row 5–8 in Table 5, it is found that in the case of labeling, the performance of the method using classification algorithm is slightly better, but the advantage of clustering is that it is more suitable for the case of no labeling, because labeling depends on expert experience and the cost of labeling is too high.

This paper also conducts algorithm verification experiments on an English data set which contains 1,300 academic articles and 6,450 keywords. This paper finds that the performance of the algorithm on the English data set is slightly better than the performance on the Chinese data set, because the F1 score of the former is 0.03 higher than the latter. Therefore, it is inferred that this phenomenon may be caused by improper word segmentation in the process of generating candidate words. We will continue to study it in subsequent research.

5   Conclusion

This paper proposes a keyword extraction algorithm based on multi-dimensional features. Combining term frequency, length, position, external knowledge-based and semantic feature, MDFKE can describe word information more comprehensively and better explain word importance. Semantic feature is defined as the similarity between candidate keyword and text topic, emphasizing that keywords need to be close to the text topic. The addition of semantic features increases the F1 value of the keyword extraction algorithm by 0.06. In addition, the use of clustering algorithms for unsupervised learning eliminates the deficiencies of supervised learning without labeling in advance and the over-reliance of supervised learning on labeling.

References

1. Kim, S.N., Medelyan, O., Kan, M.Y., Baldwin, T.: Automatic keyphrase extraction from scientific articles. Lang. Resour. Eval. 723–742 (2013). https://doi.org/10.1007/s10579-012-9210-3
2. Hassaine, A., Mecheter, S., Jaoua, A.: Text categorization using hyper rectangular keyword extraction: application to news articles classification. In: Kahl, W., Winter, M., Oliveira, J.N. (eds.) RAMICS 2015. LNCS, vol. 9348, pp. 312–325. Springer, Cham (2015). https://doi.org/10.1007/978-3-319-24704-5_19
3. Lu, G., Xia, Y., Wang, J., Yang, Z.: Research on text classification based on TextRank. In: 2016 International Conference on Communications, Information Management and Network Security. Atlantis Press (2016). https://doi.org/10.2991/cimns-16.2016.79

4. Qingyun, Z., Yuansheng, F., Zhenlei, S., Wanli, Z.: Keyword extraction method for complex nodes based on TextRank algorithm. In: 2020 International Conference on Computer Engineering and Application (ICCEA), pp. 359–363. IEEE (2020). https://doi.org/10.1109/iccea50009.2020.00084

5. Yao, L., Pengzhou, Z., Chi, Z.: Research on news keyword extraction technology based on TF-IDF and TextRank. In: 2019 IEEE/ACIS 18th International Conference on Computer and Information Science (ICIS), pp. 452–455. IEEE Computer Society (2019). https://doi.org/10.1109/icis46139.2019.8940293

6. Campos, R., Mangaravite, V., Pasquali, A., Jorge, A.M., Nunes, C., Jatowt, A.: A text feature based automatic keyword extraction method for single documents. In: Pasi, G., Piwowarski, B., Azzopardi, L., Hanbury, A. (eds.) ECIR 2018. LNCS, vol. 10772, pp. 684–691. Springer, Cham (2018). https://doi.org/10.1007/978-3-319-76941-7_63

7. Sheng, L., Xu, L.: Topic classification based on improved word embedding. In: 2017 14th Web Information Systems and Applications Conference (WISA), pp. 117–121. IEEE. https://doi.org/10.1109/wisa.2017.44

8. Hofmann, T.: Probabilistic latent semantic analysis. arXiv preprint arXiv:1301.6705 (2013)

9. Blei, D.M., Ng, A.Y., Jordan, M.I.: Latent dirichlet allocation. J. Mach. Learn. Res. 3, 993–1022 (2003)

10. Zhao, H., Chen, J., Xu, L.: Semantic web service discovery based on LDA clustering. In: Ni, W., Wang, X., Song, W., Li, Y. (eds.) WISA 2019. LNCS, vol. 11817, pp. 239–250. Springer, Cham (2019). https://doi.org/10.1007/978-3-030-30952-7_25

11. Frank, E., et al.: Domain-specific keyphrase extraction. In: Proceedings of the Sixteenth International Joint Conference on Artificial Intelligence (1999)

12. Aquino, G.O., Lanzarini, L.C.: Keyword identification in spanish documents using neural networks. Journal of Computer Science & Technology, 15 (2015)

13. Berend, G.: Exploiting extra-textual and linguistic information in keyphrase extraction. Nat. Lang. Eng. 22(1), 73 (2016)

14. Xie, F., Wu, X., Zhu, X.: Efficient sequential pattern mining with wildcards for keyphrase extraction. Knowl.-Based Syst. 115, 27–39 (2017). https://doi.org/10.1016/j.knosys.2016.10.011

15. Haddoud, M., Abdeddaïm, S.: Accurate keyphrase extraction by discriminating overlapping phrases. J. Inf. Sci. 40(4), 488–500 (2014). https://doi.org/10.1177/0165551514530210

16. Medelyan, O., Frank, E., Witten, I.H.: Human-competitive tagging using automatic keyphrase extraction. In: Proceedings of the 2009 Conference on Empirical Methods in Natural Language Processing, pp. 1318–1327(2009)

17. Zhou, X., Zhang, X., Hu, X.: MaxMatcher: biological concept extraction using approximate dictionary lookup. In: Yang, Q., Webb, G. (eds.) PRICAI 2006. LNCS (LNAI), vol. 4099, pp. 1145–1149. Springer, Heidelberg (2006). https://doi.org/10.1007/978-3-540-36668-3_150

Improvement of Short Text Clustering Based on Weighted Word Embeddings

Nan Yang, Qing Liu$^{(\boxtimes)}$, and Yaping Li

School of Information, Renmin University of China, Beijing 100872, China
{yangnan,qliu,ypli}@ruc.edu.cn

Abstract. The data sparseness problem in short text clustering will causes low clustering performance. One solution is to enrich short text according to the semantic relationship from external text corpus. A new one is neural network based text representation learning which is word embeddibngs. In this paper, we studied the methods of vector to represent a short text. But how to get a vector from word embeddings is a challenge job. One way is average sum of vectors, but it ignores the importance of the terms. TF-IDF weighted vectors is better way, but the sparseness of terms makes the local IDF not sufficient. We proposed a new method with TF-GIDF weighted vectors, which use global IDF to conquer the shortcoming. The experiments are set up to compare the new method with baselines and the results analysis shows that the proposed method outperforms baselines significantly.

Keywords: Short text clustering · Word embeddings · Term frequency · Inverse document frequency

1 Introduction

Nowadays the social media has generated vast quantities of information on the internet, for example, such as Web search snippets, book and movie summaries, product descriptions, and customer reviews. The clustering analysis on these information is very important. Due to the data sparseness problem, under the traditional Vector Space Model (VSM) and TF-IDF weight schema, the clustering performance is very low [10]. The reason is no enough word co-occurrence between short texts. There have been several studies that attempted to overcome the data sparseness problem. The solution is to enrich short text by a external information resources to enhance the semantic relationship between documents, for instance, by search engines [3], by Wordnet [6] and by Wikipedia [1]. These researches have shown positive improvement.

Recently, deep neural networks and representation learning have led to new ideas and many neural models for learning word representations have been proposed. Inspired by the ideas, we studied the method of vector to represent a short text. A short text can be represented by Bag Of Words (BOW) model and can be mapped into a sequence of vectors by using word embeddings. Calculating a

© Springer Nature Switzerland AG 2020
G. Wang et al. (Eds.): WISA 2020, LNCS 12432, pp. 260–267, 2020.
https://doi.org/10.1007/978-3-030-60029-7_24

vector to present a short text is a key problem. The simple way is average sum of vectors, but it ignores the importance of the terms. TF-IDF weighted sum is a better choose, but the IDF weight can not catch enough importance of the text due to term sparseness. To solve the problem, we proposed a new method with global IDF derived from a large external text corpus. To demonstrate the effectiveness of our method, we design experiments under two public datasets ODP239 and Searchsnippets.

The rest of the paper is organized as follows. In Sect. 2, we briefly review previous research works on neural network based text representation learning methods. Section 3 introduces the Word2Vec, a model of neural network of word embeddings. In Sect. 4, we study the methods of a vector to represent a short text and propose a new method with TF-GIDF weighted sum. In Sect. 5, we design the experiments under two public datasets for comparative studies and the results analysis. We give conclusions and future works in Sect. 6.

2   Related Work

Recently, deep neural networks and representation learning have led to new ideas for solving the data sparseness problem, and many neural models for learning word representations have been proposed [2,5,7–9]. The neural representation of a word is called word embeddings and in form of a real valued vector. The word embeddings enables us to measure word relatedness by simply using the distance between two embedding vectors. With the pre-trained word embeddings, neural networks demonstrate their great performance in many Natural Language Processing(NLP) tasks. Several deep learning based methods, such as Word2Vec and Doc2Vec, have been introduced to analyze text corpora [4]. According to these methods once the model is trained over a representative corpus, it can be readily used to analyze new text and find semantic constructs which can be useful for automated taxonomy creation. Classical Word2Vec methods are generally unsupervised requiring no domain information. In the context of NLP, distributed models are able to learn word representations in a low-dimensional continuous vector space using a surrounding word context in a sentence, such that semantically similar words are close to each other in the embedding space. There are a few literatures about short text clustering using deep neural networks [11]. But their works are different from ours. They use different text corpus to train the model to deal with different datasets. In our work, all datasets are tested under a unique pre-trained model.

3   Word Embeddings and Word2Vec

Mikolov et al. [8] proposed distributed word embeddings to catch semantic similarities between words, which resulted in Google's Word2Vec software open project[1]. A word embedding uses numeric value, in form of a vector to represent

---

[1] https://code.google.com/archive/p/word2vec/.

a word. The Word2Vec has two models: Continuous Bag of Words(CBOW) and Skip-Gram.

CBOW model predicts the current word under window size of n. Given word sequence $W = \{w_1, w_2, ..., w_M\}$, maximization the average logarithmic probabilities is defined as:

$$I(W) = \frac{1}{M} \sum_{i=L}^{M-L} \log p(w_i/w_{i-L}, ..., w_{i+L}) \tag{1}$$

where, L is the window size of context and M is the size of word sequence.

Skip-Gram model predicts the words in context in the window size of n of the current word. Given word sequence $W = \{w_1, w_2, ..., w_M\}$, maximization the average logarithmic probabilities is defined as:

$$I(W) = \frac{1}{M} \sum_{m=1}^{M} \sum_{-L \leq i < L} \log p(w_{m+i}/w_m) \tag{2}$$

where, L is the window size of context and M is the size of word sequence.

4 TF-GIDF Weighted Embedding Vectors

In this section, we will describe traditional VSM and TF-IDF weight schema. Then we studied the distributed presentation learning method and two methods to present a short text. A new method with TF-GIDF weighted way by using global IDF is proposed.

4.1 VSM and TF-IDF Weight Schema

A test dataset $DS = \{d_1, d_2, ..., d_n\}$ consists of n docs and each doc d_i is represented by a sequence of terms(words), $d_i = \{t_{i1}, t_{i2}, ..., t_{ik}\}$, k is the number of terms. In the future, we will use "words" and "terms" interchangablly. After all docs have been processed, a vocabulary including all terms $V_{voc} = \{tw_1, tw_2, ..., tw_m\}$ is created, m is the size of vocabulary. Then, the DS will be presented as a matrix A, with dimension $n \times m$ and each entry is the weight $a_{i,j} = TF \times IDF$. The value of weight has two parts: TF and IDF. TF is the frequency of terms in d_i, which reflects the importance of the term in d_i. IDF is inverse document frequency, which reflects the term importance over all docs.

4.2 Weighted Vectors of Word Embeddings

The algorithm Word2Vec [8] is originally a highly scalable predictive model for learning word embeddings from text. The word embeddings have the capability to capture linguistic regularities and patterns. Although the word embeddings has applied to many NLP tasks, it is a semantic relationships about words. In our case, a short text usually consists of several sentences so that we need to study how to represent a short text with a vector.

Average Weighted Vector. A direct way to get a vector from the vectors of n terms is to use terms count as the weight. A short text is, if the duplicate of terms is considered, the sequence of the two-tuples, $D = \{(t_1, tc_1), (t_2, tc_2), ..., (t_n, tc_n)\}$, tc_i is the count of t_i in the text. Here, we let tc_i be the weight of t_i, the weighted average of vectors is as following:

$$V_d = \frac{\sum_{i=1}^{n} tc_i \times V_m[t_i] \times \theta(t_i, V_{voc})}{\sum_{i=1}^{n} tc_i \times \theta(t_i, V_{voc})} \tag{3}$$

In this way, average weighted vector treats each term equally and not take the term importance into account. The next, we will combine the weights of terms and the vectors of terms together to produce a text vector.

TF-IDF Weighted Vector. TF-IDF weighted method gives the importance of terms and we use TF-IDF as the weights. Given a text, $D = \{t_1, t_2, ..., t_n\}$ and the weights $W = \{w_1, w_2, ..., w_n\}$. The weight w_i of term t_i is defined as TF-IDF weight in previous section. The vector V_d of a short text D is defined as:

$$V_d = \frac{\sum_{i=1}^{n} tf_i \times idf_i \times V_m[t_i] \times \theta(t_i, V_{voc})}{\sum_{i=1}^{n} tf_i \times idf_i \times \theta(t_i, V_{voc})} \tag{4}$$

Because majority of terms appears only once in a doc, co-occurrence of terms is very low. Due to the spareness problem, IDF can not catch the importance of the terms either.

TF-GIDF Weighted Vector. IDF is the term importance distribution over the docs. The more frequently a term appears in docs, the less important it is. Now we define the value of IDF derived from test dataset as local IDF. Because the number of docs in test datasets is small, local IDF can not catch the importance of the terms. For this reasons, we introduce global IDF, which catches the importance of the terms from global view. We use a large external text corpus to compute global IDF. The large text corpus contains a large amount of short texts such that the values of global IDF has statistical importance. Given a text $D = \{t_1, t_2, ..., t_n\}$ and the weights $W = \{w_1, w_2, ..., w_n\}$. The weight w_i of term t_i is defined as TF-GIDF weight, where tf_i is same as in previous section and GIDF is defined as:

$$gidf_i = \frac{n_g}{log(gdc_i + 1)} \tag{5}$$

where, n_g is number of docs and gdc_i is numbers of docs which contain term t_i in an external large text corpus. The vector V_d of a short text D is defined as:

$$V_d = \frac{\sum_{i=1}^{n} tf_i \times gidf_i \times V_m[t_i] \times \theta(t_i, V_{voc})}{\sum_{i=1}^{n} tf_i \times gidf_i \times \theta(t_i, V_{voc})} \tag{6}$$

5   Experimental Design and Results

5.1   Train Dataset

The Word2Vec model is trained under a large text corpus. We download the ODP project dump as corpus, it includes 1938099 short texts. We use the program under google's trunk project to train the corpus with dimension 200 and also use the corpus to calculate word and document frequencies.

5.2   Test Datasets

We use two public datasets to evaluate our method: ODP239[2] and SearchSnippets[3]. **ODP239** was extracted from Open Directory Project (ODP). It includes 239 topics, each with 10 subtopics and about 100 documents (about 10 per subtopic), for a total number of 25580 documents. Each document has 4 itesms: ID, url, title and snippet. **SearchSnippets** [10,11] was selected from the results of web search transaction using predefined phrases of 8 different topics. It contains training data with 10060 snippets and testing data with 2280 snippets. We sum both as our test dataset.

5.3   Evaluation Metrics

To evaluate the clustering performance, two metrics: Normalized Mutual Information(NMI) and Accuracy(ACC) are chosen as our metrics.

$$NMI = MI(X,Y)/\sqrt{H(X)H(Y)}. \tag{7}$$

where, $MI(Y,C)$ is the mutual information between Y and C, $H(\cdot)$ is entropy and the denominator $\sqrt{H(Y)H(C)}$ is used for normalizing the mutual information to be in the range of $[0, 1]$.

$$ACC = \frac{1}{n} \sum_{i=1}^{n} \delta(t_i, map(c_i)). \tag{8}$$

where, n is the total number of texts, $\delta(x,y)$ is the indicator function that equals one if $x = y$ and equals zero otherwise, and $map(c_i)$ is the permutation mapping function that maps each cluster label ci to the equivalent label from the text data.

5.4   Clustering Tool

For fairness, all methods should be applied under same clustering platform. After careful consideration, cluto[4] is chosen as our platform. Cluto is software used

---

[2] http://credo.fub.it/odp239/odp239.tar.gz.
[3] http://jwebpro.sourceforge.net/data-web-snippets.tar.gz.
[4] http://glaros.dtc.umn.edu/gkhome/views/cluto.

to high dimension clustering, developed by Minnesota University. The cluto is a partitional-based criterion-driven clustering algorithms and acts as an optimization process which seeks to maximize or minimize a particular clustering criterion function. In our experiment, the similarity method parameter sim is set as 'cosine', clustering method parameter $clmethod$ is set as 'rb' (repeated bisections) and criterion function is selected as i_2, other parameters are left as default settings.

5.5   Baseline Methods

The baseline methods that are employed for the comparison are as follows. TFIDF stands for Navie TF-IDF scema. AWE stands for Average Weighted

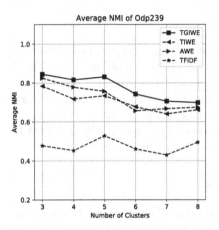

Fig. 1. Average NMI under ODP239

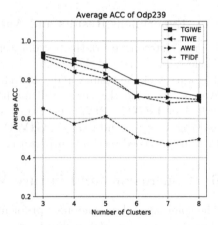

Fig. 2. Average ACC under ODP239

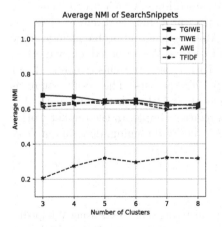

Fig. 3. Average MNI under Search-snippets

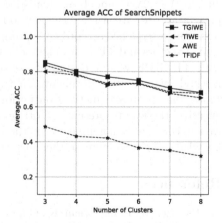

Fig. 4. Average ACC under Search-snippets

Embeddings. TIWE stands for TF-IDF Weighted Embeddings. TGIWE stands for TF-GIDF Weighted Embeddings, our method.

Table 1. Comparsion of average NMI and ACC of different methods.

Method	ODP239		SearchSnippets	
	NMI(%)	ACC(%)	NMI(%)	ACC(%)
TFIDF	47 ± 11.30	55 ± 9.15	29 ± 4.95	40 ± 4.27
AWE	73 ± 10.93	79 ± 8.78	62 ± 9.67	73 ± 7.63
TIWE	70 ± 10.63	77 ± 8.67	63 ± 10.10	73 ± 8.63
TGIWE	$\mathbf{77 \pm 10.90}$	$\mathbf{83 \pm 9.00}$	$\mathbf{65 \pm 9.57}$	$\mathbf{76 \pm 7.22}$

5.6 Experiment Results and Analysis

We run cluto on each of the test datasets. The experimental results are shown in Table 1. The curves of the results under ODP239 are shown in Fig. 1, and Fig. 2. The curves of the results under Searchsnippets are shown in Fig. 3 and Fig. 4. We can see from the curves, all methods based on word embeddings perform well than **TFIDF**. **TIWE** performs better than **AWE**. In the curves of Fig. 1 and Fig. 2, Fig. 3 and Fig. 4, **TGIWE** performs better than **AWE** and **TIWE**. In sum, proposed method outperforms others in average metrics.

6   Conclusions and Future Works

The solution of data spareness problem in short text clustering is to enrich the them by semantic relationship from a external text corpus. However, the deep neural network and distributed presentation learning has led to new ideas. Based on the Word2Vec model, also called word embeddings, which uses a vector to present a word. We studied the methods to present a short text and to present a text from a sequence of word embeddings. After considering problems in average sum of vectors and TF-IDF weighted sum of vectors. We proposed a new method with TF-GIDF weighted sum of vectors. To evaluate the effectiveness of our method, we set up experiments on two public datasets. The results analysis shows that proposed method outperforms others. Although the our method is effective, the way to present a short text is relative simple. In the further works, we will consider to combine our method with the POS tagging, domain ontology knowledge, etc., to increase the clustering performance.

References

1. Banerjee, S., Ramanathan, K., Gupta, A.: Clustering short texts using Wikipedia. In: SIGIR 2007: Proceedings of the 30th Annual International ACM SIGIR Conference on Research and Development in Information Retrieval, Amsterdam, The Netherlands, July 23–27, 2007, pp. 787–788. ACM (2007)

2. Bengio, Y., Courville, A.C., Vincent, P.: Representation learning: a review and new perspectives. IEEE Trans. Pattern Anal. Mach. Intell. **35**(8), 1798–1828 (2013)
3. Bollegala, D., Matsuo, Y., Ishizuka, M.: Measuring semantic similarity between words using web search engines. In: Proceedings of the 16th International Conference on World Wide Web, WWW 2007, Banff, Alberta, Canada, May 8–12, 2007, pp. 757–766. ACM (2007)
4. Ghosh, S., Chakraborty, P., Cohn, E., Brownstein, J.S., Ramakrishnan, N.: Characterizing diseases from unstructured text: A vocabulary driven word2vec approach. In: Proceedings of the 25th ACM International Conference on Information and Knowledge Management, CIKM 2016, Indianapolis, IN, USA, October 24–28, 2016. pp. 1129–1138. ACM (2016)
5. He, T., Lian, H., Qin, Z., Zou, Z., Luo, B.: Word embedding based document similarity for the inferring of penalty. In: Meng, X., Li, R., Wang, K., Niu, B., Wang, X., Zhao, G. (eds.) WISA 2018. LNCS, vol. 11242, pp. 240–251. Springer, Cham (2018). https://doi.org/10.1007/978-3-030-02934-0_22
6. Hotho, A., Staab, S., Stumme, G.: Ontologies improve text document clustering. In: Proceedings of the 3rd IEEE International Conference on Data Mining (ICDM 2003), 19–22 December 2003, Melbourne, Florida, USA. pp. 541–544 (2003)
7. Huang, E.H., Socher, R., Manning, C.D., Ng, A.Y.: Improving word representations via global context and multiple word prototypes. In: The 50th Annual Meeting of the Association for Computational Linguistics, Proceedings of the Conference, July 8–14, 2012, Jeju Island, Korea - Volume 1: Long Papers, pp. 873–882 (2012)
8. Mikolov, T., Chen, K., Corrado, G., Dean, J.: Efficient estimation of word representations in vector space. In: 1st International Conference on Learning Representations, ICLR 2013, Scottsdale, Arizona, USA, May 2–4, 2013, Workshop Track Proceedings (2013)
9. Mnih, A., Hinton, G.E.: Three new graphical models for statistical language modelling. In: Machine Learning, Proceedings of the Twenty-Fourth International Conference (ICML 2007), Corvallis, Oregon, USA, June 20–24, 2007. ACM International Conference Proceeding Series, vol. 227, pp. 641–648. ACM (2007)
10. Phan, X.H., Nguyen, M.L., Horiguchi, S.: Learning to classify short and sparse text & web with hidden topics from large-scale data collections. In: Proceedings of the 17th International Conference on World Wide Web, WWW 2008, Beijing, China, April 21–25, 2008. pp. 91–100. ACM (2008)
11. Xu, J., et al.: Short text clustering via convolutional neural networks. In: Proceedings of the 1st Workshop on Vector Space Modeling for Natural Language Processing, VS@NAACL-HLT 2015, June 5, 2015, Denver, Colorado, USA. pp. 62–69 (2015)

Semantic Analysis and Evaluation of Translation Based on Abstract Meaning Representation

Ying Qin[✉] and Ye Liang

Artificial Intelligence and Human Languages Lab,
Beijing Foreign Studies University, Beijing 100089, China
{qinying,liangye}@bfsu.edu.cn

Abstract. Abstract Meaning Representation (AMR) offers a novel scheme and perspective on the relation between linguistic form and semantic of various sentences. In order to explore the semantic equivalence among homologous translations, the paper analyzes the scenarios of same semantic structures covered by AMR and proposes a framework of variation in homologous translations. In addition, the framework is mapped into the annotation of AMR and used for common semantic characteristics mining in homologous translations. Accordingly AMR semantic structure matching (Smatch) score is applied to machine translation quality evaluation task. Experiments on small scale dataset preliminarily prove the effectiveness of AMR in translation quality evaluation.

Keywords: Abstract Meaning Representation · Homologous translations · Translation quality evaluation

1 Introduction

Deep learning has gained a great success in the fields of Natural Language Processing (NLP) [1]. However, Christopher Manning points out that deep learning, viewed as a black box scheme, might not be the only way to solve all the problems of computational linguistics, and the cognition and linguistic features should not be ignored [2]. The task of transforming a sentence into its meaning representation has also gained considerable attention within the community [3].

Abstract Meaning Representation (AMR) [4] is the scheme of universal dependency among the constitutions of sentence, in which the sentence is represented into a directed graph with single root node and several labeled edges and leaf nodes, as illustrated in Fig. 1. In the graph, the node denotes the concept, which can be word, frame or keyword in PropBank [5] like the type of entity, quantifier or conjunction. The directed edge indicates the relation of concepts. There are more than 100 relations defined in AMR. AMR has benefited many NLP tasks such as Machine Translation (MT) [6], Dialogue and Text Summarization. And Sembank, the English semantic database of AMR [7], has been on a certain scale and developed rapidly.

G. Wang et al. (Eds.): WISA 2020, LNCS 12432, pp. 268–275, 2020.
https://doi.org/10.1007/978-3-030-60029-7_25

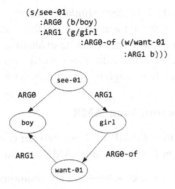

```
(s/see-01
  :ARG0 (b/boy)
  :ARG1 (g/girl
           :ARG0-of (w/want-01
                       :ARG1 b)))
```

Fig. 1. AMR graph. ARG0-1 denote the different arguments.

Usually there are a lot of acceptable translations corresponding to one source sentence, varying from word selection to syntactic structure. However the relation of linguistic form and semantic has not been thoroughly investigated. And the quality evaluation of MT is still lingering over the superficial comparison between MT and human references [8]. Evaluation based on some semantic measures provides fine-grained analysis of sentences and shows promising results in NLP tasks [3, 9], motivating us to explore the semantic features of homologous translations and new method of quality evaluation.

In the following, we first introduce the framework of AMR and induce the sentence categories with similar semantic structures but different linguistic forms (Sect. 2). Based on thoroughly analysis of homologous translations, we propose three levels of variety in translations and map them into AMR annotation (Sect. 3). In the experiment, we use Smatch [10] to calculate the similarity among AMRs of MT and reference and evaluate the quality of MT. The last section is the conclusion.

2 Sentence Categories with Similar AMR Structure

2.1 AMR

The advantage of AMR lies in semantic parsing of sentence, which is deeper than the analysis of syntactic. Three characteristics of AMR are: (1) the single root in AMR graph represents the core meaning of a sentence, (2) argument sharing can indicate complex semantic relations and (3) complementary of hidden or omitting constitute can restore the original and complete meaning [9].

AMR depicts the core meaning of sentence, that is, *who is doing what to whom*. Though having different linguistic forms, sentences with similar meaning are annotated with same AMR label. Figure 1 is the sharing AMR annotation and graph of the three sentences below.

The boy saw the girl who wanted him.
The boy saw the girl who he was wanted by.
The girl who wanted the boy was seen by him.

Obviously AMR can reflect the deep similarity among sentences, which motivates us to use AMR to analyze the translations from the same source, or homologous translations, and further to explore the semantic evaluation of translation quality. We first induce the sentence categories with same AMR structure and then apply it to analyze the relation between homologous translations with various forms.

2.2   Sentence Categories with Same AMR

According to the specification of AMR [11], we find ten categories in which sentences with different forms are annotated with same AMR structures.

Category 1. Predications in the sentences are homonyms, synonyms or near-synonyms or phrases made of such words. In AMR, these predications are stemmed then parsed into the similar subject-predicate relation. If these words or phrases have same frame in PropBank, the AMR annotations of these sentences are identical. For instance, the frame of *fear* in the following four fragments all falls into to the *fear-01* frame in PropBank, therefore, the AMR annotations of them are equivalent.

> *My fear of snakes*
> *I am fearful of snakes*
> *I fear snakes*
> *I'm afraid of snakes*

Their AMR annotation is same, represented as:

(f /fear-01
 :ARG0 (i /I)
 :ARG1 (s /snake)

Category 2. Sentences with different aspect and emphasis topic can carry the same meaning. AMR ignores the difference of verbs between active and passive. The results of annotation are also the same when the roles are reversed in order to highlight the focus or emphasize. In this scenario, the word order might be very different, as illustrated by the five sentences below, but share one AMR structure.

> *The boy wants to be believed by the girl.*
> *The boy wants the girl to believe him.*
> *The girl who was seen by the boy wants him.*
> *The boy is wanted by the girl he saw.*
> *The boy's desire is for the girl to believe him.*

Category 3. POS and inflection of non-predication word do not change AMR structure. The scenario includes a lot of types such as action or its corresponding noun and adjective, infinitive and gerund structure; adjective and its corresponding form of -*ly* adverb. For example, three sentences below share the AMR annotation.

> *The boy is responsible for the work.*
> *The boy is responsible for doing the work.*
> *The boy has the responsibility for the work.*

Category 4. Variety of modifier may not affect the AMR structure. The common varieties of modifier mainly include: a) clause vs. adjective, b) attributive noun vs. of-phrase, c) possessive vs. of-phrase, d) others reversed word order like ARG vs. ARG-of-relation. If the changes of modifier do not affect the semantic structure of a sentence, though they sometimes may differ in local word order, the AMR structure keeps unchanged as the fragments: *the white marble < - > the marble that is white.*

Category 5 Sentence meaning is similar but differs in polarity. The AMR structure may be same even though the polarity of two sentences is opposite because it is the meaning that determines the structure. For example, *I don't have any money. < - > I have no money.*

Category 6. *There-be* phrase and implied meaning of existence share the semantic structure. There are several expressions of existence, for example, the two sentences, *Four boys making pies* and *There are four boys making pies.* The former expresses the implied meaning of existence, while the later is an explicit expression.

Category 7. Various expressions of digit, date, time and currency, abbreviations and different forms of proper nouns in a sentence can own same AMR structures. For illustration, *February 29, 2012* and *29 February 2012, C$20* equals *20 Canadian dollar, and 25%* is equivalent to *twenty-five percent* or *25%.*

Category 8. Different syntactic structure may not affect the AMR if the predication keeps unchanged.

Category 9. The change of adverbial position might not affect the semantic structure. The adverbials of cause, result, condition, purpose and mode are very flexible and can appear at the beginning, middle or end of the sentence. The different position might not change the AMR structure.

Category 10. Others scenarios like same AMR in emphasis, substitution of relations in Propbank.

Besides, AMR is a kind of simplified and abstractive semantic representation of sentence. Therefore some linguistic details are not covered in the framework such as the governing of quantifier and common referential relationship between sentences.

3   AMR of Homologous Translations

3.1   Variation Level of Homologous Translations

We focus on the homologous translations to explore the variation AMR since they carry the same meaning from the source sentence. In the perspective of linguistics, we summarize four levels of variation of translations including comprehension cognition, information structure, word selection and syntactic structure. We will explain the four-level variations based on the translations of a Chinese source and corresponding English translations T1–T4 as listed below.

高新技术产品出口亮点频现，为广东对外贸易的增长做出了重要贡献。

T1: Export of high-tech products has frequently been in the spotlight, making a significant contribution to the growth of foreign trade in Guangdong.

T2: There are many bright signs in the export of new high technology products, which have significantly contributed to Guangdong's growth in foreign trades.

T3: High technology product export registers positive signs and makes great contribution to Guangdong's foreign trade growth.

T4: Hot spots of new hi-tech product export frequently appear, making significant contribution to the growth of foreign trade volume of Guangdong Province.

Level_1: Cognition Level. Due to the various understanding of the source, translators use different information structure to express the content in their mind. The longer the sentence, the richer information, and the more variation of the corresponding translations contain. Paraphrase, metaphor and anthropomorphic all belong to this level. Level_1 is the highest difference which shares less common feature among homologous translations. This level variation is not reflected in the above examples.

Level_2: Information Structure. Based on similar understanding, there may exist different information transformations in different translations. For example, translators can regulate the position of modifiers or change the order of cause and effect sub-clause to make the translation cohesive and coherent. In the above translations, T3 use the coordinate clause while the others use subordinate sentences to express the same meaning. T2 use *there-be* structure, which is different with other translations.

Level_3: Word Selection. This level of difference is reflected in subtle word selection like emotion, mood, color, pragmatic, reference, repetition, apposition and so on. In above translations, the Chinese word 亮点 is translated into four different words by translators, which are all perfect translations.

Level_4: Syntactic Structure. Translations may vary in syntactic level like choice of voice, tense, modal and aspect. In the above, T1 uses the perfect tense while the others use the present tense. Some use singular product while others prefer plural forms.

The four-level variation frame can lead us to detect and evaluate the acceptable translations for a given source.

3.2 Mapping Variation Level into AMR

We leave alone the variation of Level_1 because it is too complex to map and grasp. Translation variation level 2–4 can be mapped into three AMR structures including same-core predication, similar-core predication and different-core predication.

Different-Core Predication. The predications of homologous translations are different as well as the corresponding arguments. A mapping example is shown below.

法国总统选举举行第一轮投票
T1: France holds the first-round of Presidential election
(h / hold-01
  :ARG0 (c / country
    :name (n / name
      :op1 "France")
    :wiki "France")
  :ARG1 (e / elect-01
    :ARG2 (p / president)
    :ARG0-of (first-round)))
T2: First round vote of French presidential election begins
(b / begin-01
  :ARG0 (v / vote-01
    :ARG0 (r / round)
    :ARG1 (e / elect-01
      :ARG2 (p / person
        :ARG0-of (h / have-org-role-91
          :ARG2 (p2 / president))))
  :location "France"))

Though the root nodes of AMR are quite different, some arguments are same like *elect-01* and *president*, conveying the core meaning of sentence.

Similar-Core Predication. The third level in which the homologous translations differ is synonym or near-synonyms. It is illustrated in the examples of Sect. 3.1, in which, the Chinese phrase 做出了重要贡献 is translated into 1) *making a significant contribution*, 2) *significantly contributed*, and 3) *making great contribution* respectively. These fragments share one AMR annotation.

Same-Core Predication. If the predications of translations are identical, the difference among them is as minor as variance on syntactic.

4 Experiment

4.1 Dataset and Smatch Score

All translations are firstly parsed using JAMR [12]. To lower the effect of parsing errors, we limit the length of sentence to 25 words because the longer the sentence is, the poorer of the parser. And manual proofreading is time-consuming, which curbs the size of dataset in the experiment.

The final dataset consists of 75 source sentences, each with four human references and six machine translations. According to the core predication, corresponding translations of 24 have different predications, account for 32%, the predications of translations of 18 sentences are synonyms or near-synonyms, account for 24%, and the left 33 sentences, which accounts for the highest percentage of 44%, have same predications.

Taking all the predications of same and similar meaning into consideration, almost 68% references are parsed into similar AMR structures.

Smatch[1] score [9] is proposed to detect the similarity of AMR. Here we use the score to calculate similarity of three levels of homologous translations, shown in Table 1. The data in Table 1 reveal that the good translations always have rather high Smatch score though with different expressions.

Table 1. Smatch score of homologous translations.

	Highest	Lowest	Average
Different core	0.81	0	0.46
Similar core	0.84	0.12	0.48
Same core	1	0.38	0.88

4.2   Quality Evaluation Using Smatch Score

The Smatch score of MT is used as the quality metric by comparing MT with references. The highest score is viewed as the final if there are more references. The experiment is carried on six machine translations. Pearson correlation with human evaluation as well as the comparative result with BLEU is shown in Table 2.

Table 2. Evaluation performance and comparison with BLEU.

	Smatch	BLEU
Pearson correlation	0.292	0.205

Based on n-gram exact matching, some BLEU score might be 0 due to the sparsity of n-grams, which cannot reflect the real quality of MT. However the Smatch score is scarcely to be 0. The evaluation result based on Smatch outperforms BLEU. Further explanation of results is that semantic similarity comparison of sentences based on AMR is beyond the superficial matching of MT and references.

5   Conclusion

This paper analyzes the scenarios of same AMR structures. By combining with variance of homologous translations, we induce three categories and use Smatch score to evaluate the quality of machine translation. The result on the small dataset is promising. The research of AMR is still in primary stage. In the future work, we will further explore the common semantic features of homologous translation as well as experiments on larger scale of test dataset.

---

[1] https://pypi.org/project/smatch/.

References

1. Pan, H.-X., Liu, H., Tang, Y.: A sequence-to-sequence text summarization model with topic based attention mechanism. In: Ni, W., Wang, X., Song, W., Li, Y. (eds.) WISA 2019. LNCS, vol. 11817, pp. 285–297. Springer, Cham (2019). https://doi.org/10.1007/978-3-030-30952-7_29

2. Manning, C.D.: Computational linguistics and deep learning. Comput. Linguist. **41**(4), 701–707 (2015). https://doi.org/10.1162/COLI_a_00239

3. Alexandra, B., Abend, O. Bojar, O., Haddow, B.: HUME: human UCCA-based evaluation of machine translation. arXiv preprint arXiv:1607.00030 (2016)

4. Palmer, M., Gildea, D., Kingsbury, P.: The proposition bank: a corpus annotated with semantic roles. Comput. Linguist. J. **31**(1), 71–106 (2005). https://doi.org/10.1162/0891201053630264

5. Banarescu, L., Bonial, C., Cai, S., et al.: Abstract meaning representation (amr) 1.0 specification. In Parsing on Freebase from Question-Answer Pairs. In: Proceedings of the 2013 Conference on Empirical Methods in Natural Language Processing. ACL, pp. 1533–1544 (2014)

6. Song, L., Gildea, D., Zhang, Y., et al.: Semantic neural machine translation using AMR. Trans. Assoc. Comput. Linguist. **7**, 19–31 (2019). https://doi.org/10.1162/tacl_a_00252

7. Xue, N., Bojar, O., Hajic, J., et al.: Not an interlingua, but close: comparison of English AMRs to Chinese and Czech. In: Proceeding of LREC, vol. 14, pp. 1765–1772 (2014)

8. Qin, Y.: Review on automatic translation quality evaluation. Appl. Res. Comput. **32**(2), 326–329 (2015). https://doi.org/10.3969/j.issn.1001-3695.2015.02.002

9. Anderson, P., Fernando, B., Johnson, M., Gould, S.: SPICE: semantic propositional image caption evaluation. In: Leibe, B., Matas, J., Sebe, N., Welling, M. (eds.) ECCV 2016. LNCS, vol. 9909, pp. 382–398. Springer, Cham (2016). https://doi.org/10.1007/978-3-319-46454-1_24

10. Cai, S., Knight, K.: Smatch: an evaluation metric for semantic feature structures. Proc. ACL **2**, 748–752 (2013)

11. Bos, J.: Expressive power of abstract meaning representations. Comput. Linguist. **42**(3), 527–535 (2016). https://doi.org/10.1162/COLI_a_00257

12. Flanigan, J., Thomson, S., Carbonell, J. et al.: A discriminative graph-based parser for the abstract meaning representation. In: Proceedings of ACL (2014)

N2One: Identifying Coreference Object Among User Generated Content with Siamese Network

Wei Yuan[1], Peng Wang[2], Mengyao Yuan[3], Yue Guo[1], and Tieke He[1(✉)]

[1] State Key Laboratory for Novel Software Technology,
Nanjing University, Nanjing 210093, China
hetieke@gmail.com
[2] Jiangsu Tongxingbao Intelligent Transportation Technology Co., Ltd.,
Nanjing, China
[3] Jiangsu Communications Holding Co., Ltd., Nanjing, China

Abstract. With the boom of social network, various service is provided based on the user generated content (UGC). Some of these UGCs semantically point to the same entities in the real world. Recognizing the same real-world entity that is referred by various user contents is crucial for providing personalized service, such as recommender systems and user profiling. Recently, most recommender systems recommend content based on keywords detection or structured data. These approaches cannot fully grasp latent semantic of UGC, so it is difficult for them to discover the coreference entities behind UGCs. As a result, when there are less obvious keywords or preset user preference, the experience of such services are poor. In this paper, we formalize this problem as a N to 1 task, which means performing entity resolution among UGCs, and a Siamese Network is presented to deal with it, namely '*N2One*'. Extensive experiments are conducted on two commonly used datasets Yelp and IMDB to evaluate the effectiveness of the proposed model, and the results demonstrate its superiority.

Keywords: Natural language · Semantic analysis · Entity resolution · Recommender system · Siamese Network

1 Introduction

Recently, lots of social network contents are created by users, such as Twitter[1], Yelp[2], Reddit[3], WeChat[4] and DianPing[5]. People can express their feelings and

---

[1] https://twitter.com//.
[2] https://www.yelp.com//.
[3] https://www.reddit.com//.
[4] https://www.wechat.com//.
[5] http://www.dianping.com//.

© Springer Nature Switzerland AG 2020
G. Wang et al. (Eds.): WISA 2020, LNCS 12432, pp. 276–288, 2020.
https://doi.org/10.1007/978-3-030-60029-7_26

interests, make comments and browse other users' contents. Due to the convenience of such applications, the number of users and contents grows dramatically.

In order to improve user experience and promote commercial values, almost every social network sites provide recommendation services. The mechanisms behind these systems varies but fall broadly into three categories according to utilized factors: user character based methods [1], target content based methods [2,3] and knowledge based methods [4]. User character based methods adopt user behavior, user self-introduction or some other user information to make recommendations. Collaborative filtering is one of the typical user character based methods that recommending based on records of user behavior. Target content based methods infer results from target object's information. They generally extract features from objects using machine learning or deep learning algorithms, then compute similarity scores and generate recommendations [5]. Knowledge based methods utilize external knowledge to provide people with better suggestions. However, all these methods didn't make full use of user generated contents like user comments or posts. Although there are some studies exploiting such user generated contents [6,7], they just only focus on the semantics rather than analyzing the real world-entity referred by these UGCs.

Actually, UGCs often contain much information that can be used for improving user specific services. For example, user A posts a tweet about restaurant R, and later user B mentioned the same restaurant. If systems could recognize these two posts are referring to the same restaurant, more reasonable and intelligent services can be provided, including followers recommendation, personalized advertising, and so on. In this sense, detecting coreference entities contributes to great commercial and research values. In this paper, we formalize this problem as a N to 1 task which is a specific entity resolution (ER) task and propose a Siamese Neural Network to recognize the coreference entities among user generated contents. The Siamese Network [8] is an artificial neural network that uses the same weights while working in tandem on two different input vectors to compute comparable output vectors. Since traditional Siamese Network [9] cannot consider ER transitivity, we propose an innovative solution to tackle this problem. We conduct extensive experiments on two real-world datasets, Yelp and IMDB, and compare our approach to other methods. The experimental results show that our 'N2One' model outperforms in entities detection.

The contributions of this paper are as follows. We formalize the identification of entities in user generated content as ER problem, and propose a Siamese Network to tackle it, which considers the natural constraint of transitivity. To the best of our knowledge, it is the first time to use Siamese Network to deal with entity coreference in UGCs. Extensive experiments are conducted on two real-world datasets to prove the efficacy of the 'N2One' model.

2   Related Work

2.1   Entity Resolution

Entity Resolution (ER) identifies and links different manifestations of the same real world object. ER essentially consists of two steps [10,11]: (i) the blocking step, which determines what entities to be compared, and (ii) the filtering step, which determines whether those entities represent the same real-world object. In the past few decades, ER has received widespread attention. Prior ER work can be categorized as (a) rule-based, (b) machine-learning based, and (c) expert or crowd-based. The core of rule-based solutions is estimating similarity between entities by using various metrics and some pre-defined rules and thresholds [12–14]. The matching results of rule-based methods are usually interpretable. However, these rule-based approaches require lots of human involvement, sometimes even need domain experts to define the rules [15]. Thus, the cost of developing an efficient rule-based ER system is usually unaffordable. To address these issues faced with rule-based methods, machine-learning based solutions either model ER as a classification problem utilizing some learning algorithm (e.g. SVM or decision trees) or formalize it as a cluster issue tackled by relative methods. The limitation of machine-learning based approaches is that there is still heavy handcrafted feature engineering. Crowdsourcing ER aims at making the best use of human cognitive [16,17]. However, the quality of crowdsourcing results may be fluctuated due to the difference in crowd worker's knowledge level and cultural background. How to control the quality of crowdsourcing is still an area that need to be further researched.

With the rapid development of deep learning (DL) theories [18], many natural language processing (NLP) [19] tasks achieve state-of-the-art performance. Applying DL methods to ER is a new trend and recently becomes more and more popular. The representation based methods learn a distribution for entity representation and then estimate the similarity between particular manifestations. In this paper, our approach is more likely the representation-based method, which learns representation distribution by Siamese Network.

2.2   Recommender System

Recommender systems are widely utilized in different applications for predicting the preference of a user in a product or service. It can also be an effective strategy to overcome information overload with the growing volume of online resources. Recommendation models can be classified into three categories: collaborative filtering (CF), Content-based, knowledge-based and hybrid [20–22]. CF makes recommendations according to user historical interactions [23]. Content-based recommendation primarily compares items' and users' auxiliary information. The auxiliary information includes text, videos and images. Knowledge-based systems use some external knowledge (e.g.. relational database or knowledge graph) to infer users' preference. Hybrid models are recommender systems which integrate two or more recommendation strategies. All these model above have been faced some challenges, including sparsity, cold start and scalability, which may affect the quality of recommendation. Even so, recommender systems have been played a vital and indispensable role in predicting user preference. In this paper, our work more like content-based recommendation, but our point is at determining which content pairs identified the same entities. So except user generated contents, we don't use any other auxiliary information.

2.3 Siamese Network

The Siamese Network is an architecture for non-linear metric learning with similarity information [9]. Through the information about the similarity between object pairs, Siamese Network learns the data representations. Thus, it is different from auto-encoder which learns representations through encoding data and reproducing them conditioned on encoding results. In contrast, Siamese Network learns representations directly through similarity or dissimilarity information. Except for the vallina Siamese Networks, there are variants of Siamese Networks, one of the famous is the Pseudo-siamese network [24]. The biggest difference between Siamese Network and Pseudo-siamese network is that the former doesn't share weights in neural network but the latter does. Thus Pseudo-siamese network is more suit for input pairs that have obvious distinction [25].

Siamese Networks were first introduced to solve signature verification as image matching problem in the early 1990s [9], since then, it has been widely used in vision applications including image dimension reduction [26], visual object tracking [27], image similarity calculation [28] and one-shot image recognition [29] and so on. Not only in the vision area, Siamese Networks have been applied to lots of diverse tasks such as acoustic modeling and scene detection. Siamese Networks also have been involved in natural language processing (NLP). Siamese convolution networks have been applied to match sentences [30] and then, Siamese Networks with recurrent layers have been applied to learning sentence semantic representations [31]. In our task, to let Siamese Network learn the transitivity constraint, we propose an innovative method based on Siamese Network.

3   Problem Statement

Given a group of user generated statement $S = \{s_1, s_2, ..., s_n\}$, each statement s_i implicitly contains some entities. Some of these statements contain the same entity. We want to find the coreference entity E among these statements. More Formally:

$$E = f(s_i, ..., s_j | \theta) \tag{1}$$

where f is the model which determines the entity behind UGCs. θ are trainable parameters. $s_i, ..., s_j$ are statements from S that we want to identify the coreference entities behind them.

Obviously, the entity resolution result should obey transitivity rule. The transitivity rule means that if statement s_i and s_j describe or contain the same entity

e_m, meanwhile, statement s_j and s_k refer to e_m, then s_i and s_k refer to the same entity:

$$if\ f(s_i, s_j|\theta) = f(s_j, s_k|\theta) = E, then\ f(s_i, s_k|\theta) = E \qquad (2)$$

4   N2One

In this section, we proceed to present the details of our model, N2One, and demonstrate how to apply 'N2One' to the problem we defined in the previous section.

4.1   Model Description

Sentence Embedding Networks (SENs) transform sentences into high dimensional vectors. There are various SENs, including Convolutional Neural Network (CNN), Recurrent Neural Networks (RNNs), transformers, and so on. CNN has been widely used in Computer Vision (CV) tasks. However, in NLP tasks, it has some fatal flaws. One of the shortcomings with CNN is its inability to model long distance dependencies, which is critical for various NLP tasks. In the meanwhile, it is also difficult for CNN to process variant length sentences. Transformers are popular in recent years since BERT obtains lots of state-of-the-art performances on NLP tasks. However, training Transformers need nearly unaffordable computing power for the individual researcher. Therefore, we choose to use Gated Recurrent Units (GRU) as a basic component of our sentence embedding module.

Siamese Networks are dual-branch networks with tied weights. They usually consist of the same network copied and merged with an energy function. Siamese Network learns relationships by encoding pairs of objects to separately measure the similarity between encoded representations. In this paper, we use GRU as part of Siamese Network. The Siamese Network extracts the semantic meaning of the entity referred by UGCs. Figure 1 shows an overview of the network architecture in this work. The training set for our network consists of triplets (s_1, s_2, y), where s_1 and s_2 are UGC sentences and $y \in \{0, 1\}$ indicates whether s_1 and s_2 are refering to same entity ($y = 1$) or not ($y = 0$). Our goal of training is to minimize the vector distance between UGCs whose entity is the same, meanwhile maximize the distance between UGCs which refer to different entities in an embedding space. The strategy of measuring distance in vector space can be cosine-similarity, Manhattan distance or Euclidean distance, and so on.

4.2   Objective Function

The proposed model contains GRU, whose output is a sequence of vectors. We can use the last hidden state, compute the mean of these vectors or calculate max-pool of these vectors to predict UGC embedding. In our experiment, we choose to compute the average of the vectors.

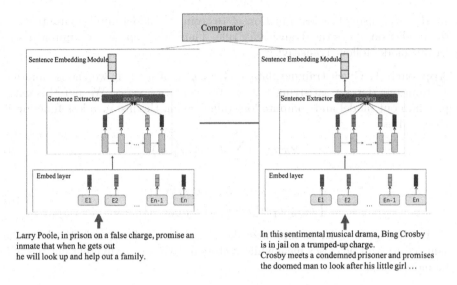

Larry Poole, in prison on a false charge, promise an inmate that when he gets out he will look up and help out a family.

In this sentimental musical drama, Bing Crosby is in jail on a trumped-up charge. Crosby meets a condemned prisoner and promises the doomed man to look after his little girl ...

Fig. 1. Model architecture

Let x_1, x_2 be the input UGC pair, e_1 and e_2 be the embedding of UGC pair. The GRU network is denoted as f_θ. The similarity between the embedding of UGCs is defined as:

$$e_{\{1,2\}} = Mean(f_\theta(x_{\{1,2\}})) \tag{3}$$

$$S_\theta(x_1, x_2) = \frac{<e_1, e_2>}{\|e_1\|\|e_2\|} \tag{4}$$

For brevity of notation, we denote $S(x_1, x_2)$ by S_θ. The loss function of our model over dataset $D = \{< x_1^i, x_2^i, y^i >\}$ is given by:

$$\mathscr{L}_\theta(D) = \sum_1^N L_\theta^i(x_1^i, x_2^i, y^i) \tag{5}$$

L_θ^i is a contrastive loss function for i'th instance. L_θ is composed of terms for $y = 1$ and $y = 0$.

$$L_\theta^i = y^i L_{pos}(x_1^i, x_2^i) + (1 - y^i)L_{neg}(x_1^i, x_2^i) \tag{6}$$

L_{pos} and L_{neg} represent the loss function of UGCs own the entity or not respectively. The details of L_{pos} and L_{neg} are as follow:

$$L_{pos}(x_1^i, x_2^i) = C(1 - S_\theta)^2 \tag{7}$$

$$L_{neg}(x_1^i, x_2^i) = \begin{cases} S_{theta}^2 & if \ S_{theta} > m \\ 0 & otherwise \end{cases} \tag{8}$$

However, using this loss function $\mathscr{L}$ to train our model won't guarantee that the prediction obeys the Transitivity rule. Thus, we propose two approaches to let our model follow Transitivity.

Approach 1. Given training data $<x_1^i, x_2^i, y^i>$, if $y^i = 1$, we choose another x_3^i from our dataset satisfying $<x_1^i, x_3^i, 0>$, and vice versa. We put this extra pair into out model and compute loss value. Then, the instance loss function is redefined as:

$$\mathscr{L}_\theta(D) = \sum_1^N L_\theta^i(x_1^i, x_2^i, y^i) \tag{9}$$

$$L_\theta^i = \begin{cases} L_{pos}(x_1^i, x_2^i) + L_{neg}(x_1^i, x_3^i) & (y^i = 1) \\ L_{neg}(x_1^i, x_2^i) + L_{pos}(x_1^i, x_3^i) & (y^i = 0) \end{cases} \tag{10}$$

Intuitively, in order to minimize the $\mathscr{L}_\theta$, the embedding of UGCs have the same entity will be closer meanwhile vectors of UGCs own different entities will be more distant.

Approach 2. Rather than sample adversed x_3^i for each pair $<x_1^i, x_2^i, y^i>$, we regulate the embedding space at the end of each epoch. Therefore, the loss of model is redefined as:

$$\mathscr{L}_\theta(D) = \sum_1^N L_\theta^i(x_1^i, x_2^i, y^i) + \gamma \tag{11}$$

where γ denotes:

$$\gamma = \sum_1^n \sum_1^k distance(S^i) \tag{12}$$

in Eq. (12), n is the number of entities our data contains, and k is the number of UGCs own the same entities. S^k is the k'th group of statements own the same entity. The Function $distance$ calculates the distance between two UGCs in S. In our paper, we use Euclidean distance as our distance function.

5   Experiments

In this section, we show our experiments to evaluate the two approaches above, on two popular NLP task datasets. The dataset description, the evaluation metrics and comparative methods are presented accordingly. We implemented our model and approaches in Python. The experiments were carried out on a PC with Intel(R) i7-8750 CPU and NVIDIA 1050Ti GPU.

5.1   Datasets

To evaluate our proposed approaches, we choose two commonly used NLP datasets, Large Movie Review Dataset V1.0 (IMDB) and Yelp. The first dataset is for binary sentiment classification containing more than $50,000$ movie reviews. All of these movie reviews come from IMDB. Each review is along with sentiment tag and corresponding movie url. In the entire collection, no more than 30 reviews are allowed for any given movi.e. In our experiment, we don't use the sentiment tag for each review. The given movie for each review is treated as entity and review as UGC. We filter out movies which have less than 10 reviews. We also remove stop words and punctuations according to NLTK's English stop word list. The reviews whose lengths is between 20 and 100 words are used in our experiments. As a result, per movie have about 15 reviews on average, and $10,966$ reviews were used for our experiment. The vocabulary size of this dataset is $40,679$. We split these $10,966$ reviews into training set and test set. The test set contains $1,585$ reviews which means per movie owns 5 reviews and the training set is the remaining $9,381$ reviews. There are a large number of classes (movies), with few examples (reviews) of each class. Thus the task on this dataset is few-shot learning task.

The Yelp dataset is a subset of Yelp businesses, reviews and user data for use in academic purposes. There are three *.json* files in Yelp dataset, *businesses.json*, *review.json* and *user.json*. The *businesses.json* file contains the basic information of restaurants such as id, review count, location, and so on. Per line in *review.json* represents a review, including review id, business id, user id and content. The *user.json* includes user name, user id and review count, and so on. We aggregate these *json* files into one file, and in the meantime remove some useless fields for our task. For simplicity, we randomly choose 10 restaurants and its related reviews whose lengths is no more than 100 words, for training and testing. Finally, there are $21,353$ records in our dataset. Each record contains business id, review id, user id, and content without stop words or punctuations. We randomly choose 60 reviews for each restaurant as a test set. Thus the test set contains 600 reviews. The vocabulary size of this dataset is $29,363$.

5.2   Experimental Setup

We use a Siamese Network as the main component of our 'N2One' model. The word embedding dimension is 100. The sentence embedding network used in this experiment is a three BGRU layers network with 64-dimensional hidden vectors h_t. There is a Dropout layer with 0.4 dropout rate between the layers for preventing overfitting. The outputs of the last layer are averaged over time and this vector is used as input to a fully connected layer. The output of this fully connected layer is 64 dimensions. The input strings are padded with token $<pad>$ according to the length of the longest sequence in an input batch. The batch size is setting to 32. The parameters of the model are optimized by Adam method and each model is trained until convergence. Start learning rate lr is 0.01. lr will be halved if the loss value in the valid set doesn't reduce. Training step wouldn't stop until the *lr* reduced more than 5 times or training epoch reaches to 50.

The previous section describes the datasets used in our experiment. After preprocessing, the datasets are divided into training set and test set. However, our model needs UGC pairs as input. If we only randomly combine every two UGCs as input, there would be data imbalance. To avoid this situation, we generate training and test pairs from training set and test set respectively. Meanwhile, let the number of positive and negative pairs roughly equal. Specifically, for each epoch, we randomly generate 20, 000 training pairs and 2, 000 ground truth pairs as training set and validation set. The ground truth pairs are generated in the beginning of program in order to keep stable results.

We conduct extensive experiments on these datasets. First, we evaluate our model with the precision of predicting whether two UGCs are referring to the same movie (or restaurant) or not. We randomly generate numbers of pairs of UGC with positive and negative tags. All models' predictions based on the same test pairs. Then, we verify if the model follows the Transitivity rule by predicting triple UGCs's relationship. Finally, we show the visual result of UGCs' distribution using T-SNE.

5.3   Result

Table 1 shows the precision of prediction in IMDB and Yelp dataset. We randomly choose 4, 000 pairs of reviews from test set as input of our model. The model predicts whether those two UGCs are referring to the same movie (or restaurant). Then we calculate the correct number of pairs and compute the precision. The result shows that predicting in IMDB dataset is more difficult than that in Yelp, because each entity in IMDB has fewer UGCs than in Yelp. Our approach 1 and approach 2 can improve Siamese Network behavior in both datasets. In IMDB dataset, the precision of Siamese Network with approach 1 and approach 2 is improved by 3.8% and 2.6%, respectively. In Yelp dataset, the precision of Siamese Network with approach 1 and approach 2 is improved by 3.2% and 2.9%, respectively. It can be noticed that Siamese Network with approach 1 and approach 2 doesn't outperform obviously than with these approaches respectively.

Table 1. Prediction precision in datasets

Model	IMDB	Yelp
rsn	0.650	0.844
rsn+approach1	0.688	0.876
rsn+approach2	0.676	0.873
rsn+approach1+approach2	0.673	0.863

Note: rsn is the abbreviation of RNN Siamese Network. The decimals in Table 2 round off to three decimal places.

Table 2 shows the Transitivity validation result in IMDB and Yelp datasets. We sample 2,000 triplets $<x_i, x_j, x_k>$ of UGCs from Yelp. The model aims to predict the relationship between x_i and x_j, x_j and x_k, x_i and x_k. Then, we determine whether the model's three times prediction follows the Transitivity rule. If the model's predictions obey Transitivity, we consider it to make correct predictions, and vice versa. Finally, we obtain the precision of our model, which is depicted in Table 2 We can see that Siamese Network with approach 1 gets the best performance among these models.

Table 2. Transitivity precision in datasets

Model	Yelp
rsn	0.86
rsn+approach1	0.92
rsn+approach2	0.89
rsn+approach1+approach2	0.88

Figure 2 shows the Yelp's UGCs distribution. We sample about 6,000 reviews from our training set and feed into model to get their sentence embedding. Then, we apply T-SNE to reduce the dimension of sentence embedding and draw them in 2 dimensional plane. From Figure 2 we can see our model can cluster those UGCs perfectly.

Fig. 2. Sentence distribution from our model

6 Conclusions

In this paper, we first formalize a user generated content coreference problem as a N to 1 task. Then we discuss its importance in recommender systems and propose the 'N2One' model to tackle it. Meanwhile, we focus on how to let our model's prediction results follow Transitivity rule which is a natural rule in entity coreference and propose two approaches. Finally, we conduct extensive experiments on two datasets to evaluate our approaches. The results demonstrate the superiority of our proposed model.

Acknowledgment. This work was partially supported by the Fundamental Research Funds for the Central Universities (14380023), and the Equipment Development Department Pre-Research Foundation of China (31511110310).

References

1. Zhao, Z.D., Shang, M.S.: User-based collaborative-filtering recommendation algorithms on hadoop. In: 2010 Third International Conference on Knowledge Discovery and Data Mining, pp. 478–481. IEEE (2010)
2. Wang, W., Yin, H., Sadiq, S., Chen, L., Xie, M., Zhou, X.: Spore: a sequential personalized spatial item recommender system. In: 2016 IEEE 32nd International Conference on Data Engineering (ICDE), pp. 954–965. IEEE (2016)
3. Chen, Y., Wang, Y., Zhao, X., Yin, H., llya, M., Rijke, M.: Local variational feature-based similarity models for recommending top-n new items. ACM Trans. Inf. Syst. (2019). https://doi.org/10.1145/3372154
4. Wang, H., Zhang, F., Zhao, M., Li, W., Xie, X., Guo, M.: Multi-task feature learning for knowledge graph enhanced recommendation. In: The World Wide Web Conference, pp. 2000–2010 (2019)
5. He, T., Lian, H., Qin, Z., Zou, Z., Luo, B.: Word embedding based document similarity for the inferring of penalty. In: Meng, X., Li, R., Wang, K., Niu, B., Wang, X., Zhao, G. (eds.) WISA 2018. LNCS, vol. 11242, pp. 240–251. Springer, Cham (2018). https://doi.org/10.1007/978-3-030-02934-0_22
6. He, T., Yin, H., Chen, Z., Zhou, X., Sadiq, S., Luo, B.: A spatial-temporal topic model for the semantic annotation of POIS in LBSNS. ACM Trans. Intell. Syst. Technol. (TIST) 8(1), 1–24 (2016)
7. Yin, H., Cui, B., Huang, Z., Wang, W., Wu, X., Zhou, X.: Joint modeling of users' interests and mobility patterns for point-of-interest recommendation. In: Proceedings of the 23rd ACM International Conference on Multimedia, pp. 819–822 (2015)
8. Bertinetto, L., Valmadre, J., Henriques, J.F., Vedaldi, A., Torr, P.H.S.: Fully-convolutional Siamese networks for object tracking. In: Hua, G., Jégou, H. (eds.) ECCV 2016. LNCS, vol. 9914, pp. 850–865. Springer, Cham (2016). https://doi.org/10.1007/978-3-319-48881-3_56
9. Bromley, J., Guyon, I., LeCun, Y., Säckinger, E., Shah, R.: Signature verification using a "Siamese" time delay neural network. In: Advances in Neural Information Processing Systems, pp. 737–744 (1994)
10. Getoor, L., Machanavajjhala, A.: Entity resolution: theory, practice & open challenges. Proc. VLDB Endowment 5(12), 2018–2019 (2012)

11. Papadakis, G., Skoutas, D., Thanos, E., Palpanas, T.: A survey of blocking and filtering techniques for entity resolution. arXiv preprint arXiv:1905.06167 (2019)
12. Fan, W., Jia, X., Li, J., Ma, S.: Reasoning about record matching rules. Proc. VLDB Endowment **2**(1), 407–418 (2009)
13. Shen, W., Li, X., Doan, A.: Constraint-based entity matching. In: AAAI, pp. 862–867 (2005)
14. Singh, R., et al.: Generating concise entity matching rules. In: Proceedings of the 2017 ACM International Conference on Management of Data, pp. 1635–1638 (2017)
15. Nie, H., et al.: Deep sequence-to-sequence entity matching for heterogeneous entity resolution. In: Proceedings of the 28th ACM International Conference on Information and Knowledge Management, pp. 629–638 (2019)
16. Wang, J., Kraska, T., Franklin, M.J., Feng, J.: Crowder: crowdsourcing entity resolution. arXiv preprint arXiv:1208.1927 (2012)
17. Marcus, A., Wu, E., Karger, D., Madden, S., Miller, R.: Human-powered sorts and joins. arXiv preprint arXiv:1109.6881 (2011)
18. Schmidhuber, J.: Deep learning in neural networks: an overview. Neural Networks **61**, 85–117 (2015)
19. Gu, B., et al.: The interaction between schema matching and record matching in data integration. IEEE Trans. Knowl. Data Eng. **29**(1), 186–199 (2016)
20. Yin, H., Wang, W., Wang, H., Chen, L., Zhou, X.: Spatial-aware hierarchical collaborative deep learning for poi recommendation. IEEE Trans. Knowl. Data Eng. **29**(11), 2537–2551 (2017)
21. Yin, H., Wang, W., Chen, L., Du, X., Nguyen, Q.V.H., Huang, Z.: MOBI-SAGE-RS: a sparse additive generative model-based mobile application recommender system. Knowl.-Based Syst. **157**, 68–80 (2018)
22. Wang, W., Yin, H., Huang, Z., Wang, Q., Du, X., Nguyen, Q.V.H.: Streaming ranking based recommender systems. In: The 41st International ACM SIGIR Conference on Research & Development in Information Retrieval, pp. 525–534 (2018)
23. He, T., Chen, Z., Liu, J., Zhou, X., Du, X., Wang, W.: An empirical study on user-topic rating based collaborative filtering methods. World Wide Web **20**(4), 815–829 (2016). https://doi.org/10.1007/s11280-016-0412-2
24. Zagoruyko, S., Komodakis, N.: Learning to compare image patches via convolutional neural networks. In: Proceedings of the IEEE Conference on Computer Vision and Pattern Recognition, pp. 4353–4361 (2015)
25. Hoffer, E., Ailon, N.: Deep metric learning using triplet network. In: Feragen, A., Pelillo, M., Loog, M. (eds.) SIMBAD 2015. LNCS, vol. 9370, pp. 84–92. Springer, Cham (2015). https://doi.org/10.1007/978-3-319-24261-3_7
26. Wang, Q., Teng, Z., Xing, J., Gao, J., Hu, W., Maybank, S.: Learning attentions: residual attentional Siamese network for high performance online visual tracking. In: Proceedings of the IEEE Conference on Computer Vision and Pattern Recognition, pp. 4854–4863 (2018)
27. Guo, Q., Feng, W., Zhou, C., Huang, R., Wan, L., Wang, S.: Learning dynamic Siamese network for visual object tracking. In: Proceedings of the IEEE International Conference on Computer Vision, pp. 1763–1771 (2017)
28. Melekhov, I., Kannala, J., Rahtu, E.: Siamese network features for image matching. In: 2016 23rd International Conference on Pattern Recognition (ICPR), pp. 378–383. IEEE (2016)
29. Koch, G., Zemel, R., Salakhutdinov, R.: Siamese neural networks for one-shot image recognition. In: ICML Deep Learning Workshop, vol. 2. Lille (2015)

30. Hu, B., Lu, Z., Li, H., Chen, Q.: Convolutional neural network architectures for matching natural language sentences. In: Advances in Neural Information Processing Systems, pp. 2042–2050 (2014)
31. Neculoiu, P., Versteegh, M., Rotaru, M.: Learning text similarity with Siamese recurrent networks. In: Proceedings of the 1st Workshop on Representation Learning for NLP, pp. 148–157 (2016)

BERT-Based Named Entity Recognition in Chinese Twenty-Four Histories

Peng Yu[1] and Xin Wang[1,2]([envelope])

[1] College of Intelligence and Computing, Tianjin University, Tianjin 300072, China
{pursuit,wangx}@tju.edu.cn
[2] Tianjin Key Laboratory of Cognitive Computing and Application,
Tianjin 300072, China

Abstract. Named entity recognition in classical Chinese plays a fundamental role in improving the ability of information extraction and constructing knowledge graphs from classical Chinese. However, due to the lack of annotated data and the complexity of grammatical rules, named entity recognition in classical Chinese has made little progress. In order to solve the problem of lack of labeled data, we propose an end-to-end solution that is not based on domain knowledge, which instead is based on the pre-trained BERT-Chinese model and integrates the BiLSTM-CRF model for classical Chinese named entity recognition. The BERT-Chinese model converts the input text into a character-level embedding vector, then the BiLSTM model is used for future training, and finally CRF is able to normalize the output of BiLSTM to obtain a globally optimal labeling sequence. We conducted fine-tuning training on ChineseDailyNerCorpus. By designing a optimized fine-tuning method, we have realized the named entity recognition task in the Chinese Twenty-Four Histories. We evaluated our model on the Chinese twenty-four histories data, and finally achieved an average F1 value of about 75%. The experimental results also show that the BERT model has a strong ability in transfer learning.

Keywords: BERT · BiLSTM-CRF · Named Entity Recognition · Chinese Twenty-Four Histories

1 Introduction

Named Entity Recognition (NER) is a basic work in the field of Natural Language Processing (NLP) [1]. After decades of development, NER has experienced rule-based and dictionary-based methods, traditional machine learning-based methods, and deep learning-based methods, as well as current mainstream transfer learning methods and hybrid methods. However, most of the named entity recognition work on the Chinese language is based on modern Chinese, and there are few studies of named entity recognition on classical Chinese. In the fields related to the processing of classical Chinese, such as digital humanities,

© Springer Nature Switzerland AG 2020
G. Wang et al. (Eds.): WISA 2020, LNCS 12432, pp. 289–301, 2020.
https://doi.org/10.1007/978-3-030-60029-7_27

there is an urgent need to identify named entities in classical Chinese, so NER on classical Chinese is a critical task in Chinese NLP.

However, compared with NER in modern Chinese, the NER task on classical Chinese is much more difficult. The main difficulty is that the grammar of classical Chinese is more complicated, and there are few labeled datasets in the field of classic Chinese, leading to little progress in NER in classical Chinese. Therefore, there is a need for a new method to deal with this task. Despite lack of data and the complexity of the grammar of classical Chinese, our research shows that with the method of transfer learning, it is possible to perform NER in classical Chinese. In this paper, we propose the BERT-BiLSTM-CRF model to implement NER in the field of classical Chinese, and evaluate the performance of the model on Chinese Twenty-Four Histories. This model is a combination of the pre-trained BERT [2] model and the BiLSTM-CRF model, where the pre-trained BERT-Chinese model is used to obtain the vector representation of each character in the input sentence, and the BiLSTM-CRF model is used for further training to get the final prediction result.

Due to the complexity of classical Chinese and the lack of available annotated data, NER of classical Chinese in the field of digital humanities has become a bottleneck in NER tasks. We use the BERT pre-training model for preliminary feature extraction, and then uses a small amount of annotation data for training on the BiLSTM-CRF model to obtain the BERT-BiLSTM-CRF model for classical Chinese NER task. In order to evaluate the performance of our model on NER task in Chinese Twenty-Four Histories, we selected five of the twenty-four books, and each book selected a long article for annotation. The time span of each two adjacent articles is approximately 450 years. Our model finally achieved an average F1 value of about 75%, and the best result is obtained in Ming History, with an F1 value of 86%. The experimental results also show certain patterns: first of all, as time goes from the ancient Qin Dynasty to the Ming Dynasty, the performance of our model shows a trend of improvement; secondly, as shown in the experimental results, the recognition accuracy of our model in Location entities is better than that of Person entities.

Our main contributions include: (1) We propose an end-to-end BERT-BiLSTM-CRF model for named entity recognition in classical Chinese, which achieves good results in Chinese Twenty-Four Histories. (2) We study the transfer learning ability of the BERT model. Our model is trained with a modern Chinese corpus, but evaluated on classical Chinese corpus, and achieves an average F1 value of about 75%. (3) Finally, based on the experimental results, we observe that the accuracy of the model on the Location entities is higher than the accuracy of the Person entities, and the closer the time is to the present time, the higher the overall accuracy of our model.

The remainder of the paper is organized as follows: Sect. 2 describes related works of named entity recognition. Section 3 shows the BERT-BiLSTM-CRF model we use in this paper. Section 4 gives detailed information about the settings of our experiments. Section 5 is the analysis of the experimental results. Finally we conclude in Sect. 6.

2   Related Work

2.1   Overview of Named Entity Recognition

Named entity recognition, also known as entity identification and entity extraction, is a subtask of information extraction that seeks to locate and classify named entities mentioned in unstructured text into pre-defined categories such as person, organization, location, time expression, etc. [3].

In the early days, NER was usually based on rules and dictionaries, which uses specific rule templates or dictionaries constructed manually by linguists based on the characteristics of the dataset. Rau et al. first proposed the method of combining manually written rules with heuristic ideas to implement automatic extraction of named entities of company name types from text [4]. However, the rule-based and dictionary-based approaches are labor-intensive and time-consuming.

In machine learning-based methods, named entity recognition is regarded as a sequence tagging problem, in which the current predicted label is not only related to the current input feature, but also related to the previous predicted label. Traditional machine learning methods mainly include: Hidden Markov Model, Maximum Entropy Markov Model, Conditional Random Field, etc. The advantage of the Conditional Random Field (CRF) is that it can make full use of internal and contextual feature information in the process of labeling a position, which makes it currently the mainstream model for NER [5].

With the development of hardware computing power, the introduction of word embedding representation, and the application of deep learning, the mainstream methods of NER have been leveraging deep neural networks. Collobert et al. first proposed a NER method based on neural networks [6], but it fail to consider the effective information between long-distance words. To overcome this shortcoming, Chiu and Nichols proposed a Bidirectional LSTM-CNN architecture that can automatically detect both word and character level features [7]. Huang et al. also proposed a Bidirectional LSTM-CRF model, which can effectively use both previous and following input features [8], and with a CRF layer, the BiLSTM-CRF model could also make full use of sentence-level tagging information. Similarly, Han et al. proposed the CNN-BiLSTM-CRF model for term extraction in Chinese Corpus [9].

With the application of NER in the fields of biology, geography, medicine, etc., while the annotated data in these specific fields is quite limited, the performance of traditional supervised learning methods is greatly undermined. In recent years, many researches have proposed solutions to the problem of insufficient labeled data. The Transformer model proposed by Ashish et al. totally abandoned traditional CNN and RNN architecture [10], whose network structure is entirely composed of attention mechanism, which can solve the problem of long-term dependence effectively. Jacob et al. proposed the BERT model, which adopts the idea of transfer learning [2]. The BERT model is pre-trained on large amounts of unlabeled data, followed by fine-tuning training on a small amount of labeled data, and finally could achieve excellent performance on various NPL tasks. The attention mechanism, transfer learning, unsupervised learning, and other methods have greatly alleviated the workload of manual annotation, and significantly improve the performance on the NLP tasks.

2.2 Applications of Named Entity Recognition

Named entity recognition is the basis of many natural language processing tasks such as relation extraction and machine translation, and has a wide range of applications. Researchers from different fields apply named entity recognition to different application scenarios.

Weidlich et al. proposed ChemSpot [11], which is a NER tool for identifying mentions of chemicals in natural language texts, such as trivial names, drugs, abbreviations, and molecular formulas. ChemSpot uses a hybrid approach combining the CRF model with the dictionary-based method to implement chemical named entity recognition. Chen et al. developed an active learning-enabled annotation system for clinical named entity recognition [12], which enables the user to evaluate the actual performance of active learning in practice. Li et al., who works in the field of biomedicine, proposed a two-phase biomedical named entity recognition method using CRFs [13], which divide the recognition task into two subtasks: Named Entity Detection (NED) and Named Entity Classification (NEC), which are finished in two phases with two respective CRF models. The NED and NEC method can reduce the training time significantly and furthermore, more features can be selected for each subtask.

In addition to the above-mentioned applications, NER is also widely used in the field of digital humanities, which can provide feasible solutions to the problems in this field. However, named entity recognition in classical Chinese is quite different from that in modern Chinese. At present, labeled data in the field of classical Chinese is extremely scarce, and data labeling in classical Chinese is much more difficult, which leads to the slow development of named entity recognition in classical Chinese. With the transfer learning approach, we can pretrain the model on unlabeled data and fine-tune the model on a small amount of labeled data to overcome the NER problems in classical Chinese.

3   BERT-Based Named Entity Recognition

With the application of deep neural network models in the field of NLP, end-to-end models that do not rely on features engeering have increasingly become the mainstream method. In this paper, the BiLSTM-CRF sequence tagging model is used as the basic model for NER. By constructing the BiLSTM-CRF model on top of the pre-trained BERT model, we implemented NER in Chinese Twenty-Four Histories. Compared with English NER, Chinese NER is more difficult. There are different granularity divisions of characters and words in Chinese, resulting in the corresponding character-based NER, word-based NER, and character-word combined NER. The BERT model uses a character-based pre-training method and is well-performed in English NER tasks. Therefore, we also use the BERT model for preliminary feature extraction to further verify its transfer learning ability in NER task in classical Chinese.

3.1   Model Architecture

In this paper, we propose the BERT-BiLSTM-CRF model to implement NER in classical Chinese. The model consists of three parts: (1) first, the BERT model is used to pre-train the vector representation of each character in the input sentence, (2) then the character vector sequence is input into BiLSTM model for further training, (3) finally the maximum probability label sequence is output through a CRF layer. The structure of our model is shown in Fig. 1.

In Fig. 1, "蒙骜攻韩" is a sentence that comes from "史记" (Records of the Grand Historian) which means that a person named "蒙骜" attacks "韩" (an ancient place name), and we use it to demonstrate how our model works.

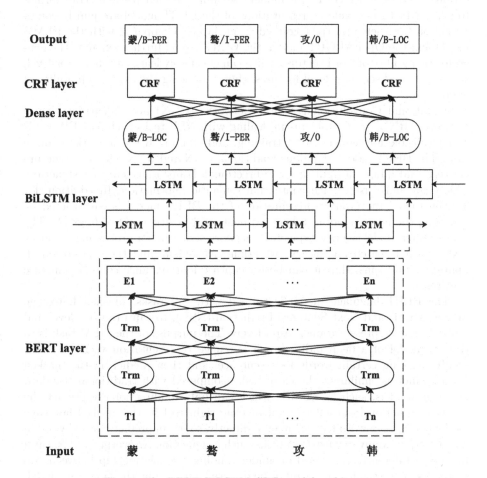

Fig. 1. The BERT-BiLSTM-CRF model architecture.

In the entire model, the pre-trained BERT model is trained by Google on a large-scale Chinese corpus [2], which can characterize the ambiguity of characters. In the process of training our model, the parameters of the BERT model are fixed, and only the parameters of the BiLSTM-CRF model need to be updated, thus the training parameters and the training time could be both reduced.

3.2   Pre-trained BERT Model

BERT is the first fine-tuning based representation model that achieves state-of-the-art performance on a large number of sentence-level and token-level tasks, outperforming a lot of task-specific architectures [2].

Unlike the ELMO [14] model, which uses a left-to-right and a right-to-left LSTM language model to extract context-sensitive features, the BERT [2] model uses masked language models to enable pre-trained deep bidirectional representations. The OpenAI GPT [15] model uses only one left-to-right Transformer to extract features, while representations of the BERT model are jointly conditioned on both left and right context in all layers. Compared with the ELMO model and the OpenAI GPT model, the BERT model further expands the generalization ability of word vectors, which can effectively learn the character-level, word-level, and sentence-level features, thus enhancing the expression ability of word vectors.

Traditional Convolutional Neural Networks (CNN) and Recurrent Neural Networks (RNN) have problems in dealing with NLP tasks such as NER: CNN is not suitable for sequence text training, and training RNN networks is quite slow. The BERT model abandons traditional CNN and RNN network structure entirely, and adopts a multi-layer bidirectional Transformer encoder structure, which can make use of rich contextual semantic information. In addition, for a given token, the input representation of the BERT model is constructed by summing the corresponding token, segment, and position embeddings [2]. The Transformer model is a new type of the network structure for processing sequence text, which is based on the self-attention mechanism, and there is no length limitation problems, thus it can better capture long-distance context semantic features.

The BERT language model is trained with two tasks, Masked Language Model (MLM) and Next Sentence Prediction (NSP), to capture word-level and sentence-level representations, respectively. The purpose of the MLM task is to train a model of a deep bidirectional language representation vector by randomly masking certain words in a sentence, and then predicting the masked words, which is similar to the cloze task. Compared with traditional standard language models, such as ELMO, which can only unidirectionally predict the objective function from left to right or from right to left, the masked language model allows each word to "see" itself indirectly, so it can predict masked word in any direction. Many downstream tasks, such as Question Answering and Natural Language Inference, are based on understanding the relationship between two sentences, while the features cannot be directly captured by the language model.

Therefore, the NSP task is used to train the model that could understand sentence level relationships.

In this paper, we use the pre-trained BERT-Chinese model provided by Google for the first step of feature extraction in Chinese Twenty-Four Histories. This pre-trained BERT-Chinese model contains 12 layers (Transformer blocks), with 768 hidden layers, and has 12 self-attention heads [2].

3.3   Fine-Tuning with BiLSTM-CRF Model

BiLSTM-CRF model is the combination of the BiLSTM model and the CRF model. The bidirectional LSTM component makes it possible for the model to use both previous and following input features. In addition, with a CRF layer, the model can make use of sentence level tag information. Due to these characteristics, the BiLSTM-CRF model could achieve better performance on sequence tagging tasks such as POS (Part-of-Speech Tagging), chunking, and NER.

In sequence tagging tasks, RNN could maintain memory based on historical information, which enables the model to predict the current output through long-distance features. Long Short-Term Memory (LSTM) [16] networks are quite similar to RNNs, but the update module of the hidden layer is replaced by special memory cells, which makes LSTM easier to discover and explore long-term dependence information. Compared with LSTM, BiLSTM can effectively make use of previous features and following features through forward and backward states. Therefore, the BiLSTM model can better capture bidirectional semantic dependencies and effectively learn contextual semantic information, thereby improving the performance on NER.

A typical LSTM network consists of the forget gate layer, the input gate layer, and the output gate layer, and its structure can be formalized as follows:

$$f_t = \sigma(W_f \cdot [h_{t-1}, x_t] + b_f) \tag{1}$$

$$i_t = \sigma(W_i \cdot [h_{t-1}, x_t] + b_i) \tag{2}$$

$$\tilde{C}_t = \tanh(W_C \cdot [h_{t-1}, x_t] + b_C) \tag{3}$$

$$C_t = f_t * C_{t-1} + i_t * \tilde{C}_t \tag{4}$$

$$o_t = \sigma(W_o \cdot [h_{t-1}, x_t] + b_o) \tag{5}$$

$$h_t = o_t * \tanh(C_t) \tag{6}$$

where h_{t-1} and x_t are information from the previous cell, σ is the logistic sigmoid function, C represents cell vectors, and f, i, o stand for the forget gate, input gate, and output gate, respectively. The forget gate layer looks at h_{t-1} and x_t to decide whether to keep this information or get rid of it or discard part of it. The input gate layer decides which values to update, and C_t is the new candidate values, scaled by how much we decide to update each state value. The output gate layer runs a sigmoid layer to decide which part of the cell state to output, and h_t is the final output value by the current cell.

Although the BiLSTM model can effectively identify the boundaries of named entities, sometimes it may fail to consider the relationship between the labeled entity sequences. Therefore, we can use the CRF [5] layer to add some constraints to ensure that the final prediction result is valid, which can be automatically learned by the CRF layer when training on datasets. The BiLSTM model does not consider the correlation between labels, and a significant advantage of CRF networks is that it can obtain a globally optimal labeling sequence by considering the relationship between adjacent labels. Therefore, a combined BiLSTM-CRF model can be used to reduce the possible erroneous prediction sequence and obtain the best named entity recognition results.

4   Experiments

4.1   Experimental Datasets

In our proposed BERT-BiLSTM-CRF model, the BERT model we use is the Google official version [2], whose parameters are fixed and do not need to be updated. However, the BiLSTM-CRF model is self-defined and requires fine-tuning training on a small amount of annotated data of named entities. The parameters of each layer of the BiLSTM-CRF model are updated during the training process.

In the process of fine-tuning the model, we use the ChineseDailyNerCorpus, which contains a large number of modern Chinese sentences and their corresponding BOI annotations. The ChineseDailyNerCorpus can be obtained at: http://s3.bmio.net/kashgari/china-people-daily-ner-corpus.tar.gz. Named entities have different categories such as location, person, and organization. In the BOI annotation method, B (beginning) indicates that the current character is the starting position of a named entity, O (outside) indicates that the current character is not a character in any entity, and I (inside) indicates that the current character is the middle position of a named entity.

We chose the ChineseDailyNerCorpus as the training data for the fine-tuning stage for two reasons: (1) first, it is difficult to find a large-scale annotated classical Chinese NER dataset, so we instead chose an easily accessible modern Chinese annotated dataset; (2) second, this training method can be used to verify the transfer learning characteristics of the BERT model. The BERT-Chinese model is trained on a large-scale Chinese corpus [2], and the BiLSTM-CRF model is trained on the ChineseDailyNerCorpus modern Chinese corpus. Although the two stages of training are not conducted on classical Chinese, but our final verification is performed by using the Chinese Twenty-Four Histories.

In the process of data preparation, we spent most of the time annotating the Chinese Twenty-Four Histories data. The Chinese Twenty-four Histories data is available at: https://github.com/quzhi1/ChineseHistoricalSource/tree/master/json, but they are merely classical Chinese texts and unlabeled data. In order to use the BERT-BiLSTM-CRF model to make predictions on the Chinese Twenty-Four Histories data, and to evaluate the precision and recall of the results, we manually labeled part of the Chinese Twenty-Four Histories data. The Chinese

Twenty-Four Histories are the Chinese official historical books covering a period from 3000 BC to the Ming dynasty in the 17th century, including "Records of the Grand Historian", "Book of Han", "Book of Song", and so on.

In order to evaluate the performance of our model for named entity recognition in Chinese Twenty-Four Histories, we selected five of the twenty-four history books, and in each book we selected a long article for annotation. They are the "Sixth of the Qin Shihuang Basic Annals" in the "Records of the Grand Historian" on year 300 BC, the "Seventh of the Emperor Xiaohuan" in the "Book of the Later Han" on year 150 AD, the "First of the Emperor Basic Annals" in the "Book of Sui" on year 600 AD, the "Ninth of the Basic Annals" in the "Song History" on year 1050 AD, and the "Fifteenth of the Basic Annals" in the "Ming History" on year 1500 AD. The procedure of manual labeling is as follows: (1) first, we find these five long articles in the Chinese Twenty-Four Histories data, (2) then find the corresponding chapters in the paper versions of these books, (3) and manually label the data according to the person and location names marked in the book. In the Chinese Twenty-Four Histories, there are few named entities of the Organization type, so we only labeled entity of Person and Location types. Although this work is very time-consuming and tedious, it is necessary for the subsequent model evaluation.

4.2 Experimental Results

As mentioned above, the parameters of the pre-trained BERT-Chinese model are as follows: the BERT-Chinese model has 12 Transformer blocks with 768 hidden layers, and 12 self-attention heads, which has 110 M parameters totally [2]. During the training process, the maximum sequence length is 128, batch size is set to 512, and hidden layer dimension of the BiLSTM model is 256.

We use the ChineseDailyNerCorpus corpus to train our model on a Windows 10 operating system and a single Inter(R) Core(TM) i7-7700HQ (2.80 GHz) CPU, and it took 6 h to complete the training process. In order to evaluate the results, this paper uses the precision, recall, and F1 value as measures, which are defined as follows:

$$Precision = \frac{TP}{TP + FP} \tag{7}$$

$$Recall = \frac{TP}{TP + FN} \tag{8}$$

$$F1 = \frac{2 * Precision * Recall}{Precision + Recall} = \frac{2 * TP}{2 * TP + FP + FN} \tag{9}$$

where TP indicates that the sample is correctly identified, FP indicates that the sample is incorrectly identified, and FN indicates that the sample is incorrectly rejected, thus $Precision$ represents the proportion of correctly predicted positive categories in the total number of all predicted positive categories, $Recall$ represents the proportion of correctly predicted positive categories in the total number of true positive categories in the sample, and $F1$ is the harmonic mean of $Precision$ and $Recall$, which can represent the overall performance.

The experimental results are shown in Table 1. Detailed information of our selected chapters are shown in Table 2.

Table 1. Experimental results in Chinese Twenty-Four Histories.

Book name	Entity type	Precision	Recall	F1 value
Records of the Grand Historian	Location	0.8657	0.8529	0.8593
Records of the Grand Historian	Person	0.6489	0.5169	0.5755
Records of the Grand Historian	Location and Person	0.7650	0.6969	0.7274
Book of the Later Han	Location	0.7778	0.7656	0.7717
Book of the Later Han	Person	0.7353	0.7905	0.7619
Book of the Later Han	Location and Person	0.7609	0.7755	0.7678
Book of Sui	Location	0.7730	0.8460	0.8078
Book of Sui	Person	0.7188	0.6788	0.6982
Book of Sui	Location and Person	0.7465	0.7642	0.7542
Song History	Location	0.7830	0.8191	0.8006
Song History	Person	0.7397	0.7500	0.7448
Song History	Location and Person	0.7691	0.7969	0.7827
Ming History	Location	0.9060	0.9164	0.9112
Ming History	Person	0.7955	0.7527	0.7735
Ming History	Location and Person	0.8674	0.8593	0.8631

Table 2. Selected chapters and detailed information.

Time	Book name in English	Selected chapter	Book name in Chinese
300 BC	Records of the Grand Historian	Volume 6	史记
150 AD	Book of the Later Han	Volume 7	后汉书
600 AD	Book of Sui	Volume 1	隋书
1050 AD	Song History	Volume 9	宋史
1500 AD	Ming History	Volume 15	明史

We calculated the *precision, recall*, and *F*1 value on each book with entity type of Location, Person, and both of them, respectively.

5   Analysis

The experimental results shown in Table 1 are not intuitive enough, so we draw a line chart to demonstrate the results of the experimental evaluation, which is shown in Fig. 2.

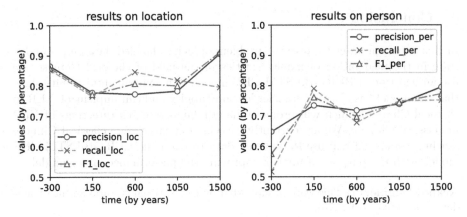

Fig. 2. Evaluation results on Chinese Twenty-Four Histories.

According to the experimental results, we can observe that the overall recognition accuracy of Location entities by our model is higher than that of Person entities. Judging from the evaluation results of five long articles in different dynasties, our model has all achieved an F1 value higher than 72%. From the experimental results in Table 1, we can see that, as time goes by, the overall F1 value becomes higher and higher, and the best result achieves an F1 value of 86%. Comparing the results of the model's prediction on Location entities and Person entities, from 300 BC to 1500 AC, the recognition accuracy of our model in these two named entity types shows an increasing trend, and the best performance is achieved for the book Ming History.

First of all, in classical Chinese, the majority of named entities of Location type less frequently change, which means that a location name in the ancient years is usually quite similar to what its name is now. However, a named entity of person type has various forms, such as given name, literary name, and nickname, resulting in the predicted results on Location entities significantly better than those on Person entities. For example, "吕不韦" (an ancient man from the Qin Dynasty) has the nickname "文信侯" (a kind of knighthood honor, something like a landlord), but the nickname rarely appears in books, thus it may not be recognized by our model.

Secondly, in ancient times, classic Chinese is more different from modern Chinese, which reduces the model performance. However, as time gets closer to current time, named entities also change over time and are more similar to what their names are currently. For instance, "湖南" (a province in China) is also called "楚" (a place name in the Spring and Autumn Period and the Warring States Period) in the Qin Dynasty, which is quite obscure. Therefore, as time goes by, these measures of NER in classical Chinese have shown improvements.

6   Conclusion

In this paper, in order to solve the problem of lack of labeled data and complex rules in the recognition of named entities in the field of classical Chinese, we use an end-to-end BERT-BiLSTM-CRF model. Our experimental results show that the model has a strong transfer learning ability and can implement NER in classical Chinese, which will promote many related research directions of NLP in classical Chinese. We are currently carrying out large-scale classical Chinese labeling work, and will use the labeled data to train the BERT-BiLSTM-CRF model, with the purpose of further improving the performance of our model.

Acknowledgments. This work is supported by the National Natural Science Foundation of China (61972275).

References

1. Natural language processing. https://en.wikipedia.org/w/index.php?title=Natural_language_processing&oldid=971205596. Accessed 16 Aug 2020
2. Devlin, J., Chang, M.W., Lee, K., et al.: Bert: pre-training of deep bidirectional transformers for language understanding. arXiv preprint arXiv:1810.04805 (2018)
3. Named-entity recognition. https://en.wikipedia.org/w/index.php?title=Named-entity_recognition&oldid=959772078. Accessed 27 June 2020
4. Xie, R., Liu, Z., Jia, J., et al.: Representation learning of knowledge graphs with entity descriptions. In: Thirtieth AAAI Conference on Artificial Intelligence (2016)
5. Lafferty, J., McCallum, A., Pereira, F.C.N.: Conditional random fields: probabilistic models for segmenting and labeling sequence data. In: 18th International Conference on Machine Learning, pp. 282–289 (2001)
6. Collobert, R., et al.: Natural language processing (almost) from scratch. J. Mach. Learn. Res. **12**, 2493–2537 (2011)
7. Chiu, J.P.C., Nichols, E.: Named entity recognition with bidirectional LSTM-CNNs. Trans. Assoc. Comput. Linguist. **4**, 357–370 (2016)
8. Huang, Z., Xu, W., Yu, K.: Bidirectional LSTM-CRF models for sequence tagging. arXiv preprint arXiv:1508.01991 (2015)
9. Han, X., Xu, L., Qiao, F.: CNN-BiLSTM-CRF model for term extraction in Chinese corpus. In: Meng, X., Li, R., Wang, K., Niu, B., Wang, X., Zhao, G. (eds.) WISA 2018. LNCS, vol. 11242, pp. 267–274. Springer, Cham (2018). https://doi.org/10.1007/978-3-030-02934-0_25
10. Vaswani, A., Shazeer, N., Parmar, N., et al.: Attention is all you need. In: Advances in Neural Information Processing Systems, pp. 5998–6008 (2017)
11. Weidlich, M., Leser, U., et al.: ChemSpot: a hybrid system for chemical named entity recognition. Bioinformatics **28**(12), 1633–1640 (2012)
12. Chen, Y., Lask, T.A., Mei, Q., et al.: An active learning-enabled annotation system for clinical named entity recognition. BMC Med. Inform. Decis. Mak. **17**(2), 35–44 (2017)
13. Li, L., Zhou, R., Huang, D.: Two-phase biomedical named entity recognition using CRFs. Comput. Biol. Chem. **33**(4), 334–338 (2009)

14. Peters, M.E., Neumann, M., Iyyer, M., et al.: Deep contextualized word representations. arXiv preprint arXiv:1802.05365 (2018)
15. Radford, A., Narasimhan, K., Salimans, T., et al.: Improving language understanding by generative pre-training (2018). https://s3-us-west-2.amazonaws.com/
openai-assets/research-covers/language-unsupervised/language_understanding_
paper.pdf
16. Hochreiter, S., Schmidhuber, J.: Long short-term memory. Neural Comput. 9(8),
1735–1780 (1997)

Machine Learning

Explainable Enterprise Rating Using Attention Based Convolutional Neural Network

Weiyu Guo[1]([⊠]), Bin Cao[2], and Zhenxing Li[2]

[1] Central University of Finance and Economics, Beijing, People's Republic of China
weiyu.guo@cufe.edu.cn
[2] HuaXia iFinance Information Technology Co., Ltd.,
Beijing, People's Republic of China
{bincao,lizhenxing}@hxifin.com

Abstract. Convolutional Neural Networks (CNNs) are being applied to identification problems in a variety of fields, and showing higher discrimination accuracy than conventional methods. However, the applications based on CNNs are rare in enterprise rating. The reason for this seems to be that most of CNNs are lacking interpretability. The users of financial industry can not accept the results providing by an artificial intelligence model which they can not understand. In this paper, we propose to model enterprise credits using CNNs with attribute and sequence attention modules, which enables an explainable and more precise enterprise credit rating. The experimental results indicate that our approach is not only achieving a higher performance compared with conventional methods, e.g., Decision Trees, Z-Score, Random Forests, linear model, but also can provide multi-granular interpretations for the rating results.

Keywords: Deep learning · Interpretability · Enterprise rating

1 Introduction

With the rapid development of the deep learning, Deep Neural Networks (DNNs) are gathering great attention in the field of artificial intelligence (AI), and even have become ubiquitous in a variety of applications, e.g., image processing [1], natural language processing [2], since they can automatically extract valuable features from original data without artificial feature engineering. However, the number of applications of deep learning in financial analysis is extremely limited except for several reports on the prediction of stock price fluctuations [3]. The reason for this phenomenon seems to be that the forecasting models based on deep neural networks are usually lacking interpretability. Fair lending is the foundation of financial industry. Thus, one significant challenge of using AI-based systems in credit evaluation, is that it is hard to provide the needed "reason codes" to users the explanation of why they were denied credit. Especially when the basis for denial is the output from an opaque machine learning algorithm.

© Springer Nature Switzerland AG 2020
G. Wang et al. (Eds.): WISA 2020, LNCS 12432, pp. 305–313, 2020.
https://doi.org/10.1007/978-3-030-60029-7_28

Financing institutions usually use linear models, e.g., logistic regression [4], Z-score [5] to evaluate credit ratings of enterprises. The advantage of these linear models is that they are easy to interpret. Users can accurately obtain the relationship between the input indicators and the assessment result in the process of credit rating. However, the accuracy of linear models is often not enough comparing with DNN based models, especially when they are suffering from high-dimension inputs and noise, their performance may decline significantly.

The present research aims to accurately evaluate credit ratings of enterprises by mining the historical data of enterprise as well as output readable explanations. We treat the credit rating as a problem of sequential modeling, and propose a DNN based model to predict the future credit rating of a firm by using its historical sequence data. In general, there are two main branches in DNN, i.e., Convolutional Neural Networks(CNN) and Recurrent Neural Networks (RNN). The former one is considered suitable for processing images and the latter for sequential data. However, instead of constructing RNN based model, a CNN based model with attributive and sequential attentions is proposed to predict the credit ratings of enterprises, which is different from RNN based model that usually require serial processing of sequential data, and can make a better use of the advantages of parallel computing hardware, such as GPU. In a nutshell, our contributions in this work can be summarized as follows:

- A CNN based model is proposed to analyze financial sequential data, which can better utilize the parallel computing power of GPU than traditional RNN based sequence modeling, and extract the high order credit features of enterprise. To the best of our knowledge, this is the first paper to use attention-based CNNs for enterprise rating.
- Dual attention modules are used in proposed model, i.e., an attribute attention module and a sequence attention module. These modules are used to select informative indicators and important time points which contribute to the enterprise rating, respectively. These modules can give the users insight to the reasons of enterprise rating prediction.
- A series of experiments are conducted to validate the proposed model. The results show that our model not only can obtain more precise enterprise ratings than conventional approaches of enterprise rating, but also output comprehensible explanations.

2   Related Work

This work devotes to achieve a deep neural network with high accuracy and explainable credit rating. Its related work can be divided into two groups, i.e., traditional credit rating methods and explainable deep neural networks.

2.1   Traditional Credit Rating Methods

Enterprise credit assessment as an intermediary service in the financial field has existed for more than one hundred years. A mount of approaches is proposed to handle this problem, e.g., factor analysis based methods [6], statistic based methods [5], machine learning based methods [4].

The factor analysis based methods [6] are usually to score the credit-related factors of the enterprise based on the experience of experts, which can be applied flexibly to qualitative analysis of enterprise credit. However, this kind of methods relies too much on the subjective judgment of experts and lacks the ability of quantitative analysis. Different from the factor analysis based methods, statistic based methods quantify the credit score of company based on its financial indicators. For example, Z-Score[5] treats a linear weighted sum of given financial indicators as the credit score of the company. The weights in Z-Score model are calculated based on historical data of companies. However, this kind of methods lacks generalization ability because the weights and score thresholds are fixed by experts.

With the development of artificial intelligence technology, some machine learning models which have interpretability, such as logistic regression [4] and decision tree [7], are introduced into credit rating. For example, logistic regression [4] is often used in place of Z-Score [5] to handle the large scale task of credit rating, and usually can obtain higher accuracy than Z-Score. The decision tree [7] is also popular in credit rating, which can provide accurate credit assessments as well as extract decision rules from dataset. However, as the feature of companies become more and more complex, the prediction performance of these models based on shallow feature representation are getting harder to be promoted. Therefore, for a higher rating accuracy, the deep nerual networks are increasingly attention in credit rating. For example, Hosaka et al. [8] use a convolutional nerual network to make bankruptcy prediction for Japanese companies, and achieve a higher accuracy than the models with shallow feature representation. However, the deep neural networks are usually thought of as black box models, which can not provide necessary explanations for the predictions.

2.2 Explainable Deep Neural Networks

Aiming at reducing the unexplainability of neural networks, some scholars use agent models which can be interpreted by common users to explain the decision process of given neural networks indirectly. For example, [9] proposed a quantitative method to explain each prediction made by pretrained convolutional neural networks, by training a decision tree to generate the specific reasons for each prediction. [10] proposed a framework to use linear models as the agents to individually explain the prediction of given deep neural network. Besides, [11] proposed local and global attentions to improve the forecasting performance as well as generate explanations with inputted features. The attention based models provide us another way to insight the forecasting process of neural networks.

3 Proposed Model

In this section, we describe our dual attentions based CNN model, SA-Attn-CNN, for enterprise credit rating. The overall architecture of SA-Attn-CNN is

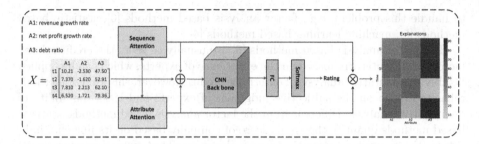

Fig. 1. The framework of dual attentions based CNN model for explainable enterprise rating. The **X** is a toy example of model input, which is expressed as a matrix. The rows of matrix indicate the time points and the columns of matrix indicate the attributes of companies. The heat map can indicate which attributes at which time points are more importance for a specific rating prediction.

shown as Fig. 1. In SA-Attn-CNN, we first regroup the inputted sequential data to be a gray picture, and use dual attention modules to mining the contributions of financial indicators and the importance of time points, respectively. Then, a CNN backbone network is used to extract the high order credit features of enterprise. Finally, a fully connected layer following the CNN backbone is utilized to predict the credit ratings.

3.1 Sequence Attention Module

The module of sequence attention aims to learn which time points are more informative in sequential data, and facilitate the following CNN backbone paying more attention to the feature of critical moment. Considering that the feature fluctuations of in adjacent time points often are key patterns for credit rating forecasting, we first apply the attention through sliding kernels to inputted sequence. Let $\mathbf{x}_i$ be the feature vector of center time point and the d be the kernel width. We compute the weights in sliding window for each time points of sequence, with a learn-able matrix $A_s \in \mathbb{R}^{d \times n}$ as follows:

$$
\begin{aligned}
\mathbf{X}_{S-Attn,i} &= (\mathbf{x}_{i-\frac{d+1}{2}}, .., \mathbf{x}_i, .., \mathbf{x}_{i+\frac{d+1}{2}}) \\
s_i &= f(\mathbf{X}_{S-Attn,i} * A_s), i \in [1, t] \\
\widehat{\mathbf{x}}_i &= s_i \mathbf{x}_i
\end{aligned}
\tag{1}
$$

where $*$ is a kind of arithmetic operation which first performs element wise multiplication, and then sum. In this work, we realize the arithmetic operation with a single channel convolution which convolutional kernel is A_s. We use the sigmoid function for the activation function $f(\cdot)$. s_i is an attention score which indicates the importance of time point i in its d-gram neighborhoods. $\widehat{\mathbf{x}}_i$ indicates weighted features of time point i, which are weighted by its volatility comparing with its neighborhoods. Hence, if a given time point i has a smaller score than its neighborhoods, then we can interpret that the features of that time point do not fluctuate much and is less important for credit rating.

For further learning the weights of each time points from a global viewpoint, the weighted sequence $(\widehat{\mathbf{x}}_1, \widehat{\mathbf{x}}_2, ..., \widehat{\mathbf{x}}_t)$ is then passed through a full connection neural network. Let $\mathbf{V} \in \mathbb{R}^{n \times m}$ and $\mathbf{W} \in \mathbb{R}^{m \times k}$ be the parameters of the network, then the global weights of each time points can be computed as follows:

$$a_i = f(tanh(\widehat{\mathbf{x}}_i \cdot \mathbf{V}) \cdot \mathbf{W}), i \in [1, t]$$
$$\widehat{\mathbf{x}}_i^s = a_i \mathbf{x}_i$$
(2)

where $f(\cdot)$ is the element wise sigmoid function. Moreover, we set $m = \frac{n}{2}$ for feature dimension reduction and $k = 1$ for getting scalar weights. $\widehat{\mathbf{x}}_i^s$ are weighted features of time point i. Hence, if a given time point i has a smaller weight than another time point, then we can interpret that the time point in current sequence is less important than the time point with larger weight for credit rating.

3.2 Attribute Attention Module

The module of attribute attention aims to learn the influence of inputted indicators on the result of credit rating, and propels the following CNN backbone to focus on those important indicators. In order to adaptively calculate the attention scores of each indicator, we assign each indicator c_i with a learn-able parameter matrix $\mathbf{W}_i \in \mathbb{R}^{t \times n}$. Specifically, we compute the weights of indicators for each sample as follows:

$$p_i = f(\mathbf{X} * W_i), i \in [1, n]$$
$$\widehat{\mathbf{x}}_i^a = \mathbf{p} \odot \mathbf{x}_i$$
(3)

where $*$ is a kind of arithmetic operation which first performs element wise multiplication, and then sum. In this work, we realize the arithmetic operation with a convolutional layer which kernel size is $t \times n$ and the number of output channels is n. $\widehat{\mathbf{x}}_i^a$ are weighted features of time point i. $\mathbf{p} = (p_1, p_2, .., p_n)$ is the weight vector of indicators. $\odot$ is an arithmetic operation that performs element wise multiplication. Hence, if a given indicator c_i has a smaller weight than another indicator, then we can interpret that the indicator in current sequence is less important than the indicator with larger weight for credit rating.

3.3 Enterprise Rating Model

As shown in Fig. 1, the proposed dual attention modules are integrated into a CNN based backbone to extract the high order credit features of enterprise. Specifically, the outputs of the sequence and attribute attention modules, i.e. $\widehat{\mathbf{x}}_i^a$ and $\widehat{\mathbf{x}}_i^s$, are first added to $\mathbf{x}_i$, and then run pass the backbone network and a fully connected layer in succession to predict credit ratings for enterprise. Therefore, our enterprise rating model can be express as:

$$
\begin{aligned}
\mathbf{Z} &= \mathbf{X} + \lambda_1 \widehat{\mathbf{X}}^s + \lambda_2 \widehat{\mathbf{X}}^a \\
\mathbf{h} &= \mathrm{CNN}(\mathbf{Z};\ [\mathbf{W}^{(1)}, \mathbf{W}^{(2)}, .., \mathbf{W}^{(l)}]) \\
y &= softmax(\mathrm{FC}(\mathbf{h};\ \mathbf{W}^{(F)}))
\end{aligned}
\tag{4}
$$

where λ_1 and λ_2 are hyper-parameters for tuning the effects of dual attention inputs. $\mathrm{CNN}(\cdot\ ;\ \cdot)$ indicates the CNN based backbone. $\mathbf{Z}$ and $[\mathbf{W}^{(1)}, \mathbf{W}^{(2)}, .., \mathbf{W}^{(l)}]$ are the input and the parameters of backbone, respectively. $\mathbf{h}$ is the embedding of input sequence. In this work, we treat the credit ratings of enterprises as categories. Therefore, given c credit ratings, we use a fully connected layer $\mathrm{FC}(\cdot\ ;\ \cdot)$ with parameter matrix $\mathbf{W}^{(F)} \in \mathbb{R}^{h \times c}$ to calculate the probabilities of different categories, and $softmax$ to infer the final credit rating y for enterprise.

3.4   Explanation of Prediction

For each prediction of SA-Attn-CNN, we can further generate the explanations for the prediction by utilizing the attention scores of the dual attention modules. If given attention score vectors $\mathbf{p}$ and $\mathbf{a}$ generate from sequential attention module and attribute attention module, respectively, then we can get the explanations by following algorithm:

$$
\begin{aligned}
\mathbf{M} &= sigmoid(\mathbf{p} \otimes \mathbf{a}) \\
(e_1, e_2, .., e_k) &= \arg(\mathrm{topK}(M))
\end{aligned}
\tag{5}
$$

where, $M \in \mathbb{R}^{t \times n}$ can be treated as the weights of input information. $sigmoid(\cdot)$ is used to normalize the elements of matrix into $[0, 1]$. $(e_1, e_2, .., e_k)$ is a set which elements indicate the locations of top-k weights in matrix. Through the e_i, we can locate the important information, i.e., we can find out which attribute at which time point is important to the prediction.

4   Experiment

4.1   Data and Experimental Settings

In this research, we crawl historical data of 7968 Chinese listed companies from multiple data sources. For each company, we obtain its historical financial data from IPO to third fiscal quarter of 2019, and extract 90 financial indicators from each data of fiscal quarter. According to the experience of the investment analysts, the situation of revenue usually indicates the credit rating of the company. Hence, we randomly split training (70%) and testing (30%) data, and predict the credit rating of the company indirectly through classifying whether its revenue increased in the third fiscal quarter of 2019.

We treat the original data of each company to be a 95×90 feature matrix D, which contains the features of company in past 95 fiscal quarters. Then, we upsample D to be a 112×112 matrix $\mathbf{X}$ leveraging the bilinear interpolation algorithm, and use $\mathbf{X}$ to be the input of our model. We initialize our CNN backbone with the corresponding layers of resnet18[1] which pretrained on imagenet, and other parts of our network with Xavier algorithm. Moreover, we utilize multistages SGD algorithm with the initial learning rate 0.001 and multiplicative factor of learning rate decay 0.1 to train our model. The training stages in this experiment are [500, 800, 1000]. Additionally, we concatenate the features of company over all time periods(95 fiscal quarters) to be the input(8550-dimensional feature vector) of compared models except Z-Score.

Table 1. The performance of enterprise credit rating. The testing dataset contains 2451 Chinese listed companies. There are 1493 positive samples, and the rest are negative samples. The learning models, i.e., decision tree, random forest, logistics regression and GBDT, are realized by scikit-learn with default settings. The coefficients of Z-Score model are [0.517, −0.460, 18.640, 0.388, 1.158].

Methods	Accuracy	Type error		AUC
		I	II	
Decision Tree	0.7458	0.3497	0.1916	0.7955
Random Forest	0.7013	0.6928	0.1285	0.7397
GBDT	0.7821	0.3256	0.1474	0.8433
Z-Score	0.5788	0.3497	0.5164	–
LR	0.6907	0.5731	**0.1099**	0.7630
SA-Attn-CNN	**0.7918**	**0.2922**	0.1567	**0.8673**

4.2 Results and Analysis

Benefit from the feature engineering capabilities of deep neural network, the proposed model achieves the best performance on the task of enterprise credit rating. As shown in Table 1, the SA-Attn-CNN can offer significant performance improvements over others in multiple criterions, i.e., Accuracy, AUC, and Type I Error. Moreover, we can observe an interesting anomaly that the performance of Random Forest is lower than decision tress on most indicators and the performance of our model is not best on the Type II Error. This phenomenon may indicate that the training samples is still not enough, the complicated model is easier to suffer from overfitting problem. That is why we do not utilize a deeper network to be the backbone of our model in this work.

Benefit from the dual attention modules, the proposed model not only can provide credit rating prediction for given company, but also can generate corresponding explanations for the prediction. As shown in Fig. 2, we can visualize the weights of the input features generated by our model to be heatmaps. The intensity of the color blocks indicate the importance of corresponding features. From the Fig. 2, we can observe that different financial features of company have different influences on the predicted results at different times.

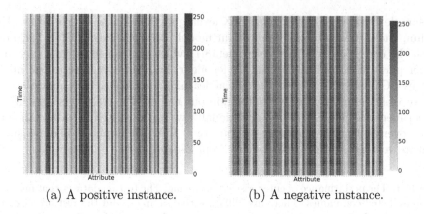

(a) A positive instance.   (b) A negative instance.

Fig. 2. The heatmap examples of inputted feature weights. The weights are first normalized to $[0, 1]$, and then mapped into the color space $[0, 255]$. (Color figure online)

5 Conclusion

To achieve a model of enterprise credit evaluation with high accuracy and explainable, we propose an attention based CNN model, i.e., SA-Attn-CNN which contains an attribute attention module and a sequence attention module. The attribute attention provides us insight on which financial indicators have significant impact on a rating prediction. The sequence attention facilitates the CNN backbone focusing on the important time points. Moreover, we evaluate the proposed model through comprehensive experiments and prove that our model not only can get higher prediction accuracy than traditional credit rating models, but also can provide explanations for the prediction outputted by SA-Attn-CNN.

However, the deep neural networks are a kind of data-driven approach, which are easy to suffer from overfitting on small training set. Therefore, we plan to extend our experimental dataset to include non-listed companies, and our model to support non-financial information of companies. Furthermore, various deeper architectures are deserved to attempt to use as the backbone of SA-Attn-CNN, when we have larger scale dataset.

References

1. He, K. et al.: Deep residual learning for image recognition. In: CVPR (2016)
2. Meng, X., et al.: Multilingual short text classification based on LDA and BILSTM-CNN neural network. In: WISA (2019)
3. Tsantekidis, A., et al.: Forecasting stock prices from the limit order book using convolutional neural networks. In: CBI (2017)
4. Bolton, C, et al.: Logistic regression and its application in credit scoring. Ph.D. thesis, University of Pretoria (2010)

5. Altman, E., et al.: Predicting financial distress of companies: revisiting the z-score and zeta models. In: Handbook of Research Methods and Applications in Empirical Finance (2010)
6. Mccrae, R.R., et al.: An introduction to the five-factor model and its applications. J. Personal. **60**(2), 175–215 (1992)
7. Xia, Y., et al.: A boosted decision tree approach using Bayesian hyper-parameter optimization for credit scoring. Expert Syst. Appl. **78**, 225–241 (2017)
8. Hosaka, T.: Bankruptcy prediction using imaged financial ratios and convolutional neural networks. Expert Syst. Appl. **117**, 287–299 (2019)
9. Zhang, Q., et al.: Interpreting CNNs via decision trees. In: CVPR (2019)
10. Ribeiro, M.T., et al.: Why should i trust you explaining the predictions of any classifier. In: KDD (2016)
11. Seo, S., et al.: Interpretable convolutional neural networks with dual local and global attention for review rating prediction. In: RecSys (2017)

Hospitalization Cost Prediction for Cardiovascular Disease by Effective Feature Selection

Wei Dai[1], Mengxing Huang[1(✉)], Qian Wu[1], Hanzhi Cai[2],
Ming Sheng[3], and Xin Li[4]

[1] Hainan University, Haikou 570228, China
{1848429933,854682796}@qq.com, huangmx09@163.com
[2] University of Sheffield, Sheffield, UK
caihanzhil996@qq.com
[3] BNRist Tsinghua University, Beijing 100084, China
shengming@tsinghua.edu.cn
[4] Beijing Tsinghua Changgung Hospital, Beijing 102218, China
Horsebackdancing@sina.com

Abstract. The burden of cardiovascular diseases is increasing, and the annual growth rate of hospitalization expenses for cardiovascular diseases is much higher than that of GDP. Therefore, researchers have developed a number of intelligent systems to predict hospitalization costs for cardiovascular disease. However, there are some problems with these methods, such as the performance of real world data sets and the differences between the feature selection and the actual selection of doctors. This paper proposes a method to construct a Medical Concept Knowledge Graph (MCKG) by combining open source knowledge graphs such as Wikidata and OpenKG, open source knowledge bases such as UMLS, and doctors' prior medical knowledge. A Medical Instance Knowledge Graph (MIKG) is constructed based on MCKG and the data of cardiovascular disease related medical records from the cooperative hospital. We conduct feature selection according to MIKG, draw feature alternatives, and combine with doctor-defined rules to arrive at final feature selection. We predict hospitalization costs with random forest algorithm. Experimental results show that the average error rate of our method is lower than that of the baseline algorithms.

Keywords: Cardiovascular diseases · Concept knowledge graph · Instance knowledge graph · Feature selection · Machine learning

1 Introduction

Cardiovascular disease is a serious threat to human beings. In China, the mortality rate of cardiovascular disease is still the highest among all diseases. Cardiovascular disease is considered to be one of the major causes of death in the world. With the aging of society and the acceleration of urbanization, the prevalence of unhealthy lifestyles among Chinese resident, the risk factors of cardiovascular disease are generally exposed. At the same time, the national burden of cardiovascular disease is growing increasingly heavy.

© Springer Nature Switzerland AG 2020
G. Wang et al. (Eds.): WISA 2020, LNCS 12432, pp. 314–325, 2020.
https://doi.org/10.1007/978-3-030-60029-7_29

Since 2004, the average annual growth rate of hospitalization expenses for cardiovascular disease is much higher than the growth rate of gross domestic product (GDP) [1]. Therefore, being able to predict hospitalization expenses in advance is of great significance to both patients and hospitals [2], and how to select features according to sample data and doctors' needs is crucial. Feature selection to improve the accuracy of prediction and combined with doctors' prior knowledge can effectively reduce the error rate of prediction is a major research of machine learning [3]. Many researchers have created different algorithms to predict the hospitalization costs of cardiovascular diseases. However, these systems have the problems of unsatisfactory accuracy when facing real world data sets [4] and different requirements from actual doctors in feature selection.

The concept of knowledge graph is proposed by Google on May 17, 2012. Google will use this as a basis to build a next-generation intelligent search engine. In essence, knowledge graph is a semantic network that reveals the relationship between entities. Formal descriptions of real-world things and their relationships can be made. With theproliferation of semantic Web resources and the publication and sharing of vast amounts of RDF data, researchers in academia and industry have spent a great deal of effort building a variety of structured knowledge bases. These knowledge bases can be roughly divided into two categories: open link knowledge base and industry knowledge base. Typical examples of open linked knowledge base are Freebase, Wikidata, OpenKG, YAGO; Typical examples of vertical industry knowledge base are: IMDB (movie data), MusicBrainz (music data), MusicBrainz (semantic knowledge network).

We apply the knowledge graph to the medical field [5], and use the knowledge graph in combination with the interaction between doctors for feature selection, and use the selected data to predict the hospitalization cost of cardiovascular diseases.

The main contributions of this paper include:

a) We create the medical health concept knowledge graph (MCKG) using the open source knowledge graph such as Wikidata, OpenKG and the open source knowledge base such as the language specification defined by UMLS.
b) Based on MCKG, we build the medical instance knowledge graph (MIKG) with real data from cooperative hospitals.
c) Based on the constructed knowledge graph, we use it to conduct feature selection and obtain feature alternatives. Doctors define rules and requirements in the alternative and further obtain the final feature selection scheme.
d) We use the selected feature data to predict the hospitalization cost of cardiovascular disease, and the experiment reduces the average error rate of the prediction.

The rest of the paper is organized as follows: In Sect. 2 we discuss the related work. Section 3 introduces the methodology about how to construct MCKG and MIKG. Section 4 shows the experiment and prediction results. At last, we conclude the paper in Sect. 5.

2   Related Work

Knowledge graph is an important part of artificial intelligence technology [6]. It has been a hot trend in the field of artificial intelligence to make use of core technologies such as knowledge extraction and knowledge representation [7] of knowledge graph to carryout relevant research. Knowledge graph has a very broad application prospect in the medical services, the technology can solve the problems of strong data professionalism and complex structure in the medical field, improve medical and health services [8] and plays an important role in clinical decision support system [9].

At present, most of the studies related to cardiovascular diseases use data sets of UCI CLEVELAND [10]. Aiming at feature selection, Senthilkumarmohan et al. proposed a method about Hybrid Random Forest with Linear Model, which uses artificial neural network model with feedback for feature selection [11]. FajrI et al. used discrete minimum wavelet method for feature selection [12]. AliL et al. explained method of exhaustion to search the best configuration of the network to select relevant features from the feature space [13], and Fatih et al. selected features based on the simplified rule library [14]. Jesmin et al. combined with medical knowledge, computing intelligently to delete clinical features [15]. Prakash, S et al. used optimality criterion feature selection method for feature selection [16]. Chandra Babu Gokulnath et al. combining genetic algorithm with support vector machine used to select features in feature space [17]. Ting-Ting Zhao et al. used discriminant minimum class locality preserving canonical correlation analysis to extract features from two data sets based on gain and entropy of motion vector [18]. Sarah P et al. used convolutional neural network to make sense of feature selection [19]. Ashirjaveed et al. employed random searching algorithm to select relevant features [20].

These feature selection methods are not combined with knowledge graph. In this paper, we used a different feature selection method. We first construct MCKG based on doctors' prior knowledge, open source knowledge base and open source knowledge graph, and then integrate the structured data of hospital database and case data to obtain MIKG. We use MIKG for feature selection and get the alternative scheme of features. Then, we further screen the alternative scheme according to the rules defined by doctors and the actual needs of doctors to get the final feature selection scheme.

3   Methodology

Most of the existing medical knowledge graphs are constructed based on medical literature published on the Internet as well as various public data sets and electronic medical records. Although such data are easy to obtain, there are some problems such as limited knowledge sources, low data purity and data redundancy. The existing feature selection methods are rarely combined with knowledge graph. Using more efficient data storage method of medical knowledge graph and combining with more authoritative medical knowledge of doctors to screen the hospitalization features of cardiovascular diseases can effectively reduce the average error of prediction costs.

To deal with these problems, this article proposes such a method: Open source knowledge graphs, such as Wikidata, OpenKG, etc. and open source knowledge base, such as medical language specifications defined by UMLS and doctors' prior medical knowledge are used to construct the medical concept knowledge graph (MCKG), the medical instance knowledge graph(MIKG) is completed using data of cardiovascular disease related cases from cooperative hospitals. Based on the constructed MIKG, we obtain a feature alternative scheme, and then combine with the actual needs of doctors and rules to generate the final feature selection scheme in the feature alternative scheme.

The knowledge graph data combined with doctor's interaction, the all features data, and the feature data selected by random search algorithm [20] are compared in three dimensions by combining the machine learning algorithm of the three schools, random forest [21], support vector machine [22], and line regression [23], the training set and the test set use a ratio of 70%: 30%, the evaluation standard is the average hospitalization cost error. The average error rate of the selected feature data combined with the random forest algorithm is reduced to 11.86%. This is a significant improvement over the feature data selected by other methods. Figure 1 is the core process of this paper:

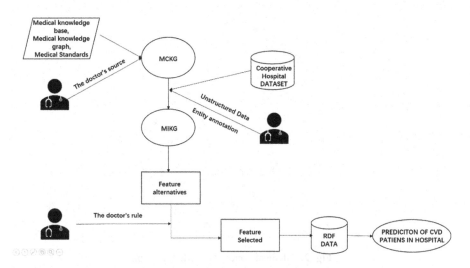

Fig. 1. Hospital cost prediction flow chart

MCKG's data sources mainly include public medical knowledge base, medical knowledge graph, and unified medical standards and specifications, which further guide the construction of MIKG. The data source of MIKG is mainly the structured data of the cooperative hospitals and the unstructured data entities marked by the doctors. The detailed process of MCKG and MIKG construction will be introduced in Sect. 3.1 and Sect. 3.2.

3.1   The Construction of MCKG

As we all know, natural language has the characteristics of polysemy and multiple synonyms, so there is a problem of concept confusion in the traditional medical knowledge graph. In this paper, open source knowledge graph such as Wikidata and OpenKG published on the Internet are combined with the prior medical knowledge of doctors in cooperative hospitals. The knowledge of doctors' dictionaries in unified standardized language provided by UMLS is imported into the conceptual knowledge graph. The knowledge graph is defined with entities as nodes and relationships and attributes as edges. Using ontology notation, that is a triplet(entity-relationship-entity) represents two associated nodes.

The MCKG constructed include the medical knowledge of Chinese and English knowledge as well as the medical specifications defined by UMLS, the main sources of data are from medical knowledge base, medical knowledge graph and doctor. MCKG includes 8,298,580 medical concepts from 116 word-lists and 51 entity words from cooperative hospital. The part of the MCKG constructed in this article is shown in Fig. 2, strictly in accordance with the UMLS definition specification, which is helpful to accurately understand the concepts and relationships between entities (only a part of the concept graph is intercepted in the figure).

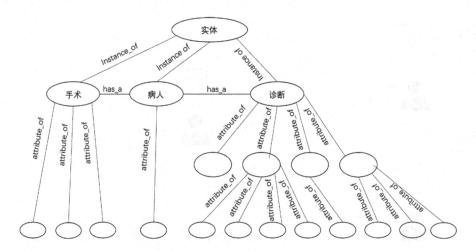

Fig. 2. Medical concept knowledge graph

It can be seen from the figure that the entity part has a surgery part, a patient part, and a diagnosis part. The attribute value part will be completed by the MIKG mentioned in the next section. The main role of the MCKG is mainly two points. First, it clarifies the relationship between the various parts of the graph, and second it guides the construction of MIKG.

3.2   The Construction of MIKG

MIKG is constructed under the guidance of the MCKG described in Sect. 3.1. The cardiovascular diseases data sets used in this paper are all from cooperative hospital. The data is divided into structured data and unstructured data.

The instantiation of the KG is mainly the process of knowledge extraction. The main process of this experiment generates a medical dictionary based on the latest cardiovascular disease diagnosis rules defined by experts, unstructured data (the medical record data of some patients) mainly adopts the method of entity annotation, defines relevant rules, extracts features related to hospitalization costs, and imports them into the MIKG. Here is an example of entity annotation of unstructured data in Table 1(Only part of a patient's case data is intercepted):

Table 1. Entity annotation sample

Annotation text	Entity tags			
Case history	Symptom	Examination	Disease	Medication
Patients were due to eight years ago, no significant incentives in paroxysmal retrosternal pain, for the stuffy pain, no radiation, for 20 to 30 min each time, postoperative oral "aspirin, wave force d" antiplatelet therapy In July 2009, coronary angiography showed that after the anterior descending branch and the first diagonal branch stenting, the intima of the stent was slightly enlarged, the wall of the anterior descending branch was irregular, and the distal segment was 90% narrow. After the operation, angina pectoris occurred several times, and the hospital treatment improved and discharged	paroxysmal retrosternal pain	irregular branch wall; coronary angiography;	angina pectoris;	stenting; aspirin;

The table shows that we divide the types of entity annotation into four categories: symptom entity, examination entity, disease entity and medication entity. The entity tags serve as the entity node of MIKG and they are imported into MIKG in the form of RDF triples.

The structured data of cardiovascular disease comes from the hospital database, including basic patient information, surgical information, diagnosis information and other information. The structured data is mapped according to the rules of relational data (ER)-mapping-RDF data. For example, If the table contains "cardiovascular diseases" and related hospital information, we can map it to an RDF triple. The goal of instantiation of MCKG is to extract the entities and relationships of cardiovascular diseases from textual data and structured data, then realize the visualization of MIKG and select features through interaction with doctors.

First, under the guidance of doctors, seven tables related to the prediction of hospitalization costs for cardiovascular diseases were extracted, as shown in Table 2:

Table 2. Related ER data

No	Table_Name	Description
1	PATIENT_VISIT	Patient's operation information
2	OPERATION	Patient's operation information
3	LAB_MASTER	Patient's test information
4	LAB_RESULT	Patient's test result
5	DIAG_TYPE	Diagnosis type
6	MASTER	Patient master index
7	ORDER	Doctor's advice information

We extract the entities of all patient records in these tables, namely patient entity, surgery entity, diagnosis entity, diagnosis result entity, diagnosis type entity, main index entity, medical order entity (the above-mentioned information related to patient's privacy has been desensitized):

1. Patient entity extraction: extract the ID and admission ID of each patient with cardiovascular disease from the patient ID (PATIENT_ID) and the patient's admission ID (VISIT_ID) as the attribute value of the patient's entity.
2. Surgery entity extraction: due to the different conditions of each patient and the different operations performed, different types of operations such as vascular exploration, coronary angiography, and coronary artery bypass grafting are extracted from the surgical entities of the patient as a subclass of surgical entities entity.
3. Diagnosis entity extraction: Each patient's examination number, examination date, and patient's basic information such as gender and age were extracted as the subclass entities of the diagnostic entity.
4. Diagnosis result entity extraction: The diagnosis results of each patient are necessarily different, and indicators such as WBC, NEUT%, RBC, etc. as well as the diagnosis result time are extracted as the subclass entities of the diagnosis result entity.
5. Diagnosis type entity extraction: Different patients have different types of diagnosis according to the needs of different types of cardiovascular diseases. Different diagnosis types such as vascular headache, carotid atherosclerosis, coronary atherosclerotic heart disease are extracted from the diagnosis entities as diagnosis type entity.
6. Main index entity extraction: The payment types of each patient, such as out-of-pocket, public expense, medical insurance, as well as entities such as place of birth and date of birth, are extracted as the subclass entities of the main index entity.
7. medical order entity extraction: Entities such as the medical examination performed by each patient, the drugs related to cardiovascular disease used, the corresponding dose, the starting time and the end time of the medication are extracted as the subclass entities of the medical order entity.

To sum up, the relationship between different entities is extracted by applying the MCKG to MIKG, for example: the relationship bet ween the patient entity and the patient entity is has_a, The relationship between the type of surgery and the surgical entity is attribute_of, The relationship between the type of diagnosis and diagnosis entity attribute_of and so on. Converting the cardiovascular disease data from the cooperative hospital into RDF data, the construction of the medical instance knowledge graph is shown in Fig. 3:

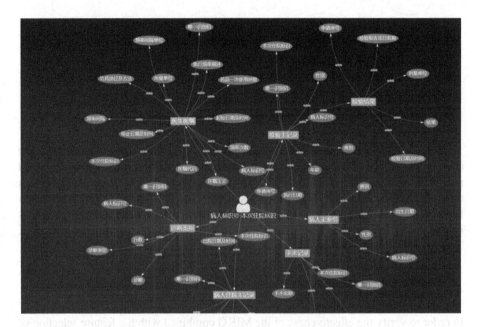

Fig. 3. Medical instance knowledge graph

The size of the constructed medical instance knowledge graph is 83.6 GB which contains 698946023 triples. Taking the patient entity as the center, different entity nodes (rectangular nodes in the figure) are connected and different nodes are connected to the corresponding instance nodes (elliptical nodes in the figure).

3.3   KG Feature Selection

Based on the already generated MCKG and MIKG, we have screened up to 189 features provided by the original database into 47 features as shown in Fig. 3. Combing the selected KG with the doctor's needs and regulations, according to the 1–2 steps closest to the patient's hospitalization information, the nine feature KG with the highest correlation with hospitalization costs are finally extracted as shown in Fig. 4:

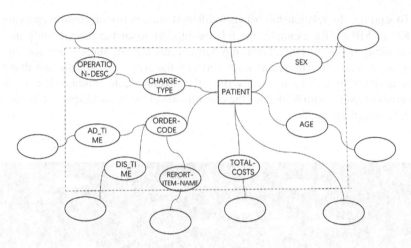

Fig. 4. Knowledge graph feature selection

The dotted frame in the figure is divided into alternative plans submitted to the doctor, who selects features based on his/her prior medical knowledge and clinical needs. ORDER_CODE is the doctor's advice code, AD_TIME is patient's admission time, DISCHARGE_TIME is patient's discharge time, CHARGE_TYPE is patient's type of payment, REPORT_ITEM_NAME is patient' examination items, SEX is patient's sex, Age is patient's age, OPERATION_DESC is patient's type of operation, TOTAL_COSTS is Total cost of patient hospitalization.

4   Experiments

In order to verify the effectiveness of the MIKG combined with the feature selection of doctors' interactive, we divide the experimental data into three groups:

The first group is DAF (data of all features), the second group is the data filtered by Random Searching Algorithm (RSA), and the third group is the data filtered by MIKG mentioned in Sect. 3 combined with the knowledge of doctors (KG-D).

The data set used in the experiment is RDF triplet data set, 10,000 of these triples are randomly selected, the training set and the test set use a ratio of 70%: 30%.

Experimental environment for this experiment: Processor: Inter® Core ™ i5-8265U CPU @ 1.8 Hz; RAM: 8.0 GB, operating system: WIN 10. This experiment uses Python3.7 software package.

The experiment uses machine learning algorithms: SVM, RF and LR. The average prediction error of hospitalization cost (Averr) is used as the evaluation index.

$$Averr = \frac{1}{n}\sum_{i=1}^{n}\left|\frac{\hat{y}_i - y_i}{y_i}\right| \times 100\% \qquad (1)$$

$\hat{y}_i$ represents the predicted value of hospitalization costs, y_i represents the actual value of hospitalization expenses, n is the total number of samples. The process of feature selection of RSA-RF [17] is shown in Fig. 5:

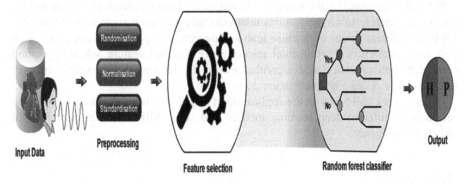

Fig. 5. RSA feature selection

The average prediction error rate of this experiment is shown in Table 3:

Table 3. Average prediction error for different feature selection methods

Feature selection method	SVM	LR	RF
DAF	36.34%	39.69%	35.12%
RSA	17.42%	21.17%	16.48%
KG-D	12.74%	14.55%	11.86%

It can be seen from the table that when all the features related to cardiovascular disease of patients are used to predict hospitalization cost, no matter which classifier is used, SVM, LR or RF, there is a high prediction error. When we use the feature data selected by the random search algorithm to predict the hospitalization cost, we can see that the prediction error is reduced. When we used MIKG in combination with the feature data of doctors' interactive selection for prediction, the prediction error of the classifier was significantly reduced, among which the best effect was achieved when RF was used, and the prediction cost error was reduced to 11.86%. This experiment proves the proposed the effectiveness of this method.

5   Conclusion and Future Work

We build MCKG using open source knowledge graphs such as Wikadata, OpenKG, etc. and open source knowledge base such as the medical language specifications defined by UMLS and the doctor's prior medical knowledge. Then we integrate the structured data in the database of the cooperative hospital and the unstructured data processed by doctors through entity annotation. We select features through the above-mentioned knowledge graph to get a feature alternative, and then combine the doctor's clinical needs and definition rules to get the final feature selection. Based on the feature selected by the above-mentioned method, we compare the corresponding RDF data with the data obtained by the features selected by the random search algorithm and the data corresponding to all the features related to hospitalization costs using SVM, RF, LR three different genres of machine learning algorithms to perform the hospitalization cost error prediction, experimental results prove that our feature selection method combined with random forest algorithm effectively reduces the prediction error of cardiovascular disease hospitalization costs.

Future work will focus on the application of MIKG to other data sets, as well as the selection of different deep learning models, and apply MIKG to a broader field of artificial intelligence.

References

1. Chinese cardiovascular disease report compilation group: Summary of Chinese cardiovascular disease report 2016. China Circul. J. **032**, 521–530 (2017)
2. Zhang, Y., Wang, S.N., Liu, Y.: Application of ARIMA model on predicting monthly hospital admissions and hospitalization expenses for respiratory diseases. China Health statistics **032**, 197–200 (2015)
3. Guyon, I.: An introduction to variable and feature selection. JMLR.org (2003)
4. Guo, K.W., Pan, H.L., Hou, A.: Classification algorithm based on feature selection and clustering. J. Jilin Univ. (Science Ed.) **056**, 395–398 (2018)
5. Ansong, S., Eteffa, Kalkidan F., Li, C., Sheng, M., Zhang, Y., Xing, C.: How to empower disease diagnosis in a medical education system using knowledge graph. In: Ni, W., Wang, X., Song, W., Li, Y. (eds.) WISA 2019. LNCS, vol. 11817, pp. 518–523. Springer, Cham (2019). https://doi.org/10.1007/978-3-030-30952-7_52
6. Sheng, M., Hu, Q., Zhang, Y., Xing, C., Zhang, T.: A data-intensive CDSS platform based on knowledge graph. In: Siuly, S., Lee, I., Huang, Z., Zhou, R., Wang, H., Xiang, Wei (eds.) HIS 2018. LNCS, vol. 11148, pp. 146–155. Springer, Cham (2018). https://doi.org/10.1007/978-3-030-01078-2_13
7. Xu, Z.L., He, L.R, Wang, Y.F.: Overview of knowledge graph technology. J. Electr. Sci. Technol. 589–606
8. Research on current situation and strategy of artificial intelligence-assisted diagnosis and treatment. Chinese Eng. Sci. 20, 1–128 (2018)
9. Sheng, M., et al.: CLMed: a cross-lingual knowledge graph framework for cardiovascular diseases. Web Inf. Syst. Appl. 512–517 (2019)
10. Uyar, K., lhan, A.: Diagnosis of heart disease using genetic algorithm based trained recurrent fuzzy neural networks. Procedia Comput. Sci. **120**, 588–593 (2017)
11. Mohan, S., Thirumalai, C., Srivastava, G.: Effective heart disease prediction using hybrid machine learning techniques. IEEE Access 1 (2019)
12. Alarsan, F.I., Younes, M.: Analysis and classification of heart diseases using heartbeat features and machine learning algorithms (2019)
13. Ali, L., Rahman, A., Khan, A., Zhou, M., Javeed, A., Khan, J.A.: An automated diagnostic system for heart disease prediction based on $\chi 2$ statistical model and optimally configured deep neural network. IEEE Access **7**, 34938–34945 (2019)

14. Basciftci, F., Eldem, A.: Using reduced rule base with Expert System for the diagnosis of disease in hypertension. Med. Biol. Eng. Comput. **51**, 1287–1293 (2013)
15. Nahar, J., Imam, T., Tickle, K.S., Chen, Y.-P.P.: Computational intelligence for heart disease diagnosis: a medical knowledge driven approach. Expert Syst. Appl. **40**, 96–104 (2013)
16. Prakash, S., Sangeetha, K., Ramkumar, N.: An optimal criterion feature selection method for prediction and effective analysis of heart disease. Cluster Comput. **22**, 11957–11963 (2019)
17. Gokulnath, C.B., Shantharajah, S.P.: An optimized feature selection based on genetic approach and support vector machine for heart disease. Cluster Comput. **22**, 1–11 (2019)
18. Zhao, T.T., Yuan, Y.B., Wang, Y.J., Gao, J., He, P.: Heart disease classification based on feature fusion. In: 2017 International Conference on Machine Learning and Cybernetics (2017)
19. Sarah, P., Ira, K.S., Enzo, F., Matthew, L., Ricardo, G., Ben, G., Daniel, R.: Disease prediction using graph convolutional networks: application to autism spectrum disorder and Alzheimer's disease. medical image analysis S1361841518303554 (2018)
20. Javeed, A., Zhou, S., Yongjian, L., Qasim, I., Noor, A., Nour, R.: An intelligent learning system based on random search algorithm and optimized random forest model for improved heart disease detection. IEEE Access **7**, 180235–180243 (2019)
21. Singh, Y.K., Sinha, N., Singh, S.K.: Heart disease prediction system using random forest. In: International Conference on Advances in Computing and Data Sciences (2017)
22. Saunders, C., et al.: Support vector machine. Comput. Sci. **1**, 1–28 (2002)
23. Allison, L.: Coding Ockham's Razor. Linear Regression, pp. 103–111. Springer, Heidelberg (2018). https://doi.org/10.1007/978-3-319-76433-7

A Classroom Student Counting System Based on Improved Context-Based Face Detector

Rong Chen, Yu Jin, and Lizhen Xu[(⊠)]

Department of Computer Science and Engineering, Southeast University,
Nanjing 21189, China
{220181702,lzxu}@seu.edu.cn, seu_yjin@163.com

Abstract. Classroom student counting based on surveillance video is the basis for important tasks such as student behavior analysis, resource optimization, school security, and intelligent management. In recent years, with the development of deep learning, the research of face detection has been greatly promoted, but it still cannot solve the problems of difficult recognition of different poses such as tiny faces in the back row, occlusion of students, and head-down in the classroom. This paper designed an intelligent student counting system based on classroom surveillance video to solve the problem of counting people in classroom scenarios. Also the paper proposed a face detector which mixes feature enhancement modules and background-based modules, using Background information including shoulders, bodies and desks, and search for unobstructed, less-occluded, and head-up targets in multiple video frames in the surveillance video based on the relative stillness of the students in the classroom scene, which finally greatly improves the accuracy of the classroom student counting statistics.

Keywords: Face detection · People counting · Context-based

1 Introduction

People counting is not only an important functional part of the modern intelligent monitoring system, but also the basis of important tasks such as crowd behavior analysis, modern security, resource optimization, business information collection, and intelligent management. With the large-scale deployment, installation and application of surveillance cameras, the application scenarios of visual people counting have also expanded.

Face detection, aiming at determining and locating the regions of faces in the natural images, is one of the fundamental steps in various face analysis, including face alignment [1], recognition [2], verification [3], people counting [4] etc. with the breakthrough of deep-learning, more and more face detection model are proposed and applied to different areas. However, for the special scene of the classroom, there are still problems that small faces are difficult to be detected or mis-detected. The paper proposed an FPN-based [5] algorithm similar to pyramid-Box [6] to exploit contextual information including shoulder, body and desks to help detect tiny or obstacle faces. At the same time, we add feature-enhanced module to dig more semantic of the face

© Springer Nature Switzerland AG 2020
G. Wang et al. (Eds.): WISA 2020, LNCS 12432, pp. 326–332, 2020.
https://doi.org/10.1007/978-3-030-60029-7_30

features. Finally, we propose a video validate module to improve student counting accuracy by exploit the relative stillness of students in limited time.

2 Related Work

Early face detection mainly relied on hand-made features, such as Harr-like features [7], control point set [8], and edge direction histogram [9]. However, hand-crafted features design is lack of guidance. With the great progress of deep learning, hand-crafted features have been replaced by Convolutional Neural Networks (CNN). At the beginning, Overfeat [10], CascadeCNN [11], and MTCNN [12] use CNN as a sliding window detector on the image pyramid to build a feature pyramid. However, the use of image pyramids is slow and memory inefficient. As a result, most two-stage detectors extract features in a single ratio. R-CNN [13] obtains region proposals through selective search [14], and then forwards each normalized image region through CNN for classification. Faster R-CNN [13], R-FCN [15] employ Region Proposal Network (RPN) to generate initial region proposals. More recently, some research indicates that multi-scale features perform better for tiny objects. FPN, a top-down architecture, integrate high-level semantic information to all scales. FPN-based methods, such as FAN, PyramidBox achieves significant improvement on detection. DSFD propose a feature enhance module that incorporates multi-level dilated convolutional layers to enhance the semantic of the features. Our paper propose FPN-based methods similar to PyramidBox and exploit feature-enhanced module to enhance the semantic of the features.

3 Proposed Model

3.1 Mixture of FEM and Context-Sensitive Predict Layers

The brief overview of our model is shown in Fig. 1. Our architecture uses the same extended VGG16 backbone and Feature enhanced model as DSFD [20]. We select conv3_3, conv4_3, conv5_3, conv_fc7, conv6_2 and conv7_2 to generate six original feature maps named of1; of2; of3; of4; of5; of6. Then, FEM transfers these original feature maps into six enhanced feature maps named ef1; ef2; ef3; ef4; ef5; ef6, which have the same sizes as the original ones and are fed into Context-sensitive Predict Layers to enhance the performance on small faces. Finally, we design a video validate layer to make use of the relative stillness of the students to improve the accuracy of student counting system.

Feature Enhance Module is able to enhance original features to make them more discriminable and robust, which is called FEM for short. Figure 2 illustrates the idea of FEM, which is inspired by FPN and DSFD. The feature enhanced model first use 1*1 convolutional kernel to normalize the feature maps. Then, we up-sample upper feature maps to do element-wise product with the current ones. Finally, we split the feature maps to three parts, followed by three sub-networks containing different numbers of dilation convolutional layers which improve the inceptive field without decreasing the feature map size.

Actually, it is clear that faces never occur isolated in the real world, usually with shoulders or bodies, and desks for students in classroom. Therefore, the network should be able to learn features for not only faces, but also contextual parts such as heads, bodies and desks. To achieve the goal, inspired by Pyramid Box, our model use a context-sensitive predict model to make use of the contextual information. At first, we use a semi-solution to generate approximate labels for contextual parts related to faces and a series of anchors related to contextual information. Next, contextual features go through Feature enhance layer learn more features. Finally, all features are used for classification and regression of faces, heads, bodies and desks.

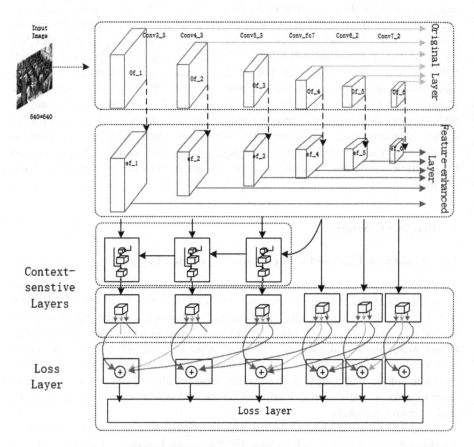

Fig. 1. The framework of our face detector

3.2   Loss Function

The loss function in our model for an image is defined as

$$L(\{p_{n,i}\}, \{t_{n,i}\}) = \sum_n \lambda_n L_n(\{p_{n,i}\}, \{t_{n,i}\}) \tag{1}$$

Where the n-th contextual anchor loss is given by

$$L_n(\{p_{n,i}\}, \{t_{n,i}\}) = \frac{\lambda}{\lambda_{n,cls}} \sum_{i_n} L_{n,cls}(p_{n,i}, p_{n,i}^*) + \frac{1}{\lambda_{n,reg}} \sum_{i_n} p_{n,i}^* L_{n,reg}(t_{n,i}, t_{n,i}^*) \tag{2}$$

Here n denotes the index of contextual anchors(n = 0,1, and 2 represents for face, head and body or desk), and i denotes the index of an anchor. $p_{n,i}$ is the predicted probability of anchor i being the n-th object(face, head and body or desk) and $p_{n,i}^*$ is the ground-truth label. Moreover,$t_{n,i}$ is vector representing the 4 parameterized coordinates of the predicted bounding box, and $t_{n,i}^*$ is that of ground-truth box associated with a positive anchor. The classification loss $L_{n,cls}c$ is log loss over two classes (face vs. not face) and the regression loss $L_{n,reg}$ is the smooth L1 loss over the 4 parameterized coordinates between predict box and ground-truth box. Finally, $\lambda_{n,cls}$ and $\lambda_{n,reg}$ are regularization parameters of the classification loss and the regression loss, and,λ and λ_n are balancing weights.

3.3   Video Validate Module

Although we enhance the feature and make use of the contextual information, there are still some students hard to detect who are hanging or turning their heads or occluded by other students. However, in real word, we all know almost every student will stay in a fixed position and will not change their position in limited time. At the same time, classroom surveillance cameras are fixed in one position too. Therefore, students have what we call relative stillness. To improve accuracy, we add one validate module which detect faces from multi frames extracted from surveillance videos and do Non-Maximum Suppression to get result predict boxes. Firstly, we divide class time into multiple five minutes and extract one frame every half minutes in the 5 min from videos that maybe the students in some frames are not blocked (default, more accurate but slower if extract more frames). Secondly, we fed all frame into our network to get multi predicts boxes which have many coincident boxes. Thirdly, we perform Non-Maximum Suppression of all boxes and get the student counting results of every 5 min. Finally, we remove the minimum value and maximum of these counting number and output the maximum counting number.

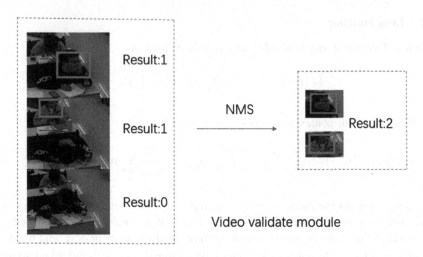

Fig. 2. General view of video validate module

4 Results

In order to verify the actual effect of the system designed and improved in this article, we conducted training on some photos of the WilderFace of the public dataset and 1384 classroom photos collected in different states, and conducted experiments in 50 classroom monitoring videos collected. After the detection, the number of faces read out is used to display the current number of people. Table 1 shows the specific experimental results.

The mAP in Table 1 stands for mean Average Precision while average Precision is a popular metric in measuring the accuracy of object detectors which computes the average precision value for recall value over 0 to 1 [16]. And the standard formula for calculating the student counting Accuracy in Table 1 is

$$P = \frac{\sum_{i=1}^{n}\left(\frac{m_i - |m_i - q_i|}{m_i}\right)}{n} \tag{3}$$

P is the final statistical accuracy of the number of people; n is the number of classrooms tested in total; i is the i-th classroom tested; mi is the actual number of classroom i;

It can be seen from experiment that compared with the original DSFD and Pyramid Box algorithms, our system has obvious advantages in student counting (Fig. 3).

Fig. 3. Detection results

Table 1. Performance of student counting layer

Method	mAP	Student counting accuracy
Pyramid Box	88.5%	92.1%
DSFD	90.3%	93%
Student counting system	91.1%	96.5%

5 Conclusion and Future Work

The present paper proposes an efficient face detection algorithm and a flexible and efficient face detection-based student counting system. Our model uses feature enhanced model and context sensitive model to make use of background information, and through extracting multiple frames from videos to improve accuracy by utilizing students' relative stillness. The method in this paper has achieved excellent detection results in classroom scenarios.

Still, much more work should be performed on face detection algorithms and the detecting strategy can be optimized for detecting speed. What's more, we will continue researching how to do auto-attendance by face recognition after face detection.

References

1. Tai, Y., et al.: Towards highly accurate and stable face alignment for high-resolution videos. In: Proceedings of the AAAI Conference on Artificial Intelligence, vol. 33, pp. 8893–8900 (2019). https://doi.org/10.1609/aaai.v33i01.33018893

2. Turk, M., Pentland, A.: Face recognition using eigenfaces. In: Proceedings. 1991 IEEE Computer Society Conference on Computer Vision and Pattern Recognition, pp. 586–587 (1991)

3. Kumar, N., Berg, A.C., Belhumeur, P.N., Nayar, S.K.: Attribute and simile classifiers for face verification. In: 2009 IEEE 12th International Conference on Computer Vision, pp. 365–372. IEEE (2009). https://doi.org/10.1109/iccv.2009.5459250

4. Hou, Y.-L., Pang, G.K.H.: People counting and human detection in a challenging situation. IEEE Trans. Syst. Man Cybernet.-Part A Syst. Hum. 41(1), 24–33 (2010). https://doi.org/10.1109/TSMCA.2010.2064299

5. Lin, T.-Y., Dollár, P., Girshick, R., He, K., Hariharan, B., Belongie, S.: Feature pyramid networks for object detection. In: Proceedings of the IEEE Conference on Computer Vision and Pattern Recognition, pp. 2117–2125 (2017). https://doi.org/10.1109/cvpr.2017.106

6. Tang, X., Du, Daniel K., He, Z., Liu, J.: PyramidBox: a context-assisted single shot face detector. In: Ferrari, V., Hebert, M., Sminchisescu, C., Weiss, Y. (eds.) ECCV 2018. LNCS, vol. 11213, pp. 812–828. Springer, Cham (2018). https://doi.org/10.1007/978-3-030-01240-3_49

7. Viola, P., Jones, M.J.: Robust real-time face detection. Int. J. Comput. Vision 57(2) (2004). https://doi.org/10.1023/b:visi.0000013087.49260.fb

8. Abramson, Y., Steux, B., Ghorayeb, H.: Yet even faster (YEF) real-time object detection. Int. J. Intell. Syst. Technol. Appl. 2(2–3), 102–112 (2007). https://doi.org/10.1504/IJISTA.2007.012476

9. Levi, K., Weiss, Y.: Learning object detection from a small number of examples: the importance of good features. In: Proceedings of the 2004 IEEE Computer Society Conference on Computer Vision and Pattern Recognition, 2004. CVPR 2004, vol. 2, pp. II-II. IEEE (2004). https://doi.org/10.1109/cvpr.2004.145

10. Sermanet, P., Eigen, D., Zhang, X., Mathieu, M., Fergus, R., LeCun, Y.: Overfeat: integrated recognition, localization and detection using convolutional networks. arXiv preprint arXiv: 1312.6229 (2013)

11. Li, H., Lin, Z., Shen, X., Brandt, J., Hua. G.: A convolutional neural network cascade for face detection. In: Proceedings of the IEEE Conference on Computer Vision and Pattern Recognition, pp. 5325–5334 (2015). https://doi.org/10.1109/cvpr.2015.7299170

12. Yin, X., Liu, X.: Multi-task convolutional neural network for pose-invariant face recognition. IEEE Trans. Image Process. 27(2), 964–975 (2017). https://doi.org/10.1109/TIP.2017.2765830

13. Girshick, R., Donahue, J., Darrell, T., Malik, J.: Rich feature hierarchies for accurate object detection and semantic segmentation. In: Proceedings of the IEEE Conference on Computer Vision and Pattern Recognition, pp. 580–587. 2014). https://doi.org/10.1109/cvpr.2014.81

14. Uijlings, J.R.R., Van De Sande, K.E.A., Gevers, T., Smeulders, A.W.M.: Selective search for object recognition. Int. J. Comput. Vision 104(2), 154–171 (2013). https://doi.org/10.1007/s11263-013-0620-5

15. Dai, J., Li, Y., He, K., Sun, J.: R-FCN: object detection via region-based fully convolutional networks. In: Advances in Neural Information Processing Systems, pp. 379–387 (2016)

16. Shang, A., Li, C., Zheng, H., Shi, M.: Extraction algorithm, visualization and structure analysis of python software networks. In: Meng, X., Li, R., Wang, K., Niu, B., Wang, X., Zhao, G. (eds.) WISA 2018. LNCS, vol. 11242, pp. 357–368. Springer, Cham (2018). https://doi.org/10.1007/978-3-030-02934-0_33

Named Entity Recognition in Aircraft Design Field Based on Deep Learning

Yubin Bao[1]([⊠]) [iD], Yuanming An[1], Zhu Cheng[1], Rimeng Jiao[1],
Chao Zhu[1], Fangling Leng[1], Shuai Wang[2], Ping Wu[2], and Ge Yu[1]

[1] School of Computer Science and Engineering, Northeastern University,
Shenyang 110819, China
{baoyb,lengfl,yuge}@mail.neu.edu.cn
[2] Shenyang Aircraft Design and Research Institute, Shenyang 110035, China
60124s@163.com

Abstract. Aircraft design is a kind of knowledge-intensive work involving multi-disciplinary integration, which needs the support of a large amount of knowledge on aircraft design field (ADF). At the same time, a large number of technical documents about AD also accumulate rich aircraft design knowledge. If this knowledge can be extracted, it can be used to guide the intelligent design and maintenance of aircraft. In this paper, we conduct our research for the named entity recognition, which is an important step of knowledge graph construction in ADF. For the problem of knowledge dispersion in ADF and lacking of training dataset, we design a platform for data acquisition and processing, and corpus annotation by crowdsourcing. And a novel deep neural network model, named AR+BiLSTM+CRF, which combines attention mechanism, Ranger optimizer, bidirectional LSTM, and CRF, is proposed for named entity recognition in ADF. The experimental results show that AR+BiLSTM +CRF model has excellent performance for named entity recognition in ADF.

Keywords: Aircraft design · Named entity recognition · Deep learning · Crowdsourcing

1 Introduction

Aviation industry is a knowledge intensive industry, and a large amount of data will be generated and collected in each stage of the full lifecycle of aircraft design, such as design documents, demonstration reports, failure reports and so on, which contain a vast amount of knowledge. If the knowledge can be effectively extracted, it can be used to guide new product design process, and improve design quality.

Therefore, the aviation industry urgently needs knowledge-driven design, manufacturing and management, and named entity recognition (NER) is an indispensable technology when we want to extract knowledge from technical documents. Generally speaking, named entity refers to entities with specific meaning, such as aircraft names, technical names and performance indicators. The purpose of NER is to identify these entities in the text [1]. It is an important task in the construction of knowledge base.

© Springer Nature Switzerland AG 2020
G. Wang et al. (Eds.): WISA 2020, LNCS 12432, pp. 333–340, 2020.
https://doi.org/10.1007/978-3-030-60029-7_31

In this paper, a NER system is developed and a new neural network model, AR +BiLSTM+CRF, is proposed for NER in aircraft design fields (ADF). The second section of this paper introduces the related research work, the third section introduces how to build the dataset, including the collection and processing of data, building domain dictionary and a crowdsourcing annotation system. The model is discussed in 4th section, experiments is in 5th section, and the last gives the conclusion.

2   Related Work

At present, NER is mainly based on rules, machine learning and deep learning. The early research work used rules defined by experts to match entities in texts. With the development of machine learning, the classification methods based on machine learning are applied to NER. Machine learning methods such as Hidden Markov Model (HMM) [2], Conditional Random Field (CRF) [3] and Support Vector Ma-chines (SVM) [4] perform well in NER. Recently, deep learning has made outstanding performance. Peng N et al. [5] proposed a LSTM+CRF model and it performed well in NER task. In addition, Convolutional Neural Network (CNN) [6] and other deep learning models have been successfully applied to NER tasks.

Due to the great difference between Chinese and English grammatical expressions, some improved models or methods were proposed. For example, Qiang B H et al. [7] proposed the neural network structure based on part of speech. Xiaowei Han et al. [8] proposed CNN-BiLSTM-CRF model to minimize the influence of different word segmentation results. However, Chinese NER models have not been widely used in ADF. In order to quickly and effectively get potential knowledge in aircraft design, the AR+BiLSTM+CRF model is proposed for Chinese NER in ADF in this paper.

3   Dictionary Construction and Dataset Annotation

3.1   The Construction of Domain Dictionary

Due to the lack of relevant data, we construct the ADF dataset for NER. In the process of dataset construction, we adopt semi-supervised strategy. Firstly, we construct entity set called *domain dictionary* for automatically pre-labeling entities in texts, and then correct the wrong annotations manually.

For obtaining semi-structured data from open knowledge base with web spider, we create different types of wrappers, like baike wrapper that extracts semi-structured data from www.baike.com, zhishi.me wrapper that extracts knowledge from zhi-shi.me and so on. For instance, based on the BFS strategy, we get entities that have HTML hyperlinks from the entries in related baike pages. At the same time, for un-structured texts, they are segmented by Jieba which is a segment tool. Then, TF-IDF and Tex-tRank algorithms are combined to extract keywords. According to the score of each word calculated by two methods, the noise is filtered and then the domain dictionary is built.

3.2   Crowdsourcing Annotation System

It is necessary to define the entity category before annotating. Based on *China Aviation Encyclopedia Dictionary* [9], *Aircraft Design Manual* [10], and the knowledge from domain experts, we divide entities into 12 categories: aircraft, standard parts, flight principle, aviation design, dynamics, aviation safety, aviation medicine, airborne weapons, flight control, avionics, aviation economy and manufacturing.

Data annotation is a necessary pre-task for NER. At present, the main methods for annotation are still manual. In order to solve the disadvantages of low efficiency and accuracy of traditional annotation methods, we implement an annotation system in ADF through the crowdsourcing cross-labeling algorithm to build training dataset.

In this algorithm, firstly, according to the domain dictionary, automatic pre-labeling is executed by locating entities and recognizing their boundaries in the text. Then, experts annotate a part of text as "authoritative data". After that, users are asked to label these authoritative data, and will be given different weight W_i according accuracy of their label compared with the experts' label. In the formal annotation, multiple people parallel annotate the text, and the score S of the category of words is counted according to the weights of different users. In the equation, L_i is a 12 dimensional one-hot vector that corresponds to 12 categories, and N is the number of annotators.

$$S = softmax(\sum_{i=1}^{N} \left(\frac{W_i \cdot L_i}{N}\right)) \tag{1}$$

If the score S is lower than the threshold $\varphi_2 = 0.6$, experts will re-label this word. The structure of the system is shown in Fig. 1. In this paper, the dataset of NER task is collected from *Aircraft Design Manual* after OCR processing, and is annotated by this crowdsourcing annotation system. We use BIO label to represent the annotation.

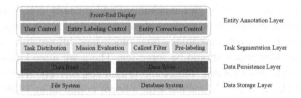

Fig. 1. The architecture of crowdsourcing annotation system

4   Named Entity Recognition Model

4.1   AR+BiLSTM+CRF Model

The data in this paper is recognized from the PDF files as image format, that has a lot of noise. What's more, some representations of professional terms in ADF are special. The traditional BiLSTM+CRF model cannot achieve the desired accuracy in such complex text. Therefore, based on the BiLSTM+CRF model, we combine the character-embedding and attention mechanism, and propose a novel model, called Attention and

Ranger Improved BiLSTM+CRF model, or AR+BiLSTM+CRF. The model structure is shown in Fig. 2. It includes the following five parts: input layer, embedding layer, BiLSTM layer, fully connected layer, CRF layer, and uses ranger optimizer.

Considering that a character is the smallest unit in Chinese, we use char-level model, and convert Chinese characters into vectors through embedding. The input of the model is a text $T = \{c_1, c_2, \ldots c_t, \ldots, c_n\}$ where c_t represents t-th character in the text. The goal is to predict the corresponding label sequence $Y = \{y_1, y_2, \ldots y_t, \ldots, y_n\}$.

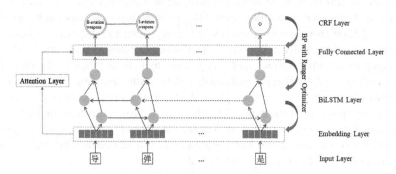

Fig. 2. AR+BiLSTM+CRF model

4.2   Character Embedding and BiLSTM Layer

The data T is input into the network through the input layer, and then enters the embedding layer for embedding operation. It uses Word2vec pre-trained model to get embedding vectors. According to Word2vec, characters in the text will be converted to vectors representation as $X = \{x_1, x_2, \ldots, x_t, \ldots, x_n\}$, and will be input to next layer.

BiLSTM is a kind of RNN (recurrent neural network), which can consider the sequence information. By using gate mechanism and memory cell, LSTM solves the problem of gradient explosion or gradient disappearance of RNN during back propagation. The gate mechanism determines how much new information is added to the memory cell, and how much should be forgotten. There are three kinds of gate structures: i, f, o, which control input, forget and output respectively. The output of these gates is obtained by a Sigmoid function with a linear combination of the current input x_t, h_t, c_t and the previous state h_{t-1}, c_{t-1}. The principle of LSTM is shown in Fig. 3.

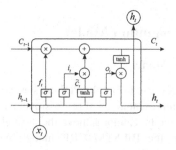

Fig. 3. Structure of LSTM

Finally, for BiLSTM, the hidden vector $\overrightarrow{h_t}$ is obtained by the forward LSTM, while $\overleftarrow{h_t}$ is obtained by the backward LSTM, and $h_t = [\overrightarrow{h_t}; \overleftarrow{h_t}]$ is the final output. This model can fully consider the feature of the bi-directional information of current element, and extract entities by the contextual semantic information of the sequence. Therefore, it is very suitable for NER task of complex data in ADF.

4.3   Fully Connected Layer with Multi-head Self-attention

The fully connected layer not only receives the output from the BiLSTM layer, but also receives the output of the embedding layer through the attention mechanism. In order to solve the problem of noise and focus on important information, we use the attention mechanism to get more accurately global feature of text and local feature of characters.

In the experiment, we use multi-head self-attention mechanism [11]. Different from the general attention mechanism, multi-head self-attention can be regarded as an information weight allocation model, which is carried out within the sequence, and can obtain the relationship between different characters.

$$head_t = softmax\left(\frac{(W_Q h_t)(W_K h_t)^T}{\sqrt{d}}\right)(W_V h_t) \tag{2}$$

W_Q, W_K and W_V are the weight matrices to be trained. $\sqrt{d}$ is the smoothing term, and d is the dimension of h_t. Finally, n single-head self-attention units are combined to obtain the multi-head vector m_t. In this paper, n is set to 3, and W_M is a weight matrix.

$$m_t = MHead_t = W_M\left(Head_t^1, Head_t^2, \ldots Head_t^n\right) \tag{3}$$

In this model, the attention mechanism is to directly operate on the vectors obtained by the embedding layer. In Word2vec, the vector of each character is the same in different context, and it cannot reflect its position feature. Especially, in ADF, the length of text is usually so long, so that the effect may have some limitation in Word2vec and BiLSTM. Through the use of attention, the model can obtain the global information of each Chinese character in different positions on the basis of Word2vec, and then get its global feature vector, which is helpful to solve this problem. At the same time, neither Word2vec nor attention can make good use of the sequence information. Fortunately, BiLSTM can make up for this shortcoming and achieve complementary effect. In the fully connected layer, we concatenate h_t and m_t obtained from BiLSTM and attention by a parameter β, so as to optimize the information carried by the feature vector z_t.

$$\beta = \sigma(W_s[h_t; m_t]) \tag{4}$$

$$z_t = (1 - \beta) \cdot h_t + \beta \cdot m_t \tag{5}$$

4.4   CRF Layer and Ranger Optimizer

Based on features extracted from the fully connected layer, the CRF layer can obtain a good prediction sequence by the relationship between adjacent tags, which can make up for the deficiency of BiLSTM. For the input sequence $Z = \{z_1, z_2, \ldots, z_t, \ldots, z_n\}$ and the output sequence $Y = \{y_1, y_2, \ldots, y_t, \ldots, y_n\}$, the conditional probability of CRF is:

$$P(y|z) = \frac{1}{\sum_y exp \sum_{t=1}^{n} w_t f_t(y, z)} \left(exp \sum_{t=1}^{n} w_t f_t(y, z) \right) \tag{6}$$

Where $f_t(y, z)$ is the function of transition and state characteristic, and w_t represents the weight. The conditional probability distribution can be obtained by Viterbi algorithm, and the final output is obtained by softmax to get the annotation sequence.

In addition, we choose Ranger optimizer to make the system have higher speed and stronger stability, and to solve the warm-up problem of optimizing the learning rate. Ranger is a combination of RAdam and LookAhead optimizer. RAdam [12] has the advantages of both Adam and SGD, and can automatically adjust the adaptive learning rate according to the dispersion of variance. LookAhead [13] reduces the number of hyper-parameters that need to be adjusted. It allows faster weight sets to be explored forward, while slower weights are left behind to provide longer-term stability.

For Ranger optimizer, based on Lookahead, a synchronization parameters θ has been created from initial parameters $\emptyset$, and then it searches $k = 0.5$ batches through Radam optimizer. After that, Lookahead multiplies the difference between θ and the latest weight of Radam with slow weigths step size $\alpha = 5$, and updates $\emptyset$ to $\emptyset^*$, so as to iterate parameters in *group['params']* of the model.

5   Experiments

The dataset contains 104,731 named entities, while the training test set are randomly divided according to the ratio of 8:2. The evaluation index includes precision(P), recall (R) and F1-score(F1).

In order to compare the impact of different character vector dimensions, we first use the random character vector model in embedding layer, and dimensions of the vector are 100, 200, and 300. The results are shown in Table 1. We can find that the performance of the model increases with the increase of the vector dimensions. Therefore, the dimension of the vector is selected as 300. In addition, when dimensions of the character vector increase continually, the performance is no longer rise continually.

Table 1. Performance of character vector dimension.

Vector dimension	P(%)	R(%)	F1(%)
100	82.30	81.46	82.30
200	83.14	82.20	83.14
300	83.85	82.93	83.85

In order to verify whether the pre-trained character vector can improve the model performance, we designed a comparative experiment between the random character vector and the Word2vec pre-trained vector. The results are shown in Table 2.

Table 2. Performance of pre-training char vector

The way of embedding	P(%)	R(%)	F1(%)
random character vector	83.85	82.93	83.39
Word2vec	85.63	84.17	84.89

It can be seen that the Word2vec pre-trained model has a significant effect on the model performance. This is because the vectors pre-trained by Word2vec already contain some semantic information. The semantic information is of great help in identifying named entities, while the random character vectors only transforms the character symbol representation into the vector representation, which is equivalent to losing some semantic information in disguise, so the performance will be slightly worse.

What's more, Table 3 gives the performance compared with other models. We set $hidden\_size = 300$, $lr = 1e-3$, and $dropout = 0.5$. It shows that the AR+BiLSTM +CRF model performs outstanding performance in the task of NER in ADF.

Table 3. Performance comparison with other NER models

Model	P(%)	R(%)	F1(%)
BiLSTM	81.82	80.54	81.17
BiLSTM+CRF	85.63	84.19	84.90
Attetion+BiLSTM+CRF	88.14	87.37	87.75
AR+BiLSTM+CRF	**89.47**	**88.35**	**88.91**

We can see that BiLSTM has a natural advantage in NER, but this advantage may be lost while the length of sentence gradually increases, and attention can make up for this information loss to a certain extent. In addition, using the Ranger optimizer can dynamically turn on or turn off the adaptive learning rate by the variance dispersion, without preheating the adjustment parameter learning rate. It reduces the number of hyper-parameters, and makes the model difficult to fall into local optimal solution.

6   Conclusion

In this paper, we design a system framework for NER in the ADF, including data collection, dictionary construction, entity annotation and other functions, and propose a new neural network model named AR+BiLSTM+CRF, which introduces multi-head self-attention and ranger optimizer. The experimental results show that the model proposed in this paper has excellent performance in ADF, and its F1 can reach 88.91%.

Acknowledgments. This research is supported by the National Defense Basic Scientific Research Program of China (JCKY2018205C012).

References

1. David, N., Satoshi, S.: A survey of named entity recognition and classification. Logist. Invest. **30**(1), 3 (2003). https://doi.org/10.1075/li.30.1.03nad
2. Bikel, D.M.: An algorithm that learns what's in a name. Mach. Learn. **34** (1999). https://doi.org/10.1023/A:1007558221122
3. McCallum, A., Li, W.: Early results for named entity recognition with conditional random fields, feature induction and web-enhanced lexicons. In: Proceedings of the Seventh Conference on Natural Language Learning at HLT-NAACL, vol. 4, pp. 188–191 (2003). https://doi.org/10.3115/1119176.1119206
4. Isozaki, H.: Efficient support vector classifiers for named entity recognition. In: Proceedings of COLING (2002). https://doi.org/10.3115/1072228.1072282
5. Peng, N.: Improving named entity recognition for chinese social media with word segmentation representation learning. In: The 54th Annual Meeting of the Association for Computational Linguistics (2016). https://doi.org/10.18653/v1/P16-2025
6. Dong, X., Qian, L., Guan, Y.: A multiclass classification method based on deep learning for named entity recognition in electronic medical records. I: 2016 New York Scientific Data Summit (NYSDS). IEEE (2016). https://doi.org/10.1109/NYSDS.2016.7747810
7. Qiang, B.H.: Research on Chinese named entity recognition using combined boundary-PoS feature. In: The International Conference on Design, Manufacturing and Mechatronics (2015). https://doi.org/10.1142/9789814730518_0098
8. Han, X., Xu, L., Qiao, F.: CNN-BiLSTM-CRF model for term extraction in Chinese corpus. In: Meng, X., Li, R., Wang, K., Niu, B., Wang, X., Zhao, G. (eds.) WISA 2018. LNCS, vol. 11242, pp. 267–274. Springer, Cham (2018). https://doi.org/10.1007/978-3-030-02934-0_25
9. Xuanchao, C.: China Aviation Encyclopedia Dictionary, 1st edn. Aviation industry Press, Beijing (2000)
10. Editorial Board of the Aircraft Design Manual: Aircraft Design Manual. 1st edn. Aviation Industry Press, Beijing (2000)
11. Vaswani, A., Shazeer, N., Parmar, N., et al.: Attention is All You Need. In: Advances in Neural Information Processing Systems, pp. 5998–6008 (2017). https://arxiv.org/abs/1706.03762
12. Liu, L.: On the variance of the adaptive learning rate and beyond. In: Proceedings of the Eighth International Conference on Learning Representations (2020). https://arxiv.org/abs/1908.03265
13. Zhang, M.R.: Lookahead Optimizer: k steps forward, 1 step back. In: Advances in Neural Information Processing Systems, pp. 9593–9604 (2019). https://arxiv.org/abs/1907.08610

Interpretable Text-to-SQL Generation
with Joint Optimization

Mingdong Zhu[1,2(✉)], Xianfang Wang[1], and Yang Zhang[1]

[1] School of Computer Science and Technology, Henan Institute of Technology,
Xinxiang, China
{zhumingdong, zhangyang}@hait.edu.cn,
2wangfang@163.com
[2] Intelligent Industrial Big Data Application Engineering Technology Research
Center of Xinxiang, Xinxiang, China

Abstract. The purpose of Text-to-SQL is to obtain the correct answer for a textual question from the database, which can take advantage of advanced database system to provide reliable and efficient response. Existing Text-to-SQL methods generally focus on accuracy by designing complex deep neural network models, and hardly consider interpretability, which is very important for serious applications. To address this, in this paper we propose a novel framework for Interpretable Text-to-SQL Generation (ITSG) with joint optimization, which achieves state-of-the-art accuracy and possesses two-level interpretability at the same time. The framework mainly consists of three layers: a sequence encoder which encodes questions, table headers and significant table contents, an attention-based LSTM layer which generates SQL queries and a reinforcement learning layer which boosts the execution accuracy. Comparing with state-of-the-art methods on benchmark datasets, the experimental results show the effectiveness and interpretability of our ITSG framework.

Keywords: Text-to-SQL · Interpretability · Attention-based LSTM · Reinforcement learning

1 Introduction

Nowadays Question Answering (QA) is one of the most active research areas in natural language processing [1, 2]. In this paper, we mainly focus on question answering over databases, which can take advantage of advanced database system to provide reliable and efficient response. This is usually done by Text-to-SQL, which maps natural language question to a corresponding SQL query, followed by executing the SQL query against databases to obtain the answer. Figure 1 shows an example of question answering over databases.

G. Wang et al. (Eds.): WISA 2020, LNCS 12432, pp. 341–351, 2020.
https://doi.org/10.1007/978-3-030-60029-7_32

Place	Player	Country	Score	To par
T1	"greg norman"	"Australia"	"63.0"	"−9"
T2	"phil mickelson"	"United States"	"65.0"	"−7"
T3	"scott hoch"	"United States"	"67.0"	"−5"
T4	"bob tway"	"United States"	"67.0"	"−5"
T5	"lee janzen"	"United States"	"68.0"	"−4"
T6	"david gilford"	"England"	"69.0"	"−3"

- **QUESTION:**
  What country does david gilford play for?

- **SQL:** SELECT Country FROM table_2_16514480_1 WHERE Player == "david Gilford"

- **ANSWER:** "England"

Fig. 1. An example of question answering over databases

Early works adopt a deep neural sequence-to-sequence model, while recent researches try to incorporate the SQL syntax into neural models [3, 4], for example [4] uses a slot filling approach where syntactic correctness is ensured by predefined sketches. Existing works mainly focus on accuracy while tend to overlook interpretability which is vital for serious applications. By combining the deep neural sequence-to-sequence model and the feature of SQL structures, we propose a novel framework for Interpretable Text-to-SQL Generation (ITSG) with joint optimization, which achieves state-of-the-art accuracy and possesses two-level interpretability at the same time.

The key challenges for an efficient and adaptable framework for Text-to-SQL generation are three folds. First, it's insufficient to just consider the textual questions and table headers, lots of useful semantic information are hided in table contents. Second, a general structure model for generating SQL programs should be designed, instead of limiting to just a few frequent subtasks. Third, the natural language is evolving and the database will be constantly updated, the framework should process a certain degree of adaptability. In the face of these challenges, our framework consists of three layers: sequence encoder layer, attention-based Long Short-Term Memory (LSTM) layer and reinforcement learning layer. In order to improve the accuracy for semantic parsing, similarities between questions and significant table contents are incorporated in the sequence encoder layer. And then the attention-based LSTM layer constructs the foundation for interpretability by computing impact weights of input tokens for each output. At last, the reinforcement learning layer is integrated to ensure the adaptability of the framework. In summary, our main contributions are:

(1) A framework with two-level interpretability for Text-to-SQL generation is proposed. The first level is the SQL query program for each question answer, which provides the process rather than only the result. The second level is the attention weights of the LSTM layer for each output, which show the fragment relations between questions and generated SQL query programs.

(2) In the sequence encoder layer, semantic similarities between questions and significant table contents are incorporated to enhance the accuracy.

(3) In the attention-based LSTM layer, a general structure for generating SQL programs is designed, instead of limiting to just a few frequent SQL structures.

2   Related Work

Semantic parsing is the task of translating natural language to machine interpretable meaning representations, which is highly correlated to this work. In terms of logical form annotations, related works can be divided into two categories, i.e., depending on logical form annotations or not.

2.1   Semantic Parsing Without Logical Form Annotations

Two common learning paradigms for semantic parsing are maximum marginal likelihood (MML) and reward-based methods [1]. In MML, given an answer pair (x, y), the objective maximizes $\sum_{l \in \hat{L}} P(l|x)$, where $\hat{L}$ is an approximation of a set of logical forms output y [5, 6]. In reward-based methods, a reward function is defined as a prior, and the model parameters are updated with respect to it [7, 8]. Discrete Hard EM [1] develops a hard EM learning scheme that computes gradients relative to the most likely solution at each update. These methods highly rely on large data sets, especially for questions with rich semantics, and therefore becomes unfeasible for complex questions demanding high accuracy.

2.2   Text-to-SQL with Logical Form Annotations

[9] uses rewards from in-the-loop query execution over databases to learn a policy to generate the query and leverages the structure of SQL to prune the space of generated queries and significantly simplify the generation problem. Annotated Seq2seq [10] utilizes a sequence-to-sequence model after automatic annotation of input natural language. Execution guided decoding is suggested in [11], in which non-executable (partial) SQL queries candidates are removed from output candidates during decoding step. Basing on BERT-style pre-training model, X-SQL [3] enhances the structural schema representation with the contextual output. Existing works mainly focus on accuracy while tend to overlook interpretability. Our framework belongs to this category and is compared with two most relevant methods in the experiments.

3   Our Approach

Text-to-SQL is to generate a SQL query over certain databases from a natural language question. The overall framework of our approach mainly consists of three layers: sequence encoder, attention-based LSTM and reinforcement learning, as shown in Fig. 2. Sequence encoder layer generates embedding representations of questions and tables, which try to catch semantic relationships between questions and table contents. And then attention-based LSTM layer generates SQL query programs of textual questions. At last, reinforcement learning layer tries to optimize the parameters of networks by iteratively executing generated SQL programs on databases.

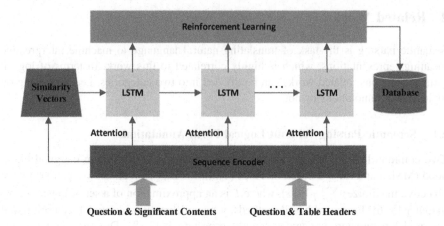

Fig. 2. The framework of our approach. It mainly consists of three parts: sequence encoder (blue color), attention-based LSTM (yellow color) and reinforcement learning (green color) (Color figure online)

3.1 Sequence Encoder

For sequence encoder, we use a model similar to BERT (Bidirectional Encoder Representation from Transformers) to generate the embedding representations of questions and database table. Previous works mainly utilize questions and table headers as input of the sequence encoder, however, in our framework significant contents of columns in terms of TF-IDF(Term Frequency–Inverse Document Frequency) are considered. Specifically, the embedding representations are divided into two parts. The first part is token vectors of questions and table headers, denoted as qh, and the second part is similarity matrix of questions and significant contents, denoted as qc.

qh is set as $\{h_{[CLS]}, h_{q_1}, h_{q_i} \cdots, h_{[SEP]}, h_{c_i}, h[SEP], h_{c_j}, h[SEP], \cdots\}$, which denotes the output from the encoder with dimension d. Each question token is encoded as h_{q_i}, followed by h_{c_j} which encodes the j-th column name. $h_{[CLS]}$ is the global embedding representation of the input and $h_{[SEP]}$ is the embedding of separators. qh is sent to each cell of LSTM layer for attention model.

qc is the semantic similarity matrix between questions and significant content of columns. the significant content vector SC_i of column i is obtained by TF-IDF measure, and is computed by $SC_i = \sum_j^{cn_i} TF\text{-}IDF_{ij} * h_{c_{ij}}$, where cn_i is the number of tuples of column i, $TF\text{-}IDF_{ij}$ is TF-IDF value for the j-th tuple of column i, and $h_{c_{ij}}$ is the output from the encoder for the j-th tuple of column i. For large datasets, we just utilize items with top k highest TF-IDF values for approximation.

qc is denoted in formula (1) as follows,

$$qc = \{qc_{ij} | qc_{ij} = f(U * h_{q_i}, V * SC_i)\}, \tag{1}$$

where both $U, V \in R^{m \times d}$, and f is simple dot product. qc is sent to the LSTM layer as an encoder.

Please note that we initialize our encoder with ELECTRA [12], which has the similar architecture as generative adversarial networks, but trained text encoders as discriminators rather than generators. ELECTRA has less parameters but is more efficient than BERT for general NLP tasks.

3.2 Attention-Based LSTM

The attention-based LSTM layer composes the SQL query from the similarity matrix qc and embedding encoder qh. To ensure the generality of our model, we divide the sub-tasks of generating SQL queries in three categories, instead of limiting the model by specific subtasks [9, 13]. Three categories are Finite-Task, Semi-Task and Infinite-Task. The Finite-Task means the sub-task with the definite and finite output, such as choosing aggregation operator which contains totally 6 possible choices (COUNT, MIN, MAX, SUM, AVG and NONE). The Semi-Task denotes the sub-task which outputs column names defined in tables, such as SELECT column names tasks. The Infinite-Task means the sub-task which generates the output which is semantically relative with questions and table contents, such as the condition value in WHERE.

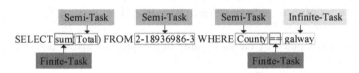

Fig. 3. An example for the categories of subtasks

Each SQL query is composed of serval keywords, each keyword has the fixed pattern of subtasks. Figure 3 is an example for the categories of subtasks. The pattern of SELECT is Finite-task * {Finite-task, Semi-Task}, where the first Finite-task is to predict the number of columns and corresponding to the example in Fig. 3, the output should be 1 * {sum, Total}. The pattern of FROM is Finite-task * {Semi-Task}, where the first Finite-task indicates the number of referred tables, and for Fig. 3 the output should be 1 * {2-18936986-3}. The pattern of WHERE is Finite-task * {Finite-task, Semi-Task, Finite-task, Infinite-task}, where the first Finite-task shows how many conditions and {Finite-task, Semi-Task, Finite-task, Infinite-task} try to predict logical operator between conditions, column name, comparison operator and condition value, respectively, and for Fig. 3 the output is 1 * {NULL, County, ==, galway}, where logical operator is NULL because there is only one condition.

The attention-based LSTM layer is based on Seq2Seq mode [14]. We denote c_i is the context vector of step i, computed by two layers of bidirectional LSTM layers with attention, as shown in formula (2).

$$c_i = \sum\nolimits_{j=1}^{NT} \alpha_{ij} E_j, \qquad (2)$$

where E_j is the ELECTRA embedding of the j-th token, and α_{ij} is the corresponding attention weight, NT is the numbers of tokens which are used in this step.

The hide state s_i for each step is computed by formula (3).

$$s_i = LSTM(s_{i-1}, y_{i-1}, c_i),$$ (3)

where y_{i-1} is the output of last step, $y_0 = qc$.

For the Finite-task, the output is computed by formula (4).

$$p(y_i) = \text{softmax}(W_f \tanh(W_s s_i; W_c c_i),$$ (4)

where $[\cdot; \cdot]$ denotes the concatenation of two vectors, W_f, W_s, W_c are affine transformation.

For the Semi-Task, the output is computed by formula (5).

$$p(y_i) = \text{softmax}(W \tanh(W_s s_i; W_c c_i, W_t h_{[CLS]}),$$ (5)

where $h_{[CLS]}$ contains global semantic and c_i is computed by the embedding of the column names by formula (2).

For the Infinite-task, the output is computed by formula (6).

$$p(y_i) = \text{softmax}(W \tanh(W_s s_i; W_c c_i, W_t h_{[CLS]}; W_q qc),$$ (6)

where c_i is computed by the embedding of the question tokens and column names by formula (2).

Cross entropy loss is utilized for training parameters in the attention-based LSTM layer, denoted as L_l.

3.3 Reinforcement Learning

In order to improve the adaptability of our framework, a reinforcement layer is integrated. The attention-based LSTM layer mainly tries to get the correct SQL query program, while reinforcement layer mainly focuses on obtaining right query result by optimizing parameters according to execution results of generated SQL query programs.

We divide SQL executions into two types: nonexecutable statements and executable statements. For nonexecutable statements, a SQL program throws a run-time error which may be caused by mismatch between operator and operands, and local repair approach [11] is adopted to mitigate this. For executable statements, let y denote the sequence of generated SQL program. Let $q(y)$ denote the query generated by the model and t denote the ground truth query result. Then we define the reward $R(q(y), t)$ as

$$R(q(y), t) = 2(\text{JACCARD } (q(y), t) - \frac{1}{2}),$$ (7)

where JACCARD is Jaccard similarity function. It can tell that the more similar it is between $q(y)$ and t, the closer reward $R(q(y), t)$ to 1, on the contrary, $R(q(y), t)$ is close to -1. The loss $L_r = -E_y[R(q (y), t)]$, is the negative expected reward over possible results.

According to [15], by Monte-Carlo sample, the policy gradient for training parameters is defined as

$$
\begin{aligned}
\nabla L_r &= -\nabla\left(E_y[R(q(y),t)]\right) \\
&= -E_y[R(q(y),t)\nabla\sum_i \log p(y_i;\theta)] \\
&\approx -R(q(y),t)\nabla\sum_i \log p(y_i;\theta)
\end{aligned}
\tag{8}
$$

where $p(y_i;\theta)$ is the probability of choosing the i-th token of y.

3.4 Joint Optimization

The proposed framework consists of three parts: Sequence Encoder, Attention-based LSTM and Reinforcement Learning. The training process is empirically divided into two stages, i.e. Attention-based LSTM training followed by joint fine tuning with reinforcement learning.

For the first stage, as our framework is implemented based on the ELECTRA [12], we initialize the parameters with those of the corresponding layers of ELECTRA. As the Sequence Encoder is shared by all training objects, its parameters are kept fixed at this stage. we train the Attention-based LSTM layer using the stochastic gradient descent (SGD) algorithm with a batch size of 32, a momentum of 0.9 and a weight decay of 0.00005. The initial learning rate is set as 0.001, and it is divided by 10 when the error plateaus.

For the second stage, after the attention-based LSTM layer is trained, we jointly finetune the entire framework by combining the loss terms over all layers $L = L_l + L_r$ and set smaller initial learning rate as 0.0001.

4 Experiments

4.1 Experimental Setting

To validate the effectiveness and the interpretability of our framework, we have conducted experiments on two benchmark datasets: WikiSQL dataset [9] and ATIS dataset [16]. WikiSQL dataset contains 56,324 training pairs, 8,421 dev pairs, and 15,878 test pairs. ATIS dataset has more diverse SQL structures, following the preprocessing method of [17], the dataset consists of 933 examples, with 714/93/126 examples in the train/dev/test split, respectively.

In our experiments, accuracy is used to evaluate effectiveness of our framework, including accuracy of query results and accuracy of SQL programs as in [9]. Let N denotes the number of examples in the dataset, N_{res} the number of queries that, when executed, result in the correct result, and N_{sql} the number of queries has exact match with the ground truth SQL query. By integrating result accuracy $Acc_{res} = \frac{N_{res}}{N}$ and program accuracy $Acc_{sql} = \frac{N_{sql}}{N}$, we set accuracy metric as $Acc = \frac{N_{res} + N_{sql}}{2N}$.

Our framework, denoted as ITSG, is compared with the current state-of-the-art SQLova [18] and X-SQL [3], and for fair comparation, execution guidance is applied during inference. Our framework is implemented by PyTorch 1.1, and all the experiments are conducted on a server with Intel E5-2650v3, 128 GB RAM, 2 * NVIDIA V100 and Ubuntu 16.04 OS.

4.2   Performance of Accuracy

Performance of accuracy are presented in Table 1 and Table 2. As shown in Table 1 for Wiki-SQL dataset, our framework achieves comparable results with the current state-of-the-art SQLova and X-SQL models. The error rate of ITSG is 24.6% less than SQLova, and 11.1% less than X-SQL. Typically Acc_{sql} is slightly less than Acc_{res}, because different SQL programs can lead to same query results, such as different SQL programs with only different orders of WHERE conditions. Table 2 shows the performance of ITSG on ATIS, ATIS has more diverse SQL structures, and not applicable to SQLova and X-SQL which are limited to six sub-module of SQL structure. Due to the much smaller data quantity, Test accuracy is less than Dev accuracy. Table 2 shows generality of ITSG.

Table 1. Performance of accuracy on Wiki-SQL

Model	Acc_{res}	Acc_{sql}	Acc
SQLova	88.9%	83.5%	86.2%
X-SQL	90.3%	86.2%	88.3%
ITSG (Our)	91.3%	87.9%	89.6%

Table 2. Accuracy of ITSG on ATIS

ITSG	Acc_{res}	Acc_{sql}	Acc
Dev	88.4%	86.2%	87.3%
Test	75.8%	72.2%	74%

4.3   Ablation Study

To understand the importance of each part of ITSG, we evaluate ablations in Table 3. In Sequence Encoder layer, we replace ELECTRA with BERT_Base, and it shows that ELECTRA contributes to the overall accuracy by 1.3%. In order to verify the value of significant table contents, we get rid of the input of similarity vectors in the attention-based LSTM layer, and overall accuracy decreases 3.7%. And then we remove the reinforcement layer, result accuracy, program accuracy and overall accuracy decrease 3.7%, 4.7% and 4.2%, which validate the necessity of reinforcement layer.

Table 3. The results of ablation study

Model	Acc_{res}	Acc_{sql}	Acc
ITSG	91.3%	87.9%	89.6%
-ELECTRA + BERT_Base	90.1%	86.5%	88.3%
-Similarity Vector	87.8%	84.1%	85.9%
-Reinforcement Layer	87.6%	83.2%	85.4%

4.4 Interpretability

From textual questions to query results in databases, our framework is provided with two levels of interpretability. The first level is the SQL query programs. Given a textual question, the framework doesn't only care the result, but also emphasizes the process, that's is, the corresponding SQL query program, execution of which outputs correct answer. The second level is the attention weights of the LSTM layer, which explain which token is mainly responsible for the current output.

Figure 4 is an illustration of two-level interpretability. For the question "What is the administrative division that has an area of 30 km^2?", Not only the answer "macau" is returned, but also the SQL query program for the returned answer, "SELECT (Administrative Division) FROM table_1_171666_1 WHERE (Area (km \u00b2) == 30", is provided, which shows the process of obtaining the answer as the first level of interpretability. Further, rectangles with the same color and line type show the close relationship between question tokens and fragments of SQL query programs and column headers, which illustrate the second level of interpretability.

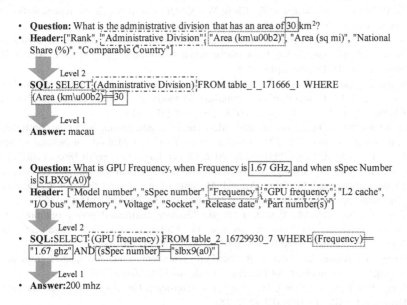

Fig. 4. Illustration of two-level interpretability (Color figure online)

5   Conclusion

In this paper, we proposed ITSG, a deep neural network for interpretable Text-to-SQL generation with joint optimization. ITSG leverages similarity matrices between questions and table contents as inputs of our model to enhance the probability of semantic matches between textual questions and structural SQL query programs, utilize attention-based LSTM to obtain impact weights of input tokens for each output which provides foundation of interpretability, and integrates reinforcement learning to improve accuracy and generality. Our evaluation on benchmark datasets shows that our model achieves state-of-the-art accuracy with two-level interpretability. In our future work, we will study more self-supervised learning based Text-to-SQL approaches to further reduce dependence on large training data of certain area.

Acknowledgments. This work is supported by the National Natural Science Foundation of China (61802116), the Science and Technology Plan of Henan Province (192102210113, 192102210248, 202102210372).

References

1. Min, S., Chen, D., Hajishirzi, H., Zettlemoyer, L.: A discrete hard EM approach for weakly supervised question answering. In: 2019 Conference on Empirical Methods in Natural Language Processing & International Joint Conference on Natural Language Processing (EMNLP-IJCNLP), pp. 2851–2864. ACL (2019). https://doi.org/10.18653/v1/d19-1284
2. Han, Z., Jiang, X., Li, M., et al.: An integrated semantic-syntactic SBLSTM model for aspect specific opinion extraction. In: 15th International Conference of Web Information Systems and Applications, pp. 191–199. WISA (2018). https://doi.org/10.1007/978-3-030-02934-0_18
3. He, P., Mao, Y., Chakrabarti, K., Chen, W.: X-SQL: reinforce schema representation with context. CoRR abs/1908.08113 (2019)
4. Dong, L., Lapata, M.: Coarse-to-fine decoding for neural semantic parsing. In: 56th Annual Meeting of the Association for Computational Linguistics, pp. 731–742. ACL (2018). https://doi.org/10.18653/v1/p18-1068
5. Berant, J., Chou, A., Frostig, R., et al.: Semantic parsing on freebase from Question-Answer pairs. In: 2013 Conference on Empirical Methods in Natural Language Processing (EMNLP), pp. 1533–1544. ACL (2013). Doi:10.1.1.408.319
6. Krishnamurthy, J., Dasigi, P., Gardner, M.: Neural semantic parsing with type constraints for semi-structured tables. In: 2017 Conference on Empirical Methods in Natural Language Processing (EMNLP), pp. 1516–1526. ACL (2017). https://doi.org/10.18653/v1/d17-1160
7. Iyyer, M., Yih, W., Chang, M.: Search-based neural structured learning for sequential question answering. In: 2017 Annual Meeting of the Association for Computational Linguistics, pp. 1821–1831. ACL (2017). https://doi.org/10.18653/v1/p17-1167
8. Liang, C., Norouzi, M., Berant, J., et al.: Memory augmented policy optimization for program synthesis and semantic parsing. In: 32rd Conference on Neural Information Processing Systems (NeurIPS), pp. 10015–10027. (NIPS) (2018)
9. Zhong, V., Xiong, C., Socher, R.: Seq2sql: generating structured queries from natural language using reinforcement learning. CoRR, abs/1709.00103 (2017)
10. Wang, W., Tian, Y., Xiong, H., et al.: A transfer-learnable natural language interface for databases. CoRR, abs/1809.02649 (2018)

11. Wang, C., Huang, P., Polozov, A., et al.: Execution-Guided neural program decoding. CoRR, abs/1807.03100 (2018)
12. Clark, K., Luong, M., Le, Q., et al.: ELECTRA: pre-training text encoders as discriminators rather than generators. In: 8th International Conference on Learning Representations (ICLR) (2020)
13. Wang, C., Tatwawadi, K., Brockschmidt, M.: Robust Text-to-SQL generation with execution-guided decoding. CoRR, abs/ 1807.03100 (2018)
14. Zhou, P., Shi, W., Tian, J., et al.: Attention-based bidirectional long short-term memory networks for relation classification. In: 2016 Annual Meeting of the Association for Computational Linguistics, pp. 207–212. ACL (2016). https://doi.org/10.18653/v1/p16-2034
15. Schulman, J., Heess, N., Weber, T., et al.: Gradient estimation using stochastic computation graphs. In: 29th Annual Conference on Neural Information Processing Systems, pp. 3528–3536. NIPS (2015). https://doi.org/10.5555/2969442.2969633
16. Dahl, D., Bates, M., Brown, M., et al.: Expanding the scope of the ATIS task: the ATIS-3 corpus. In: Proceedings of the workshop on Human Language Technology(HLT), pp. 43-48. ACM (1994). https://doi.org/10.3115/1075812.1075823
17. Shi, T., Tatwawadi, K., Chakrabarti, K., et al.: IncSQL: training incremental Text-to-SQL parsers with non-deterministic oracles. CoRR abs/1809.05054 (2018)
18. Hwang, W., Yim, J., Park, S., et al.: A comprehensive exploration on WikiSQL with table-aware word contextualization. CoRR abs/1902.01069 (2019)

Ranking-Based Fuzzy Min-Max Classification Neural Network

Lingli Xue[1,2], Wei Huang[1,2(✉)], and Jinsong Wang[1,2]

[1] Tianjin Key Laboratory of Intelligence Computing and Novel
Software Technology, Tianjin University of Technology, Tianjin 300384, China
xuelingli7@126.com, huangwabc@163.com,
jswang70@126.com
[2] School of Computer Science and Engineering,
Tianjin University of Technology, Tianjin 300384, China

Abstract. The performance of fuzzy min-max classification neural network (FMM) is affected by the input sequence of training set patterns. This paper proposes a ranking-based fuzzy min-max Classification Neural Network (RFMM) to overcome this shortcoming. RFMM improves FMM through the following three aspects. First, RFMM ranks the input order of the training set patterns according to their membership degree to the center point of same class, so that the finally constructed network is fixed and does not depend on the input order of the training set. Second, a new membership function based on Manhattan distance is constructed, which overcomes the problem that the membership degree obtained by the membership function in the FMM will not decrease steadily with the increase of the distance between the input pattern and the hyperbox. At last, RFMM uses the method based on individual contour coefficient to classify the patterns in overlapping regions, which overcomes the problem that when the FMM eliminates the overlapping region by shrinking hyperboxes, the membership degree of the patterns in the contracted region to the class they belong is changed. Experimental results show that RFMM has better learning ability, and compared with other FMM methods, RFMM shows higher classification accuracy and lower network complexity.

Keywords: Pattern recognition · Fuzzy Min-Max neural network · Hyperbox fuzzy set

1 Introduction

With its powerful learning ability, machine learning has been widely used in data mining, medical diagnosis, pattern recognition and other fields [1–5]. Pattern recognition uses a series of mathematical methods to make computers realize human recognition capabilities. However, in pattern recognition, traditional set theory describes clear events without intermediate positions. Zadeh proposed the concept of fuzzy sets, which measures the degree of occurrence of events. As a method of dealing with uncertain or fuzzy data, it emphasizes the importance of fuzzy logic for pattern recognition [6–8]. Since then, many researchers have studied the combination of fuzzy logic and pattern recognition.

© Springer Nature Switzerland AG 2020
G. Wang et al. (Eds.): WISA 2020, LNCS 12432, pp. 352–364, 2020.
https://doi.org/10.1007/978-3-030-60029-7_33

Simpson proposed a fuzzy min-max classification neural network (FMM) in [9], learned the data structure by constructing hyperbox fuzzy sets for each class, and classified the linear and inseparable multidimensional data easily according to the membership degree of the finally formed hyperbox fuzzy sets. It has the advantages of online learning and one pass. Although FMM has obvious advantages, it also has certain limitations. Nowadays, many researchers have improved FMM for different purposes [10].

A general fuzzy min-max neural network for clustering and classification (GFMM) is proposed in [11], supervised learning and unsupervised learning are combined in the training algorithm, the membership function, constraint conditions and network structure in the FMM method are modified. The methods proposed in [12] and [13] mainly improve the performance of FMM by creating new neurons to process overlapping regions. An inclusion/exclusion fuzzy hyperbox classifier (EFC) is proposed in [12], which uses inclusion hyperboxes to include patterns from the same class, and uses exclusion hyperboxes to include patterns that located in the overlapping regions, instead of shrinking hyperboxes to eliminate overlapping. A fuzzy min-max neural network classifier with compensatory neuron architecture (FMCN) is proposed in [13], which uses compensation neurons to deal with the overlap between different classes of hyperboxes. Data-core-based fuzzy min-max neural network for pattern classification (DCFMM) is proposed in [14], which defines a new membership function that takes into account noise, the geometric center of the hyperbox, and the data core. Instead of FMM using contraction to eliminate overlap, DCFMM proposes an overlapping neuron to represent overlapping regions of different classes of hyperboxes. Multi-level fuzzy min-max neutral network classifier (MLF) is proposed in [15], which adopts a multi-layer tree structure and creates overlapping boxes in subnets to deal with patterns located in the overlapping areas of the upper layer. However, almost all the above methods have the problem that the constructed network depends on the input order of training set patterns. If the input sequence changes, the constructed network will change.

In this paper, a ranking-based fuzzy min-max classification neural network (RFMM) is proposed. RFMM determines the input order of patterns in training set according to the membership degree of the input pattern to the center point of each class, which overcomes the problem of FMM's strong dependence on input order. Furthermore, this paper proposes a new membership function based on Manhattan distance, which overcomes the problem that the membership degree value obtained by the membership function in the FMM will not decrease steadily with the increase of the distance between the input pattern and the hyperbox. Different from FMM, RFMM uses a method based on individual contour coefficients to deal with the pattern classification problem in overlapping regions, which overcomes the problem that the membership degree of the patterns in contracted regions to the class they belong is changed when FMM contracts the hyperboxes. At the same time, the network complexity is low.

This paper is organized as follows: In the second section, fuzzy min-max classification neural network is analyzed. In the third section, the ranking-based fuzzy min-max classification neural network proposed in this paper is described in detail. The fourth section describes the experimental results of the proposed method and other FMM methods under different data sets. The fifth section gives the conclusion.

2 Preliminaries

2.1 Fuzzy Min-Max Classification Neural Network

The fuzzy min-max classification neural network (FMM) consists of multiple fuzzy hyperbox sets. A hyperbox includes two points V and W, respectively corresponding to the minimum and maximum points of the hyperbox. Simpson specifies that the value range of each dimension of the hyperbox is 0 to 1. Each hyperbox defines an region in the *n*-dimensional pattern space. All patterns contained in the hyperbox have full membership to the hyperbox. The fuzzy min-max classification neural network includes three layers, as shown in Fig. 1.

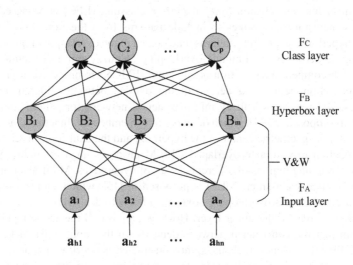

Fig. 1. Structure of FMM

When an untrained input pattern enters, FMM first checks whether there is a hyperbox of the same class in the network. If it does not exist, a new hyperbox is created, the minimum point and maximum point of the hyperbox are the corresponding points of the input pattern; If it exist, it is judged whether the pattern is included in a hyperbox of the same class. If included, no processing is performed on the pattern and then the next pattern is trained; otherwise, the expansion and overlap detection and processing steps are performed.

1) Expansion: In this step, first find the hyperbox of the same class as the input pattern and with the highest degree of membership. Simpson defines membership function as

$$b_j(A_h) = \frac{1}{2n} \sum_{\Delta=1}^{n} \left(\begin{array}{c} max\big(0, 1 - \big(max\big(0, \gamma min\big(1, a_{h\Delta} - w_{j\Delta}\big)\big)\big)\big) + \\ max\big(0, 1 - \big(max\big(0, \gamma min\big(1, v_{j\Delta} - a_{h\Delta}\big)\big)\big)\big) \end{array} \right) \quad (1)$$

Where $A_h = (a_{h1}, a_{h2}, \ldots, a_{hn}) \in I^n$ is the hth input pattern in the training set, Δ refers to the dimension, γ is the sensitivity coefficient, it is used to adjust the degree to which A_h belongs to the hyperbox B_j as the distance between A_h and hyperbox B_j increases. The value is defined by the user and ranges from 0 to 1.

Then it is judged whether the hyperbox satisfies the expansion rule. The size of the hyperbox is limited by the user-defined expansion parameter θ, where the expansion rule is as follows

$$\sum_{\Delta=1}^{n} \left(max(w_{j\Delta}, a_{h\Delta}) - min(v_{j\Delta}, a_{h\Delta}) \right) \leq n \cdot \theta \tag{2}$$

If the expansion condition is satisfied, the hyperbox is expanded; if it is not satisfied, a new hyperbox is created.

2) Overlap detection and processing: In this step, it is detected whether the extended hyperbox constitutes overlap areas with other classes of hyperboxes. If each dimension belongs to one of the following four cases, it is considered that the two hyperboxes overlap. According to the corresponding overlapping case, the smallest overlapping dimension is selected to perform the shrinking operation to eliminate the overlapping.

Case 1: $V_{j\Delta} < V_{k\Delta} < W_{j\Delta} < W_{k\Delta}$

$$W_{j\Delta}^{new} = V_{k\Delta}^{new} = \frac{1}{2} \left(W_{j\Delta}^{old} + V_{k\Delta}^{old} \right)$$

Case 2: $V_{k\Delta} < V_{j\Delta} < W_{k\Delta} < W_{j\Delta}$

$$V_{j\Delta}^{new} = W_{k\Delta}^{new} = \frac{1}{2} \left(V_{j\Delta}^{old} + W_{k\Delta}^{old} \right)$$

Case 3a: $V_{j\Delta} < V_{k\Delta} < W_{k\Delta} < W_{j\Delta}$ and $\left(W_{k\Delta} - V_{j\Delta} \right) < \left(W_{j\Delta} - V_{k\Delta} \right)$

$$V_{j\Delta}^{new} = W_{k\Delta}^{old}$$

Case 3b: $V_{j\Delta} < V_{k\Delta} < W_{k\Delta} < W_{j\Delta}$ and $\left(W_{k\Delta} - V_{j\Delta} \right) > \left(W_{j\Delta} - V_{k\Delta} \right)$

$$W_{j\Delta}^{new} = V_{k\Delta}^{old}$$

Case 4a: $V_{k\Delta} < V_{j\Delta} < W_{j\Delta} < W_{k\Delta}$ and $\left(W_{k\Delta} - V_{j\Delta} \right) < \left(W_{j\Delta} - V_{k\Delta} \right)$

$$W_{k\Delta}^{new} = V_{j\Delta}^{old}$$

Case 4b: $V_{k\Delta} < V_{j\Delta} < W_{j\Delta} < W_{k\Delta}$ and $\left(W_{k\Delta} - V_{j\Delta} \right) > \left(W_{j\Delta} - V_{k\Delta} \right)$

$$V_{k\Delta}^{new} = W_{j\Delta}^{old}$$

After the training is completed, hyperboxes are expanded to cover almost the entire pattern space. Each class is composed of many hyperboxes, and hyperboxes of different classes do not overlap each other. For each pattern in the test set, its membership degree to all hyperboxes in the network is calculated, and the hyperbox with the highest membership degree is selected to classify it.

2.2 Limitations of Fuzzy Min-Max Classification Neural Network

The FMM method has the following three limitations: Firstly, the construction of network depends on the input sequence of training sets. Secondly, the membership value will not decrease steadily with the increase of the distance between the input pattern and the hyperbox. Thirdly, shrinking hyperboxes will lead to the change in membership degree of shrinking areas to corresponding hyperboxes.

1) FMM constructs the network according to the input sequence of training set patterns. For the newly emerging input pattern, FMM performs the steps of creating or expanding a hyperbox. Once the overlap region is created, it is immediately contracted to eliminate the overlap region, and then train the next untrained pattern. The training results of the previous patterns affect the training of the next untrained pattern, which makes FMM rely on the order of input patterns in the training set. If the order is changed, the hyperbox set in the finally constructed network will also change, which will affect the accuracy of classification.

Figure 2 shows the training results of three 2-dimensional input patterns A1 : $(0.6, 0.7)$, A2 : $(0.1, 0.4)$, A3 : $(0.1, 0.9)$ under different input sequences for the same extended parameter $\theta = 0.4$.

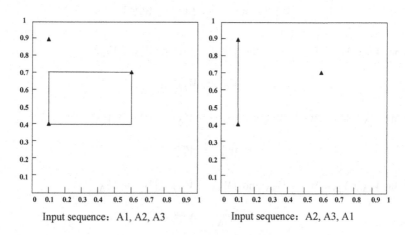

Fig. 2. The training results of input patterns A1, A2, A3 for different input sequences ($\theta = 0.4$)

2) The membership degree value obtained by membership function in FMM will not decrease with the increase of the distance between input pattern and hyperbox [11].

3) FMM eliminates the overlap region by shrinking the hyperboxes, but this will causes the membership degree of the patterns in contracted region to the class they belong is changed. In Fig. 3, the three input patterns A, B, and D come from class 1, and the two input patterns C and E come from class 2. Figure 3(a) shows the training results of these five input patterns, and the two hyperboxes of different classes overlap. The membership degree of A, B, D to class 1 is 1, the membership degree of C, E for class 2 is 1. Figure 3(b) shows the result of eliminating the overlapping region by shrinking the hyperbox, but because C and D fall outside the hyperboxes, the membership degree of D to class 1 is less than 1, and the membership degree of C to class 2 is less than 1.

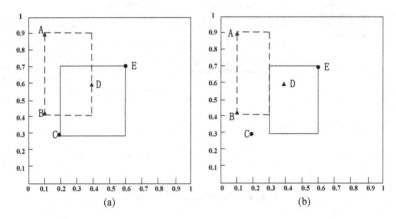

Fig. 3. Shrink hyperboxes to eliminate overlapping regions

3 Ranking-Based Fuzzy Min-Max Classification Neural Network

In the following, we will introduce the RFMM in detail from two aspects of training process and classification process. The architecture of RFMM is shown in Fig. 4.

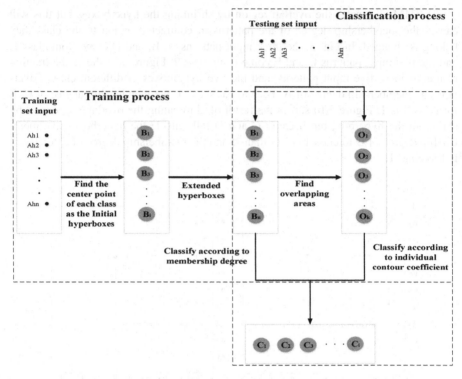

Fig. 4. The structure of RFMM

3.1 Training Process

In the method proposed in this paper, the training process is divided into two stages. First, rank the input order of the patterns in the training set, then the extended hyperbox operation is performed on other patterns in the training set.

1) Ranking. For each pattern in the training set, calculate the sum of the Euclidean distances to other patterns of the same class, find out the center point of each class as the initial hyperboxes in the network. For the non-central point patterns in the training set, the membership degree to hyperboxes of the same class in the network is calculated, and the order of their input into the network is ranked from high to low according to the membership degree.

This method uses a Manhattan distance-based method to calculate membership degree. The membership degree of the input pattern to the hyperbox is inversely proportional to its Manhattan distance to the hyperbox. The Manhattan distance from the input pattern to the hyperbox is calculated by (3).

$$\begin{cases} distance = \sum_{\Delta=1}^{n} distance_\Delta \\ where \\ distance_\Delta = \begin{cases} 0, \ if \ V_\Delta \leq a_{h\Delta} \leq W_\Delta \\ |a_{h\Delta} - V_\Delta|, \ if \ |a_{h\Delta} - V_\Delta| \leq |a_{h\Delta} - W_\Delta| \\ |a_{h\Delta} - W_\Delta|, \ if \ |a_{h\Delta} - W_\Delta| \leq |a_{h\Delta} - V_\Delta| \end{cases} \end{cases} \quad (3)$$

Where a_h represents the h th input pattern, and V and W represent the minimum and maximum points of the hyperbox.

In this way, hyperbox fuzzy sets are constructed and expanded from the central area of each class, which can not only reduce the number of hyperboxes finally created and reduce the complexity of the network, but also fix the finally constructed hyperbox fuzzy sets so that they do not change with the input pattern sequence in the training set.

2) Extended hyperboxes. After entering the input pattern, check whether it is located in an existing hyperbox of the same class, if it exists, do not need processing, and continue to train the next input pattern; If there is no such hyperbox, the hyperbox with the highest membership degree of the same class is found to expand. The extended rule is the same as FMM. If the extension rule is not satisfied, a new hyperbox is created.

Different from FMM performing overlap detection and processing immediately after an input pattern expansion hyperbox is completed, the method proposed in this paper performs overlap processing after the hyperboxes creation and expansion of all input patterns in the training set are completed, thus reducing the time complexity.

After the steps of creating and expanding the hyperboxes of all patterns in the training set are completed, the overlapping area constructed in the previous step is identified to classify the patterns in the test set that fall in the overlapping regions. Since the overlapping cases listed in the FMM are incomplete, the overlapping cases proposed in [16] are used in this method as follows.

Case 1: $V_{j\Delta} < V_{k\Delta} < W_{j\Delta} < W_{k\Delta}$ Case 2: $V_{k\Delta} < V_{j\Delta} < W_{k\Delta} < W_{j\Delta}$

Case 3: $V_{j\Delta} = V_{k\Delta} < W_{j\Delta} < W_{k\Delta}$ Case 4: $V_{j\Delta} < V_{k\Delta} < W_{j\Delta} = W_{k\Delta}$

Case 5: $V_{k\Delta} = V_{j\Delta} < W_{k\Delta} < W_{j\Delta}$ Case 6: $V_{k\Delta} < V_{j\Delta} < W_{k\Delta} = W_{j\Delta}$

Case 7: $V_{j\Delta} < V_{k\Delta} \leq W_{k\Delta} < W_{j\Delta}$ Case 8: $V_{k\Delta} < V_{j\Delta} \leq W_{j\Delta} < W_{k\Delta}$

Case 9: $V_{k\Delta} = V_{j\Delta} < W_{k\Delta} = W_{j\Delta}$

3.2 Classification Process

After the patterns of the test set comes in, first check whether it is located in the overlapping area. If it is located in the overlapping region, the individual contour coefficients to the two overlapping hyperboxes are calculated according to (4).

$$\begin{cases} b_i = \frac{m(i)-n(i)}{max(m(i),n(i))} \\ where \\ m(i) = \frac{1}{n_1}\sum_{j\in H_1} d(i,j) \\ n(i) = \frac{1}{n_2}\sum_{j\in H_2} d(i,j) \end{cases} \tag{4}$$

Since the overlapping area is composed of two hyperboxes of different classes, in the above formula, H_1 represents one of the hyperboxes constituting the current overlapping area, n_1 represents the number of input patterns of the same class as hyperbox H_1 contained in hyperbox H_1, $m(i)$ represents the average distance between the current pattern to be classified and all the input patterns of the same class as hyperbox H_1 contained in hyperbox H_1; H_2 represents another hyperbox that constitutes the current overlapping area, n_2 represents the number of input patterns of the same class as hyperbox H_2 contained in hyperbox H_2, and $n(i)$ represents the average distance between the current pattern to be classified and all the input patterns of the same class as hyperbox H_2 contained in hyperbox H_2. b_i contains the average distance between the two hyperboxes that constitute the overlapping area, which is used to evaluate the rationality that a pattern to be classified in the overlapping area is divided into a certain class, and its value is between 1 and -1. If the value is close to 1, it means that the average distance of the patterns in hyperbox H_1 is much smaller than the average distances in hyperbox H_2, indicating that the pattern is more in line with the distribution pattern of the class of hyperbox H_1, and should be classified into the class to which the hyperbox H_1 belongs.

If the pattern is not in the overlapping area, the hyperbox class with the highest degree of membership is selected to be assigned to the pattern.

4 Experimental Studies

In order to verify the performance of the method, three different standard data sets Iris, Wine and Breast Cancer are used for experiments to compare the performance of the proposed method and other FMM methods. In this experiment, the classification performance of each method is judged by comparing its classification accuracy and network complexity.

Since the pattern space is a unit cube In, all data sets are normalized before the experiment. Two parts of experiments are carried out on three data sets: (1) all samples in the data set are taken as training sets and test sets for experiments, and the learning ability of RFMM is judged through comparison with FMM; (2) Through K-fold cross-validation experiments on three different data sets and comparison with other FMM methods, the classification performance of RFMM are judged.

4.1 Training and Testing the Complete Data Set

In this part, all samples of each data set are used for training and testing to compare the learning ability of RFMM and FMM. Figures 5, 6 and 7 show the classification accuracy and network complexity of Iris, Wine and Breast Cancer respectively. With the gradual increase of θ until the number of hyperboxes in the network is very small,

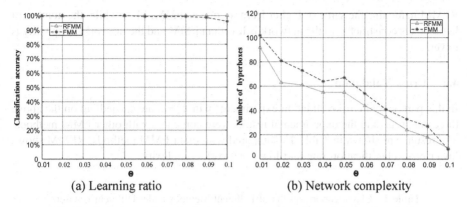

(a) Learning ratio   (b) Network complexity

Fig. 5. Experimental results for Iris (complete data for training and testing)

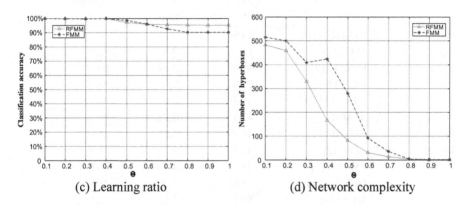

(c) Learning ratio   (d) Network complexity

Fig. 6. Experimental results for Breast Cancer (complete data for training and testing)

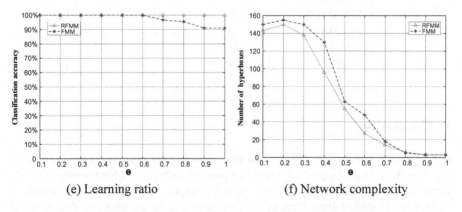

(e) Learning ratio   (f) Network complexity

Fig. 7. Experimental results for Wine (complete data for training and testing)

RFMM maintains the same or higher learning rate as FMM. Meanwhile, the number of hyperboxes constructed in the network is less than FMM and the network complexity is lower.

4.2   Cross Validation

In this part, k-fold cross-validation experiments with k = 3 are carried out on the above three data sets, and the experimental results were compared with other FMM methods. The results are shown in Table 1. Compared with other FMM methods, RFMM shows lower error classification rate and lower network complexity.

Table 1. Classification Results of Different Methods under Different Datasets.

Data Set	Method	Miss Classification (%)		Number of hyperboxes
		Max	Min	
Iris	GFMM [11]	18.67	3.33	38
	EFC [12]	11.33	3.33	42
	FMCN [13]	10.00	2.67	64
	DCFMN [14]	7.89	2.67	38
	MLF [15]	4.67	2.67	57
	RFMM	4.44	2.22	36
Breast Cancer	GFMM [11]	44.49	4.26	154
	EFC [12]	41.56	5.73	183
	FMCN [13]	31.42	4.70	244
	DCFMN [14]	15.85	4.26	190
	MLF [15]	6.31	3.52	227
	RFMM	6.63	3.54	138
Wine	GFMM [11]	15.00	2.22	129
	EFC [12]	7.78	2.22	133
	FMCN [13]	7.78	2.22	183
	DCFMN [14]	7.78	2.22	124
	MLF [15]	7.78	2.22	133
	RFMM	6.94	0	98

5   Conclusion

In this paper, a ranking-based fuzzy min-max classification neural network (RFMM) is proposed. In the training process, the center point of each class in the training set is first found as the initial hyperbox in the network, and the input order of the remaining training set is ranked from high to low according to the membership degree to the hyperboxes of same class in the network, which overcomes the problem that the FMM training process depends heavily on the input order of the training set patterns.

Moreover, a new membership degree function based on Manhattan distance is proposed. A new method is adopted to deal with the classification of patterns falling in overlapping areas, which overcomes the problem of membership change of contracted areas caused by contraction process of FMM. Experiments show that RFMM has lower network complexity while maintaining higher classification performance.

According to RFMM's advantages of high classification accuracy and low complexity, we plan to apply it to speech recognition and face recognition in the future.

Acknowledgements. This work was supported by the National Natural Science Foundation of China under Grant 61673295, by the National Key R&D Program of China under Grant 2018YFC0831405, by the Natural Science Foundation of Tianjin for Distinguished Young Scholars under Grant 19JCJQJC61500, by the Natural Science Foundation of Tianjin (Key Program) under Grant 18JCZDJC30700, and by the Natural Science Foundation of Tianjin (General Program) under Grant 18JCYBJC85200, by the Major Project for New Generation of AI Grant 2018AAA0100400.

References

1. Zheng, H., Shi, D.: Using a LSTM-RNN based deep learning framework for ICU mortality prediction. In: Meng, X., Li, R., Wang, K., Niu, B., Wang, X., Zhao, G. (eds.) WISA 2018. LNCS, vol. 11242, pp. 60–67. Springer, Cham (2018). https://doi.org/10.1007/978-3-030-02934-0_6

2. Wang, D., Wan, S., Guizani, N.: Context-based probability neural network classifiers realized by genetic optimization for medical decision making. Multimedia Tools and Applications 77(17), 21995–22006 (2018). https://doi.org/10.1007/s11042-018-5631-3

3. Xu, Z., He, T., Lian, H., Wan, J., Wang, H.: Case facts analysis method based on deep learning. In: Ni, W., Wang, X., Song, W., Li, Yukun (eds.) WISA 2019. LNCS, vol. 11817, pp. 92–97. Springer, Cham (2019). https://doi.org/10.1007/978-3-030-30952-7_11

4. Wang, D., Xiong, C.C., Zhang, X.K.: An opposition-based group search optimizer with diversity guidance. Math. Prob. Eng. 2015(546181), 1–12 (2015)

5. Zong, W., Zhou, H., Huang, G.-B., Lin, Z.: Face recognition based on kernelized extreme learning machine. In: Kamel, M., Karray, F., Gueaieb, W., Khamis, A. (eds.) AIS 2011. LNCS (LNAI), vol. 6752, pp. 263–272. Springer, Heidelberg (2011). https://doi.org/10.1007/978-3-642-21538-4_26

6. Zadeh, L.A.: Fuzzy sets. Inf. Control 8, 338–353 (1965)

7. Bellman, R.E., Kalaba, R., Zadeh, L.A.: Abstraction and pattern classification. J. Math. Anal. Appl. 13, 1–7 (1966)

8. Wang, D., Oh, S.K., Kim, E.H.: Design of space search-optimized polynomial neural networks with the aid of ranking selection and l2-norm regularization. J. Electric. Eng. Technol. 13(4), 1724–1731 (2018)

9. Simpson, P.K.: Fuzzy min-max neural networks – Part1: classification. IEEE Trans. Neural Netw. 3(5), 776–786 (1992)

10. Sayaydeh, O.N.A., Mohammed, M.F., Lim, C.P.: Survey of Fuzzy Min-Max Neural Network for Pattern Classification Variants and Applications. IEEE Trans. Fuzzy Syst. 27(4), 635–645 (2019)

11. Gabrys, B., Bargiela, A.: General fuzzy min-max neural network for clustering and classification. IEEE Trans. Neural Netw. 11(3), 769–783 (2000)

12. Bargiela, A., Pedrycz, W., Tanaka, M.: An inclusion/exclusion fuzzy hyperbox classifer. Int. J. Knowl. based Intell. Eng. Syst. **8**(2), 91–98 (2004)
13. Nandedkar, A.V., Biswas, P.K.: A fuzzy min-max neural network classifier with compensatory neuron architecture. IEEE Trans. Neural Netw. **8**(1), 42–54 (2007)
14. Zhang, H., Liu, J., Ma, D., Wang, Z.: Data-core-based fuzzy min-max neural network for pattern classification. IEEE Trans. Neural Netw. **22**(12), 2339–2352 (2011)
15. Davtalab, R., Dezfoulian, M.H., Mansoorizadeh, M.: Multi-level fuzzy min-max neural network classifier. IEEE Trans. Neural Netw. Learn. Syst. **25**(3), 470–482 (2014)
16. Mohammed, M.F., Lim, C.P.: An enhanced fuzzy min-max neural network for pattern classification. IEEE Trans. Neural Netw. Learn. Syst. **26**(3), 417–429 (2015)

A Multi-Stages Chromosome Segmentation and Mixed Classification Method for Chromosome Automatic Karyotyping

Chengchuang Lin[1,3,4(✉)], Gansen Zhao[1,3,4], Aihua Yin[2], Bichao Ding[1,3,4],
Li Guo[2], and Hanbiao Chen[2]

[1] School of Computer Science, South China Normal University,
Guangzhou 510631, China
{chengchuang.lin,gzhao,dingbichao}@m.scnu.edu.cn
[2] Guangdong Women and Children Hospital, Guangzhou 511400, China
yinaiwa@vip.126.com, guoli3861@163.com, chenhanbiao2000@126.com
[3] Key Lab on Cloud Security and Assessment technology of Guangzhou,
Guangzhou, China
[4] SCNU & VeChina Joint Lab on BlockChain Technology and Application,
Guangzhou 510631, China

Abstract. The chromosome karyotyping task is vital and indispensable but tedious work for birth defect diagnosis and biomedical research. In this work, we tackle chromosome automatic karyotyping using a multi-stages chromosome segmentation and mixed classification method. Firstly, we apply a global binary threshold-based method to segment the metaphase chromosome microscope grayscale image into several image slices, consisting of chromosome instances and chromosome clusters. Afterward, we propose a mixed chromosome classification method for identifying a given image is a chromosome cluster or corresponding instance label. After that, we use a deep learning-based approach to segment chromosome cluster images into chromosome instances and apply the mixed chromosome classification model to recognize their corresponding labels. Finally, we synthesize a chromosome karyotype from all corresponding instances and labels. In the mixed classification stage, the proposed method yields $99.53 \pm 0.23\%$ classification accuracy on the clinical dataset. In segmentation stages, the proposed method achieves 90.81% comprehensive segmentation accuracy and 85.00% instance segmentation accuracy with 90.63% AP_{50} precision. The experimental results show that our proposed method is promising for solving chromosome segmentation and classification task of the clinical chromosome automatic karyotyping.

This work was supported by Key-Area Research and Development Program of Guangdong Province(No.2019B010137003), NationalKey-Area Research and Development Program of China (2018YFB1404402), Guangdong Science and Technology Fund (No.2016B030305006, No.2018A07071702, No.201804010314), Guangzhou Science & Technology Fund (No.201804010314), VeChain Foundation (No.SCNU-2018-01).

© Springer Nature Switzerland AG 2020
G. Wang et al. (Eds.): WISA 2020, LNCS 12432, pp. 365–376, 2020.
https://doi.org/10.1007/978-3-030-60029-7_34

Keywords: Chromosome segmentation · Chromosome classification ·
Chromosome automatic karyotyping

1   Introduction

Human chromosomes contain human genetic information, which is commonly
used for analyzing human genetic diseases. In general, there are 23 pairs of chromosomes in a healthy human body, including 22 pairs of autosomes and a pair
of sex chromosomes (X and Y chromosome in male cells and double X in female
cells) [1]. Karyotype analysis, illustrated by *Fig*.1, is a fundamental approach
for clinical cytogeneticists to diagnose human chromosomes genetic diseases and
birth defects, which is generated by arranging these chromosomes after extracting them from the metaphase chromosome images. For cytogeneticists, karyotyping is laborious work, many researchers have dedicated to auto-karyotyping
using computation techniques [2–6] for years.

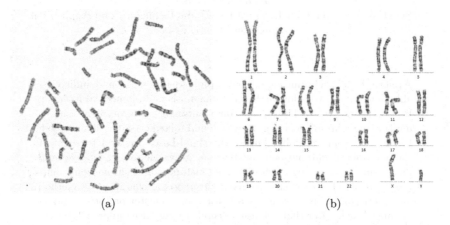

(a)   (b)

Fig. 1. Example of the chromosome karyotype analysis. The Fig. 1(a) is an grayscale
image illustrating a G-band metaphase chromosome cell microphotograph. The Fig.
1(a) is a corresponding chromosome karyotype of Fig. 1(a).

In general, researches on chromosome auto-karyotyping follow the sequential
procedure of chromosome classification [1,5] and segmentation [7,8]. Although
the above research has advanced some progress, there are still some limitations
in the automatic chromosomes karyotyping.

Challenges: The most significant challenge of automatic karyotyping is the
segmentation of overlapping and touching chromosomes. There are three mainstream methods of chromosome segmentation. The threshold-based segmentation method is the most primal method for overlapping and touching chromosome segmentation. The most notable strength of the threshold-based segmentation method is high running efficiency and the most noticeable weakness is

poor segmentation effects. The geometric features-based approach is the most commonly used for overlapping and touching chromosomes segmentation in the last decade. The geometric features-based approach makes a tradeoff between the running efficiency and outcome effects. As it is a rule-based approach, it may not work when encountering more complicated overlapping and touching chromosome clusters. The instance segmentation method based on deep learning is the most promising way to solve the task of overlapping and touching chromosomes segmentation currently. These methods can deal with a variety of complicated overlapping and touching chromosome clusters accurately. However, training and running these models require a large scale of the labeled dataset and enormous computational resources.

Motivations: Motivated by the strengths and weaknesses of the above methods, we propose a novel approach to tackle the chromosome auto-karyotyping problem. We utilize a global threshold method to segment the full-size original metaphase image into several image slices at extremely low running costs. After that, we design a mixed chromosome classifier to identify chromosome cluster images that are required to segment by instance segmentation model. And the classifier will give the corresponding labels of the rest images. The most significant advantage of our proposed method is that it compounds the benefit of threshold-based segmentation and instance segmentation methods.

Results: To evaluate the overall performance and clinical application effect of our proposed method, we train and test our method on the clinical dataset which is constructed by skilled cytogeneticists. In the chromosome classification stage, our proposed method achieves mixed classification performance with $(99.53 \pm 0.23)\%$ accuracy, which is a state-of-the-art classification result in previously reported literature. In segmentation stages, the proposed method achieves 90.81% comprehensive segmentation accuracy and 85.00% instance segmentation accuracy with 90.63% AP_{50} precision, which is promising on the clinical dataset.

Contributions: According to the mentioned explorations, we proposed a multiple-stages method, illustrated in Fig.2, for the chromosome auto-karyotyping task. First, to tackle the chromosome auto-karyotyping task, we decompose the task into multiple-stages. After that, we propose a classification model and a segmentation model based on the deep neural network. Second, to evaluate our proposed model in clinical application, we build a clinical chromosome classification dataset and label a chromosome instance segmentation dataset. Finally, we conduct experiments to demonstrate the effectiveness of the proposed method.

2   Related Work

Chromosome Segmentation
As the chromosome segmentation plays a vital role in chromosome automatic karyotyping, so it attracts numerous researches to try it out.

Enea et al. [2] reviewed and implemented a variety of thresholding strategies applied to human chromosome segmentation. According to the conclusion of [2], over-segmentation and under-segmentation are common phenomena in thresholding strategies methods because it is very difficult to set the threshold value for those methods applied in various situations.

Shervin et al. [3] proposed a geometric-based method for separation chromosome clusters when touching and overlapping chromosome clusters were detected automatically. The authors applied their method to a database containing 62 touching and partially overlapping chromosomes and a success rate of 91.9% is achieved. However, their method is a customized method based on the geometric features of the existing chromosome database, which means the doubtful effectiveness of the clinical application.

R. Lily et al. [4] proposed a neural network-based image segmentation method to the problem of distinguishing between partially overlapping chromosomes. Their method achieved intersection over union (IOU) scores of 94.7% for the overlapping region and 88.94% on the non-overlapping chromosome regions. However, their training and testing chromosome images are semi-synthetically generated. The authors did not report the effectiveness of their method in clinical chromosome application.

Tanvi Arora [9] proposed a human metaphase chromosome images segmentation approach of using region based active contours. The author claimed that this method has been tested on Advanced Digital Imaging Research (ADIR) dataset and yielded quite good performance. However, their method is based on the features active contours in ADIR. Therefore, the clinical effectiveness is still uncertain.

Chromosome Classification

In previous studies, there are some works [1,5] for chromosome classification. In these studies, the best chromosome classification accuracy was 98.9% reported by Yulei [5] in their private dataset. However, the author did not disclose their dataset for others to verify their results and also did not report sensitivity and specificity which are two important metrics for chromosome classification.

Additionally, all the above methods are 24-classification which are designed for identifying each of 22 autosomes, X, and Y sex chromosomes.

Chromosome Automatic Karyotyping

Reem et al. [10] proposed an automatic segmentation method for chromosome cells using difference of gaussian (DoG) as a sharpening filter. However, all chromosome cell images appeared in this paper have no overlapping chromosomes or touching chromosomes, which is impractical in clinical applications.

Yirui et al. [8] proposed an end-to-end chromosome karyotyping method based on generative adversial networks. Their approach can automatically detect, segment and classify chromosomes from original cell images. However, according to their reported paper, the chromosome brands are too nebulous to recognize by geneticists. It is extremely incorrect and irresponsible to make genetic diseases and birth defect diagnoses based on virtually generated karyotypes.

So far, studies that can effectively tackle the problem of chromosome automatic karyotyping are still very scarce. Motivating by the limitations of the above methods for solving the chromosome auto-karyotyping task, we proposed our chromosome auto-karyotyping framework.

3   Proposed Method

The proposed method, illustrated by Fig. 2, consists of four stages, the global binary threshold segmentation, mixed classification, chromosome cluster instance segmentation, and karyotyping operation stage. In this section, we go through the detail of the above stages.

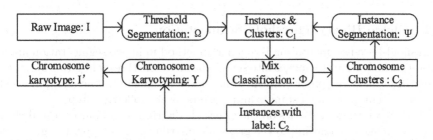

Fig. 2. The plot depicts the operation flow of the proposed method. Forgiven a chromosome cell microphotograph raw image I, the global threshold-based segmentation operation Ω separates I into image slices set C_1 including instances and clusters. Whereafter, the chromosome mixed classifier Φ classifies all instances from C_1 into chromosome instance set C_2 with their corresponding labels, and chromosome clusters C_3. All image slices from C3 are segmented into C1 by the instance segmentation operation Ψ. When C_1 and C_3 are empty, it means that are clusters have been segmented into instances with corresponding labels. Finally, the chromosome karyotyping operation Υ generates the karyotype I' from C_2.

3.1   Problem Formalization

We use the symbol I to denote a metaphase cell image depicted by Fig.1(a) Similarly, we apply symbol I' to denote a chromosome karyotype illustrated by Fig. 1(b). We let symbols $C1$, $C2$, and $C3$ represent the images set of chromosome instances & clusters, chromosome instances, and chromosome clusters. We use symbols Ω, Ψ to denote the operation of the global binary threshold segmentation and chromosome instance segmentation. Furthermore, we use symbols Φ and Υ to denote mixed chromosome classification and karyotyping operation.

Global Binary Threshold Segmentation: According to the above symbols, we depict the global binary segmentation procedure of I as $C1 = \Omega(I)$.

Mixed Chromosome Classification: The chromosome mixed classification operation Φ identifies the chromosome instance images in $C1$, and labels these instance images with corresponding chromosome number tags. All labeled chromosome instance images will be moved from $C1$ to $C2$, and the rest images in $C1$ are seen as chromosome clusters, illustrated by $C2 = \Phi(C1)$. Therefore, the mixed chromosome classification operation for $C1$ can be described as $C3 = C1 - C3$.

Instance Segmentation: The chromosome instance segmentation operation Ψ is used for separating a chromosome cluster images into multiple chromosome instance images. The instance segmentation process can be described as $\forall C \in C3 \to C1 = C1 \cup \Psi(C), C3 = C3 - C$.

Chromosome karyotyping: The chromosome karyotyping operation Υ refers to the process of generating a karyotype image I' by arranging chromosome instances of $C2$ which can be formalized as $I' = \Upsilon(C2)$.

3.2 Global Binary Threshold Segmentation

Threshold-based segmentation is a general method in image segmentation applications, which separate images into several parts by the threshold grayscale value. We utilize the global binary threshold method for segmenting the metaphase cell image into chromosome instances and clusters by the following steps.

Firstly, we make statistics on the grayscale distribution of our collected chromosome images. According to the grayscale distribution, we apply the grayscale threshold value 250 to segment a chromosome image into slices. When the total pixels of a slice less than 260, the slice will be seen as non-chromosomal objects and removed. Finally, we save the rest image slices as chromosome instance and cluster images for further processing.

3.3 Mixed Chromosome Classification

We build the mixed chromosome classification model based on the ResNeXt [11] backbone with weakly supervised pre-train weights [12].

The ResNeXt backbone is an orderly stack of regular neural network modules called building blocks as Fig. 3 depicted. A usual building block is assembled by multiple paths of convolutional neurons. The total quantity of assembled paths is termed by *cardinality*. The channel amount of the convolutional neurons in the building block is the width of the block, termed by the symbol d. We use F to denote the input and output tensor dimensions of the building block. In our mixed chromosome classifier, we apply ResNeXt 101-32 × 16d as the concrete backbone network, which means the *cardinality* of the building block is 32 and the *width* is 16. To improve the applicability of the model in the chromosome classification application, we design 9 layers of neurons as the customized header adding to the backbone network as the output of the classifier. Since there are 22 pairs of autosomes and a pair of sex chromosomes (X and Y), we have 24 categories of chromosome instances. Meanwhile, we take all chromosome clusters as an extra category, therefore the output of the classification model is an one-hot vector with the shape of 25 × 1. Table 1 shows the detail of the mixed chromosome classification model.

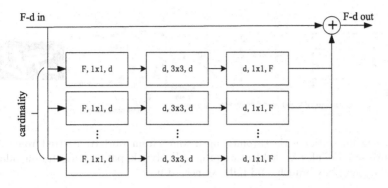

Fig. 3. This figure describes a standard building block assembled by multiple paths of convolutional neurons. The symbol F denotes the dimension of the input x_i and output y_i while the d represents the total input or output channels number of each inner convolution neuron. The *cardinality* means the total number of aggregated neuronal paths.

Table 1. The complete architecture of the chromosome classification model

Block name	Detail	Output shape
conv1	conv2d(7×7,64,stride=2)	$112 \times 112 \times 3$
conv2	maxpooling(3×3,stride=2) block(F=128,d=16,c=32)$\times$3	$56 \times 56 \times 128$
conv3	block(F=256,d=16,c=32)$\times$4	$28 \times 28 \times 256$
conv4	block(F=512,d=16,c=32)$\times$23	$14 \times 14 \times 512$
conv5	block(F=1024,d=16,c=32)$\times$3	$7 \times 7 \times 1024$
customized header	Pooling2d() Flatten() BatchNormal1d(4096) Dropout(0.25) Linear(in=4096,out=512) Relu() BatchNormal1d(512) Dropout(0.5) Linear(in=512,out=25)	25×1

3.4 Chromosome Instance Segmentation

Motivated by the achievements of the PANet [13] in the COCO instance segmentation challenge competition, we transfer this model for solving the task of chromosome instance segmentation, illustrated by Fig. 4. The PANet model consists of five sub-modules: a feature proposal network(FPN), a path augmentation sub-module, an adaptive feature pooling sub-module, a box predict branch and a fully-connected fusion sub-module.

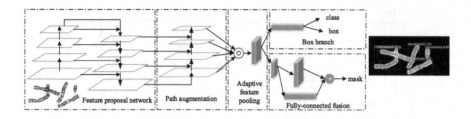

Fig. 4. The illustration is the chromosome instance framework transferred from PANet. It has five sub-modules named feature proposal network, path augmentation, adaptive feature pooling, box branch, and fully-connected fusion.

The input of the chromosome instance segmentation framework is an entire chromosome cluster image with 300 × 300 pixels. The output of the framework consists of two parts. The output of the framework consists of two parts. The first part of the output is the chromosome instances of proposal boxes and their corresponding probabilities belong to the chromosome. Another output is a pixel mask for each chromosome instance. We separate chromosomes through their respective masks.

3.5   Chromosome Karyotyping

After the preceding processing, we have gathered all chromosome instances in a cell image and their corresponding categories. In the chromosome karyotyping stage, we only need to arrange chromosome instances into a karyotype according to an international system for cytogenetic nomenclature (ISCN) criterion [14].

4   Experiments

4.1   Implementation Details

We implement the global binary threshold segmentation using OpenCV-Python library. According to statistics of the clinical chromosome grayscale, we set up the grayscale segmentation threshold to 225. Meanwhile, we set up the threshold of the minimal pixels filter to 260 which will filter candidate images whose pixels are lower the threshold.

We build the mixed chromosome classification model with PyTorch which is an open source machine learning framework. We adapt the discriminative learning rate [15] to help the model obtain better convergence. First, we load the WSL pre-trained weights [12] as the initial weights of the ResNeXt backbone. Second, we freeze the weights of the backbone and train our customized header with the learning rate at *slice(1e-5, 4e-5)*, 100 maximum epochs. We monitor the training loss and apply early-stopping when the training is no more decrease in the last 5 epochs. Finally, we unfreeze the weights of the backbone and fine-tune the whole model in 100 epochs at *slice(1e-6, 4e-6)* learning rate. Furthermore, we apply the early-stopping strategy at training loss indicator with 5 epochs patience.

We transfer the chromosome instance segmentation framework and its implementation from PANet [13] where is originally designed for the COCO instance segmentation task. We adapt ResNet50 [16] as the FPN backbone and load the COCO pre-trained wights to the model as initial weights for chromosome instance segmentation task. To alleviate the overfitting of the model, we add a *Dropout* layer with a 0.25 drop out rate before the last fully-connected fusion layer.

4.2 Dataset Description

This study is motivated by the challenges of the clinical work on genetic disease diagnostic at Guangdong Women and Children Hospital where we obtain 500 privacy-removal clinical metaphase cell images and corresponding karyotypes manually done by skilled geneticists.

We construct a mixed classification dataset with 130 images of each autosome labeled from 0 to 21, and 98 X chromosome images labeled to 22, and 32 Y chromosome images labeled to 23. At the same time, we add 4876 chromosome cluster images with label 24 into this dataset. All chromosome images are padded to 224 × 224 pixels.

We annotate 882 chromosome cluster images as the chromosome instance segmented dataset manually. In the chromosome instance segmented dataset, we randomly select 20% images as test image and other as training data. To mitigate the overfitting of the instance segmentation model, we conduct a series of data augmentations in training images.

Firstly, We do a horizontal and vertical flip for individual image and its mask in the training set. Secondary, we rotate all images and their masks in the training dataset every 15°.

4.3 Experimental Evaluation Metrics and Results

As the global binary threshold segmentation stage and chromosome karyotyping stage are engineering tasks, and their operation results are determined. Therefore, the experimental evaluation focus on the chromosome mixed classification and the instances segmentation stages.

Evaluation of Mixed Chromosome Classification: In the mixed classification stage, we apply *accuracy, F1, sensitivity,* and *specificity* general metrics to quantitatively evaluate the overall performance of the classification model.

To evaluate the stability of the proposed classification model, we conduct all evaluation experiments utilizing cross-validation by K-Folds. The original chromosome mixed classification dataset is divided into 5 folds by random stratified sampling. One fold data is used as validation data and the other four folds as training data in turn. Table 2 gives the classification evaluation results of the proposed framework on the clinical dataset in five runs.

According to Table 2, the proposed classification model yields $99.53 \pm 0.23\%$ *accuracy*, $98.85 \pm 0.89\%$ *F1* value, $98.82 \pm 0.86\%$ true positive rate (*sensitivity*), and $99.98 \pm 0.01\%$ true negative rate (*specificity*), which means that our proposed classification method promising for solving the mixed classification problem. Meanwhile, the *kappa* value with $99.23 \pm 0.47\%$ means that the proposed model works well even in the unbalanced dataset.

Table 2. Summary results of chromosome mixed classification.

	$Acc(\%)$	$F1(\%)$	$TPR(\%)$	$TFR(\%)$	$kappa(\%)$
1	99.68	98.88	**98.81**	99.48	99.48
2	99.36	98.37	98.46	98.96	98.96
3	99.36	98.33	98.25	98.96	98.96
4	**99.68**	**99.41**	98.38	**99.99**	**99.48**
5	99.56	99.26	99.21	99.98	99.26
mean	99.53	98.85	98.82	99.98	99.23
std	0.23	0.89	0.86	0.01	0.47

Evaluation of Chromosome Instance Segmentation. To evaluate the performance of the instance segmentation model, we adopt $Accuracy_m$ as the major evaluation metric which is calculated by the total number of correct segmented instances dividing by the number of total instances in the chromosome cluster set. When the bounding box of the predicted instance has equal or more than 50% intersection over union (IoU) with a ground-truth instance bounding box, we regard this instance as a correct segmented instance. Meanwhile, we follow the metrics of AP_{50}, one of standard evaluation metrics for instance segmentation tasks [13], for quantitatively evaluating the precision of the instance segmentation. Finally, we introduce $Accuracy_t$ as the general segmentation evaluation metric for our proposed multiple stages segmentation method where $Accuracy_t$ is computed by the number of all correct segmented instances dividing by the number of all ground-truth instances in both global binary threshold segmentation stage and instance segmentation stage.

Table 3 summarizes the quantitative evaluation results. According to the results, we can draw the following conclusions. First, without data augmentation, the instance segmentation model yields a promising result in the verify set while performs poorly in the test set, which demonstrates the overfitting of the model and the necessity of data augmentation. Second, though small performance gaps in verify set and test set still exist, the proposed method obtains promising results with 90.81% $Accuracy_t$, 85% $Accuracy_m$, and 90.63% AP_{50} precision.

To better depict the effectiveness of the instance segmentation model, we show several examples in Fig. 5. Chromosome instances in various cluster images are separated precisely by the instance segmentation model.

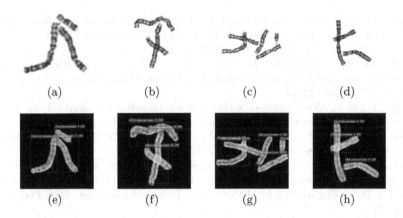

Fig. 5. The plots in the first row are original chromosome cluster images while the plots in the second row are their corresponding instance masks predicted by the instance segmentation model.

Table 3. Summary results of chromosome instance segmentation.

	$Accuracy_t$	$Accuracy_m$	AP_{50}
Verify Set(No Aug)	98.35%	96.80%	97.8%
Test Set (No Aug)	61.52%	55.33%	65.86%
Verify Set (Aug)	92.25%	87.50%	96.25%
Test Set (Aug)	90.81%	85.00%	90.63%

5   Conclusion

In this paper, we proposed a multi-stages chromosome segmentation and mixed classification method for chromosome automatic karyotyping. According to the experimental results on the clinical chromosome dataset, the proposed classification model obtains a promising result ($99.53 \pm 0.23\%$ accuracy) for solving the mixed chromosome classification problem. Meanwhile, the chromosome segmentation result is quite encouraging (90.81% accuracy). The above results demonstrate the promises for solving the clinical chromosome auto-karyotyping problem currently.

References

1. Lin, C., et al.: CIR-Net: automatic classification of human chromosome based on inception-ResNet architecture. IEEE/ACM Trans. Comput. Biol. Bioinform. (2020). https://doi.org/10.1109/TCBB.2020.3003445
2. Poletti, E., Zappelli, F., Ruggeri, A., Grisan, E.: A review of thresholding strategies applied to human chromosome segmentation. Comput. Methods Prog. Biomed. **108**(2), 679–688 (2012)

3. Shervin, M., Mehran, F., Babak, H.K.: A geometric approach to fully automatic chromosome segmentation. In: 2014 IEEE Signal Processing in Medicine and Biology Symposium (SPMB), pp. 1–6. IEEE (2014)
4. Lily Hu, R., Karnowski, J., Fadely, R., Pommier, J.-P.: Image segmentation to distinguish between overlapping human chromosomes. arXiv preprint arXiv:1712.07639 (2017)
5. Qin, Y., et al.: Varifocal-net: a chromosome classification approach using deep convolutional networks. IEEE Transactions on Medical Imaging (2019)
6. Ni, W., Wang, X., Song, W., Li, Y. (eds.): WISA 2019. LNCS, vol. 11817. Springer, Cham (2019). https://doi.org/10.1007/978-3-030-30952-7
7. Khan, S., DSouza, A., Sanches, J., Ventura, R.: Geometric correction of deformed chromosomes for automatic karyotyping. In: 2012 Annual International Conference of the IEEE Engineering in Medicine and Biology Society, pp. 4438–4441. IEEE (2012)
8. Wu, Y., Yue, Y., Tan, X., Wang, W., Lu, T.: End-to-end chromosome karyotyping with data augmentation using GAN. In: 2018 25th IEEE International Conference on Image Processing (ICIP), pp. 2456–2460. IEEE (2018)
9. Arora, T., Dhir, R.: A novel approach for segmentation of human metaphase chromosome images using region based active contours. Int. Arab J. Inf. Technol. **16**(1), 132–137 (2019)
10. Bashmail, R., Elrefaei, L.A., Alhalabi, W.: Automatic segmentation of chromosome cells. In: Hassanien, A.E., Tolba, M.F., Shaalan, K., Azar, A.T. (eds.) AISI 2018. AISC, vol. 845, pp. 654–663. Springer, Cham (2019). https://doi.org/10.1007/978-3-319-99010-1_60
11. Xie, S., Girshick, R., Dollár, P., Tu, Z., He, K.: Aggregated residual transformations for deep neural networks. In: Proceedings of the IEEE Conference on Computer Vision and Pattern Recognition, pp. 1492–1500 (2017)
12. Dhruv, M., et al.: Exploring the limits of weakly supervised pretraining. In: Proceedings of the European Conference on Computer Vision (ECCV), pp. 181–196 (2018)
13. Shu, L., Lu, Q., Qin, H., Shi, J., Jia, J.: Path aggregation network for instance segmentation. In: Proceedings of the IEEE Conference on Computer Vision and Pattern Recognition, pp. 8759–8768 (2018)
14. Shaffer, L.G., Jean, M.-J., Schmid, M., et al.: ISCN 2013: an international system for human cytogenetic nomenclature (2013). Karger Medical and Scientific Publishers (2013)
15. Yang, Y., Igor, G., Boris, G.: Large batch training of convolutional networks. arXiv preprint arXiv:1708.03888 (2017)
16. He, K., Zhang, X., Ren, S., Sun, J.: Deep residual learning for image recognition. In: Proceedings of the IEEE Conference on Computer Vision and Pattern Recognition, pp. 770–778 (2016)

An End-to-End Deep Neural Network
for Truth Discovery

Huafeng Chen[1,2], Yongquan Dong[1,2(✉)], Qing Gu[1,2], and Yali Liu[1,2]

[1] School of Computer Science and Technology, Jiangsu Normal University,
Xuzhou 221000, China
s_dimple@163.com, tomdyq@163.com, gdq@qq.com,
1204240062@qq.com
[2] Xuzhou Engineering Research Center of Cloud Computing,
Xuzhou 221000, China

Abstract. With the rapid growth of web data, information about the same target gathered from multiple sources often exhibits conflicts. This problem motivates the need for truth discovery, which is to automatically resolve conflicts and find the truth from multiple conflicting claims. Existing truth discovery methods are mainly based on iterative updates or probability models. A common limitation of these methods is that their models are complex to be built. In this paper, we propose a concise end-to-end deep neural network for truth discovery, which regards the task as a classification problem. Firstly, for each target, we extract a unique claim, and for each unique claim, we construct a source-unique-claim vector depending on whether the source provides this value. Then on the training dataset, we label the vector as true/false according to the ground truth. Finally, we use a deep neural network to build a classification model for each target to judge which claim is the truth. Experimental results on two real-world datasets show that our proposed model has better performance than existing state-of-the-art methods.

Keywords: Truth discovery · Deep learning · Data fusion · Information extraction

1 Introduction

At present, the Web has become a significant way for most people to obtain information. However, the information collected from multiple sources is usually noisy and conflicting. For example, several web sites have provided several different numbers on the death of the Paris terrorist attacks, some of which are vague. Another example is that even for the same flight, different flight booking websites provide different boarding gates on the same date [1, 2]. As the unreliability in declared claims and sources is increasing, the need to derive the true claims on conflicting data from multiple sources is becoming an urgent task. It has stimulated the generation and research of truth discovery technology. Truth discovery focuses on getting the most reliable information from multi-source data. It has facilitated many NLP tasks, such as knowledge representation, question answering, and information retrieval. These tasks are all based on reliable information extracted from multiple sources. In the era of big

© Springer Nature Switzerland AG 2020
G. Wang et al. (Eds.): WISA 2020, LNCS 12432, pp. 377–387, 2020.
https://doi.org/10.1007/978-3-030-60029-7_35

data, truth discovery is of great significance for extracting credible information and improving data quality.

A variety of methods have been proposed to solve truth discovery. According to a survey on truth discovery [3], many of the previous algorithms are based on manually defined rules to iteratively calculate and update the reliability of the source and the credibility of the claim. We also find that all the proposed truth discovery models utilized a voting mechanism, which conforms to the general principle of truth discovery. These methods can be divided into the following four categories: 1) Iterative methods [1, 4] design truth discovery as an iterative process, in which the steps of claim credibility estimation and source reliability calculation are iterated until convergence. 2) Probabilistic graph methods [5–8] use hyper-parameters to capture external knowledge. 3) Optimization-based methods [9–12] rely on setting an optimization function that can learn the relations between sources' qualities and claims' credibility with an iterative method to compute two sets of parameters jointly. 4) Machine learning methods [13–15] use different estimators to address truth discovery with large-scale data.

Though showing certain effectiveness in some applications, there are some limitations in existing truth discovery methods. First, most truth discovery methods begin with a uniform weight among all sources. The source reliability initialization strategies are inaccurate because they do not use existing prior knowledge to calculate the reliability of sources, which will lead to suboptimal results. Second, the calculation methods and iteration rules of these traditional models are too complicated and depend on many hyper-parameters. In particular, the variation of the parameters of each truth discovery algorithm has a significant influence on the quality of the algorithm. Third, previous works usually return the claim confidence score as a direct output, but people prefer to know what the true claim is.

In this paper, we propose a model based on an end-to-end neural network (EENN) to address the above limitation. We assume that a source is likely to provide true information with equal probability to all targets. Most truth discovery methods are also based on this consistency assumption. Firstly, for each target, we extract a unique claim, and for each unique claim, we construct a source-unique-claim vector. Then on the training dataset, we label the vector as true/false according to the ground truth. Finally, we use an end-to-end neural network to build a classification model for each target to infer which claim is the truth. Experiments on real-world datasets show that our EENN model outperforms the state-of-the-art methods.

2 Problem Statement

In this paper, we consider a general truth discovery problem for numerical data. Before introducing the model, we first introduce important definitions and define the truth discovery problem. Then, we use an example on the stock database (Table 1) to illustrate these concepts.

Definition 1. A *target* is a thing of interest, and the truth of a *target* is defined as its accurate information, which is unique.

Definition 2. A *source* describes the place where the information about targets can be collected.

Definition 3. A *claim* is the data describing a target from a source, and such a claim can be either true or false.

Definition 4. The truth discovery problem includes targets, sources, and claims. For a set of targets T that we are interested in, related claims can be collected from a set of sources S. The truth discovery aims to find the true claim for each target $t \in T$ by resolving the conflicts among the claims from different sources $s \in S$.

As an example, let us consider two targets: the opening price and the rate of change of stock symbol AAPL on a specific day. The claims for these two targets provided by the four sources are shown in Table 1.

Table 1. The opening price and the rate of change of AAPL on a specific day

Sources	t1 AAPL-Change%	t2 AAPL-Open Price
s1	0.25	22.26
s2	0.24	22.29
s3	0.25	22.26
s4	0.25	21.03

The sets of unique claims for two targets are {22.26, 22.29, 21.03} and {0.24, 0.25}. For each target, the truth discovery aims to infer which unique claim is true automatically. Figure 1 shows a graphical representation of the truth discovery problem based on the example mentioned above.

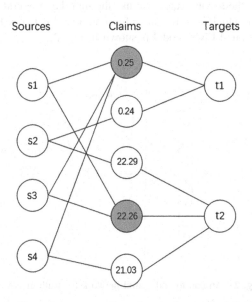

Fig. 1. A graph representation of truth discovery of Table 1.

3   An End-to-End Neural Network Model

In this section, we first introduce an EENN model for truth discovery. Then we introduce the central functions of the proposed model.

3.1   EENN Framework

Suppose there is a target t whose data are collected from M sources denoted by $s_1, s_2, \ldots, s_M$. These sources provide N unique claims denoted by $c_1, c_2, \ldots, c_N$. For each unique claim $c_j (1 \leq j \leq N)$, there is a corresponding source-unique-claim vector $v_j = [v_{1j}, v_{2j}, \ldots, v_{Mj}]$ where $v_{ij} = 1$ if source s_i provides unique claim c_j and $v_{ij} = 0$ otherwise. If the unique claim c_j is the ground truth, the label of c_j is marked as "true", otherwise it is marked as "false". For example, Table 2 provides vector representations of the data in Table 1.

Table 2. Representations of unique claims of Table 1

Targets	Unique claims	Source-unique-claim vector	Label
t1	c1 = 0.25	[1, 0, 1, 1]	True
	c2 = 0.24	[0, 1, 0, 0]	False
t2	c1 = 22.260	[1, 0, 1, 0]	True
	c2 = 22.290	[0, 1, 0, 0]	False
	c3 = 21.030	[0, 0, 0, 1]	False

In the first stage, we mainly construct a vector for each unique claim in the manner described above. In the second stage, we use the multilayer neural network to build a classification model, which is called an end-to-end neural network(EENN) model. The main framework of our EENN model is shown in Fig. 2.

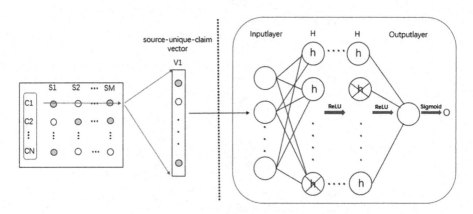

Fig. 2. An end-to-end neural network for truth discovery

3.2 Model Principle

In our EENN model, from the input layer to the first hidden layer, we first use a fully connected layer to learn the dependence between the unique claims and multiple sources. The first hidden layer is of bigger sizes than the input layer, and the second hidden layer has fewer neuron nodes than the previous layer. Then to prevent the model from overfitting, we utilize the dropout layer. In the hidden layers, we use ReLU as the activation function, which provides simple derivatives during the error backpropagation process, allows deeper networks to converge faster, and prevents the vanishing gradient effectively. The ReLU function is shown in Eq. (1).

$$ReLU = \begin{cases} x & x > 0 \\ 0 & x \leq 0 \end{cases} \tag{1}$$

The core of our model is to train the weight matrix W between each layer. Through forward propagation computation, we obtain the output of the model, which is the credibility of the unique claim, expressed as a probability value by activation function sigmoid. The output of our EENN model is as follows:

$$Z = w^T x + b \tag{2}$$

$$output = sigmoid(Z) = \frac{1}{1 + \exp(-Z)} \tag{3}$$

where Z is obtained by the output layer and we use sigmoid to get the final output. The w^T represents the weight matrix from the last hidden layer to the output layer, x represents the output of the previous layer, and b represents the bias. We use a binary classification cross-entropy loss function to measure the error between the predicted value and the truth:

$$loss(o,t) = \frac{1}{n} \sum_i (t_i * \log(o_i) + (1 - t_i) * \log(1 - o_i)) \tag{4}$$

where t_i is the label of the ith sample in the training set and o_i is the output of our model said above with respect to the ith sample.

The most important thing is that, through training, the weight matrix W and bias b of each layer are adjusted continuously along the direction of gradient descent until the training of the model is completed and the parameters converge. We choose a stochastic gradient descent optimization strategy to speed up the training of the model. Finally, for the set $\{c_1, c_2, \ldots, c_N\}$ which describes the same target, a unique claim with the highest probability value will be regarded as the truth. We think that the complex dependence between the reliability of the source and the credibility of the claim can be learned based on the end-to-end neural network.

4   Experiment and Analysis

In order to verify the effectiveness of our end-to-end neural network for truth discovery, experiments are conducted on public datasets. We first introduce the datasets and evaluation criteria, then describe the baseline approaches briefly. Finally, giving the experimental results and analysis.

4.1   Datasets

Stock Dataset [16]
Li et al. [1] collected trading data of 1000 stock symbols from 55 sources on every workday in July 2011. They also provided the gold standard of 100 kinds of stock data. The dataset contains 1000 stock symbols with 16 properties. Li et al. [15] considered the attributes of Volume, Shares outstanding, and Market Cap as a continuous type, and other attributes are classified as categorical type. We follow their example, and only the categorical data are selected for our experiment. Note that a target is a property of a specific stock symbol on a particular day in this dataset.

Population Dataset [17]
The dataset provides demographic data collected by multiple sources for different U.S. cities in different years. The gold standard contains 301 values.

Table 3 shows the statistics of the datasets.

Table 3. Statistics of the two real-world datasets

	Sources	Claims	Targets	Ground truths
Stock Dataset	55	12056684	336000	30870
Population Dataset	4264	49955	41196	308

4.2   Evaluation Criteria

In order to compare with the previous truth discovery methods, we choose the same evaluation criteria as Li et al. [15]. We focus on five measures. Note that we finally multiply the four measures of precision, recall, accuracy, and F1 by 100%, respectively, for better representation in the Figure.

- **Error rate:** The error rate is the percentage of the wrong prediction of categorical data. The lower the error rate, the better a method performs.
- **Precision**: The percentage of the predicted claims that are consistent with a gold standard.
- **Recall**: The percentage of the claims in the ground truth table being output as correct.
- **Accuracy**: The percentage of true claims and false claims correctly predicted.
- **F1**: $\frac{2 \times Precision \times Recall}{Precision + Recall}$

4.3   Baseline Methods

We compare the following baseline methods with our model:

Voting: This method chooses the claim with the highest occurrence as the predicted truth.

Investment [18]: In this algorithm, a source 'invest' its reliability in the claims it offered uniformly.

PooledInvestment [18]: Compared with Investment, PooledInvestment differs in that it scales the credibility linearly.

2-Estimates [19]: This approach uses a single truth hypothesis. It takes complementary voting by assuming that 'each target has only one real value'.

3-Estimates [19]: It increases by 2- estimate by considering the difficulty of obtaining the credibility of each claim.

TruthFinder [4]: TruthFinder adopts Bayesian analysis iteration to estimate source reliability and infer truth. This method also puts forward the hypothesis of source consistency and the concept of 'implication', which is used to adjust the vote of value by considering the influences between claims.

AccuSim [20]: It also uses Bayesian analysis. In order to obtain the similarity of claims, the implication function is introduced.

CRH [11]: CRH is a model for dealing with heterogeneous datasets with categorical and continuous data, and the estimation of source reliability can be performed jointly between all data types.

FFMN [15]: This method incorporates memory mechanisms and resolves the truth discovery by using a memory network-based model to learn the reliability of sources and predict the credibility of claims.

FBMN [15]: It is similar to the FFMN model. The difference is that FBMN is a backpropagation memory network. This model considers source reliability as latent knowledge and merges it in the composition of the hidden layer representation.

4.4   Experiment Results and Analysis

In experiment 1, we first carry out strict data cleaning and standardization on the stock dataset. For each target, we normalize claims to the same format. It is vital to normalize duplicate records from multiple sources [21]. Due to the enormous redundant data in the stock dataset, we eliminate repeated claims, building source-claim vectors with unique claims only. As we find that the ground truth also has invalid information, such as null values, we discard it and then use it to label each the source-unique-claim vector. Finally, our EENN model acts as a classifier to predict that the probability of each unique claim may be true.

In the training process, the main parameter settings of our 4-layer EENN model are as follows: activation = 'relu', activation = 'sigmoid'(output), dropout = 0.2, learning rate = 0.001, loss function = "BCEloss", momentum = 0.78. In order to ensure that the evaluation is effective, we conducted three sets of experiments by dividing the 21 days data into three groups randomly, and one group contains seven days of data. In particular, one group is reserved as the test set of the verification model, and the others are used for training. Finally, the average value of the three experimental results is taken as the performance of the model. We compare the performance of all algorithms by assessing their error rates in classifying true and false claims. The results are shown in Table 4.

Table 4. Performance comparison on stock dataset

Categorical	Method	Error rate
Previous method	Voting	0.0817
	Investment	0.0983
	PooledInvestment	0.0990
	2-Estimates	0.0726
	3-Estimates	0.0818
	TruthFinder	0.1194
	AccuSim	0.0726
	CRH	0.0700
Memory NN	FFMN	0.0207
	FBMN	0.0644
Our Model	**EENN**	**0.0026**

Experiment 2 studies the impact of different data scales on the model. We sample the data from 21 days for 1 day, 5 days, 10 days, 15 days, and all the data. We conduct a series of experiments to calculate the average results. The experimental results are shown in Fig. 3.

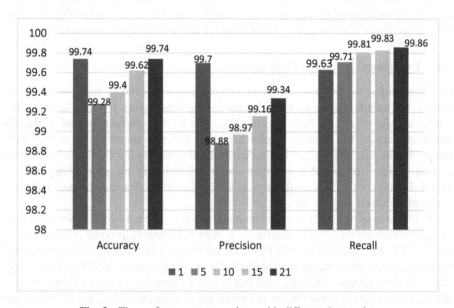

Fig. 3. The performance comparison with different data scales

In order to analyze the effect of the number of iterations on the results of our model, we try epochs from 2 to 20 in experiment 1. We observe that our EENN model takes only a few iterations to converge. The results are shown in Fig. 4.

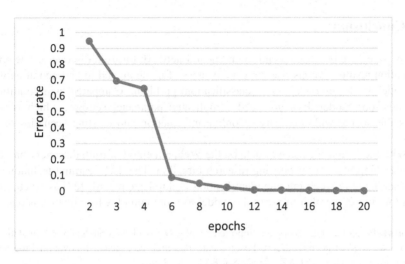

Fig. 4. Results under different iterations

In experiment 3, we use the population dataset for comparison to illustrate the applicability of the data of different data distribution of our model. The baseline method refers to the homepage [22]. Figure 5 shows the results.

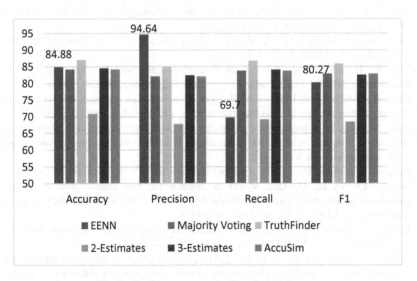

Fig. 5. Performance on the Population dataset.

We can observe that our proposed model EENN has the best performance on the stock dataset in Table 4 and shows the highest precision on the population dataset in Fig. 5, which verifies that our method outperforms the baseline methods.

5 Conclusion

In this paper, we propose an end-to-end neural network for truth discovery. The above evaluation results demonstrate the effectiveness of using an end-to-end neural network to model truth discovery as a classification problem. Compared with traditional methods, our model does not need complicated parameter settings, which is more applicable to the wide diversity of online information and existing scenarios on the Web.

With the increasing scale of data on the Web, a method of truth discovery based on deep learning is a very promising research direction. The deep learning technique has also facilitated research into the identification of duplicate records [23] associated with it. In the future, we plan to apply the model to more complex heterogeneous datasets.

Acknowledgments. This work is supported by the National Natural Science Foundation of China (No. 61872168, 61702237), Postgraduate Research & Practice Innovation Program of Jiangsu Province (No. KYCX20_2382, No. KYCX20_2396).

References

1. Li, X., Dong, X.L., Lyons, K., Meng, W., et al.: Truth finding on the deep web: is the problem solved? Proc. VLDB Endowment **6**(2), 97–108 (2012)
2. Dong, X.L., Saha, B., Srivastava, D.: Less is more: selecting sources wisely for integration. Proc. VLDB Endowment **6**(2), 37–48 (2012)
3. Li, Y., Gao, J., Meng, C., Li, Q., et al.: A survey on truth discovery. Proc. ACM SIGKDD Explorations Newsletter **17**(2), 1–16 (2016)
4. Yin, X., Han, J., Philip, S.Y.: Truth discovery with multiple conflicting information providers on the web. IEEE Trans. Knowl. Data Eng. **20**(6), 796–808 (2008)
5. Zhao, B., Rubinstein, B.I., Gemmell, J., Han, J.: A Bayesian approach to discovering truth from conflicting sources for data integration. Proc. VLDB Endowment **5**(6), 550–561 (2012)
6. Zhao, B., Han, J.: A probabilistic model for estimating real-valued truth from conflicting sources. In: Proceedings of the 10th International Workshop on Quality in Databases, pp. 1–7. In conjunction with VLDB (2012)
7. Pasternack, J., Roth, D.: Latent credibility analysis. In: WWW, pp. 1009–1020 (2013)
8. Ma, F., Li, Y., Li, Q., Qiu, M., et al.: FaitCrowd: fine grained truth discovery for crowdsourced data aggregation. In: Proceedings of the 21th ACM SIGKDD International Conference on KDD, ACM, pp. 745–754 (2015)
9. Li, Q., Li, Y., Gao, J., Su, L., et al.: A confidence-aware approach for truth discovery on long-tail data. Proc. VLDB Endowment **8**(4), 425–436 (2014)
10. Aydin, B.I., Yilmaz, Y.S., Li, Y., Li, Q., Gao, J., et al.: Crowdsourcing for multiple-choice question answering. In: Proceedings of the Twenty-Sixth Annual Conference on Innovative Applications of Artificial Intelligence, Association for the Advancement of Artificial Intelligence. pp. 2946–2953 (2014)

11. Li, Q., Li, Y., Gao, J., Zhao, B., et al.: Resolving conflicts in heterogeneous data by truth discovery and source reliability estimation. In: Proceedings of the ACM SIGMOD Int. Conf. Manage. Data, pp. 1187–1198 (2014)
12. Li, Y., Li, Q., Gao, J., Su, L., et al.: On the discovery of evolving truth. In: Proceedings of 21th ACM SIGMOD International Conference on KDD, ACM, pp. 675–684 (2015)
13. Xiao, H., Gao, J., Li, Q., Ma, F., et al.: Towards confidence in the truth: a bootstrapping based truth discovery approach. In: Proceedings of the 22th ACM SIGKDD International Conference on KDD, pp 1935–1944 (2016)
14. Lyu, S., Ouyang, W., Wang, Y., Shen, H., et al.: Truth discovery by claim and source embedding. IEEE Trans. Knowl. Data Eng. 1 (2019)
15. Li, L., Qin, B., Ren, W., Lin, T.: Truth discovery with memory network. Tsinghua Sci. Technol. 22(6), 609–618 (2017)
16. Luna Homepage, http://www.lunadong.com/fusionDataSets.htm. Accessed 10 June 2020
17. Pasternack, J., Roth, D.: Knowing what to believe (when you already know something). In: COLING.ACL, pp. 877–885 (2010)
18. Pasternack, J., Roth, D.: Making better informed trust decisions with generalized fact-finding. In: International Joint Conference on Artificial Intelligence (2011)
19. Galland, A., Abiteboul, S., Marian, A., Senellart, P.: Corroborating information from disagreeing views. In: ACM International Conference on Web Search & Data Mining, pp. 131–140 (2010)
20. Dong, X.L., Berti-Equille, L., Srivastava, D.: Integrating conflicting data: the role of source dependence. Proc. VlDB Endowment 2(1), 550–561 (2009)
21. Dong, Y., Dragut, E.C., Meng, W.: Normalization of duplicate records from multiple sources. IEEE Trans. Knowl. Data Eng. 31(4), 769–782 (2019)
22. Dafna Homepage. http://da.qcri.org/dafna/#/dafna/home_sections/home.html. Accessed 05 June 2020
23. Gu, Q., Dong, Y., Hu, Y., Liu, Y.: A method for duplicate record detection using deep learning. In: Ni, W., Wang, X., Song, W., Li, Y. (eds.) WISA 2019. LNCS, vol. 11817, pp. 85–91. Springer, Cham (2019). https://doi.org/10.1007/978-3-030-30952-7_10

Sliding Covariance Matrix: Co-learning Spatiotemporal Geometry Feature for Skeleton Based Action Recognition

Guan Huang, Qiuyan Yan[✉], and Guan Yuan

School of Computer Science and Technology,
China University of Mining and Technology, Xuzhou, China
yanqy@cumt.edu.cn

Abstract. The covariance matrix is a generic feature representation in vision applications. It can accurately and efficiently capture geometric features of Riemannian manifold especially in the condition of data size is medium-scaled. When the covariance matrix is applied in describing skeleton data, how to represent spatial and temporal relations of skeleton joints, meanwhile ensuring the matrix is nonsingular is a challenging problem. In this work, we first propose a sliding window-based frame appending model acquiring a nonsingular covariance matrix descriptor for all skeleton frames. Then, sliding covariance matrixes for all sliding windows are sequentially fed to the modified Long Short-Term Memory (LSTM) network for extracting the spatiotemporal characteristics and action recognition. The proposed method is verified by the experiments on five medium-sized skeleton datasets and the results show that the proposed method improves the accuracy by 6%–20% compared to the state-of-the-art models. Meanwhile, the experiment results clarify that when the data size is not so large, our proposed method can describe spatiotemporal characters of skeleton data more accurately and efficiently than deep network methods.

Keywords: Sliding covariance matrix · Long short-term memory network · Human action recognition · Skeleton data

1 Introduction

Human action recognition refers to the description and understanding of human action by computers. Besides, human action recognition represents a multidisciplinary cross-research topic, and it has been widely applied to human-computer interaction, monitoring systems, and medical diagnosis. The existing human action recognition technology still has many limitations caused by different action scenes, highly similar actions, and changes in light intensity, thus making the research on human action recognition technology an urgent and challenging problem that needs to be addressed.

With the development of the depth sensor technology, such as Kinect sensors, the three-dimensional (3D) skeleton-based human action recognition has attracted great research attention anew. The human body can be considered as an articulated system consisted of connected rigid segments, and the skeleton activity data are in a Riemannian manifold. Since the deep learning networks have achieved state-of-the-art

© Springer Nature Switzerland AG 2020
G. Wang et al. (Eds.): WISA 2020, LNCS 12432, pp. 388–400, 2020.
https://doi.org/10.1007/978-3-030-60029-7_36

learning ability in the non-Euclidean space, several recent works suggested employing the deep learning methods to learn the spatial-temporal features of the skeleton automatically. However, the deep networks require abundant data to train the model for obtaining the optimal values of hyperparameters, which is unfeasible in some practice.

The covariance matrix has recently received increasing attention in computer vision by leveraging Riemannian geometry symmetric positive-definite (SPD) matrices. It can effectively fuse multiple geometry features and be efficiently calculated via integral images. Compared with application in traditional RGB images, the covariance matrix is applied in skeleton data rarely during to only represent the spatial relationship of joints. Nevertheless, kinds of the hierarchical structure described for time property of skeleton activity are proposed. However, the covariance matrix has shortcomings of being prone to be singular when the number of feature vectors smaller than the dimensions, which might result in lower accuracy in modeling complicated features. In that case, to utilize Riemannian metrics, a small-scaled matrix has to be appended. To this end, few hierarchical structures consider the appending schema.

In this paper, a sliding window-based appending model is proposed for obtaining the nonsingular covariance matrices to represent the manifold geometry of skeleton data in a continuous-time dimension. The covariance matrices are then sequentially fed into the Long Short-Term Memory (LSTM) network to learn the temporal characteristics of the activity and conduct the skeleton-based action recognition. As shown that when the data size is not so large, our proposed method can describe spatiotemporal characters of skeleton data more accurately and efficiently than state-of-the-art compared methods.

2   Related Work

In this section, we briefly review the existing literature that closely relates to the proposed model. Wang et al. [1] use the spatial and temporal dictionaries of the parts to represent actions, which can capture the spatial structure of human body and movements. Vemulapalli et al. [2] utilize rotations and translations to represent the 3D geometric relations of body parts in Lie group, and then employ Dynamic Time Warping (DTW) and Fourier Temporal Pyramid (FTP) to model the temporal dynamics. Chaudhry et al. [3] encode the skeleton structure with a spatial-temporal hierarchy, and exploit Linear Dynamical Systems to learn the dynamic features. Luo et al. [4] develop a novel dictionary learning method combined with Temporal Pyramid Matching, to keep the temporal dynamics. Wang et al. [5] extract the local occupancy patterns from the appearance around skeleton joints, and then process with FTP to obtain temporal structure. Cho and Chen [6] perform action recognition with a hybrid multi-layer perception. In the above methods, the local temporal dynamics is generally represented within a certain time window or differential quantities, it cannot capture the relationships between joints.

The combination of neural network and RNN can classify sequences directly. Baccouche et al. [7] propose a LSTM-RNN to recognize actions by obtaining sequential representations with a 3D convolutional neural network. Grushin et al. [8] use LSTM-RNN for action recognition by regarding the histograms of optical flow as inputs.

Lefebvre et al. [9] propose a bidirectional LSTM-RNN with one forward hidden layer and one backward hidden layer for gesture classification. Moreover, except the study of Du et al. [10], all the work above just uses RNN-LSTMs as a classifier. While we propose a solution including both feature leaning and sequence classification.

3 Proposed Method

In the field of action recognition based on the human skeleton data, the depth-sensing devices, such as Kinect sensors, are commonly used to obtain 2D or 3D coordinates of human skeletons to represent skeleton sequences. Early methods for motion recognition from skeletal data are based on a modeling of the skeleton's movement as time series, however, these methods have limited capabilities to represent the high correlations existing between the movement of two adjacent gestures.

The SPD matrices have been shown effective in various vision tasks. The covariance matrix is a second-order descriptor by leveraging Riemannian geometry symmetric positive-definite (SPD) matrices. However, the covariance matrix has shortcomings of prone to be singular and limited capability in modeling temporal feature relationships. In this paper, a sliding window-based frame appending model is proposed for obtaining the nonsingular covariance matrices to represent the manifold geometry of skeleton data in a continuous-time dimension. The covariance matrices are then sequentially fed into the Long Short-Term Memory (LSTM) network to learn the temporal characteristics of the activity and conduct the skeleton-based action recognition.

3.1 Sliding Window-Based Frame Appending Model

Different from the RGB dataset, a skeleton dataset usually has the dimensions of 20, while the number of skeleton frames per action instance only ranges from 15 to 120, might far from being enough to estimate a reliable covariance matrix, i.e. incur the singularity of the covariance matrix. When the scale of the dataset is small, this problem is serious. To ensure the covariance matrix of samples in each sliding window nonsingular, we propose a sliding window-based skeleton frame appending model. The model structure is clarified in Fig. 1.

Assume the length of the skeleton sequence is L frames, the number of skeleton joints is n, the length of the sliding window is w, the step size of the window movement is t. We select each skeleton joint as a feature, to ensure the number of feature vector larger or equal to the feature dimension, the w is usually assigned equal to n. When frames in the current window are not enough, the p frames are randomly select from this current window, the appending number p is assigned to $p = w \times C - (w - t) \times (C - 1) - L$, C is the number of sliding windows $C = L / t$. Because the window size is relatively small, randomly selected frames have little influence on the accuracy and contain the spatial and temporal relationship of joints.

For example, there is an instance that has $L = 30$ frames and the number of skeleton joints is 20. Originally, this instance only can produce one window and corresponds to one covariance matrix satisfying nonsingular. When using our sliding window-based frame appending model, and sliding step $t = 2$, this instance generates

$C = 15$ sliding windows and corresponding covariance matrices. For the last window, $p = 18$ frames will be randomly selected and appended to this window. When action dataset sale is small, this model not only can increase sample size but also ensure the non-singularity of the covariance matrix.

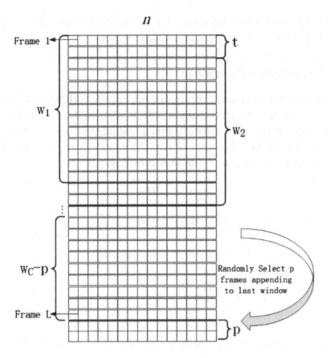

Fig. 1. The structure of sliding window-based frame appending model

3.2 Sliding Covariance Matrix Computation

The feature-based covariance has been introduced into object monitoring and texture classification methods in the computer vision field. It has also been successfully applied to object tracking, shape modeling, and face recognition. However, restricted by the limited representation ability on temporal attribution, it rarely is applied to action recognition in realistic and dynamic scenes, especially in 3D skeleton data.

In this work, the original skeleton data represented by a three-dimensional coordinate sequence, i.e. each skeleton joints of the human body is represented by (x, y, z) coordinate. To take the value of x, y, z as a whole to describe a skeleton joint, we select the coordinate of each joint as a feature to compute the covariance matrix, and the number of feature vectors equals the width of the sliding window, the calculation formula is as Eq. 1. This covariance matrix depicts the spatial distribution of all joints in an action frame, and the matrix sequence generated by the sliding window-based frame appending model maintains the spatial and temporal property of skeleton data simultaneously. The covariance matrix computed by the sliding window-based frame appending model is named as Sliding Covariance Matrix (SCM).

$$C = \frac{1}{n-1}\sum\nolimits_{i=1}^{n} (J_{mi} - \mu_m)(J_{mi} - \mu_m)^T \tag{1}$$

Where J_{mi} represents the mth joints of the ith skeleton frame, and μ_m represents the mean of the mth joints of all frames in a sliding window.

3.3 SCM-Based Action Recognition

The LSTM represents an improved recurrent neural network (RNN) architecture, which mitigates the vanishing gradient effect of the RNN. An LSTM neuron contains a memory cell c_t which has a self-connected recurrent edge of weight. At each time step t, the neuron can choose to write, reset, or read the memory cell governed the input gate i_t, forget gate f_t, and output gate o_t, as shown in Eqs. (2)–(7).

$$i_t = \sigma_g(W_i x_t + U_i h_{t-1} + b_i) \tag{2}$$

$$f_t = \sigma_g\left(W_f x_t + U_f h_{t-1} + b_f\right) \tag{3}$$

$$\tilde{c}_t = \sigma_h(W_c x_t + U_c h_{t-1} + b_c) \tag{4}$$

$$o_t = \sigma_g(W_o x_t + U_o h_{t-1} + b_o) \tag{5}$$

$$c_t = f_t * c_{t-1} + i_t * \tilde{c}_t \tag{6}$$

$$h_t = o_t * \sigma_h(c_t) \tag{7}$$

Where the initial values are $c_0 = 0$ and $h_0 = 0$ and the operator $*$ denote the Hadamard product (element-wise product). The subscript t indexes the time step. The input vector to the LSTM unit is x_t. The forget gate's activation vector is f_t. The input gate's activation vector is $i_t \in R^h$. The output gate's activation vector $o_t \in R^h$. The hidden state vector also known as output vector of the LSTM unit is $h_t \in R^h$. $\tilde{c}_t \in R^h$ is the cell input activation vector. $c_t \in R^h$ is the cell state vector. $W \in R^{h \times d}$, $U \in R^{h \times h}$ and $b \in R^h$ refer to weight matrices and bias vector parameters that need to be learned during training. The superscripts d and h refer to the number of input features and the number of hidden units, respectively. σ_g is the sigmoid function. σ_c is the hyperbolic tangent function. σ_h is the hyperbolic tangent function. To take full advantage of the human skeleton data in action recognition, an optimized LSTM model is proposed as in Fig. 2. The SCMs extracted from sliding windows are fed to the LSTM sequentially. In the optimized LSTM network architecture, there are three LSTM networks, six fully-connected layers, and four dropout layers.

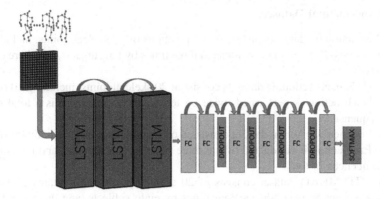

Fig. 2. Optimized LSTM Network Architecture

Several LSTM networks can extract more temporal features of SCMs. Fully-connected layers are added to the traditional LSTM network, which would produce the higher-order representation of temporal features that are more easily separable into different classes. However, a fully-connected layer occupies most of the parameters. Neurons develop the co-dependency on each other during the training, which limits the individual ability of each neuron, thus leading to the overfitting. To prevent overfitting, some dropout layers are added to the network structure. During this phase, a set of neurons is chosen randomly. Besides, the incoming and outgoing edges to a dropped-out node are also removed. In the experiment, we verify that the architecture of three LSTMs, six Fully-connected layers, and four dropout layers has the best performance on experiment datasets.

4 Experiments

For the convenience of expression, in the following, the proposed LSTM model based on the **SCM** is denoted as **CMSW-LSTM**. Each fully connected layer with a ReLU activation function. A dropout rate of 0.5 was used on the Dropout layers. The LSTM layer contains 100 units. The fully connected layers contain 2048, 1024, 512, 128, 128, and 32 neurons, respectively. During the phase of training, the proposed method is trained for 80 epochs. The proposed method was evaluated on five datasets including MSR-Action3D (MSR for short), UTKinect-Action3D (UTK for short), Florence 3D (Florence for short), UTD-MHAD (UTD for short) and a real classroom student action dataset (Classroom for short). We use ADAM optimizer for training, with an initial learning rate of 1×10^{-4} and a minibatch size of 16. To verify the effectiveness of the **SCM**, the proposed model is compared with the LSTM model that used the original skeleton sequence as an input. In order to evaluate the performance of the **CMSW-LSTM** model, it is compared with some state-of-the-art skeleton data-based models: RA-GCN [11], SPDNet [12], Lie Group [2], Co-LSTM [13], and STA-LSTM [14].

4.1   Experimental Datasets

The MSR-Action3D dataset consists of 3D positions of 20 skeletal joints, including 20 types of actions. Each action is conducted three times by ten subjects, and there are 557 action sequences.

The UTKinect-Action3D datasets consist of 20 skeletal joints, including 10 types of actions. Each action is performed twice by ten subjects, and there is a total of 199 action sequences.

The Florence 3D Action dataset consisted of 15 skeletal joints, including 9 types of actions. Each action is conducted two to three times by ten subjects, and there are 215 action sequences in total.

The UTD-MHAD dataset consists of 20 skeletal joints and contains 27 types of actions. Each action is conducted four times by eight subjects, and there is a total of 816 action sequences.

To test the effect of the proposed method in real scenes, we collected data on the actions of students in the classroom through kinect2, as shown in Fig. 3. The classroom dataset consists of 25 skeletal joints, including 7 actions, and each action is done three times. There are ten experimental subjects, and each action lasts about 13 to 15 s.

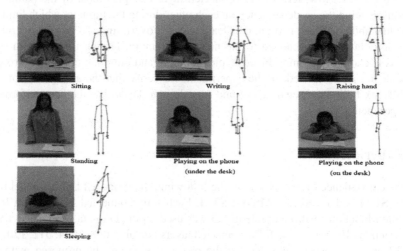

Fig. 3. Human actions in the classroom dataset

The specific method of collecting classroom data is as follows. The data was obtained by a fixed Kinect device. The Kinect base was placed approximately 90 cm from the ground, and the horizontal distance from the Kinect camera to the subject was 120 cm, as shown in Fig. 4. The desk height of the student participants was 76 cm, and the chair height was 45 cm, as shown in Fig. 5. The dataset consists of seven actions performed by ten participants, and each participant performed each action three times. All seven movements that were captured denote the movements with a higher frequency in a classroom, including sitting, writing, raising a hand, standing, using a mobile phone under the table, using a mobile phone on the table, sleeping on the table.

Fig. 4. Data collection process. The horizontal distance between the Kinect camera and a student was 120 cm

Fig. 5. Illustration of the data collection setup. The desk height was 76 cm, and the chair height was 45 cm

4.2 Experimental Results

Firstly, we represent the recognition accuracy of the CMSW-LSTM model on every action of all datasets in confusion matrices. The confusion matrix represents a summary of the prediction results of a classification model. After the training, the confusion matrices of the CMSW-LSTM model are presented in Figs. 6, 7, 8, 9 and 10.

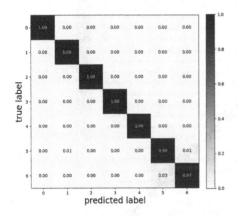
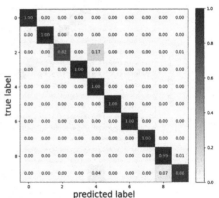

Fig. 6. Confusion matrix on the Classroom Dataset

Fig. 7. Confusion matrix on the UTK Dataset

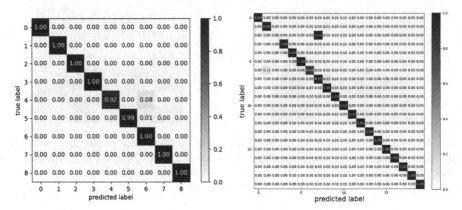

Fig. 8. Confusion matrix on the Florence Dataset

Fig. 9. Confusion matrix on the MSR Dataset

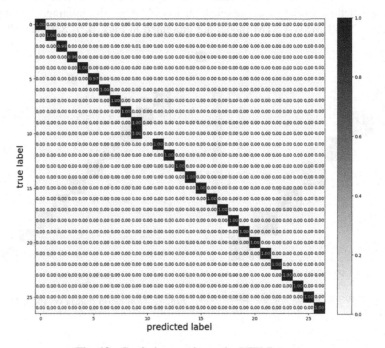

Fig. 10. Confusion matrix on the UTK Dataset

From the results are presented in confusion matrices, it is clear that the **CMSW-LSTM** method has the best accuracy in the diagonal of all confusion matrix, which means almost all actions in the dataset can be classified accurately.

Secondly, to compare the performance of **CMSW-LSTM** with state-of-the-art skeleton-based classification model, the compared results of CMSW-LSTM and other models on different datasets are given in Table 1. The boldface of the values means the best-obtained results for each dataset. In this experiment, the structure of **CMSW-LSTM** is set as in Fig. 2, i.e. three LSTM layers, six fully-connected layers, and four dropout layers, which is testified the best structure on recognition accuracy.

Table 1. Accuracy of different models on different datasets

Dataset	CMSW-LSTM	RA-GCN	SPDNet	LSTM	Lie Group	Co-LSTM	STA-LSTM
Florence	**97.21**	81.36	49.296	63.38	91.40	78.24	80.29
UTD	**99.36**	79.26	50.617	81.28	92.36	76.35	79.56
UTK	**97.69**	89.23	67.742	61.29	97.20	80.21	79.26
MSR	**98.56**	56.74	47.749	66.83	90.37	77.32	81.16
Classroom	**98.39**	76.51	32.258	43.67	56.40	62.58	68.37

As shown in Table 1, the **CMSW-LSTM** model achieved the best results on all datasets. The conclusion is that the LSTM model inputted with the **SCM** can effectively extract the spatial and temporal relations of skeleton joints by efficiently capture geometric features of Riemannian manifold, much better than the LSTM without the covariance and time window (LSTM, Co-LSTM, STA -LSTM), geometry learning model (Lie Group, SPDNet), and graph learning model (RA-GCN), thus achieving a significant accuracy improvement compared to the state-of-the-art models.

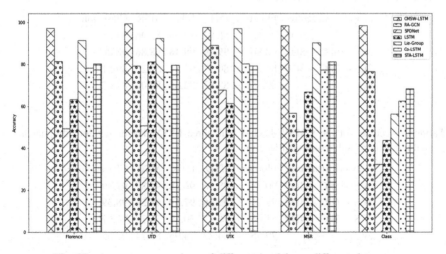

Fig. 11. Accuracy comparison of different models on different datasets

To show the results more intuitively, we illustrate the compared results in Table 1 in Fig. 11. As shown in Fig. 11, the accuracy of the **CMSW-LSTM** model on all five datasets was significantly higher than the compared methods. The second place is the Lie Group method. The results verify that learning the spatial distribution of skeleton data is more effective and feasible than a deep learning network when data size is small or medium scale.

Lastly, to show that different network structures of the proposed model yield different performances, we clarify the accuracy of the **CMSW-LSTM** with the different structures on different datasets in Table 2, 3 and 4. LS represents the LSTM layer, FC represents the fully connected layer, and DP represents the dropout layer. For example, 3LS + 5FC + 4DP represents three LSTM layers, five fully connected layers, and four Dropout layers. From Table 2, 3 and 4 we can see that the network with three LSTM layers, six fully connected layers, and four dropout layers has the best accuracy in three out of five datasets.

Table 2. Accuracy of the **CMSW-LSTM** with six fully-connected layers and four dropout layers (varied numbers of LSTM layers)

Network structure	Florence	UTD	UTK	MSR	Classroom
2LS + 6FC + 4DP	93.31	83.08	93.94	93.83	94.41
3LS + 6FC + 4DP	**97.21**	**99.36**	**97.69**	**98.56**	**98.39**
4LS + 6FC + 4DP	74.48	73.96	98.90	60.48	96.50

Table 2 shows the accuracy of the network with six fully connected layers, four dropout layers, and the varied number of LSTM layers (2,3 and 4). Three LSTM layers have the best accuracy in all five datasets.

Table 3. Accuracy of the **CMSW-LSTM** with three LSTM layers and three dropout layers (varied numbers of fully-connected layers)

Network structure	Florence	UTD	UTK	MSR	Classroom
3LS + 5FC + 3DP	84.66	86.82	87.93	85.96	74.65
3LS + 6FC + 3DP	**93.86**	**99.49**	**98.46**	**98.86**	**98.86**
3LS + 7FC + 3DP	89.54	80.17	91.55	82.25	89.51
avg	89.35	88.83	92.65	89.02	87.67

Table 4. Accuracy of the **CMSW-LSTM** with three LSTM layers and four dropout layers (varied numbers of fully-connected layers)

Network structure	Florence	UTD	UTK	MSR	Classroom
3LS + 5FC + 4DP	90.38	88.56	96.38	94.87	97.89
3LS + 6FC + 4DP	**97.21**	**99.36**	**97.69**	**98.56**	**98.39**
3LS + 7FC + 4DP	51.05	66.25	90.78	75.78	96.67
avg	79.55	84.72	94.95	89.74	97.65

Then we analyze the accuracy of the network with three LSTM layers, three (Table 3) or four (Table 4) dropout layers, and varied numbers of fully-connected layers (5, 6 and 7). As shown in Table 3 and Table 4, compared with five and seven fully-connected layers in two groups, the network contains four fully-connected layers outperforms on five datasets.

Table 5. Accuracy of the **CMSW-LSTM** with three LSTM layers and six fully-connected layers (varied dropout layers)

Network structure	Florence	UTD	UTK	MSR	Classroom
3LS + 6FC + 3DP	93.86	**99.49**	**98.46**	**98.86**	**98.86**
3LS + 6FC + 4DP	**97.21**	99.36	97.69	98.56	98.39
3LS + 6FC + 5DP	79.50	66.15	93.30	46.39	85.76

Lastly, we compare the accuracy of the network with a different number of dropout layers. From Table 2, 3 and 4, we find three LSTM layers and six fully-connected layers outperform other network architectures. To investigate the effect of dropout layers, we compare the accuracy of the different number of dropout layers. As shown in Table 5, the network contains three fully-connected layers outperforms on four out of five datasets. However, to make the overall performance comparison more intuitionistic, we compute the average accuracy of the network with 3DP and 4DP for each data set in Table 3 and Table 4. We find that the network with 4DP has the best overall performance.

5 Conclusion

The paper studies the skeleton-based human action recognition using the covariance matrix. A covariance matrix computed model by sliding window-based frame appending called **SCM** and an optimized LSTM architecture called **CMSW-LSTM** is proposed. The **SCM** is used to extract the short-term temporal and spatial features, which are then passed to the LSTM network to perform temporal modeling. The results show that the proposed method achieves a significant improvement in accuracy compared to the state-of-the-art methods.

References

1. Wang, C., Wang, Y., Yuille, A.L.: An approach to pose-based action recognition. In: Proceedings of the IEEE Conference on Computer Vision and Pattern Recognition, pp. 915–922 (2013)
2. Vemulapalli, R., Arrate, F., Chellappa, R.: Human action recognition by representing 3d skeletons as points in a lie group. In: Proceedings of the IEEE Conference on Computer Vision and Pattern Recognition, pp. 588–595 (2014)

3. Chaudhry, R., Ofli, F., Kurillo, G., Bajcsy, R., Vidal, R.: Bio-inspired dynamic 3d discriminative skeletal features for human action recognition. In: Proceedings of the IEEE Conference on Computer Vision and Pattern Recognition Workshops, pp. 471–478 (2013)

4. Luo, J., Wang, W., Qi, H.: Group sparsity and geometry constrained dictionary learning for action recognition from depth maps. In: Proceedings of the IEEE international conference on computer vision, pp. 1809–1816 (2013)

5. Wang, J., Liu, Z., Wu, Y., Yuan, J.: Mining actionlet ensemble for action recognition with depth cameras. In: 2012 IEEE Conference on Computer Vision and Pattern Recognition, pp. 1290–1297. IEEE (2012)

6. Cho, K., Chen, X.: Classifying and visualizing motion capture sequences using deep neural networks. In: 2014 International Conference on Computer Vision Theory and Applications (VISAPP), pp. 122–130. IEEE (2014)

7. Baccouche, M., Mamalet, F., Wolf, C., Garcia, C., Baskurt, A.: Sequential deep learning for human action recognition. In: Salah, A.A., Lepri, B. (eds.) HBU 2011. LNCS, vol. 7065, pp. 29–39. Springer, Heidelberg (2011). https://doi.org/10.1007/978-3-642-25446-8_4

8. Grushin, A., Monner, D.D., Reggia, J.A., Mishra, A.: Robust human action recognition via long short-term memory. In: The 2013 International Joint Conference on Neural Networks (IJCNN), pp. 1–8. IEEE (2013)

9. Lefebvre, G., Berlemont, S., Mamalet, F., Garcia, C.: BLSTM-RNN based 3D gesture classification. In: Mladenov, V., Koprinkova-Hristova, P., Palm, G., Villa, A.E.P., Appollini, B., Kasabov, N. (eds.) ICANN 2013. LNCS, vol. 8131, pp. 381–388. Springer, Heidelberg (2013). https://doi.org/10.1007/978-3-642-40728-4_48

10. Du, Y., Wang, W., Wang, L.: Hierarchical recurrent neural network for skeleton based action recognition. In: Proceedings of the IEEE Conference on Computer Vision and Pattern Recognition, pp. 1110–1118 (2015)

11. Song, Y.-F., Zhang, Z., Wang, L.: Richly activated graph convolutional network for action recognition with incomplete skeletons. In: 2019 IEEE International Conference on Image Processing (ICIP), pp. 1–5. IEEE (2019)

12. Huang, Z., Van Gool, L.: A riemannian network for spd matrix learning. In: Thirty-First AAAI Conference on Artificial Intelligence (2017)

13. Zhu, W., et al.: Co-occurrence feature learning for skeleton based action recognition using regularized deep LSTM networks. In: Thirtieth AAAI Conference on Artificial Intelligence (2016)

14. Song, S., Lan, C., Xing, J., Zeng, W., Liu, J.: Spatio-temporal attention-based LSTM networks for 3D action recognition and detection. IEEE Trans. Image Process. **27**, 3459–3471 (2018)

Web Table Column Type Detection Using Deep Learning and Probability Graph Model

Tong Guo, Derong Shen[(⊠)], Tiezheng Nie, and Yue Kou

School of Computer Science and Engineering, Northeastern University,
Shenyang 110004, China
Shirley_GT@163.com, {shenderong,nietiezheng,
kouyue}@cse.neu.edu.cn

Abstract. The rich knowledge contains on the web plays an important role in the researches and practical applications including web search, multi-question answering, and knowledge base construction. How to correctly detect the semantic types of all the data columns is critical to understand the web table. The traditional methods have the following limitations: (1) Most of them rely on dictionary lookup and regular expression matching, and are generally not robust to dirty data; (2) They only consider character data besides numeric data which accounts for a large proportion; (3) Some models take the characteristics of a single column and do not consider the special organizational structure of the table. In this paper, a column type detection method combining deep learning and probability graph model is proposed, taking the semantic features of a single column and the interaction between multiple columns into account to improve the prediction accuracy. Experimental results show that our method has higher accuracy compared with the state-of-the-art approaches.

Keywords: Web tables · Column type detection · Deep learning · Probability graph model

1 Introduction

Web tables are web content displayed in the form of tables, which can provide a large amount of high-quality data and have a wide range of topics. The effective use of massive amounts of high-quality table data is of great significance. There are currently many practical applications that utilize the rich semi-structured data resources of web tables, including question answering, table search, table expansion, knowledge base (KB) completion, semantic retrieval and the creation of linked open data (LOD) [1] and so on. The usefulness of web table data depends largely on the semantic understanding. One way to gain the semantics of a table is to match it with a KB, and the process is also called "table interpretion" [2]. The correct detection of the semantic types of table columns is vital for the task of table interpretation.

There have been many works for column type detection. Traditional methods based on search and ontology matching, which find the corresponding entities of the cells contained in the column and use the types corresponding to these entities as the result. These works are only applicable to the named-entity columns (NE-columns) and can't

© Springer Nature Switzerland AG 2020
G. Wang et al. (Eds.): WISA 2020, LNCS 12432, pp. 401–414, 2020.
https://doi.org/10.1007/978-3-030-60029-7_37

deal with the large amount of dirty data in the real-world data set. Another category of prior method uses a probabilistic model which is still based on the ontology matching of single cells, so it has the same disadvantages. Most of the research in recent years is based on feature engineering. A variety of features in web tables and knowledge bases are extracted to build supervised or unsupervised models. However, most of the features contains metadata like external web page information and column header information which are manually extracted, time-consuming and labor-intensive. At the same time, the accuracy is not high due to the neglect of the semantic association between the cells.

For example, Fig. 1 shows an example of one web table with unknown column types. The tags on the top of headers are the correct results after semantic type detection. The third header is missing, so the header description message cannot be used to determine the type of data in the column; the city "Ottawa" in the second row of the third column is incorrectly spelled "Ottaw". If only use the data of single column, it is possible to get the incorrect semantic type <location> instead of the correct semantic type <city> .

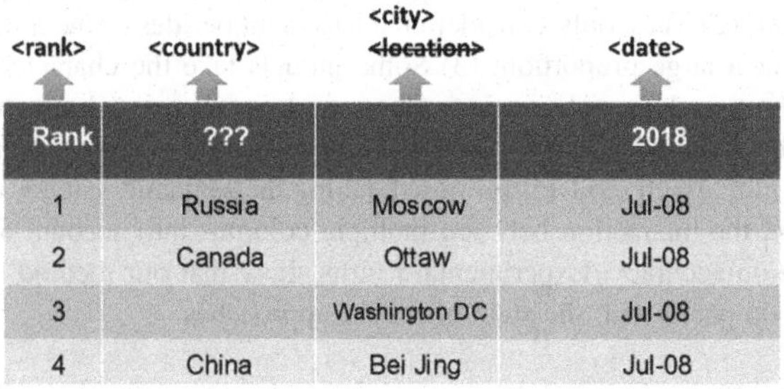

<rank>	<country>	<city> <del><location></del>	<date>
Rank	???		2018
1	Russia	Moscow	Jul-08
2	Canada	Ottaw	Jul-08
3		Washington DC	Jul-08
4	China	Bei Jing	Jul-08

Fig. 1. An instance of Web Table Column Type Detection

In view of the above reasons, this paper first synthesizes the contextual semantics of cells in a single column and uses the deep learning model to build a single-column classification model to pre-classify the semantic type, then uses the relationship between columns to comprehensively detect all the columns' types, hence further improve the accuracy of column type detection. Our contributions can be summarized as follows:

1. A method for detecting column types of web tables combining deep learning with probability graph model is proposed. This method can detect both of character data columns and numeric data ones at the same time without any metadata.
2. A single-column type detection model based on hybrid neural network is proposed. The deep learning model of BiGRU + Attention is used to further obtain deep level semantic relationship for improving prediction performance.
3. A multi-column type detection model based on the probability graph model is proposed, which comprehensively considers the co-occurrence relationship of various semantic types in the knowledge base and the real data set, and better utilizes the implicit semantic relationship between columns.

4. Through experiments, the proposed method performs better than other column type detection algorithms in applications.

2   Related Work

Column type detection of web tables constitutes table semantic interpretation tasks like attribute annotations and foreign key discovery [2]. Its research results can also provide labeled table data for tasks such as question answering [3], knowledge base completion [4] and table expansion [5]. The current methods can be divided into three categories: ontology-based methods, probabilistic methods, and featured-based methods.

The traditional methods take ontology matching as the core. [6] and [7] focus on instance-level matching. They find the matched candidate entity classes for each cell in the target column, then chooses the type with the highest frequency as the result. [8] uses word vectors to represent entities in cells and knowledge bases and then uses majority voting algorithm. However, these methods are limited to the low coverage of knowledge base and ignores the great differences between data from different sources, such as different naming and abbreviations.

Another kind of studies use probabilistic model. [9] used a probabilistic graphical model to express different degrees of matching relationships and annotate the table with a series of related random variables according to a suitable joint density distribution. Based on this work, [10] used a more lightweight graphical model. [11] weakened the assumption that there is a knowledge base correspondence between columns and entity sets, but strengthened the relationship with Wikipedia documents. [12] proposed a maximum likelihood inference model. This kind of methods ignore possible associations between cells.

In recent years, some researchers based on feature engineering have been presented. [13] used Kolmogorov-Smirnov test and TF-IDF to describe these types, [14] used more features, including Mann-Whitney test for numerical data and Jaccard similarity for text data to train logistic regression and random forest models. These methods rely on some external data such as the table title, so the detect results depend on the size of the table and the number of entity columns.

Compared with the existing work, this paper has the following advantages: (1) More consideration is given to the semantic information and don't rely on external data and metadata since this kind of data is often missing; (2) In order to solve the problems of dirty data, this article combines word embedding and character embedding to learn the distribution information of short sequences in words to a certain extent; (3) By selecting the appropriate deep learning model to better extract the deep semantic information of the text; (4) Based on the single-column classification model, our model combined with the probability graph model to extract the contextual characteristics and relationship characteristics between columns make the classification results more accurate.

3   Problem Statement and Definition

We first give the relevant definitions, and then elaborate on the problems.

Definition 1. Web table. The web table x consists of m rows and n columns, $\{C_1, C_2, \ldots, C_n\}$ represents the set of all columns, where each column is composed of a column header C_i^h and a group of cells, marked as $C_i = \{v_{i1}, v_{i2}, \ldots, v_{im}\}$, the data in the cell exists in the form of characters or numbers. Column headers and cells are allowed to be null.

Definition 2. KB. Given a knowledge base, define it as a six-tuple form: $KB = \{E, T, R, B, G, F\}$, where E, T, R represent a group of instances, a collection of types and the relationship between two entities; B is a collection of $<e, T>$ binary tuples, used to indicate that instance e belongs to a certain type T; G is a classification graph, representing the hierarchical structure between types, each directed edge is from a more general type to a more specific type, there is a subclass-of relationship between the parent class and the subclass; F is a collection of fact triples, representing the relationship fact or attribute fact between two entities.

Problem 1. Given a KB, for each table in the web table corpus (WTC), find the semantic type $Y_i \in T$ in KB for each column C_i in the table that best capture C_i's content.

The model proposed in this paper is mainly divided into two parts: single-column pre-classification model and multi-column type detection model. The overall framework is shown in Fig. 2. Firstly, the columns in the web table are divided into character and numeric columns, character column type detection model and numeric column type detection model are constructed respectively; The linear chain conditional random field in the graph model is used to construct the undirected graph model, where the state features are represented by the classification probability obtained by the single-column classification model, and the transition features are represented by the matrix obtained from the existing latent semantic relationship in the corpus and KB; Finally, the final classification result is obtained through calculation.

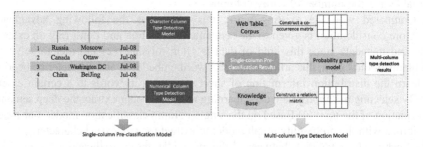

Fig. 2. The overall framework

4 Single-Column Pre-classification Model

We first propose a semantic classification model for single columns, which detects the semantic type based on the content of cells contained in the target column. This task can be formally defined as a multi-classification problem, assuming that a set of K predefined KB types which do not intersect with each other is given as $\{P_1, P_2, \ldots, P_k\}, P_i \in T$. Taking all cells in the column as input, we assign that each category P_i has an actual value score S_{P_i} through the classification model, so that the correct type of the target column has the highest score. We separate the columns into numeric and character ones and then build two classification models.

4.1 Character Column Type Detection Model

Embedding Layer. Existing research shows that there is inter-cell correlation in the web table [15]. In order to better extract the local features and context features of the data in the column, we use a sliding window of size H to randomly collect adjacent H cells on the target column and combine all the word sequences in these cells. The initial text is formed together, then M initial texts are obtained after collecting M times. Because there is no uniform standard for the structure of the network table, we give a fixed size value N as the specified text length.

Considering the diversity and irregularity of the sources of web tables and the low frequency of most entity words, large-scale data sets will inevitably have a large number of OOV (Out-Of-Vocabulary) words, so this paper uses a vector that combines character embedding and word embedding as input to the subsequent neural network model(see Fig. 3). Compared with the latest research such as the Sherlock model [16], which uses character statistics as a feature method, character-level representation is more flexible in handling spelling errors and rare word problems [17].

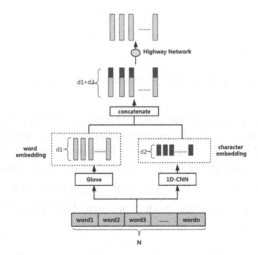

Fig. 3. Embedding model combined word embedding with character embedding.

For each initial text of length N, we use Glove and 1D-CNN (one-dimensional convolutional neural network) to get word embedding and character embedding about it, and then connect the two vertically to generate a matrix, and then pass it through Highway-NN (High Speed Neural Network) to get the embedding result. Only a small part of the input will be affected by Eq. (2), which is a single-layer feedforward neural network, and the rest is allowed to pass through the untransformed network.

$$z = t \odot g(W_H y + b_H) + (1 - t) \odot y \tag{1}$$

$$z = g(Wy + b) \tag{2}$$

Model Building Layer. Existing studies have used simple feed-forward NN [16], CNN [15] and BiRNN [19] to capture the features of text. However, these methods still can't fully learn the context information of the text and extract the deep-level information in the table content. Therefore, we propose a column type detection method based on BiGRU-Attention hybrid neural network model, as shown in Fig. 4. The model is divided into three parts: input layer, hidden layer and output layer.

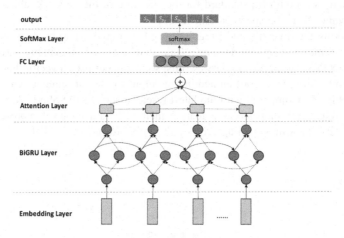

Fig. 4. BiGRU + Attention network model.

In embedding layer, through the word embedding and character embedding fusion model introduced above, a fixed-size matrix is obtained as the vectorized representation of the text of the target column, and then the text features are extracted through the hidden layer. BiGRU layer contains two unidirectional, opposite GRUs, respectively $(t-1)$ the output of the hidden layer state forward $\overrightarrow{h_t}$ and the output of the reverse hidden layer state $\overleftarrow{h_t}$, it also contains a neural network model composed of GRUs determined by the output of the two.

$$\begin{cases} \overrightarrow{h_t} = GRU\left(x_t, \overrightarrow{h_{t-1}}\right) \\ \overleftarrow{h_t} = GRU\left(x_t, \overleftarrow{h_{t-1}}\right) \\ h_t = w_t\overrightarrow{h_t} + v_t\overleftarrow{h_t} + b_t \end{cases} \tag{3}$$

We then put the output vector of the BiGRU layer into the Attention layer, which can further highlight the key information of the text to improve the quality of the feature extraction of the data layer of the hidden layer. The attention mechanism calculates the probability weight of the word vector, performs automatic weighted transformation on the data, and highlights the important words. The weight coefficient is specifically calculated, where u_w represents a randomly initialized attention matrix, h_t is the output vector of the BiGRU in the previous layer, and w_w is the weight coefficient.

$$\alpha_t = \frac{\exp\left(u_t^T u_w\right)}{\sum_t \exp\left(u_t^T u_w\right)}$$

$$u_t = \tan h(w_w h_t + b_w) \tag{4}$$

$$s_t = \sum_{i=1}^{n} \alpha_t h_t$$

Next we further stacked with a fully connected (FC) layer that learns the nonlinear relationship between input and output. Finally, the SoftMax function is used to perform the corresponding calculation on the input of the output layer to perform text classification. Get the output vector C_i of the target column $\overrightarrow{C_i} = \left\{S_{p_1}, S_{p_2}, \ldots, S_{p_k}\right\}$, K is the number of types in the predefined KB, each bit of C_i represents the probability that the column belongs to a certain semantic type.

4.2 Numerical Column Type Detection Model

Most of the cells in the numeric column only have a single numeric value, and generally do not contain too many context features, so we use a simple classification method based on statistical features. For example, statistics such as the mean, variance, median, mode, maximum, minimum, peak, skewness, and standard deviation of values.

In addition to numbers, the numeric column also has the possibility of characters. Taking Fig. 1 as an example, the second column has a cell of "Jul-08". These text messages greatly influence the judgment of column types, so we also extract features from these text messages. Statistics such as the frequency of occurrence of each letter, the mean and variance of the character length and the proportion of cells with characters. After extracting these values, they are modeled using classic machine learning algorithms such as random forest, the vector representation of probability $\left\{S_{p_1}, S_{p_2}, \ldots, S_{p_k}\right\}$ is also obtained.

5   Multi-column Type Detection Model

Through the single-column classification model proposed in Sect. 4, we can get a preliminary judgment on the semantic type. In order to make the result more accurate, we also need to consider the relationship between the semantic types of adjacent columns, and use the local context features between the columns to detect the column type.

5.1   CRF Model

In view of the above situation, this paper proposes a multi-column type detection based on a probability graph model to model the correlation between structural link variables to perform joint prediction. Considering that the interaction between the two columns does not have directionality, we use linear-CRF (linear chain conditional random field) on the entire network table, an undirected graph model that directly models the conditional probability.

$$score(C, y) = \sum_{k,i}^{n} \lambda_k \phi_{single}(C_i, y_i) + \sum_{l,i}^{n} \mu_l \phi_{multi}(y_{i-2} y_{i-1}, y_i) \tag{5}$$

$$P(y|C) = \frac{e^{score(C,y)}}{\sum_{y \in Y_c} e^{score(C,y)}} \tag{6}$$

C represents the sequence of the input table columns $\{C_1, C_2, \ldots, C_n\}$, Y represents the type sequence corresponding to each column of the output $\{Y_1, Y_2, \ldots, Y_n\}$, $Y_i \in T$ in KB, λ_k and μ_l are weight coefficients. $P(y|C)$ obtained after normalizing $score(C, y)$ represents the conditional probability that the predicted output sequence is Y when the input sequence C is given. The model defines the multi-column type detection problem as maximizing the joint conditional probability when the content in the table column is given. When the probability $P(y|C)$ reach the maximum, the corresponding Y is the semantic type sequence of the table columns. Figure 5 shows an instance.

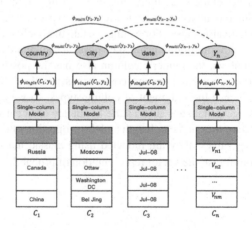

Fig. 5. Multi-column type detection model based on CRF.

5.2   Potential Functions

We define two potential functions for this model:

Status Feature Function $\phi_{single}(C_i, y_i)$. We use it to measure the influence of the feature obtained based on the column value on the semantic type detection of a column when a column is given, here we use the type probability value corresponding to the column $\{S_{p_1}, S_{p_2}, \ldots, S_{p_k}\}$ to represent. This probability can be obtained by the single-column model proposed in Sect. 4. The character sequence corresponds to the hybrid network model in 4.1, and the numerical sequence corresponds to the random forest model in 4.2.

Transfer Feature Function $\phi_{multi}(y_{i-2}, y_{i-1}, y_i)$. It depends on the status of the current column y_i, y_{i-2} and y_{i-1}, which is used to indicate the influence of the existing relationship between the columns on its semantic type detection. When the semantic relationship is closer, more likely to occur at the same time, $\phi_{multi}(y_{i-2}, y_{i-1}, y_i)$ has a higher value. In order to quantify this relation, we define a matrix Q of size K × K. This paper will calculate this value from two aspects.

First we consider to quantify the co-occurrence relationship extracted from the data set: give a marked web table, take $\{Y_1, Y_2, Y_3\}$ = {country, language, currency} as an example, co-occurrence type pairs are <country,language> , <language, currency > and <country, currency > . Traverse all tables to get matrix P_{corr}. In order to increase the accuracy and versatility of the algorithm, we use the semantic relationship marked in KB to further calculate P_{rela} (see Algorithm 1). Then Q_{ij} is the probability value obtained after normalizing the $P_{corr}(Y_i, Y_j)$ and $P_{rela}(Y_i, Y_j)$.

Algorithm 1: *ComputeRelation*
Input: KB, $P = \{ P_1, P_2, \ldots, P_k \}$, $P_i \in T$
Output: matrix P_{rela}
1. for each P_i in P
2.   { Properties} $\leftarrow$ search for P_i. $(is\ rdfs\text{:}domain\ of\ )$ /* Query the type's attribute */
3.   for each property in { Properties }
4.   {classes} $\leftarrow$ search for property. $(rdfs\text{:}range)$
5.   if $\exists\ P_j \in P$ in {classes}/* classes contains predefined semantic types */
                    $P_{rela}(P_i, P_j) = 1$
6.   { Properties} $\leftarrow$ search for P_i. $(is\ rdfs\text{:}range\ of\ )$
7.   for each property in {Properties }
8.   { classes} $\leftarrow$ search for property. $(rdfs\text{:}domain)$
9.   if $\exists\ P_l \in P$ in {classes}
                    $P_{relation}(P_i, P_l) = 1$
10. return P_{rela}

---

6   Experiments

6.1   Experiment Dataset

Two real-world data sets T2Dv2 and Limaye derived from the Gold Standard are used as the web table corpus in the experiment. The data set is json files obtained from various types of web pages. These tabular data manually marked by previous researchers which shows diversity in size, existence and sparsity of relationships. The specific information is shown in Table 1. The KB used in this article is DBPedia which is at the core of the LOD project.

Table 1. Statistics of Web Tables datasets.

Name	Tables	Avg. cells	Rows	Columns	Types	Coverage of entity
T2Dv2	779	124	84	5	65	30.5%
Limaye	428	23	34	4	421	25.8%

6.2   Comparative Experiment

In order to verify the feasibility of the model proposed in this paper, compare it with the following models: Lookup-Vote [5], T2K Match [18], ColNet [15], and Sherlock [16]. T2K Match is a collective matching method which first selects a series of candidate instances from the KB and then performs attribute matching based on the repeated pattern. ColNet integrates KB reasoning and lookup with machine learning and trains CNN model. Sherlock is a multi-layer neural network model.

6.3   Evaluation Metrics

In order to evaluate the performance of the model, we use the precision and recall rates obtained by comparing the real and predicted values and F1 score value as evaluation indicators to measure the effect of different models.

$$F1 = \frac{2 \times \text{precision} \times \text{recall}}{\text{precision} + \text{recall}} \tag{7}$$

6.4   Experimental Results

Overall Performance: We run the proposed model on T2Dv2 and Limaye repeatedly, and then compared the classification results with other models. The specific data under the three measurement indicators are shown in Table 2 and Table 3. It can be seen that in these methods, the overall classification effects of Lookup-Vote and T2K Match are the most unsatisfactory, because these two methods based on ontology matching rely heavily on whether the cell in the web table can find the corresponding entity in the KB. It can be seen from Table 1 that the entity coverage of the cells in the web table is very low, and the difference in performance on different data sets is significant.

Table 2. Results of models on T2Dv2.

Methods	Precision	Recall	F1 score
Lookup-Vote	0.862	0.821	0.841
T2K Match	0.624	0.727	0.671
ColNet	0.765	0.811	0.787
Sherlock	0.850	0.834	0.842
Proposed	0.903	0.871	0.887

Table 3. Results of models on Limaye.

Methods	Precision	Recall	F1 score
Lookup-Vote	0.732	0.660	0.694
T2K Match	0.560	0.408	0.472
ColNet	0.763	0.820	0.791
Sherlock	0.810	0.859	0.833
Proposed	0.864	0.841	0.852

ColNet and Sherlock, two feature-based methods, achieved better results by introducing convolutional neural networks and feedforward neural networks, but only considered the characteristics of single-column data. Our method takes the semantic characteristics of the single column and the influence of the relationship between the columns into account, so the overall performance on both data sets is significantly better than several other methods.

Relationship Between Columns Impact: In order to further verify the influence of the relationship between columns on the overall classification results, the experiment compared the results of the model proposed by Sherlock and the model proposed in this paper. In order to comprehensively measure the classification effect of the model, we separately calculate the evaluation indicators of Micro-F1, Macro-F1 and weighted-F1.

As shown in Fig. 6, the fusion of the multi-column detection model proposed in this paper can improve the classification accuracy, and the fusion of the multi-column detection model on the single-column model proposed in this paper can also improve the classification accuracy. It shows that the relationship between columns can affect the final classification effect.

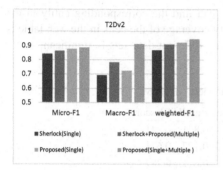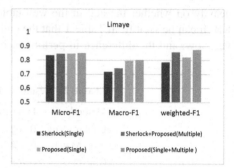

Fig. 6. Comparison of F1 scores of our algorithm and other algorithms on T2Dv2 and Limaye.

Sliding Window Size Impact: When embedding the text data in the character column, we use the sliding window method to combine adjacent cells together to form a fixed-size text. We set 5 different sizes to determine the effect of the sliding window size on the result. As can be seen from Fig. 7, the larger the sliding window, the higher the accuracy of the classification. This situation shows that combining multiple cells together for embedding is more conducive to model building local features. However, when the sliding window value is too large, considering the small size of some web tables, random data may be introduced to reduce the accuracy.

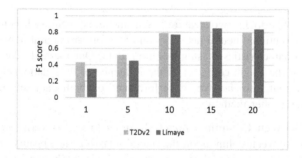

Fig. 7. The effect of sliding window size on character data column type detection

7   Conclusion and Outlook

In view of the problem of semantic type detection of data columns in web tables, this paper proposes a method that combines deep learning and probability graph model. This method uses deep learning model to fully extract the semantic relationship features that exist in single column. The probability graph model is used to take the relationship between columns into account. Experimental results show that the method we proposed can achieve satisfactory results on two real-world data sets, and is significantly better than other comparison methods in accuracy. Next, on the basis of this article, we will conduct a deeper study on how the relationship between the rows and columns of the table affects the classification results. Furthermore, it is an important direction in the future to develop an iterative process during the classification.

Acknowledgment. This work is supported by the National Key R&D Program of China (2018YFB1003404) and the National Natural Science Foundation of China (61672142).

References

1. Shuo, Z., Krisztian, B.: Web table extraction, retrieval, and augmentation: a survey. ACM. Trans. Intell. Syst. Technol. **11**, 2, Article 13, 35 (2020)
2. Michael, C., Hongrae, L.: Ten Years of Web Tables. PVLDB, **11**(12), 2140–2149 (2018). http://doi.org/10.14778/3229863.3240492
3. Sun, H.: Table cell search for question answering. In: Proceedings of the 25th International Conference on WWW, pp. 771–782 (2016). https://doi.org/10.1145/2872427.2883080
4. Ritze, D., Lehmberg, O.: Profiling the potential of web tables for augmenting cross-domain knowledge bases. In: Proceedings of the 25th International Conference on World Wide Web, pp. 251–261 (2016). https://doi.org/10.1145/2872427.2883017
5. Yoones, A., Paolo, M.: Knowledge base augmentation using tabular data. In: Prof. of WWW 2014 (2014)
6. Zwicklbauer, S., Einsiedler, C., Seifert, C.: Towards disambiguating web tables. In: International Semantic Web Conference, pp. 205–208 (2013)
7. Zhang, Z.: Effective and efficient semantic table interpretation using tableminer + . Semantic Web, **8**(6), 921–957 (2017). https://doi.org/10.3233/sw-160242
8. Efthymiou, V., Hassanzadeh, O., Rodriguez-Muro, M., Christophides, V.: Matching web tables with knowledge base entities: from entity lookups to entity embeddings. In: d'Amato, C., Fernandez, M., Tamma, V., Lecue, F., Cudré-Mauroux, P., Sequeda, J., Lange, C., Heflin, J. (eds.) ISWC 2017. LNCS, vol. 10587, pp. 260–277. Springer, Cham (2017). https://doi.org/10.1007/978-3-319-68288-4_16
9. Limaye, G., Chakrabarti, S.: Annotating and searching web tables using entities, types and relationships. Proc. VLDB Endowment 3(1–2), 1338–1347 (2010)
10. Mulwad, V.: Using linked data to interpret tables. Proc. First Int. Workshop Consum. Linked Data (2010). https://doi.org/10.13016/M2NS0M24R
11. Bhagavatula, C., Noraset, T., Downey, D.: Tabel: entity linking in web tables. In: International Semantic Web Conference, pp. 425–441 (2015)
12. Venetis, P., Halevy, A., Wu, C.: Recovering semantics of tables on the web. In: Proc. VLDB, pp. 528–538 (2011). https://doi.org/10.14778/2002938.2002939
13. Krishnamurthy, S., Pedro, S.: Assigning semantic labels to data sources. In European Semantic Web Conference. Springer, pp. 403–417(2015)
14. Minh, P., Suresh, A., and Pedro, S.: Semantic labeling: a domain-independent approach. In International Semantic Web Conference. Springer, pp. 446–462(2016)
15. Jiaoyan, C., Ernesto, J.: ColNet: embedding the semantics of web tables for column type prediction. AAAI (2018). https://doi.org/10.1609/aaai.v33i01.330129
16. Hulsebos, M., K. Z. Hu.: Sherlock: a deep learning approach to semantic data type detection. In: KDD, pp. 1500–1508 (2019). https://doi.org/10.1145/329250
17. Quoc, L., Tomas, M.: Distributed representations of sentences and documents. In: International Conference on Machine Learning. pp. 1188–1196 (2014)

18. Ritze, D., Lehmberg, O., Bizer, C.: Matching html tables to dbpedia. In: Proceedings of the 5th International Conference on Web Intelligence, Mining and Semantics, p. 10. ACM (2015). https://doi.org/10.1145/2797115.2797118
19. Xu, B., Yan, S., Yang, D.: BiRNN-DKT: transfer bi-directional LSTM RNN for knowledge tracing. In: Ni, W., Wang, X., Song, W., Li, Y. (eds.) WISA 2019. LNCS, vol. 11817, pp. 22–27. Springer, Cham (2019). https://doi.org/10.1007/978-3-030-30952-7_3

An Evolutionary Algorithm Based on Compressed Representation for Computing Weak Structural Balance in Large-Scale Signed Networks

Xingong Chang$^{(\boxtimes)}$ ⓘ and Fei Zhang

Information College, Shanxi University of Finance and Economics,
Taiyuan, China
19951001@sxufe.edu.cn

Abstract. With the integration of sentimental information, signed networks have a wide range of applications. The calculation of the degree of weak unbalance, which reflects the tension between positive and negative relations, is an NP-hard problem. In this paper, an evolutionary algorithm EAWSB for computing the degree of weak unbalance is proposed, where an indirect individual representation based on compression is designed to reduce the space complexity of the algorithm. In addition, a rotation operator is proposed to increase the population diversity. Experimental results show the effectiveness and efficacy of EAWSB. A thorough comparison show that EAWSB outperforms or is comparable to other state-of-the-art algorithms.

Keywords: Weak structural balance · Signed networks · Evolutionary algorithms · Compressed representation

1 Introduction

Networks are common models of many complex systems. Represented as graphs, networks use vertices to represent entities and edges to represent relationships between entities. Usually, networks fall into two broad categories: unsigned and signed. In unsigned networks, the edges typically indicate such relationships as friendship, collaboration, and sharing information, which usually have positive connotations, while in signed networks relationships can be friendly or hostile. For example, in an on-line discussion site Slashdot, users can tag others as "friends" or "foes", and in Wikipedia, a signed network is used to construct a voting network where a signed link indicates a positive or negative vote by one user on the promotion to admin status of another. In recent years, signed networks have attracted numerous research, including trust propagation [1, 2], link prediction [3, 4], recommender systems [5], and community detection [6], etc.

Perhaps the most basic theory applicable to signed networks but does not appear in the study of unsigned networks is that of social structural balance [7, 8]. The theory states that relationships in friend-enemy networks tend to follow patterns such as "an enemy of my friend is my enemy" and "an enemy of my enemy is my friend". A notion

© Springer Nature Switzerland AG 2020
G. Wang et al. (Eds.): WISA 2020, LNCS 12432, pp. 415–427, 2020.
https://doi.org/10.1007/978-3-030-60029-7_38

called weak balance [9] further generalizes structural balance by arguing that in many cases an enemy of one's enemy can indeed act as an enemy. Some research [3, 10] show that weak balance is more common in realistic signed networks than (strong) balance. For a signed network, the degree of unbalance, i.e., a distance to exact balance or weak balance, indicates a kind of instability or disharmony of it. Although the computation of structural balance is a well-known NP-hard problem [11], many algorithms [12–20] have been proposed to solve it. The block modelling method [16, 17] is an early method for the problem of weak structural balance. Because it is based on matrix representation and permutation operation, it has high space and time complexities and therefore is not suitable for large signed networks. MLMSB [18], ILS-CC [19], and VNS [20] are currently the three state-of-the-art algorithms to solve the problem of weak structural balance, which will be described in detail in Sect. 4.3. In this paper, we propose an Evolutionary Algorithm for Weak Structural Balance (EAWSB) where an indirect individual representation method based on compression is designed, which can effectively reduce the size of the genotype space and make the search of the algorithm more adequate. In addition, a rotation operator is proposed to improve the diversity of the population without affecting the fitnesses of individuals.

2   Related Background

2.1   Structural Balance and Weak Structural Balance

The theory of balance goes back to Heider [7] who asserted that a social system is balanced if there is no tension resulting in a tendency to change. The most appropriate model for structural balance is that of signed networks. Formally, a signed network is a graph $G = (V, E)$ together with a function $\sigma: E \rightarrow \{+, -\}$, which associates each edge with a sign of positive (+) and negative (-). In 1956, Cartwright and Harary [8] obtained the following important result which states an equivalent definition of balance.

Theorem 1. A signed graph is balanced if and only if its vertex set can be partitioned into two classes so that every edge joining vertices within a class is positive and every edge joining vertices between classes is negative.

For weakly balance case, we also have the following Theorem [21].

Theorem 2. A signed graph is weakly balanced if and only if its vertex set can be partitioned into multiple classes (more than two) so that every edge connecting two vertices that belong to the same class is positive and every edge connecting two vertices that belong to different classes is negative.

2.2   Computation of Weak Structural Balance

There are three major methods for the computation of weak structural balance. Firstly, Doreian and Mrvar [15] studied this problem earlier by proposing a block modelling method and then improved the method [16, 17] by exploring the evolutionary mechanism of international relations. Recently, Brusco and Doreian [20] proposed an variant of the variable neighborhood search for weak structural balance and obtained excellent experimental results.

Secondly, There is a large body of research work under the name of Correlation Clustering which is very relevant to weak structural balance. Correlation clustering [23] is a very useful and flexible framework for unsupervised learning in the setting: given a graph $G = (V, E)$ with weights $w_{ij} \in R$ on the edges which may be both positive and negative. The goal is to partition the vertices into clusters so that the sum of the weights of edges within a cluster is maximized (or the sum of the weights of edges between clusters is minimized). As a matter of fact, weak structural balance is a special case of correlation clustering when the weights are -1, 0 and 1. Levorato *et al.* [19] put forward an efficient local search algorithm ILS-CC to compute the degree of weak unbalance by solving the problem of correlation clustering.

Thirdly, Evolutionary Computation [24] is a powerful search and optimization technique inspired by the process of natural evolution, which are parallel in nature and do not require differentiability of objective functions and constraints. Considering the good performance of evolutionary algorithm in solving many difficult NP problems, we propose EAWSB for solving the problem of weak structural balance.

3   The Proposed Algorithm EAWSB

3.1   Fitness Function

Consider a signed network $G = (V, E)$ where $V = \{v_1, v_2, ..., v_n\}$ and $E \subseteq V \times V = \{(v_i, v_j) \mid v_i, v_j \in V$ and $i \neq j\}$ are the set of nodes and edges respectively. Let $A = (a_{ij})_{n \times n}$ be an adjacency matrix where $a_{ij} \in \{-1, 0, 1\}$, $a_{ij} = 1$ if the relationship between v_i and v_j is positive, $a_{ij} = -1$ for negative relationship and $a_{ij} = 0$ while the relationship is unknown or missing. Let $S = \{s_1, s_2, ..., s_n\}$ be a status set, where $s_i \in \{0, ..., k-1\}$ for $i = 1, ..., n$ denotes that the *i*th vertex belongs to the s_ith class, and k is a prespecified value to denote the maximum possible number of class labels that G can be partitioned.

According to Theorem 2, we can define the energy function as

$$E(S) = \frac{1}{2} \sum_{(v_i,v_j) \in E} (1 - a_{ij}\, \delta(s_i, s_j)) \tag{1}$$

where $\delta(s_i, s_j) = 1$ if $s_i = s_j$, -1 otherwise. The summation runs over all adjacent pairs of vertices. $E(S)$ is the sum of the number of all negative edges within same classes and the number of all positive edges between different classes. Its minimum value is the degree of weak unbalance of the signed network G. In this paper, the fitness function of the EAWSB algorithm is defined as:

$$F(S) = \sum_{(v_i,v_j) \in E} a_{ij}\, \delta(s_i, s_j) \tag{2}$$

Obviously, $E(S) = (m - F(S))/2$, where $m = |E|$ is the number of edges of G. So, minimizing $E(S)$ is equivalent to maximizing $F(S)$.

3.2   Natural Representation and Compressed Representation

Generally speaking, an individual representation can be divided into direct representation and indirect representation [24]. Naturally an individual *ind* in the population can be encoded as a status sequence $ind = s_1 s_2 \ldots s_n$, where $s_i \in \{0, \ldots, k\text{-}1\}$, which means that the ith vertex belongs to the s_ith class for $i = 1, \ldots, n$. We call this kind of representation a natural representation. In this case, the length of an individual is $n = |V|$. For large scale networks, the above representation will require significant amount of storage space and the operations, such as selection, crossover and mutation will consume lots of computation time. To address these issues, we propose a compressed representation based on following theorem. In the following, we define

$$s_{opt}(G, h) = \underset{s_h \in \{0, \ldots, k-1\}}{\arg\max} \sum_{v_j \in N(v_h)} a_{hj} \, \delta(s_h, s_j) \tag{3}$$

where $N(v_i) = \{v_k | (v_i, v_k) \in E\}$ is the neighborhood of v_i.

Theorem 3. Given a signed network $G = (V, E)$, $S^* = \{s_1^*, s_2^*, \ldots, s_n^*\}$ is the optimal solution of Eq. (3), then $\forall i \in \{1, \ldots, n\}$, $s_i^* = s_{opt}(G, i)$.

Proof. For a certain index h, $s_h^* \neq s_{opt}(G, h)$ if we have

$$\sum_{(v_i, v_j) \in E} a_{ij}\delta(s_i, s_j) = \sum_{(v_h, v_j) \in E} a_{hj}\delta(s_h, s_j) + \sum_{(v_i, v_j) \in E \wedge i \neq h} a_{ij}\delta(s_i, s_j)$$

$$= \sum_{v_j \in N(v_h)} a_{hj}\delta(s_h, s_j) + \sum_{(v_i, v_j) \in E \wedge i \neq h} a_{ij}\delta(s_i, s_j)$$

note h only appears in the first part and not in the second part of the expression above. So, we can define $s_h^\# = s_{opt}(G, h)$, $s_j^\# = s_j^*$ for $j \neq h$. Then $S^\# = \{s_1^\#, s_2^\#, \ldots, s_n^\#\}$ is more optimal than $S^* = \{s_1^*, s_2^*, \ldots, s_n^*\}$, because

$$\sum_{(v_i, v_j) \in E} a_{ij}\delta(s_i^\#, s_j^\#) = \sum_{v_j \in N(v_h)} a_{hj}\delta(s_h^\#, s_j^\#) + \sum_{(v_i, v_j) \in E \wedge i \neq h} a_{ij}\delta(s_i^\#, s_j^\#)$$

$$> \sum_{v_j \in N(v_h)} a_{hj}\delta(s_h^*, s_j^*) + \sum_{(v_i, v_j) \in E \wedge i \neq h} a_{ij}\delta(s_i^\#, s_j^\#) = \sum_{v_j \in N(v_h)} a_{hj}\delta(s_h^*, s_j^*) + \sum_{(v_i, v_j) \in E \wedge i \neq h} a_{ij}\delta(s_i^*, s_j^*)$$

$$= \sum_{(v_i, v_j) \in E} a_{ij}\delta(s_i^*, s_j^*)$$

that is in contradiction with the optimality of $S^* = \{s_1^*, s_2^*, \ldots, s_n^*\}$. ∎

Theorem 3 indicates that the best status value of v_h can be calculated from its neighborhood $N(v_h)$. Thus, we only need to encode the vertices in U which is a proper subset of V satisfying the condition: $\forall v \in V \backslash U$ (set difference), $N(v) \subseteq U$. The reason is that if the status values of vertices in U are known, the status values of vertices in $V \backslash U$ can be calculated from their neighborhood. The immediate question is how to find the subset U. We give the Algorithm 1 to solve the problem.

Algorithm 1. the procedure to generate the compressed representation

1: **Input**: The adjacency matrix of $G(V, E)$: $A=(a_{ij})_{n \times n}$
2: Calculate the array of vertex degrees: degree[0.. n-1]
3: original_degree[0..n-1]= degree[0.. n-1]
4: Initialize each element of array selNode[0..n-1] to be 0
5: for each vertex i with original_degree[i]=1 and degree[i]>0 do
6:   j=the neighbor of i
7:   selNode[j]=1, degree[j]=0
8:   for each j's neighbor p do
9:     if(degree[p]>0)degree[p]= degree[p]-1
10: endfor
11: endfor
12: repeat
13:   Randomly select a vertex j with degree[j]>0
14:   selNode[j]=1, degree[j]=0
15:   for each j's neighbor p do
16:     if(degree[p]>0)degree[p]= degree[p]-1
17:   endfor
18: until degree[i]=0 for all $i \in \{0, ..., n$-1$\}$
19: **Output**: all is with selNode[i]=1

Algorithm 1 consists of three parts. In Part 1 (line 2–line 4), three arrays are defined. Array original_degree[] and degree[] hold the information of vertex degrees, and they are identical initially. Through the algorithm, values in degree[] decrease while original_degree[] remains unchanged. When all values in degree[] become zero the algorithm stops. All elements in the third array selNode[] are initialized to 0. Subsequently every time when a vertex j is selected we set selNode[j] = 1. Part 2 (line 5–line 11) deals with leaf vertices (vertices with degree one), the only neighbor of a leaf vertex is selected if its degree has not been decreased to zero. Like fitness-proportional selection [24], Part 3 (line 12–line 18) uses degree-proportional selection strategy, *i.e.* degree value is used to associate a probability of selection with each vertex and a vertex with a higher degree will be more likely to be selected. Both in Part 2 and Part 3, every time when a vertex i is selected degree[i] is set to zero indicating that it will not be selected again. All its neighbors' degree will be reduced by one so that their probabilities being selected decrease because one of their neighbours vertex i has been just selected. When all neighbours of a vertex i are selected, degree[i] becomes zero, and can not be selected later. This is exactly what Theorem 3 tells us.

3.3   Initialization

First, each individual of the population is generated randomly. Then, we adjust the gene values according to the Homophily theory [21] which says that objects tend to be similar to those surrounding them. Thus, we select a gene (vertex) position in an individual randomly and set the same class label of the individual to all its neighbors if the neighbor has positive edge (its friend) link to it. We repeat this operation for *iniK* times, where *iniK* is an integer parameter used to control the intensity of homophily.

3.4   Evolutionary Operators

Crossover. The one-way crossover operator introduced in Tasgin *et al.* [25] is used. The main idea is as follows: providing two individual inputs *ind1* and *ind2*, randomly selecting a gene from *ind1*, iteratively searching the genes with the same status value as that of the gene selected above in *ind1* and then transferring the status values of those genes in *ind1* to the corresponding genes in *ind2*. The modified *ind2* is returned as the crossover result. Figure 1 shows the details of one-way cross-over scheme in our algorithm. From the crossover process, we can see that, unlike the common single point, two-point and uniform crossover operators that emphasize the combination of excellent building blocks, one-way crossover operators emphasize the retention and accumulation of excellent building blocks. Therefore, its "tearing" effect is much weaker than the former, and is more suitable for use in the network evolution environment.

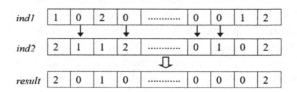

Fig. 1. Illustration of the one-way crossover

Rotation. Any gene value is in $\{0, \ldots, k\text{-}1\}$. In the course of evolution, we let each individual rotate *i.e.*, the status value $0 \to 1$, $1 \to 2$, $\ldots$, $k\text{-}1 \to 0$, with a lower probability (0.05 in our research). Note that this rotation does not change the fitness of an individual, because those vertices that were in the same class after rotation are still in the same class, and those who were not in the same class after rotation are still not in the same class. However, the benefit of this operation is that the diversity of the population is maintained, because each gene position has the opportunity to obtain a new genetic value without changing the individual's fitness. Its time complexity is $O(n)$.

In addition, our research adopts tournament selection and elitist policy [24] and mutation is one-point mutation which are useful to prevent "tearing" of good building blocks, just like the above crossover operators.

3.5   Local Search

Based on Theorem 3, the local search can be designed as follows: for a given individual, we first randomly select a gene from it and then change the value of the gene with the best status value $\in \{0, ..., k-1\}$ obtained from Eq. (3), and use the changed individual as a result. We can repeat this procedure $locK$ times, where $locK$ is a positive integer that represents local search intensity. It is generally accepted that the performance of evolutionary algorithms depends on the tradeoff between exploration and exploitation. For local search, increase the value of $locK$ if we want to encourage exploitation, or reduce its value if we want to encourage exploration.

3.6   The Framework of EAWSB

Algorithm 2 describes the framework of EAWSB where we first use Algorithm 1 to generate the compressed individual representation. Then the population is initialized. In the main loop of EA, the selection procedure is responsible for selecting parental population for mating in EA and elitism and tournament selection are used to maintain the diversity of population. The crossover, mutation, rotation and local search are executed in order. The steps in Lines 5–9 are repeated until some termination criterion is met. Finally, the fittest individual in population is returned as a result.

Algorithm 2. The framework of EAWSB
___

1: **Input**: Adjacency matrix of $G(V, E)$: $A=(a_{ij})_{n \times n}$, Population size: $popSize$, Initialization intensity: $iniK$, Local search intensity: $locK$, Maximum number of generations: $maxGen$, Tournament size: $tourSize$, Crossover probability: p_c, Mutation probability: p_m, Rotation probability: p_r

2: Use algorithm 1 to generate a compressed representation of G

3: P←Generate_initial_population($popSize$, $iniK$)

4: repeat

5:    P_{parent}←Selection(P, $tourSize$)

6:    P_{child}←Crossover(P_{parent}, p_c)

7:    P_{child}←Mutation(P_{child}, p_m)

8:    P_{child}←Rotation(P_{child}, p_r)

9:    P←LocalSearch(P_{child}, $locK$)

10: until Termination_criterion($maxGen$)

11: **Output**: the fittest individual in P
___

The time complexity of EAWSB can be obtained by synthesizing the time complexities of the above components. The fitness of an individual can be calculated according to Eq. (2), and its time complexity is $O(m)$. The overall time complexity of the EAWSB algorithm is $O(popSize*maxGen*(n + m))$.

4 Experiments

4.1 Datasets

The experiments used five large signed network datasets: Epinions, Slashdot, WikiElec, SlashSCC and WikiRfa which are downloaded from Stanford Network Analysis platform [26]. Epinions (epinions.com) is a product review site, where members can rate others' reviews as "useful" or "useless" and then label commenters with "trust" or "distrust" thus makeing a network of trust/distrust. Slashdot (slashdot.com), a technology news website, allows users to mark authors as "friends" or "enemies" for their articles, and its data also makes a signed network with friend/enemy labels. WikiElec is about Wikipedia users' voting records for administrators and contains support/opposition relationships. SlashSCC is the largest strongly connected component of the dataset "soc-sign-Slashdot 081106". The WikiRfa network records the statistical requests from Wikipedia members for Wikipedia editors to become administrators. A request can be represented by a supporting, neutral, or opposing vote. We preprocessed the data sets by removing the conflicting edges, multiple edges, loops and outliers, and turning them into undirected graphs. For WikiRfa, we discarded all its neutral votes. As a result, we obtained 5 large undirected signed networks for our experiments, as shown in Table 1, where n and m are the number of vertices and edges of undirected networks, and m^- and m^+ are the number of negative and positive edges of networks.

Table 1. Preprocessed datasets

Experimental datasets	n	m	m^+	m^-
Epinions	131513	708507	589888	118619
Slashdot	82062	498532	380933	117599
WikiElec	7114	99862	78371	21491
SlashSCC	21369	273646	206087	67559
WikiRfa	11253	168722	131769	36953

4.2 Experimental Results

Parameter Settings. The running parameters of EAWSB algorithms are set as follows: Population size $popSize = 100$, Initialization intensity $iniK = 500$, Local search intensity $locK = 500$, Maximum number of generations $maxGen = 100$, Tournament size $tourSize = 2$, Crossover probability $p_c = 0.9$, Mutation probability $p_m = 0.1$, and Rotation probability $p_r = 0.05$. The algorithm EAWSB is programmed with JAVA. All experiments were conducted on a Window 10 machine with an Intel(R) Core(TM) i7-7700HQ CPU and 16.0 GB Memory. In the experiments, the reported data are the statistical results based on 10 independent runs on each dataset.

Table 2. Running results of degrees of unbalance on the 5 signed networks. 'N' and 'C' denote natural and compressed representation respectively.

k	EAWSB	Epinions	Slashdot	WikiElec	SlashSCC	WikiRfa
2	N	52682	74773	14214	47335	25895
	C	52467	74728	14201	47286	25870
3	N	48976	70826	13876	45282	25289
	C	48639	70396	13843	45244	25265
4	N	48842	69743	13821	44598	25202
	C	48655	69446	13806	44578	25157
5	N	48793	69488	13809	44468	25161
	C	48604	69309	13797	44469	25142
6	N	48692	69268	13798	44390	25151
	C	48622	69226	13791	44277	25134

Running Results of EAWSB with Different Representation. Table 2 shows the results of EAWSB running on 5 datasets stated above respectively, where "N" means natural representation, "C" means compressed representation. The operation was carried out in five cases according to the number of status values $k = 2, 3, 4, 5, 6$, in which $k = 2$ is the case of structural balance and can be regarded as a special case of weak structural balance. It can be seen from the table that EAWSB with compressed representation usually gives the better solutions because the compressed representation reduces the length of the individuals, and then reduce the size of the search space which leads to a more sufficient search and better solutions can be obtained.

4.3   Comparisons with Similar Algorithms

EAWSB is compared with four classical algorithms, namely MEMESB [12], MLMSB [18], ILS-CC [19] and VNS [20]. Table 3 records the statistical results which are the best experimental results when the number of status values k is taken from 2 to 6. When EAWSB runs, we try our best to keep consistent with the original algorithms in data preprocessing and parameter settings.

Table 3. Statistic results of degrees of unbalance on the 5 real-world signed networks. '#' represents that the result was not provided in the original paper.

Algorithms	Epinions	Slashdot	WikiElec	SlashSCC	WikiRfa
MEMESB	56799	75022	13850	61581	25762
MLMSB	#	#	13841	62120	25974
ILS-CC	55901	69994	14300	#	#
VNS	#	**67782**	14142	#	#
EAWSB	**48594**	69224	**13799**	**44426**	**25143**

MEMESB is an evolutionary algorithm for structural balance. We generalize it to solve the problem of weak unbalance. It uses natural representation, two-point crossover, single point mutation, and performs local search only for the optimal individual of the current population. As shown in the table, EAWSB is better than algorithm MEMESB comprehensively.

MLMSB is a multilevel learning based memetic algorithm, which incorporates network-specific knowledge such as the neighborhoods of node, community and partition. MLMSB employs natural representation, two-way crossover, and neighbourhood-based mutation. In the experimental analysis of its original paper three large-scale signed networks SlashSCC, WikiElec, and WikiRfa were used and the experimental results are shown in Table 3. With the same parameter settings EAWSB get better results as shown in Table 3.

Both ILS-CC and VNS are local search algorithms which usually start from an initial solution, and then select the best neighbor from its neighborhood, meanwhile take it as the current solution of the next iteration, repeating the process until a local optimal solution is reached. ILS is a metaheuristic based on stochastic multi-restart search, which iteratively applies local search to perturbations of the current best solution, generating a random walk in the space of local optima. Each new solution obtained is then refined using an embedded heuristic. ILS-CC considers the weak structural balance problem as the problem of correlation clustering and obtains the degree of weak unbalance of the network through ILS. The authors of ILS-CC reported the experimental results of three large-scale signed networks Epinions, Slashdot and WikiElec. It is easy to see that the experimental results of EAWSB are better than those of ILS-CC.

Based on iterated local search, VNS enforces greater control on the size of the neighborhood around the current solution wherein the search takes place. Brusco and Doreian [20] adapt the variable neighborhood search procedure to refine the solutions obtained by the multistart relocation heuristic for weak structural balance and obtained excellent experimental results. As the number of clusters k grows from 2 to 14 the best results they gave were 67782 on Slashdot and 14142 on WikiElec. The former is better than that of EAWSB, but in the latter case on WikiElec, EAWSB performs better. In addition, for Epinions, a very large data set with more than 130,000 nodes, VNS does not provide corresponding experimental results, while EAWSB provides many good experimental results. So we can say that EAWSB is comparable to VNS.

In summary, in comparison with the other four state-of-the-art algorithms on the five large-scale signed networks, algorithm EAWSB performs poorly in only one of the 14 scenarios, and outperforms its opponents in all the other 13 scenarios.

4.4   Experimental Analysis of EAWSB Submodules

Experimental Analysis of Compressed Representation. We conduct experiments to measure the compressing performance of Algorithm 1 by computing the compression rate which is the ratio of the length of the compressed individual over the length of the individual before compression. On Epinions, Slashdot, WikiElec, SlashSCC and WikiElec, the compression rate are 32.73%, 44.01%, 50.62%, 89.58% and 47.16%

respectively. A good compression rate can lead to a smaller compression degree, smaller genotype space. Consequently, search can be more sufficient, and solution quality is better.

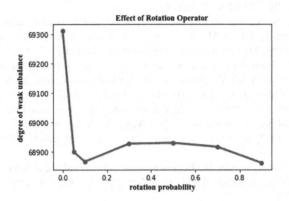

Fig. 2. Effect of Rotation Operator on Slashdot.

Experimental Analysis of Rotation Operator. Using EAWSB with *popSize* = 100, *maxGen* = 500, *iniK* = 500, *locK* = 500 and *k* = 6, experiments were conducted to study the influence of rotation operator. Figure 2 shows the change of the minimum degree of weak unbalance obtained on Slashdot with the increase of rotation probability (p_r). We can see (1) using rotation operator (where $p_r \neq 0$) can obtain better solution than not using rotation operator (where $p_r = 0$); (2) the mean minimum value is obtained at $p_r = 0.1$, and we observed that the absolute minimum value is usually obtained at $p_r = 0.05$, that's why we usually set the rotation probability p_r to 0.05 in this paper; and (3) The case at p_r is basically the same as the case at 1-p_r, it's approximately symmetric, because $100p_r$ percent of the individuals in a population are rotated, there must be $100(1$-$p_r)$ percent of the individuals that are not rotated, and it doesn't matter in which group a node is, what matters is with which nodes it is in the group.

5 Conclusion

The degree of weak unbalance is an important metric to measure the tension of signed networks. This paper proposes an evolutionary algorithm EAWSB to compute the degree of weak unbalance. We propose an individual indirect representation method based on compression to effectively reduce the size of genotype space. As a result, the search of the algorithm is more sufficient and the better solution can be obtained. In order to keep the diversity of population and avoid premature convergence, a rotation operator is proposed to improve population diversity without affecting individual fitnesses. Experiments on large scale signed networks - Epinions, Slashdot, WikiElec, SlashSCC and WikiRfa show that the proposed method is effective. In the future, we will study the weak unbalance calculation method based on deep learning.

References

1. Guha, R., Kumar, R., Raghavan, P., Tomkins, A.: Propagation of trust and distrust. In: Proceedings of the 13th International Conference on World Wide Web, pp. 403–412 (2004). https://doi.org/10.1145/988672.988727
2. Tang, J., Chang, S., Aggarwal, C., Liu, H.: Negative link prediction in social media. In: Proceedings of the 8th ACM International Conference on Web Search and Data Mining, pp. 87–96 (2015). https://doi.org/10.1145/2684822.2685295
3. Leskovec, J., Huttenlocher, D., Kleinberg, J.: Signed networks in social media. In: Proceedings of the Sigchi Conference on Human Factors in Computing Systems, pp. 1361–1370. ACM (2010). https://doi.org/10.1145/1753326.1753532
4. Chang, X., Shi, W., Zhang, F.: Signed network embedding based on noise contrastive estimation and deep learning. In: Ni, W., Wang, X., Song, W., Li, Y. (eds.) WISA 2019. LNCS, vol. 11817, pp. 40–46. Springer, Cham (2019). https://doi.org/10.1007/978-3-030-30952-7_5
5. Tang, J., Aggarwal, C., Liu, H.: Recommendations in signed social net-works. In: The 25th International Conference on World Wide Web, pp. 31–40 (2016). https://doi.org/10.1145/2872427.2882971
6. Alessia, A., Clara, P.: Community mining in signed networks: a multiobjective approach. In: Proceedings of the 2013 IEEE/ACM International Conference on Advances in Social Networks Analysis and Mining, pp. 95–99. Niagara, Canada (2013). https://doi.org/10.1109/icde.2018.00031
7. Heider, F.: Attitudes and cognitive organization. J. Psychol. 21(1), 107–112 (1946). https://doi.org/10.1080/00223980.1946.9917275
8. Dorwin, C., Frank, H.: Structure balance: a generalization of Heiders theory. Psychol. Rev. 63(5), 277–293 (1956). https://doi.org/10.1037/h0046049
9. Davis, J.A.: Clustering and structural balance in graphs. Soc. Netw. 20(2), 27–33 (1977). https://doi.org/10.1177/001872676702000206
10. Leskovec, J., Huttenlocher D., Kleinberg J.: Predicting positive and negative links in online social networks. In: Proceedings of the 19th International Conference on World Wide Web, pp. 641–650. ACM (2010). https://doi.org/10.1145/1772690.1772756
11. Barahona, F.: On the computational complexity of Ising spin glass models. J. Phys. A 15 (10), 3241–3253 (1982). https://doi.org/10.1088/0305-4470/15/10/028
12. Sun, Y., Du, H., Gong, M., et al.: Fast computing global structural balance in signed networks based on memetic algorithm. Phys. A Stat. Mech. Appl. 415, 261–272 (2014). https://doi.org/10.1016/j.physa.2014.07.071
13. Zheng, X., Zeng, D., Wang, F.-Y.: Social balance in signed networks. Inf. Syst. Front. 17(5), 1077–1095 (2014). https://doi.org/10.1007/s10796-014-9483-8
14. Aref, S.: Balance and frustration in signed networks under different contexts. arXiv:1712. 04628v2 [cs.SI] 21 Apr (2018)
15. Doreian, P., Mrvar, A.: A partitioning approach to structural balance. Soc. Netw. 18(2), 149–168 (1996). https://doi.org/10.1016/0378-8733(95)00259-6
16. Doreian, P., Mrvar, A.: Partitioning signed social networks. Soc. Netw. 31(1), 1–11 (2009). https://doi.org/10.1016/j.socnet.2008.08.001
17. Doreian, P., Mrvar, A.: Structural balance and signed international relations. J. Soc. Struct. 16, 1–49 (2015). https://doi.org/10.21307/joss-2019-012
18. Ma, L., Gong, M., Du, H., et al.: A memetic algorithm for computing and transforming structural balance in signed networks. Knowl.-Based Syst. 85, 196–209 201 (2015). https://doi.org/10.1016/j.knosys.2015.05.006

19. Levorato, M., Rosa, F., Yuri, F., et al.: Evaluating balancing on social networks through the efficient solution of Correlation Clustering problems. EURO Journal on Computational Optimization, Springer, Cham, **5**(4), pp. 467–498 (2017). http://doi.org/10.1007/s13675-017-0082-6
20. Michael, J.B., Patrick, D.: Partitioning signed networks using relocation heuristics, tabu search, and variable neighborhood search. Soc. Netw. **56**, 70–80 (2019). https://doi.org/10.1016/j.socnet.2018.08.007
21. David, E., Jon, K.: Networks, Crowds, and Markets: Reasoning about a Highly Connected World. Cambridge University Press, London (2010)
22. Srinivasan, A.: Local balancing influences global structure in social networks. Proc. Nat. Acad. Sci. **108**, 1751–1752 (2011). https://doi.org/10.1073/pnas.1018901108
23. Bansal, N., Blum, A., Chawla, S.: Correlation clustering. In: Proceedings of the 43rd Annual IEEE Symposium on Foundations of Computer Science, pp. 238–247 (2002). https://doi.org/10.1109/sfcs.2002.1181947
24. Jong, K.D.: Evolutionary Computation: A Unified Approach. MIT Press, Cambridge (2016)
25. Tasgin, M., Herdagdelen, A., Bingol, H.: Community detection in complex networks using genetic algorithms. arXiv:0711.0491 [physics.soc-ph] (2007)
26. Stanford Network Data. http://snap.stanford.edu/data/#signnets. Accessed 10 June 2020

Collaborative Filtering: Graph Neural Network with Attention

Yanli Guo$^{(\boxtimes)}$ and Zhongmin Yan

Shandong University, Jinan, China
1789213513@qq.com, yzm@sdu.edu.cn

Abstract. In recent years, graph neural networks have been widely used in natural language processing and speech recognition. However, there is relatively little exploration of graph neural networks in recommendation systems. In this work, we strive to develop neural network based technology to solve the problem of collaborative filtering recommendation based on implicit feedback. Some recent work use deep learning for recommendation, but they mainly use it for auxiliary information modeling. When modeling the key elements of collaborative filtering (the interaction between users and item features), they still use matrix factorization and apply inner products to the potential features of users and items. The collaboration signal(interaction information between users and items) will not be encoded during the embedding process. At the same time, the high-order connectivity of user-item cannot be fully utilized, making the interactive information not comprehensive enough. Therefore, the final embedding may not be enough to capture the collaborative filtering effect. By replacing the inner product with a neural architecture that can learn arbitrary functions from data, we propose a general method called "Graph Neural Network with Attention" (GNNA). GNNA captures CF signals based on neural networks and uses high-order connectivity to obtain neglected interactive information, thereby improving the embedding of users and items. By introducing attention mechanism and high-order connectivity, it can learn user vectors and item vectors on the user item interaction graph, collect neighbor interaction information for coding, and spread it on the user-item interaction graph. This can comprehensively inject user-item collaboration signals into the embedding process. We have conducted extensive experiments on two public benchmarks. Further analysis verified the importance of using attention mechanism and high-order connectivity to learn user and item representations, and proved the rationality and effectiveness of GNNA.

Keywords: Recommender systems · Information filtering · Neural networks · Attention mechanism

1 Introduction

The idea of collaborative filtering is to use the past behaviors or opinions of the existing user group to predict what the current user is most likely to like

© Springer Nature Switzerland AG 2020
G. Wang et al. (Eds.): WISA 2020, LNCS 12432, pp. 428–438, 2020.
https://doi.org/10.1007/978-3-030-60029-7_39

or interested in, and to analyze the similar preferences of users with similar behaviors to make efficient recommendations.

In order to predict user preferences from key (and widely available) user behavior data, a large amount of research work has been invested in collaborative filtering (CF) [7,8,20]. Generally speaking, a learnable CF model has two key components: 1) embedding, which transforms users and items into vectorized representations, and 2) interaction modeling, which reconstructs historical interactions based on the embeddings. Many research efforts have been devoted to enhancing these two parts. For example, matrix factorization (MF) directly embeds user/item ID as vectors and uses inner products to model user-item interactions [11]; translation-based CF model uses Euclidean distance metric as interaction function [18]; collaborative deep learning expands the embedding function by integrating information obtained from item edges [19].

Since the CF method lacks the coding of key collaboration signals, it may not accurately show the similarity between the user and the item. More specifically, since existing methods use descriptive functions (such as ID and attributes)[9] to build embedded functions, they do not consider high-order user-item interactions. Therefore, using high-order relations to capture interactive information between users and items is particularly important for improving the quality of recommendations. In this work, we introduced the attention mechanism [10] and high-order connectivity, summarized weights between different items of the same user, learned the vector representation of users and items, and improved the quality of suggestions.

To summarize, this work makes the following main contributions:

- The importance of high-order connectivity signals in embedded functions was emphasized.
- A new recommendation framework named GNNA based on graph neural network was proposed. This framework assigns different weights to the messaging of different items of the same user, and encodes potential interactive signals through embedd propagation.
- Extensive experiments have been conducted on two real world datasets, it proved the effectiveness of the GNNA method and the prospect of collaborative filtering.

2   Related Work

2.1   Collaborative Filtering

Collaborative filtering methods play an important role in recommendation systems. Now traditional collaborative filtering methods are gradually being combined with machine learning. For example, graph neural networks are used for embedded propagation [20]; recommendations based on storage networks [23]; neural networks are used to solve the problem of collaborative filtering implicit feedback [8]; conference-based recommendations [12] and one-class recommendation [16].

2.2   Implicit Feedback

Although the early literature on recommendations has focused on displaying feedback [13], recent attention is shifting to implicit data [4]. The user does not interact with the item, which does not mean that the user does not like the item; the user interacts with the item does not mean that the user likes the item. Collaborative filtering of implicit data is usually expressed as a recommendation problem. Expressing the implicit interaction data between users and items is more effective and more challenging for recommendation problems [4]. For this reason, recent literatures [1,8] focus on using deep learning techniques to capture the non-linear interactions between users and items.

2.3   Attention Mechanism

The attention mechanism is similar to human visual attention and both focus on important parts. At present, attention mechanism has been shown in various machine learning, such as image/video [14], machine translation [2], and natural language processing [6]. Recently, attention mechanism has been used in recommendation systems [15]. For example, [3] proposed an attention-based approach to align codec frameworks for machine translation. [17] proposed an attention CNN network. The author designed two attention mechanisms to learn user and item representations, namely local attention and global attention.

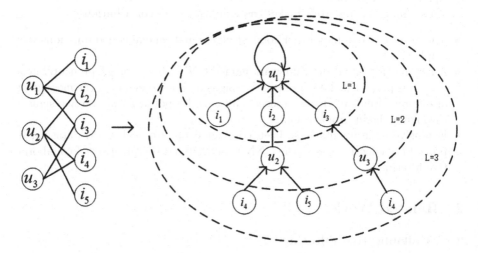

Fig. 1. User-item transformed interaction graph.(The figure on the left is the user-item bipartite graph, which represents the interaction between users and items. The figure on the right is the transformed graph structure, and L represents different orders.)

3   Graph Neural Network with Attention

The architecture of the GNNA model is shown in Fig. 2. The framework consists of three parts: (1) the embedding layer, which is responsible for the initial input of user embedding and item embedding; (2) the attention propagation layer, which introduces attention mechanism and high-order connectivity to obtain high-order interactive information. Different items of users are assigned different weights, and message embeddings are carried out; (3) the prediction layer, which aggregates embeddings from different dissemination layers, and outputs the user preference for items.

3.1   Embedding Layer

The embedding layer is used to initialize user embedding and item embedding. Following mainstream recommender models [8], it describes a user u (an item i) with an embedding vector $e_u \in R^d(e_i \in R^d)$, where d denotes the embedding size. This can be seen as an embedding look-up table:

$$E = [e_{u_1}, \cdots, e_{u_N}, e_{i_1}, \cdots, e_{i_M}] \tag{1}$$

This embedding table is used as the initial state for user embedding and item embedding, it is optimized in an end-to-end manner. In traditional recommender models (such as MF), these ID are embedded and fed directly into the interaction layer to obtain prediction score. However, in GNNA framework, the embedding is referenced by propagating the embedding on the user-item interaction diagram.

3.2   Attention Propagation Layer

Next, a GNN-based messaging architecture was established [21] to obtain collaborative filtering signals along the graph structure and optimize user and item embeddings, as shown in Fig. 2.

Message Construction. The interaction between a user and an item directly reflects the user preference [22]. Similarly, an item (user) that interacts with the user(item) can be regarded as a feature of the user(item), and two users(items) can be measured. Based on the similarity between users, the attention mechanism can be used to calculate the weight of message passing between the same user and different items, and then perform the embeddings propagation. The message during the propagation is divided into two parts: (1) user-item interaction message (2) self-connection message of u, as shown in Fig. 1. For a connected user-item pair (u, i) messages, the message from i to u as

$$m_{u \leftarrow i} = f(e_u, e_i, p_{ui}). \tag{2}$$

where $m_{u \leftarrow i}$ is the message embedding (i.e., the information to be propagated). $f(\cdot)$ is a message encoding function, which takes embeddings e_u and e_i as input, and p_{ui} is the strength of the side message. In this work , $f(\cdot)$ as:

$$m_{u \leftarrow i} = p_{ui}(W_1 e_i + W_2(e_i \otimes e_u)). \tag{3}$$

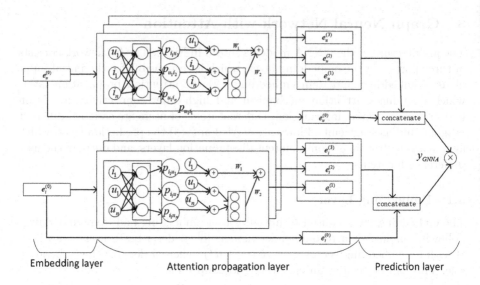

Fig. 2. GNNA model structure (The embedding layer initializes the initial embedding of users and items, then updates the embeddings of users and items on the attention propagation layer, and finally predicts the score on the prediction layer.).

where $W_1, W_2 \in R^{d_l \times d_{l-1}}$ are the trainable weight matrices to distill useful information for propagation. The interaction between e_i and e_u was passed via $e_i \otimes e_u$, where $\otimes$ denotes the element-wise product.

Attention Mechanism. Although $p_{ui} = 1/\sqrt{|N_i \parallel N_u|}$ (N_u and N_i are neighbor user set and neighbor item set), all edges have not same type or same strength, an attention mechanism is added to define the strength of user-item interaction based on the similarity between items connected with the user.

The similarity between items connected to users can be defined as:

$$cos(u_{i1}, u_{ij}) = u_{i1}^T u_{ij} / \parallel u_{i1} \parallel \parallel u_{ij} \parallel \qquad (4)$$

where u_{ij} represents the user neighbor item set. Define the attention of users interacting with the item as::

$$p_{ui} = e^{cos(u_{i1}, u_{ij})} / C \qquad (5)$$

Here C as:

$$C = \sum_{j=1}^{M} e^{cos(u_{i1}, u_{ij})} \qquad (6)$$

At this time, p_{ui} represents different weights with item edges. j represents the Mth item. By continuously learning p_{ui}, different item strengths of the same user can be obtained.

Message Aggregation. With the representations augmented by first-order connectivity modeling, it can stack more embedding propagation layers to explore the high-order connectivity information. High-order connectivities are crucial to encode the collaborative signal to estimate the relevance score between a user and item. By stacking L embedding propagation layers, a user (and an item) is capable of receiving the messages propagated from its l-hop neighbors. As Fig. 2 displays, in the l-th step, the representation of user u is recursively formulated as:

$$e_u^l = LeakyReLU(m_{u \leftarrow u}^l + \sum_{i \in N_u} m_{u \leftarrow i}^l) \tag{7}$$

where $m_{u \leftarrow u}^l$ represents user's self-connection, $\sum_{i \in N_u} m_{u \leftarrow i}^l$ represents user interactions. The LeakyReLU activation function is used to encode positive and negative signals. In addition to the message propagated from neighbors, a self-connected closed loop is also considered, which retains information about the original features. Similarly, the representation of e_i can be obtained by propagating the information of items. All in all, the advantage of this section is to get the relations between users and items as much as possible.

The message is defined as follows:

$$m_{u \leftarrow i}^{(l)} = p_{ui}(W_1^{(l)} e_i^{(l-1)} + W_2^{(l)} (e_i^{(l-1)} \otimes e_u^{(l-1)})) \tag{8}$$

$$m_{u \leftarrow u}^{(l)} = W_1^{(l)} e_u^{(l-1)} \tag{9}$$

where $W_1, W_2 \in R^{d_l \times d_{l-1}}$ represents a trainable transformation matrix. This step stores messages from neighboring nodes.

By increasing the attention mechanism to calculate the message passing weight, it focus on important neighbor side messages. User-item message strength can be enhanced.

3.3   Model Prediction

After propagating the L layer, multiple representations of users and items can be obtained, namely $\{e_u^{(1)}, e_u^{(2)}, \cdots, e_u^{(l)}\}\{e_i^{(1)}, e_i^{(2)}, \cdots, e_i^{(l)}\}$. Since different layers can represent messages delivered by different connections, they can reflect user preference well. Therefore, the final representation of the corresponding item can be obtained:

$$e_u^* = e_u^{(0)} \| \cdots \| e_u^{(l)} \tag{10}$$

$$e_i^* = e_i^{(0)} \| \cdots \| e_i^{(l)} \tag{11}$$

Where $\|$ is the connection operation [21]. The connection operation does not involve learning other parameters, so it is very convenient. In this way, not only can the initial embedding be enriched by propagation embedding, but also the propagation range can be controlled by adjusting L.

Finally, the inner product of users and items are used to evaluate user preference:

$$y_{GNNA}(u, i) = e_u^{*T} e_i^* \tag{12}$$

4 Experiments

Experiments were conducted on two real datasets to evaluate the proposed method, especially the attention propagation layer. The aim is to answer the following questions:

- RQ1:How does GNNA perform as compared with state-of-the-art CF method?
- RQ2:How do different hyper-parameter settings (e.g., number of layers) affect GNNA?

4.1 Dataset Description

Table 1. Statistics of the datasets.

Dataset	Users	Items	Interactions	Density
Gowalla	29,858	40,981	1,027,370	0.00084
Amazon-Book	52,643	91,599	2,984,108	0.00062

To evaluate the performance of GNNA, experiments were conducted on two benchmark datasets: Gowalla and Amazon-Book, which are publicly available datasets.

Gowalla: This is a check-in dataset obtained from Gowalla [20], where users share their locations via check-ins. To ensure data quality, a 10-core setting was used, which keeps at least ten interactive users and items.

Amazon-Book: Amazon Reviews is widely used in product recommendations [20]. Set up 10 cores(each user and each item have at least ten interactions).

Dataset composition: For each data set, randomly select 80% of the user's historical interactions as the training set and the remaining 20% as the test set. In the training set, randomly select 10% of the interactions as the validation set to adjust the hyper-parameters.

4.2 Experimental Settings

Evaluation Indicators. All items that do not interact with the user are treated as negative items. Each method outputs the user preference score for the item, except for the active items in the training set. The performance of GNNA was evaluated by evaluating the recall and ndcg.

Baselines. In order to prove the effectiveness of GNNA, it is compared with following methods:

- MF [11]: This is a matrix factorization optimized by Bayesian Personalized Ranking (BPR) loss optimization, which uses the user's direct interaction term as the target value of the interaction function.
- NeuMF [8]: The method is a state-of-the-art neural CF model which uses multiple hidden layers above the element-wise and concatenation of user and item embeddings to capture their nonlinear feature interactions.
- GC-MC [5]: This model uses a GCN encoder to generate user and item representations, where only first-order neighbors are considered.
- CMN [23]: It is a state-of-the-art memory-based model, where the user representation attentively combines the memory slots of neighboring users via the memory layers. Note that the firstorder connections are used to find similar users who interacted with the same items.
- HOP-Rec [13]: This is a state-of-the-art graph-based model, where the high-order neighbors derived from random walks are exploited to enrich the user-item interaction data.

Parameter Settings. In the experiment, the embedding size of all models was fixed to 64. In terms of hyper-parameters, a grid search was used to the hyper-parameters: adjusting the learning rate between 0.0001, 0.0005, 0.001, 0.005. In addition, node discard technology was used for GC-MC and GNNA, and the ratio is adjusted to 0.0, 0.1, ... 0.8. Xavier was used to initialize the model parameters and execute the early stopping strategy, that is, if the recall does not increase on 50 consecutive epochs on the verification data, it will stop prematurely. Discard technology was used to prevent overfitting of the neural network. Specifically, two techniques was used: node discard and message discard. The drop rate of the node was set to 0.1 and the message loss rate was set to 0.1.

Table 2. Performance Comparison.

	Gowalla		Amazon-Book	
	recall	ndcg	recall	ndcg
MF	0.1342	0.1178	0.0273	0.0232
NeuMF	0.1402	0.1242	0.0299	0.0254
GC-MC	0.1399	0.1238	0.0346	0.0331
CMN	0.1433	0.1276	0.0383	0.0330
HOP-Rec	0.1462	0.1285	0.0407	0.0352
GNNA	0.1623	0.1355	0.0487	0.0379

4.3   Performance Comparison(RQ1)

Overall Comparison. Table 2 shows the performance comparison of different methods, the results are as follows: MF performed poorly on both datasets, NeuMF consistently outperforms MF across all cases. This indicates that the inner product is not sufficient to capture the complex relationship between the user and the item. Compared with the performance of MF and NeuMF, the performance of GC-MC verifies that merging first-order neighbors can improve representation learning. CMN generally achieves better performance than GC-MC in most cases. Such improvement might be attributed to the neural attention mechanism, which can specify the attentive weight of each neighboring user, rather than the equal or heuristic weight used in GC-MC. The performance of HOP-Rec is better than that of CMN, because HOP-Rec uses high-order neighbors to enrich training data, while CMN only considers similar users. The performance of GNNA is always better than other methods, which proves the importance of different neighborhood weights for user-item interaction. It further illustrates that the use of attention mechanism and high- order connectivity can enhance the presentation of users and items. Table 2 shows the recall@20 and ndcg of different methods. It can be seen that the recall@20 and ndcg of GNNA are significantly higher than other methods, which shows the effectiveness of GNNA.

4.4   Study of GNNA(RQ2)

Since the embedding representation plays an important role in the attention propagation layer, the impact of different layers on performance was studied.

Table 3. Effect of attention propagation layer numbers(L).

	Gowalla		Amazon-Book	
	recall	ndcg	recall	ndcg
GNNA-1	0.1534	0.1299	0.0336	0.0229
GNNA-2	0.1557	0.1321	0.0339	0.0355
GNNA-3	0.1623	0.1355	0.0487	0.0379

Effect of Layer Numbers. In order to investigate whether the attention propagation layer can improve GNNA performance, the depth of the model was changed. The layer number in the range of 1,2,3 was explored. Table 3 summarizes the experimental results, the conclusions were as follows:

Increasing the depth of GNNA can enhance the recommendation effect. Changing the number of propagation layers can obtain corresponding additional interactive signals, which also shows that the attention propagation layer can control the performance of the model.

Table 4. Effect of graph convolution layers.

	Gowalla		Amazon-Book	
	recall	ndcg	recall	ndcg
GNNA-1	0.1534	0.1299	0.0336	0.0229
GNNA-GC-MC-1	0.1474	0.1133	0.0299	0.0199
GNNA-MF-1	0.1470	0.1009	0.0294	0.0146

To investigate how graph convolutional layers affect performance, different layers of GNNA-1 variants were considerd. In particular, the representation interaction between the node and its neighbors were removed from the message passing function. It is set to represent interaction between GC-MC and MF, which are called GNNA-GC-MC-1 and GNNA-MF-1, respectively. Table 4 shows that GNNA-1 is consistently superior to all variants. The improvement is attributed to the role of the attention mechanism. Therefore, the rationality and effectiveness of the attention transmission layer are verified.

5   Conclusions and Future Work

In this work, the attention mechanism was incorporated into the propagation process to improve the quality of recommendation. A new method called GNNA is designed, which uses the attention mechanism to capture different weights for user-item interactions. The key to GNNA is the newly proposed attention propagation layer. A large number of experiments on two real-world datasets prove the rationality and effectiveness of GNNA. In the future, we will further improve GNNA by introducing a knowledge graph to learn more interactions between users and items. We hope that GNNA in the future can help users reason about online behaviors in order to achieve more efficient recommendations.

References

1. Almaghrabi, M., Chetty, G.: A deep learning based collaborative neural network framework for recommender system (2018)
2. Bahdanau, D., Cho, K., Bengio, Y.: Neural machine translation by jointly learning to align and translate. Comput. Sci. (2014)
3. Bahdanau, D., Cho, K., Bengio, Y.: Neural machine translation by jointly learning to align and translate. Comput. Ence (2014)
4. Bayer, I., He, X., Kanagal, B., Rendle, S.: A generic coordinate descent framework for learning from implicit feedback (2017)
5. Berg, R.V.D., Kipf, T.N., Welling, M.: Graph convolutional matrix completion (2017)
6. Gao, Z.Y., Chen, C.P.: deepsa2018 at semeval-2018 task 1: Multi-task learning of different label for affect in tweets (2018)

7. He, X., He, Z., Song, J., Liu, Z., Jiang, Y.G., Chua, T.S.: Nais: neural attentive item similarity model for recommendation. IEEE Trans. Knowl. Data Eng. **30**(12), 2354–2366 (2018)
8. He, X., Liao, L., Zhang, H., Nie, L., Hu, X., Chua, T.S.: Neural collaborative filtering, pp. 173–182 (2017)
9. Huang, S., Huang, M., Zhang, Y., Li, M.: Under water object detection based on convolution neural network (2019)
10. Kakanakou, M., Xie, H., Qiang, Y.: Double attention mechanism for sentence embedding. In: International Conference on Web Information Systems and Applications (2018)
11. Koren, Y., Bell, R.M., Volinsky, C.: Matrix factorization techniques for recommender systems. Computer (2009)
12. Li, J., Ren, P., Chen, Z., Ren, Z., Ma, J.: Neural attentive session-based recommendation (2017)
13. Li, Q., Tang, X., Wang, T., Yang, H., Song, H.: Unifying task-oriented knowledge graph learning and recommendation. IEEE Access PP(99), 1–1 (2019)
14. Long, C., Zhang, H., Xiao, J., Nie, L., Chua, T.S.: SCA-CNN: Spatial and channel-wise attention in convolutional networks for image captioning (2017)
15. Lu, Y., Dong, R., Smyth, B.: Coevolutionary recommendation model: Mutual learning between ratings and reviews (2018)
16. Perera, P., Patel, V.M.: Learning deep features for one-class classification. IEEE Trans. Image Process. (2018)
17. Seo, S., Huang, J., Yang, H., Liu, Y.: Interpretable convolutional neural networks with dual local and global attention for review rating prediction (2017)
18. Tay, Y., Luu, A.T., Hui, S.C.: Latent relational metric learning via memory-based attention for collaborative ranking (2017)
19. Wang, H., Wang, N., Yeung, D.Y.: Collaborative deep learning for recommender systems (2014)
20. Wang, X., He, X., Wang, M., Feng, F., Chua, T.S.: Neural graph collaborative filtering (2019)
21. Xu, K., Li, C., Tian, Y., Sonobe, T., Kawarabayashi, K.I., Jegelka, S.: Representation learning on graphs with jumping knowledge networks. Arxiv (2018)
22. Xue, F., He, X., Wang, X., Xu, J., Liu, K., Hong, R.: Deep item-based collaborative filtering for top-n recommendation. ACM Trans. Inf. Syst. **37**(3), 1–25 (2019)
23. Zhou, X., Liu, D., Lian, J., Xie, X.: Collaborative metric learning with memory network for multi-relational recommender systems (2019)

Graph Query

Dynamic Partition of Large Graphs Combining Local Nodes Exchange with Directed Dynamic Maintenance

Xiaohuan Shan[1], Xiyi Shi[1], Yulong Song[2], Menglin Zhang[1],
and Baoyan Song[1(✉)]

[1] School of Information, Liaoning University, Shenyang, China
bysong@lnu.edu.cn
[2] Department of Computer Science and Engineering, Shanghai Jiao Tong
University, Shanghai, China

Abstract. Graph partition is the key preprocessing of query and analysis for graphs. In the era of big data, graphs have the characteristics of large scale and dynamic evolution. For such large dynamic graphs, the existing graph partition methods have the problems of too slow partition speed, unable to realize dynamic update, and uneven load caused by dynamic changes of graphs. Regarding to the above problems, in this paper, a processing technique combining initial graph partition with incremental dynamic maintenance is proposed. In the initial partition stage, a multi-level local nodes exchange partition algorithm is proposed, which is composed of graph compression, local node exchange partition and restoration optimization. Then an optimization adjustment mechanism is proposed to eliminate redundant modification and reduce computing cost. In the dynamic maintenance stage, several update strategies for different changes are executed. And a directed dynamic maintenance strategy is proposed to avoid frequent or circular exchange caused by two-way movement, so as to improve the efficiency of dynamic graph partition. The experiments show that our proposed method is quite efficient in dynamic partition of large graphs, which is performed both on real and synthetic data.

Keywords: Dynamic graphs · Graph partition · Multilevel division · Load balance · Incremental update

1 Introduction

Graph is an abstract representation of data structure commonly used in computer science. It has been widely applied in social network [1], biological information network [2], economic and military fields [3], etc. With the rapid development of emerging networks, the sizes of graphs are exploding. The original methods have been unable to meet the processing requirements of such a large graph, which use a single high-performance computer to process. So the partitioning of the graph has become a key pre-processing in the graph calculation [4].

In addition, most complex networks in real world are dynamic evolution, and the changes of networks will inevitably lead to the dynamic evolutions of the graph [5].

© Springer Nature Switzerland AG 2020
G. Wang et al. (Eds.): WISA 2020, LNCS 12432, pp. 441–453, 2020.
https://doi.org/10.1007/978-3-030-60029-7_40

For example, in Facebook, such behaviors as account registration, account cancellation and friend relationship change happen all the time. The insertion and deletion of nodes or edges and the update of edge weights in the graph will reflect the dynamic characteristics of the network.

In the context of graph size explosive increase and dynamic, existing graph partition methods are faced with challenges as follows: (1) The original algorithms have problems such as too slow partition speed, poor timeliness and inability to realize dynamic update. (2) The relationship between nodes becomes more complex, how to ensure the characteristics of high cohesion and low coupling have become a difficulty in large-scale graph partition. (3) The dynamic changes lead to uneven load, which directly cause the aggravation of "bucket effect" and affect the efficiency of query [6].

To address these challenges, we make an in-depth study of partition on large dynamic graph. And it is divided into two stages, namely the initial graph partition stage and the incremental dynamic maintenance stage in this paper. we make our contributes as follows:

- In the initial graph partition stage, a multi-level local nodes exchange partition (MLEP) method was proposed based on the idea of hierarchical partition. It includes the compression of original graph according to the proposed breadth-first search (BFS) heavy-edge matching principle; local nodes exchange utilizing the node gain; the optimized restoration of partition results.
- An optimization adjustment mechanism (OAM) for the graph change operations is proposed to eliminate the redundant update of graph data within the time threshold, so as to reduce the computation cost and communication overhead.
- In the incremental dynamic maintenance stage, firstly, we introduce several update strategies for different changes. Then this paper proposes a directed dynamic maintenance strategy, which avoids frequent or circular exchange caused by two-way movement. It improves the efficiency of dynamic graph partition by reducing unnecessary movement.

The rest of this paper is organized as follows. In Sect. 2, we review related works. The details of the initial graph partition method and incremental dynamic maintenance are discussed in Sect. 3 and Sect. 4. Experimental results and analysis are shown in Sect. 5. We finally conclude the work in Sect. 6.

2   Related Work

A good graph partition algorithm will play a good foundation role in query, analysis, and other operations. Since graph partition problem is an NP-hard [4], it is difficult to obtain an accurate solution within a feasible time, so most researchers solve this problem from the perspective of optimization and heuristics.

As classical heuristic algorithms, FM [7] can produce a good approximate optimization solution for the partition on small static graphs. Donath and Fiedler put forward the spectral method [8] to realize the two-way partition based on the eigenvalues of Laplace matrix. But its complex operation makes it impossible to achieve partition on large graphs. The geometric partition methods [9, 10] realized the partition rapidly by the information of geometric coordinate. Although the partition speed of this kind of method is faster than the spectral method, but the partition quality is poor. In recent years, the algorithms based on the idea of multi-layer partition have been widely studied [11, 12], that is because they can effectively deal with the partition of large graphs. JA-BE-JA [13] assigns a random partition to each node firstly. Then it will try to swap their nodes with other nodes belong other partitions. An advantage of this approach is that the partitions cannot change their size across the computation, therefore their balance can be guaranteed. BA framework [14] is a scatter - gather local search scheme, which can process the bulk data efficiently. However, it ignores the dynamic of graphs, so working on dynamic graphs is inefficient.

At present, the dynamic of the large graph results in that the original static partition algorithms can be applied to the initial partition, but there is no update mechanism for the dynamic changes. Therefore, the update can only be realized through re-partition, which needs to massive migration and communication overhead [15]. DynamicDFEP [16] is proposed to deal with the partition on large dynamic graph. It firstly uses DEEP [17] to finish the initial partition. It randomly assigns a partition for each node and gives an initial funding. Each node uses its funding to "buy" its neighbors to expand its partition, and the principle of random selection has a certain influence on the stability of the partition quality. On this basis, several different updating strategies are proposed, but the load balancing problem after dynamic updating is neglected.

3   Multi-level Local Nodes Exchange Partition

3.1   Heavy-Edge Matching Graph Compression Based on BFS

It is easy to ensure the efficiency and quality of partition in medium or small graph relatively. Therefore, this paper proposes a Heavy - edge Matching Graph Compression based on BFS (HMGC-BFS), which is effective to reduce the size of the graph.

The main process of HMGC-BFS is as follows: (1) it selects the node with the largest edge weight as the initial node to carry out BFS in G. (2) it compares all edge weights connected with the initial node and selects the edge with the largest edge weight. Such edge is marked and called inner compressed edge. (3) HMGC-BFS continues to traverse the other nodes to find all inner compressed edges connected by two unmarked nodes and mark them. (4) each pair of nodes connected the inner compressed edges are merged in traversal order. Meanwhile, the edges connected by either compressed nodes are also merged as the external compressed edges and the sum of the weights of external compressed edges is the new weight. At the same time, we store all compressed information for use in subsequent restoration phases.

For example, in Fig. 1, (a) is the initial graph G. We select v_1 as the initial node, which is one of endpoints of edge with the largest weight. According to the principle of BFS, the traversal order is $v_1 \rightarrow v_2 \rightarrow v_4 \rightarrow v_3 \rightarrow v_5 \rightarrow v_7 \rightarrow v_6 \rightarrow v_9 \rightarrow v_{10} \rightarrow v_8$. An ordered set of inner compressed edges that have been marked is $\{(v_1, v_2), (v_3, v_4), (v_7, v_{10}), (v_6, v_8), (v_5, v_9)\}$. The dotted edges are shown in Fig. 1 (b). According to the order of the set of inner compressed edges, the two nodes of each inner compressed edge in the set are merged and weights of inner compressed edges are recorded to ensure the correctness of the subsequent restoration. The final compression graph is shown in Fig. 1 (c).

3.2 Initial Partition of the Compression Graph

In order to realize the efficient partition of the compressed graph, a partition algorithm Local Nodes Exchange based on Weight (W-LNE) is proposed in the initial partition phase. W-LNE firstly carries out initial pre-partition on the compressed graph, which divides all nodes into K subgraphs on the basic average to ensure the relative balance and avoid large deviation. Second, we fully consider the effect of weights on the partition to keep high cohesion and low coupling relatively. We define the concept of node gain, which is used to evaluate a node whether needs to move. The formal definition is as follows.

Definition 1 (Node Gain): For a node u, its node gain is the reduction in cost of exchange to other subgraphs. N_u is the neighbor set of u, $P(v)$ is the partition of v.

$$\Delta_u(P_x) = |\{|v| * \sum W_v | v \in N_u \wedge P(v) = P_x\}| - |\{|v| * \sum W_v | v \in N_u \wedge P(v) = P(u)\}| \tag{1}$$

We use the node gain to judge whether adjusting u to subgraph P_x is better than its current. If $\Delta_u(P_x) > 0$, we call it a positive gain, which means that moving u to P_x is better. Otherwise, it is a negative gain, and node u is better to maintain the status quo.

Therefore, we calculate the node gain of nodes in each subgraph. We select the same number of nodes from different subgraphs for exchange that is to maintain load balancing. W-LNE specifies that the exchange number of nodes is the minimum number of positive gain. To obtain better partition results, the algorithm compares the external cut-edge weights W_{out} with an average weight of each partition $\bar{w} = \sum w(u)/k$ (k is the number of partitions) to determine the final exchange nodes. Through multiple exchanges, a better initial partition results can be achieved. The specific algorithm is shown in Algorithm 1.

Algorithm 1 W-LNE Algorithm
Input: $G(V,E,W)$, k
Output: k-way partition results
1   Assign each node to some partition and form k subsets of V randomly, and initialize all nodes in V unmarked;
2   compute $\overline{w} = \sum w(u)/k$;
3   for each combination $<V_a \in P_a, V_b \in P_b>$ do
4   for each $u \in V_a$ do
5   compute $\Delta_u(P_b)$;
6       $D_a \leftarrow \{u \mid u \in V_a \wedge \Delta_u(P_b) \geq 0\}$;
7   for each $v \in V_b$ do
8   compute $\Delta_v(P_a)$;
9         $D_b \leftarrow \{v \mid v \in V_b \wedge \Delta_v(P_a) \geq 0\}$;
10   if($|D_a| < |D_b|$)
11   for each $v \in V_b$ do
12   compute W_{out};
13   if $W_{out} > \overline{w}$
14   delete v form D_b;
15   for each $u \in V_a$ do
16   compute W_{out};
17   if $W_{out} > \overline{w}$
18   delete u form D_a;
19   for each $u \in V_a, v \in V_b$ do
20       $P(u) = P_b$; $P(v) = P_a$; state(u)= state(v)= marked;
21   if $|Da|=0$ or $|Db|=0$ then break;
22   return the k-way partition results;

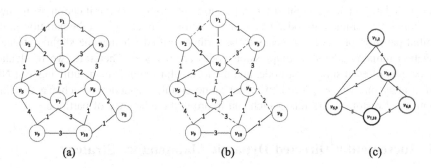

(a)   **(b)**   **(c)**

Fig. 1. Initial graph as in (a), the marked inner compressed edges in (b), and final result in (c).

Take the compressed graph G_c in Fig. 1 (c) as an example for initial partition, and we set $k = 2$. Firstly, the nodes of G_c are randomly divided into two partitions, $P_{shadow} = \{v1, 2, v7, 10\}$ and $P_{white} = \{v_{3, 4}, v_{5, 9}, v_{6, 8}\}$, as shown in Fig. 2 (a). Then we

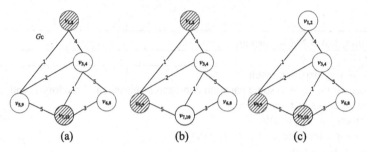

Fig. 2. Example of local nodes exchange

calculate the node gain of 5 nodes belong two partitions, such as $\Delta v_{1,\ 2}(P_{\text{white}}) = 10$, $\Delta v_{3,\ 4}\ (P_{\text{shadow}}) = -4$. According to the above calculation, the gains of nodes in P_{white} are both greater than zero, while in P_{shadow} there is only one greater than zero, so only one node will be exchanged. There are two cases for node exchange, case 1 as in Fig. 2 (b), and we swap $v_{5,\ 9}$ and $v_{7,\ 10}$. The cut edge weight after swapping is W_out = 11, and $\bar{w}$ = 10.5, W_out > $\bar{w}$. Case 2 as in Fig. 2 (c), and we swap $v_{1,\ 2}$ and $v_{5,\ 9}$, which W_out < $\bar{w}$. Therefore, swapping $v_{1,\ 2}$ and $v_{5,\ 9}$ makes better partition results.

3.3   Optimized Restoration

The task of the restoration phase is to restore the initial partition results of the compressed graph to the original graph. In this paper, the compressed nodes are restored quickly utilizing the compression information table stored in the compression phase. Then the cut-nodes will be readjusted, which are connected with the edge of other partitions. The cut-nodes are further divided into composite cut-nodes (nodes by compression expanding) and simple cut-nodes (uncompressed nodes). We compress the edges with larger weights in the compression phase, so the composite cut-nodes still retain a larger edge weight after expansion, so no further verification is needed. Although the simple cut-nodes has a slightly stronger relation with the partition in the initial partition phase, but it needs to be further verified whether the slightly stronger relation remains after the compression nodes are restored. Therefore, we recalculate node gain of each simple cut-node, and the local adjustment is made utilizing W-LNE. In conclusion, we only need to adjust some simple cut-nodes to achieve the final optimized restoration results, which can improve the efficiency of partition.

4   Incremental Directed Dynamic Maintenance Strategy

4.1   Graph Change Operation Optimization Adjustment

In real applications, the dynamics of networks are abstracted to insertion and deletion of nodes or edges, and the changes of edge weights. In this paper, we take the time threshold T as a unit, and the initial partition results are optimized and maintained by using the changes of the graph in T time. Compared with repartition cost for each change, it greatly reduces the cost to achieve the partition on large dynamic graphs. For the convenience of expression, this paper defines the above dynamic change as graph change operation.

Definition 2 (Graph Change Operation, GCO): GCO can be represented by $<op, value>$, thereinto $op = ins/del/upd$, represents insert/delete/weight update respectively, and $value$ represents the information of the node or edge corresponding to op.

It is worth noting that we ignore the case of inserting an isolated node (that is, no edges are connected to other nodes), which is lacks practical significance.

Definition 3 (Graph Change Operation Set, GCOS): GCOS is a set of GCO_t within a time threshold T. It can be represented as $GCOS_T = \{GCO_1, GCO_2..., GCO_t...\}$, and GCO_t is the GCO of some time.

According to the actual situation, within a T, GCO with different timestamps may have a corresponding relationship, that is, it may be an operation for the same node or edge. We consider that the communication and processing overhead of assigning operations to the corresponding partitions is much greater than the cost of optimizing it. So in this paper, the Optimization Adjustment Mechanism (OAM) is proposed to realize the adjustment of merging or deleting the operation for the same node or edge. Its specific process is as follows, which gives different strategies for different GCO. For ease of description, GCO with a previous timestamp is defined as GCO_{tbef}, and other GCO that arrives later and is associated with it is defined as GCO_{taft}.

- GCO_{tbef} is inserting a node. GCO_{taft} can be considered in the following cases: (1) if GCO_{taft} is inserting an associated edge with GCO_{tbef}, then insert it into the value of GCO_{tbef} and delete it. (2) if GCO_{taft} is deleting the same node with GCO_{tbef}, then delete both GCO_{tbef} and GCO_{taft}. And if GCO_{taft} is deleting the connected node with GCO_{tbef}, then delete the edge in value of GCO_{tbef}. (3) if GCO_{taft} is deleting an associated edge with GCO_{tbef}, then delete the edge from value of GCO_{tbef} and GCO_{taft}. (4) if GCO_{taft} is updating the weight of some edge in GCO_{tbef}, then update it in GCO_{tbef} and delete GCO_{taft}.
- GCO_{tbef} is inserting an edge. (1) If GCO_{taft} is inserting a node connected with GCO_{tbef} or the same edge with GCO_{tbef}, then delete GCO_{tbef}. (2) If GCO_{taft} is deleting a node connected with GCO_{tbef}, delete GCO_{tbef}. (3) If GCO_{taft} is deleting the same edge with GCO_{tbef}, delete GCO_{tbef} and GCO_{taft}. (4) If GCO_{taft} is updating the weight of the edge in GCO_{tbef}, update it in GCO_{tbef} and delete GCO_{taft}.
- GCO_{tbef} is deleting a node. If GCO_{taft} is deleting the same with GCO_{tbef}, then delete GCO_{tbef}.
- GCO_{tbef} is deleting an edge. (1) If GCO_{taft} is inserting a node connected with GCO_{tbef}, then delete GCO_{tbef}. (2) If GCO_{taft} is inserting the same edge with GCO_{tbef}, then delete GCO_{tbef} and change the op of GCO_{taft} to upd. (3) If GCO_{taft} is deleting the same node with GCO_{tbef}, then delete GCO_{tbef}. (4) If GCO_{taft} is deleting the same edge with GCO_{tbef}, then delete GCO_{tbef}.
- GCO_{tbef} is updating a weight. (1) If GCO_{taft} is deleting a node or edge associated with GCO_{tbef}, then delete GCO_{tbef}. (2) If GCO_{taft} is updating the weight of the same edge in GCO_{tbef}, then delete GCO_{tbef}.

Through the analysis of the above situations, it can be seen that the number of GCO in GCOS can be reduced by merging or deleting the relevant nodes or edges with

OAM, thus reducing the number of adjustments. It reduces the communication and computing costs effectively without affecting the evolutionary outcome.

4.2 Incremental Dynamic Update

Based on the optimized GCOS, The main idea of DDM is to number and sort for each partition according to the size of the load. The nodes that need to adjust will move in one direction according to the partition number. The purpose of this is to avoid frequent or circular exchange between fixed nodes caused by two-way movement and reduce unnecessary movement. Different update strategies are given according to the different GCO, as follows.

(1) Node/Edge insertion. In this paper, weight is used to express the closeness of the relation between nodes. Therefore, we firstly define a concept of closeness as a function to evaluate the quality of partitions.

Definition 4 (Closeness): For a given graph G, the closeness of it is $C(G)$. W_{in} is the sum of weights inside the G and W_{out} is the sum of weights of cut-edges about G.

$$C(G) = W_{in}/(W_{in} + W_{out}) \tag{2}$$

In addition, in order to determine whether a node u can be inserted into G, the concept of closeness gain is defined as follows:

$$C_G = C(G \cup \{u\}) - C(G) \tag{3}$$

It can be seen that the greater the closeness gain is, the more closeness of the partition after adding u. It indicates the closer relation between u and the partition.

For the inserted new node u, if it is connected with only one node, the node is directly added to the partition where its neighbor is. In addition, it may be related to multiple nodes, which can be represented by inserting a set of edges. If nodes connected with u belong the same partition, we add u into the partition too. Else we calculate closeness gain between u and several partitions to obtain the final partition of u.

For the inserted new edge $<u, v, w>$, if u and v have existed, the new edge is added into the edge set. If either u or v isn't belongs to any partition, which means they are new nodes, so the information of nodes and edge both update to the partition.

(2) Node/Edge deletion. If the deleted node is connected with the cut-node of the partition, we calculate the node gain of cut-node to judge whether the cut-node need to be moved. Similarly, if the deleted edge is connected an inner node and cut-node, we also calculate the node gain of cut-node. Else the deleted edge/node will be deleted directly.

(3) Weight update. If the edge need to update weight is the inner edge of a partition, we alter the information directly. But if the edge is the cut-edge, it affects the cut-node. Therefore, the gain of the involved cut-nodes should be calculated.

4.3 Directed Dynamic Maintenance

With the insertion and deletion of nodes or edges, there will be load imbalance among different partitions and it makes the task load of each partition has a large deviation, which leads to the increase of time cost and the "bucket effect".

To avoid this phenomenon, this paper proposes a directed dynamic maintenance strategy (DDM). Firstly, in order to measure the load capacity of each partition, we define the load factor of partition P as follows.

$$Load(P) = \frac{|V^P| + |E^P|}{|V| + |E|} \qquad \overline{Load(P)} = \frac{\sum_{i=1}^{|P|} Load(P_i)}{|P|} \qquad (4)$$

$Load(P)$ is load factor of P, $|V^P|$ and $|E^P|$ represent the number of nodes and edges in P. $|V|$ and $|E|$ are the number of nodes and edges in original graph. We measure the load balancing degree of each partition. The $\overline{Load(P)}$ is the average of the load factors. If the load factor of some partition is greater than the average, it indicates that this partition has excessive load, and dynamic adjustment is considered.

In addition, in order to avoid frequent exchange turbulence caused by the same two partitions exchanging the same node, this paper utilizes a directed dynamic maintenance strategy. It ranks the partitions according to $Load(P)$, and a partition with a high $Load(P)$ has a low partition id. It moves nodes of the partition with low partition id to the partition with high id. Then the nodes are moved in the opposite direction.

5 Experiments

5.1 Experimental Settings

All experiments are implemented in JAVA and built it on Giraph, which is a popular graph distributed computation framework based on Hadoop. We deployed on four machines, each is equipped with a 3.2 GHZ Intel(R) Core(TM) i5-4460 CPU, 8 GB memory, 1T hard disk and runs Ubuntu 16.04.

Datasets. We use three different real datasets and two synthetic datasets to evaluate our method. Real datasets include Amazon, Youtube and LiveJournal. The details of the real datasets above are given in Table 1. Synthetic datasets G_1, and G_2 are generated by R-MAT [16]. The number of nodes is 10^6 and 10^7 respectively, and the number of edges is 8 times nodes'. Each edge of above datasets is randomly assigned a weight between 1 and 100 to indicate the closeness of two nodes.

Table 1. Statistics of Real Datasets

Datasets	Abbr.	$|V|$	$|E|$
Amazon	AM	334863	925872
Youtube	YT	1134890	2987624
LiveJournal	LJ	4847571	68993773

Since the evolution of graphs over time is a smooth process, in order to simulate dynamism in each dataset, the datasets in this paper are divided into two parts according to the ratio of 9:1. We use 90% of the graphs in initial partition and simulate graphs change dynamically by the remaining 10%. Unless stated otherwise, we compute a 4-way partitioning of the input graphs. Each experiment was run 5 times and the average is reported here.

5.2   Performance Analysis of Initial Graph Partition

In this section, the experiments are compared with BS and DynamicDFEP algorithms from two aspects, which are partition time and the number of edge-cuts.

Figure 3 shows the comparison of the initial partition time on real and synthetic datasets. As shown, each partition time increase with the growth of graphs. There is a small gap on the smaller graphs, and with the increasing of graphs, the gap is more and more obvious. Among them, BS divides the graph into two partitions randomly, and then it swaps nodes to achieve good result according to the setting of parameters. It is affected by the parameters. DynamicDFEP assigns a partition for each node randomly and gives an initial funding. Each node uses its funding to expand its partition from its neighbors. Our MLEP compresses the graph in a smaller size, and it partition the compressed graph, then the partition result is restored to the original graph.

Figure 4 shows the comparison of the edge-cut on different datasets. Edge-cuts are the number of edges that have endpoints in different partitions. We can validate the partition quality by the edge-cuts. In order to ensure the partition principles of high cohesion and low coupling, the less edge-cuts proof that partition quality is better indirectly. BS swaps the nodes to improve the partition result, but it ignores the influence of weights for the closeness. DynamicDFEP utilizes the principle of random selection, which has a certain influence on the partition quality. MLEP considers the effect of weights and degrees for closeness, so we exchange the nodes to improve the closeness of inner and reduce it of outer.

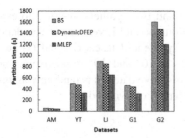

Fig. 3. Comparison of the partition time.

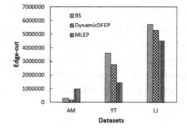

Fig. 4. Comparison of the edge-cut.

5.3   Performance Analysis of Incremental Dynamic Maintenance

We evaluate the update efficiency on different number of partitions. Due to BS is the algorithm on static graphs, then it only repartitions the graphs to achieve the update, which is much lower than the incremental update methods. So we compare our update strategy IDU with DynamicDFEP, which includes three kinds of update strategies.

As shown in Fig. 5, the proposed update strategy IDU in this paper is better than other algorithms. That is because we utilize OAM to eliminate the unnecessary operations before update, and we only update a smaller set of graph change operations incrementally. We note that the UB-Ins provides better results than other methods in terms of running time. It can be explained by that face that it only needs one round to complete. We also note that the Com-Ins is faster than Part-Ins. It is because that the Com-Ins needs to fewer rounds than the Part-Ins.

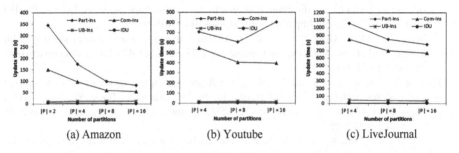

(a) Amazon   (b) Youtube   (c) LiveJournal

Fig. 5. Comparison of the update strategy on different datasets

6   Conclusion

In this paper, we proposed a partition method for large dynamic graphs, which includes initial partition and incremental dynamic maintenance. Firstly, we propose a multi-level local nodes exchanges partition method (MLEP) to partition the graph initially at some static moment. Secondly, an optimization adjustment mechanism (OAM) for the graph change operations is proposed to eliminate the redundant update of graph data, so as to reduce the computation cost and communication overhead. Finally, we proposed a directed dynamic maintenance strategy, which avoids frequent or circular exchange caused by two-way movement. Experiments show that the proposed method can effectively reduce the computation and communication overhead under the premise of partition efficiency. It effectively maintains the cohesion of each partition, and realizes the dynamic partition of large graphs.

Acknowledgements. This work was supported by the National Key Research and Development Program of China (2019YFC0850103); National Natural Science Foundation of China under Grant (No. 61472169, 61502215, 61802160, 51704138); The Key Research and Development Program of Liaoning Province (No. 2017231011); The National Natural Science Foundation of China (No. U1811261); China Postdoctoral Science Foundation Funded Project (No. 2020M672134);

Science Research Fund of Liaoning Province Education Department (No. LJC201913); Liaoning Public Opinion and Network Security Big Data System Engineering Laboratory (No. 04-2016-0089013).

References

1. Calle, J., Rivero, J., Cuadra, J., et al.: Extending ACO for fast path search in huge graphs and social networks. Expert Syst. Appl. **86**, 292–306 (2017). https://doi.org/10.1016/j.eswa.2017.05.066
2. Sonmez, A.B., Can, C.: Comparison of tissue/disease specific integrated networks using directed graphlet signatures. BMC Bioinform. **18**(4), 41–50 (2017). https://doi.org/10.1186/s12859-017-1525-z
3. Aydin, B., Henning, M., Llya, S, et al.: Recent advances in graph partitioning. Algorithm Eng. 56, 117–158 (2016). http://doi.org/10.1007/978-3-319-49487-6_4
4. Mazaheri Soudani, N., Fatemi, A., Nematbakhsh, M.: An investigation of big graph partitioning methods for distribution of graphs in vertex-centric systems. Distrib. Parallel Databases, 1–29 (2019). http://doi.org/10.1007/s10619-019-07256-z
5. Shi, Z., Li, J.H., Guo, P.F., et al.: Partitioning dynamic graph asynchronously with distributed FENNEL. Fut. Generat. Comput. Syst. **71**, 32–42 (2017). https://doi.org/10.1016/j.future.2017.01.014
6. Shan, X., Ma, J., Gao, J., Xu, Z., Song, B.: A subgraph query method based on adjacent node features on large-scale label graphs. In: Ni, W., Wang, X., Song, W., Li, Y. (eds.) WISA 2019. LNCS, vol. 11817, pp. 226–238. Springer, Cham (2019). https://doi.org/10.1007/978-3-030-30952-7_24
7. Fiduccia, C.M., Matteyses, R.A.: Linear-time heuristic for improving network partitions. In: Proceedings of the 19th Design Automation Conference. Institute of Electrical and Electronics Engineers, pp. 175–181 (1988). https://doi.org/10.1145/62882.62910
8. Pothen, A., Simon, H.D., Liou, K.: Partitioning sparse matrices with eigenvectors of graphs. SIAM J. Matrix Anal. Appl. **11**(3), 430–452 (1990). https://doi.org/10.1137/0611030
9. Berger, M.J., Bokhari, S.H.: A partitioning strategy for nonuniform problems on multiprocessors. IEEE Trans. Comput. **100**(5), 570–580 (2006). https://doi.org/10.1109/TC.1987.1676942
10. Lisser, A., Rendl, F.: Graph partitioning using linear and semidefinite programming. Math. Program. **95**(1), 91–101 (2003). https://doi.org/10.1007/s10107-002-0342-x
11. Filippidou, I., Kotid, Y.: Online partitioning of multi-labeled graphs. In: Proceedings of the GRADES 2015, ACM, 1–6 (2015). https://doi.org/10.1145/2764947.2764950
12. Preen, R.J., Smith, J.: Evolutionary n-level hypergraph partitioning with adaptive coarsening. IEEE Trans. Evol. Comput. **23**(6), 962–971 (2019). https://doi.org/10.1109/TEVC.2019.2896951
13. Rahimian, F., Payberah, A.H., Girdzijauskas, S., et al.: JA-BE-JA: a distributed algorithm for balanced graph partitioning. In: IEEE International Conference on Self-adaptive & Self-organizing Systems. IEEE, pp. 51–60 (2013). https://doi.org/10.1109/saso.2013.13
14. Chen, T., Li, B.: A distributed graph partitioning algorithm for processing large graphs. In: IEEE Symposium on Service-oriented System Engineering. IEEE, 53–59(2016). http://doi.org/10.1109/SOSE.2016.48
15. Osaba, E., et al.: Dynamic partitioning of evolving graph streams using nature-inspired heuristics. In: Rodrigues, J.M.F., et al. (eds.) ICCS 2019. LNCS, vol. 11538, pp. 367–380. Springer, Cham (2019). https://doi.org/10.1007/978-3-030-22744-9_29

16. Chayma, S., Sabeur, A., Alessio, G., et al.: DynamicDFEP: a distributed edge partitioning approach for large dynamic graphs. In: Proceedings of the 20th International Database Engineering & Applications Symposium, pp. 142–147. ACM (2016). https://doi.org/10. 1145/2938503.2938506

17. Guerrieri, A., Montresor, A.: DFEP: distributed funding-based edge partitioning. In: Träff, J.L., Hunold, S., Versaci, F. (eds.) Euro-Par 2015. LNCS, vol. 9233, pp. 346–358. Springer, Heidelberg (2015). https://doi.org/10.1007/978-3-662-48096-0_27

Link Prediction Based on Smooth Evolution of Network Embedding

Hao Dong, Yue Kou$^{(\boxtimes)}$, Derong Shen, and Tiezheng Nie

Northeastern University, Shenyang 110004, China
2431752775@qq.com, {kouyue,shenderong,
nietiezheng}@cse.neu.edu.cn

Abstract. The problem of link prediction in dynamic heterogeneous information networks has been widely studied in recent years. The technique of network embedding has been proved extremely useful for link prediction. However, the existing methods lack the close combination between deep-level features and temporal features of networks, which affects the accuracy of prediction and makes it difficult to adapt to the dynamic networks. In this paper, a Smooth Evolution model for Network Embedding (called SENE) is proposed, which considers both deep-level features and temporal features to obtain the embedded representations of the network structure, and uses the transformer mechanism to effectively obtain the smooth evolution of network embedding. Also an SENE-based link prediction algorithm is proposed, which can effectively guarantee the accuracy of link prediction. The feasibility and effectiveness of the proposed key technologies are verified by experiments.

Keywords: Link prediction · Network embedding · Smooth evolution · Transformer

1 Introduction

With the increasing popularity of the Internet, a large number of users use the Internet every day to understand the changing development and entertainment of the world. While using the Internet, users also leave a lot of data on it. And these different types of data can be combined into different networks. For example: the information from WeChat can be abstracted into a social network, treating each user as a node, and the friend relationship between the users as an edge. Compared to homogeneous networks, heterogeneous networks contain more information, so data mining for heterogeneous networks is more valuable. But, due to the inherent complexity of the heterogeneous network, how to better integrate the features contained in the heterogeneous network is still a challenge.

At the same time, link prediction [1], has attracted increasing attention in the research community, due to its importance in many real-world application. Link prediction is to analyze the known network topology and construct a prediction method to find the probability of edges between node pairs that do not yet have formed in the network. It can be divided into static link prediction and dynamic link prediction according to the time when the edge appears. Static link prediction is to predict the

© Springer Nature Switzerland AG 2020
G. Wang et al. (Eds.): WISA 2020, LNCS 12432, pp. 454–466, 2020.
https://doi.org/10.1007/978-3-030-60029-7_41

topology of an unobserved network based on the observed partial network topology of the current network. Dynamic link prediction [2] is to predict the network structure at the next moment based on the observed changes in the network topology. Because the real world networks such as e-commerce networks, social networks, and user travel networks often change dynamically, dynamic link prediction has more practical value.

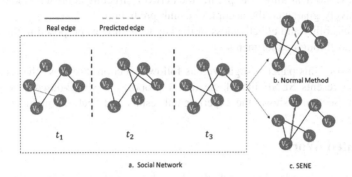

Fig. 1. An example of link prediction in social network

Let us consider the following motivating scenarios.

Scenario 1: Yesterday, employee e_1 sent an email to employee e_2, but today employee e_1 sent an email to employee e_3. This is, the real social networks are not static and will change over time. Generally, the current interaction information of employees is more dependent on recent behavior. Therefore, we need find a way to capture the temporal feature in the network.

Scenario 2: Figure 1(a) depicts a social network which includes a set of users (v_1-v_6). Each node represents a person and the edge between nodes represents the interaction information between people. t_1, t_2, t_3 represents the topological structure of the network in three adjacent time slices respectively. Next, the connection of v_1 in the next time slice will be predicted.

From **Scenario 2**, we can see the general link prediction method tend to predict between the edges that formed in previous time slice. As shown in Fig. 1 (b), it is incorrectly predicted that the edge between v_1 and v_4 is more likely to form at the next moment. Considering the network evolution information, we can predict the edge that never formed before. As shown in Fig. 1 (c), after considering the evolution information, the connection of v_1 in the next time slice can be correctly predicted.

The current link prediction methods for heterogeneous information networks [3] face some challenges: First, how to consider the inherent features and temporal features of the network itself when embedding the network? Secondly, how to effectively capture the evolution information embedded in the networks?

For the above problems, we propose a link prediction method based on smooth evolution of network embedding. The main contributions are as follows:

(1) A Smooth Evolution model for Network Embedding (called SENE) is proposed. Different from traditional network embedding models, SENE not only considers the topology feature and context feature of networks, but also makes full use of the temporal feature to capture the smooth evolution of network embedding.

(2) A SENE-based link prediction algorithm is proposed, which uses the smooth evolution information of network embedding to predict the network embedding at the next moment, further to predict the network structure at the next moment. It can effectively guarantee the accuracy of link prediction.

(3) The feasibility and effectiveness of the key technologies proposed in this paper are verified through experiments.

The rest of this paper is organized as follows. Section 2 reviews the related work. Section 3 presents SENE model. Section 4 proposes a SENE-based link prediction algorithm. Section 5 shows the experimental results and Sect. 6 concludes.

2 Related Work

Various approaches for link prediction have been studied over the years. First, we briefly review the techniques for them. Then we analyses how our work differs from them. Current link prediction methods can be divided into network topology-based link prediction and network embedding-based link prediction [4].

Topology-based link prediction [5] starts from the topology of the network and calculates the similarity between the nodes in the network. Generally based on common neighbors, Jaccard, path similarity, etc. For example, Mitzenmache et al. [6] proposed a domain-based link prediction method. The essence is that two nodes have more common neighbors, and it is more likely that there is an edge. Brin et al. [7] proposed a method based on path similarity, using random walk to obtain the path of the node, and then calculating the path similarity. Nodes with high similarity are more likely to have edge connections.

Network embedding-based link prediction [8] mainly uses a low-dimensional vector to represent the nodes, and then calculates the similarity between the node representations. For example, Perozzi et al. [9] proposed a network embedding algorithm Deepwalk, which introduces deep learn into network representation learning, uses random walk to obtain sequence information, and then uses the Skip-gram model for network embedding. Grover et al. [10] proposed the node2vec algorithm to further develop the Deepwalk algorithm, and reached a good balance between the depth and breadth of random walk. Subsequently, Dong et al. [11] proposed a network embedding algorithm matepath2vec for heterogeneous information networks, using random walk based on a given element path to capture rich context information in heterogeneous information networks. Tang et al. [12] proposed another embedding algorithm LINE, and optimized it with negative sampling, and achieved a good balance between the accuracy and time complexity of the algorithm. Chen et al. [13] proposed PME, a link prediction algorithm for heterogeneous information networks, to decompose a heterogeneous information network into multiple networks, and then embedding each network separately to complete the link prediction task.

The differences between our work and existing work are as follows:

First, the link prediction method proposed in this paper differs from the above-mentioned techniques in that the traditional link prediction technology mainly acts on the prediction on homogeneous information networks, although some papers have studied the link prediction technology on heterogeneous information networks. But usually only consider the characteristics of the network itself, and ignore the timing information.

Second, this paper propose a SENE-based Link Prediction Algorithm. When embedding nodes, it also considers the deep-level features and temporal features of the network, and finally obtains the evolution information of network embedding by using the transformer mechanism.

3 SENE Model

3.1 Problem Definition

This section first introduces several related concepts, and then formalizes the definition of link prediction for dynamic heterogeneous information networks.

Definition 1 (heterogeneous information network): A heterogeneous information network is an information network with multiple types of objects and/or multiple types of links, formally defined as $G = (V; E; T; R)$. Among them, V is the union of different types of vertices, E is the union of different types of edges, T denotes the node type set, and R denotes the edge type set.

Given a heterogeneous information network $G = (V; E; T; R)$, where $T = \{t_1, t_2, \cdots, t_n\}$, representing n consecutive time intervals. At the same time, $G_t = (V_t, E_t, T, R)$ is used to represent the network topology at time t, where V_t is a subset of V, representing the set of nodes at time t, and Et is a subset of E at time t, representing the set of edges at time t. At the same time, We assumed that the total number of nodes at each moment does not change, so G_t can represented by $G_t = (V_t, E_t, T, R)$. In total, the heterogeneous information network G can represented by $G = G_1 \cup G_2 \cup \cdots \cup G_n$ or $G = \{G_t \mid t = 1, 2, \cdots, n\}$. Based on this we give the definition of dynamic link prediction.

Definition 2 (Dynamic Link Prediction): Given the network topology at time slice $0 - T$, predicting the network connection at time $T + 1$, that is, predicting the network connection at the next time based on network history information as follows:

$$G = \{G|G_1, G_2, \cdots, G_t\} \rightarrow G_{t+1}$$

3.2 Model Overview

In order to better capture the deep-level and temporal features of heterogeneous information networks, we proposes Smooth Evolution model of Network Embedding— SENE. This section first introduces how to consider the deep-level and temporal feature when embedding the network, and then introduces the method of using the transformer mechanism to obtain network evolution information.

We builds the SENE model based on the heterogeneous network G. The model can be divided into two phases: network embedding phase and smooth evolution phase (see Fig. 2).

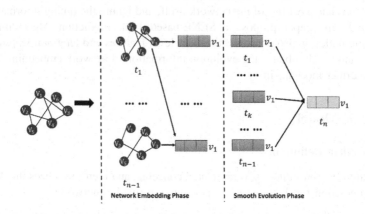

Fig. 2. Overview of SENE

Network Embedding Phase: The network is divided into sub-networks of different time slices at certain time intervals, and then each sub-network is embedded separately. When embedding, not only the topology information and context information of the node itself, but also the historical behavior information of the node must be considered to facilitate the capture of the evolution of network embedding.

Smooth Evolution Phase: In the previous network embedding stage, the network embedding that time slices from 0 to $n-1$ can already be obtained. By exploring the underlying rules in the evolution information of the network embedding from 0 to $n-1$ time slices, the network embedding at Nth time slice can be obtained. Then, we uses the transformer mechanism to capture network evolution information.

3.3 The First Phase: Network Embedding

In order to get a good embedding, we use Tang's LINE method to embedding the network, and make some modifications to the LINE algorithm to make it more suitable for our model. The LINE method is show as follows:

$$p_1\left(v_i, v_j\right) = \frac{1}{1 + \exp\left(-u_i \cdot u_j\right)} \tag{1}$$

$$O = -\sum_{(i,j) \in E} log p_1\left(v_i, v_j\right) \tag{2}$$

Where u_i is low dimensional vector representation of node v_i, p_1 (.,.) is a distribution in the vector space of $V * V$, O is the objective function.

Since the LINE does not distinguish the different of links, LINE is only applicable to homogeneous information networks. Therefore, we decompose a heterogeneous information network into multiple homogeneous information networks. At the same time, in order to ensure the consistency of embedding, we add the objective function of each homogeneous network to obtain the learning objective function O as follows:

$$O = \sum_{r \in R} \sum_{(i,j) \in E} \sum_{(i,k) \notin E} \left[s(v_i, v_j) - s(v_i, v_j) \right] \tag{3}$$

Where $s(.,.)$ represents similarity of node pairs, in this paper we use Euclidean distance to measure the similarity of node pairs.

In addition, in order to distinguish between different links, we assign different weights w to different type of links when embedding nodes, and use w_r represents the weight of edge type r. The attention [14] mechanism can achieve a corresponding weight for each different links type. Therefore, we use the attention mechanism for weight learning. The learning objective function after adding weight is shown in (4):

$$O = \sum_{r \in R} w_r \sum_{(i,j) \in E} \sum_{(i,k) \notin E} \left[s(v_i, v_j) - s(v_i, v_j) \right] \tag{4}$$

The calculation of the minimum value of Eq. (4) takes a huge time cost, and the number of non-linked pairs is extremely large compared to the linked pairs. Therefore, we adopt a negative sampling strategy to optimize the model. The general negative sampling method is to sample non-links pairs with the same probability, and does not take into account temporal feature, so we proposes a negative sampling strategy based on temporal feature.

If the i and k nodes are connected in the previous time slice, but not connected in the current time slice, the possibility of the i and k nodes connecting at the next moment is still very large, so the embedding between the nodes should be relatively similar. Based on the above analysis, we should sample nodes that have never been connected or nodes that have not been connected for a long time before.

At the same time, in order to simplify the problem, we only consider the distance between the time slice last connected between two nodes and the current time slice as follows:

$$P(j) = \frac{t - t_{ij}}{\sum_{k \notin E_t} (t - t_{ik})} \tag{5}$$

Where P (j) represents the probability that node j is sampled, and t_{ij} represents the last time slice connected between node i and node j, E_t represents the union of edge in time slice t.

At the same time, in order to ensure that the vectors embedded in different time slices belong to the same vector space, we will embed the nodes of the $t-1$ time slice as the initial value of the t time slice node embedding, and set the initial value of the 0 time slice to be a random value.

3.4   The Second Phase: Smooth Evolution

The network embedding of each time slice has been obtained through the above. Next, we introduce how to capture the evolution information of the network embedding through the network embedding of each time slice.

The change of the network topology with time is generally smooth and contains some rules. At the same time, because the network embedding we obtain implies the topology feature of the network, the evolution of the network embedding over time should also be smooth and have some rules. In order to better capture this hidden rule, we use the transformer [15] mechanism proposed by Google to capture the smooth evolution information of network embedding.

Transformer consists of a 6-layer coding layer and a 6-layer decoding layer, where each coding layer is composed of multi-head attention and a single-layer fully connected network. This article uses Multi-head attention to get the next time slice network Embedded, the specific model is shown in Fig. 3.

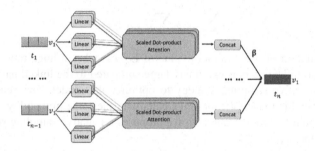

Fig. 3. Smooth Evolution Model

The use of the model in Fig. 3 to capture embedded network evolution information has the following advantages: (1) The number of operations required by the model to calculate the association between two time slices is independent of the distance between the two time slices, so it can capture long distances network evolution information. (2) The model can be calculated in parallel during training. (3) Multi-head attention can capture the evolution information embedded in the network from multiple dimensions, making the results more accurate.

4   SENE-Based Link Prediction Algorithm

The link prediction based on SENE includes the following steps (as shown in Algorithm 1):

Step 1: Dataset division. Divide the network into n different sub-networks according to a certain time interval, respectively: G_1, G_2... G_n, select the $0-n-1$ time slices as the training set, and the nth time slice network as the test set.

Step 2: Embedding the network of the $0-n-1$ time slices. According to Eq. (4), the $0-n-1$ time slices of the network are embedding, and the negative sampling strategy based on temporal feature is used to optimize the network embedding.

Step 3: Obtain the network embedding of the nth time slice. Use the transformer mechanism to capture the evolution information of network embedding, and finally capture the network embedding of the nth time slice. The network embedding of each node in the $0-n-1$ time slices is input into the evolution model introduced earlier, and the resulting output is the network embedding of the nth time slice.

Step 4: Generate prediction results. First, for each node n that to be predicted in the graph G_n, the similarity between the remaining nodes and node n is calculated. Then, according to the similarity, the top-k nodes are selected as the prediction result of n nodes. Finally, the prediction results of each node in G_n are aggregated and returned.

Algorithm 1. SENE-based link prediction algorithm

```
Input:G
output:{N₁,…, N|v|│ Nᵢ•V ∩│Nᵢ│=k∩ i=1~│Gₙ│}
1 Divide G into G₁,G₂···Gₙ;
2 For g∈{Gᵢ│ i>0 && i<=n-1}
3   REPEAT:
4     loss_g = compute_loss(g); //compute loss of g, and to loss_g
5     updata(g,loss_g); //updata embedding of g by loss_g
6   UNTIL Convergence
7 End for
8 Foreach v ∈ G
9   compute vₙ by v₁~vₙ₋₁;
10End for
11Foreach n in Gₙ
12  Sim=ComputeSim(n,Gₙ);
13 Nᵢ=sort(Sim,k);
14End for
```

---

5 Experiments

5.1 Dataset

We extracts part of the data from the Bibsonomy [16] dataset as an experimental dataset. This dataset contains the tags information that users did to the publications during 2009. There are three types of nodes in the network: user nodes, publication nodes and tag nodes. The statistic of the dataset is shown in Table 1.

Table 1. Bibsonomy network statistics

Type of Edges(A-B)	Number of Node A	Number of Node B	Avg. Degree of A	Avg. Degree of B	Number of Edge
User-Publication	1426	299269	422.8	2.1	601400
User-Tag	1426	116639	341.4	4.2	486787

When dividing the data set, we divide the data set in months. At the same time, we select the data from January to November as the training dataset, and select the data from December as the test dataset. And then, we capture the embedding evolution from January to November to predict the network embedding in December.

5.2 Evaluation Metrics

In this paper, Precision, Recall and AUC [17] curves are used to measure the link prediction results.

In definition, precision is the proportion of real positive examples in the forecast results to the whole forecast results. Recall is the proportion of positive examples in the prediction results to the real results. For this experiment, the real positive example is the exist edges in the test dataset.

AUC curve is defined as the area surrounding the coordinate axis under ROC curve. In practice, the approximate calculation can be made by formula (6). Where, k is the number of comparisons, k' is the number that the similarity of the selected edge in the positive example is greater than similarity that in the negative example, and k'' is the number that the similarity of the selected edge in the positive example is less than or equal to the similarity of negative example.

$$\text{AUC} = \frac{k' + 0.5 * k''}{k} \quad (6)$$

5.3 Performance Evaluation

First, from the perspective of temporal, we compare different link methods based on temporal feature, including:

FT: Link prediction method based on frequency time(FT) [18], if two nodes are connected more frequently in the past time slice, the possibility of connection in the next time slice is higher;

GT: Link prediction method based on generation time (GT) [19], if the node pairs are over connected in the closer time slice, the more likely the node pairs are to be connected;

SENE: SENE-based Link Prediction Algorithm proposed by this paper.

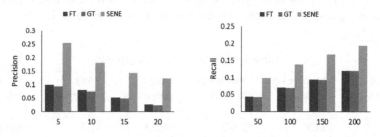

Fig. 4. Performance evaluation of temporal feature-based link prediction methods

The experimental results are shown in Fig. 4. It can be seen from Fig. 4 that the accuracy and recall rate of the model proposed in this paper are much higher than that of the comparison method no matter how much k value is taken. The reasons may be as follows: FT and GT only consider the temporal feature of the network, but ignore the topological structure feature and context feature of the network itself. Therefore, the precision and AUC values are not high. The SENE combines the deep-level features of the network with the temporal feature, which can better predict the upcoming connections. AUC values are shown in Table 2.

Table 2. The values of AUC of temporal feature-based link prediction methods

Method	AUC
FT	0.625
GT	0.587
SENE	**0.971**

Secondly, from the perspective of network embedding, we compare different network embedding methods, including:

Node2vec: Node2vec [10] is a graph embedding model proposed by Grover et al. Its core idea is to generate random walk sequences, and then use word vector model to represent and learn the nodes in random walk sequences. It forms a good balance between depth and breadth when generating random walk sequences;

Metapath2vec: Metapath2vec [11] is a node embedding algorithm for heterogeneous information networks proposed by Dong et al. It uses the random walk algorithm based on meta-path to construct the domain nodes of each node, and then uses skip-gram model to complete the node embedding;

LINE: LINE [12] is a graph embedding algorithm proposed by Tang et al. The goal is to embed large-scale information network into low dimensional space, and use edge sampling algorithm to solve the problem of gradient descent.

SENE: SENE-based Link Prediction Algorithm proposed by this paper.

Before the experiment, we explain some parameters of the network embedding method. In the experiment, both the baseline and the method proposed in this paper need to set some parameters. In this paper, we choose the embedded dimension as 16, and set the learning rate as 0.001. For node2vec, metapath2vec and LINE, we choose 100 training times, and the training times of SENE is 50. Then, all the data from January to November will be used as the training set of the comparison method, and the data from December will be used as the test set.

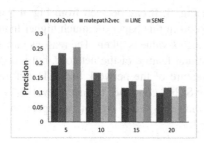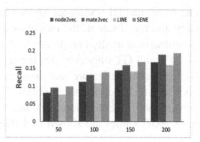

Fig. 5. Performance evaluation of network embedding-based link prediction methods

The experimental results are shown in Fig. 5, which shows that the SENE based link prediction method has the best performance. Node2vec, metapath2vec and LINE considered the topological structure feature and context feature, but ignored the temporal feature of the network. In SENE, the inherent features and temporal feature of the network are considered. The AUC value of the experiment is shown in Table 3. It can be seen from the table that even compared with the best performing method, the method proposed in this paper has nearly 5% improvement.

Table 3. The values of AUC of network embedding-based link prediction methods

Method	AUC
Node2vec	0.826
Metapath2vec	0.924
LINE	0.813
SENE	**0.971**

6   Conclusion

In view of the shortcomings of the existing link prediction technology, this paper proposed a dynamic link prediction method on heterogeneous information networks. In order to predict the edge in the next slice more accurately, this paper fully considers the deep-level and temporal features of heterogeneous information networks, and proposes the Smooth Evolution model for Network Embedding (SENE). In the model, context feature and topology feature are added at the same time, and a negative sampling strategy based on temporal feature is used to optimize the embedding, and transformer mechanism is used to obtain the smooth evolution of the network embedding, making full use of the rich features of heterogeneous information network. In addition, we propose a SENE-based Link Prediction Algorithm, by calculating the similarity of embedding between nodes to evaluate the possibility of edge formed, effectively ensuring the accuracy of link prediction. In the next step, we will study the capture high-order similarity and algorithm optimization strategy.

Acknowledgment. This work is supported by the National Key R&D Program of China (2018YFB1003404) and the National Natural Science Foundation of China (61672142).

References

1. Chen, C., et al.: Unsupervised Adversarial Graph Alignment with Graph Embedding (2019). ArXiv, abs/1907.00544
2. Mutinda, F.W., Nakashima, A., Takeuchi, K., Sasaki, Y., Onizuka, M.: Time series link prediction using NMF. IEEE International Conference on Big Data and Smart Computing (BigComp) **2019**, 1–8 (2019). https://doi.org/10.1109/BIGCOMP.2019.8679502
3. Sun, Y., Han, J., Yan, X., Yu, P.S., Wu, T.: PathSim: meta path-based Top-K similarity search in heterogeneous information networks. Proc. VLDB Endow. **4**, 992–1003 (2011)
4. Martínez, V., Galiano, F.B., Cubero, J.C.: A survey of link prediction in complex networks. ACM Comput. Surv. **49**, 69:1–69:33 (2016). https://doi.org/10.1145/3012704
5. Liu, Y., Shen, D., Kou, Y., Nie, T.: Link prediction based on node embedding and personalized time interval in temporal multi-relational network. In: Ni, W., Wang, X., Song, W., Li, Y. (eds.) WISA 2019. LNCS, vol. 11817, pp. 404–417. Springer, Cham (2019). https://doi.org/10.1007/978-3-030-30952-7_40
6. Mitzenmacher, M.: A brief history of generative models for power law and lognormal distributions. Internet Math. **1**, 226–251 (2003). https://doi.org/10.1080/15427951.2004.10129088
7. Brin, S., Page, L.: The anatomy of a large-scale hypertextual web search engine. Comput. Netw. **30**, 107–117 (1998). https://doi.org/10.1016/S0169-7552(98)00110-X
8. Wang, D., Cui, P., Zhu, W.: Structural deep network embedding. In: Proceedings of the 22nd ACM SIGKDD International Conference on Knowledge Discovery and Data Mining (2016). https://doi.org/10.1145/2939672.2939753
9. Perozzi, B., Al-Rfou, R., Skiena, S.: DeepWalk: online learning of social representations. KDD 2014 (2014). https://doi.org/10.1145/2623330.2623732
10. Grover, A., Leskovec, J.: node2vec: scalable feature learning for networks. In: Proceedings of the 22nd ACM SIGKDD International Conference on Knowledge Discovery and Data Mining (2016). https://doi.org/10.1145/2939672.2939754
11. Dong, Y., Chawla, N.V., Swami, A.: metapath2vec: scalable representation learning for heterogeneous networks. In: Proceedings of the 23rd ACM SIGKDD International Conference on Knowledge Discovery and Data Mining (2017). https://doi.org/10.1145/3097983.3098036
12. Tang, J., Qu, M., Wang, M., Zhang, M., Yan, J., Mei, Q.: LINE: Large-scale Information Network Embedding (2015). ArXiv, abs/1503.03578
13. Chen, H., Yin, H., Wang, W., Wang, H., Nguyen, Q.V., Li, X.: PME: projected metric embedding on heterogeneous networks for link prediction. In: Proceedings of the 24th ACM SIGKDD International Conference on Knowledge Discovery & Data Mining (2018). https://doi.org/10.1145/3219819.3219986
14. Luong, T., Pham, H., Manning, C.D.: Effective Approaches to Attention-based Neural Machine Translation (2015). ArXiv: abs/1508.04025
15. Vaswani, A., et al.: Attention is All you Need (2017). ArXiv, abs/1706.03762
16. Benz, D., et al.: The social bookmark and publication management system bibsonomy. VLDB J. **19**, 849–875 (2010). https://doi.org/10.1007/s00778-010-0208-4

17. Norton, M., Uryasev, S.P.: Maximization of AUC and Buffered AUC in binary classification. Mathematical Programming, **174**, 575–612 (2019). https://doi.org/10.1007/s10107-018-1312-2
18. Divakaran, A., Mohan, A.: Temporal link prediction: a survey. New Gener. Comput. **38**(1), 213–258 (2019). https://doi.org/10.1007/s00354-019-00065-z
19. Li, D., Shen, D., Kou, Y., Lin, M., Nie, T., Yu, G.: Research on a link-prediction method based on a hierarchical hybrid-feature graph. Sci. Sin. Inform. **50**, 221–238 (2020). https://doi.org/10.1360/N112018-00223

An Experimental Study of Time Series Based Patient Similarity with Graphs

Kalkidan Fekadu Eteffa, Samuel Ansong, Chao Li$^{(\boxtimes)}$, Ming Sheng,
Yong Zhang, and Chunxiao Xing

BNRist, DCST, RIIT, Institute of Internet Industry, Tsinghua University,
Beijing 100084, China
{ajq18,ssm18}@mails.tsinghua.edu.cn
{li-chao,shengming,zhangyong05,xingcx}@tsinghua.edu.cn

Abstract. Finding similarities between patients has been used to effectively and reliably predict diagnoses and guide treatments. However, Electronic Health Records (EHRs) contain characteristics that make analysis and application difficult. Firstly, it is difficult to compare two patients' time series. Also, EHRs contain a vast amount of data, which proves to be a significant barrier to developing efficient systems for the widespread use of patient similarity. In this paper, we introduce a novel graph representation of time series EHRs. Our method compresses a patient's time series medical records to reduce the storage required by more than 50%. Our paper also presents similarity metrics that can be applied to vector and graph representations of patient's time series medical records and assesses the general performance for suggested metrics.

Keywords: Patient similarity · Graph similarity computation · EHR

1 Introduction

Patient similarity computation produces a similarity score that reflects the correlation between medical histories of patients. Researchers have used patient similarity to identify patients who are at low risks of IS (Ischemic Stroke), uncover the unique characteristics of cohorts, predict mortality in intensive care units (ICUs) and distinguish patients who experience adverse drug events. It also has potential applications in medical education [1,6], as it can be a useful resource for medical students, professors, and researchers [20].

In a hospital database, tables contain information like chart events, lab events, and more that are all associated with a timestamp. Considering the timestamp helps to understand the progression of diseases and whether it is a chronic problem or not. That said, determining similarity among time-series data to better understand the information contained in the time component is challenging [8]. Besides this problem, raw EHRs are vast and contain large amounts of information per patient [4]. The analysis on EHR data is two-fold. On the one hand, it requires involvement with the timestamps in the streaming [2,5]

© Springer Nature Switzerland AG 2020
G. Wang et al. (Eds.): WISA 2020, LNCS 12432, pp. 467–474, 2020.
https://doi.org/10.1007/978-3-030-60029-7_42

and sequential [21] manner. On the other hand, it also needs to make extensive analysis on the texts [14,15].

To advance the study of patient similarity, we propose a novel graph structure to represent patients' records. The graph structure will be designed to reflect time-series data, by including relevant data points with their associated timestamp. Besides considering all the critical variables needed for analysis, this representation will be designed to reduce the storage requirement, making it an ideal format for storing EHRs.

In this paper, Sect. 2 briefly discusses related work. Section 3 introduces our graph structure and explains similarity metrics that can be used in determining vector and graph similarity. Section 4 presents and discusses the experiment. Finally, Sect. 5 concludes the work and discusses future research directions.

2   Related Work

The structure of this section will follow the layout shown in Fig. 1. Starting from the preprocessing stage of EHRs, this section discusses what researchers use for finding the representation for patients, picking similarity metrics and evaluating the computation.

The format of health records obtained from different hospitals vary. For this reason in the preprocessing stage [4] develops a unified format before analyzing the data any further. As EHRs contain vast information, preprocessing also includes extracting relevant features. [17] selects 21 medical features to filter patient records within the MIMIC-III dataset.

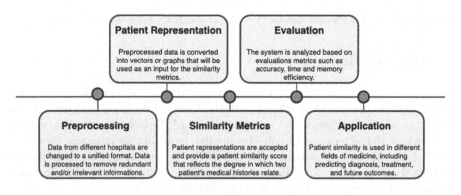

Fig. 1. Patient similarity computation layout

The patient representation stage retrieves features like diagnostic information, lab tests, and additional variables, which are later used as inputs for the patient similarity metric. Then patients can be represented as concatenations of their medical records according to their timestamps. Graph patient representations are also mentioned in some papers. [9] have constructed temporal graphs to represent the event sequences, while [17] uses nodes and edges to represent variables and the dependency of variables in adjoining time-windows, respectively.

After the patient representation has been selected, the similarity metrics function defines the patient similarity score. Measurements like cosine similarity, RV-coefficient and dCov(distance co-variance) [22] are used to find patient similarity scores between vectors of patient representations. For graph representations, GED (Graph Edit Distance) is a similarity metric that is largely used in finding similarity scores between graphs. As the Graph Neural Network approaches an increase in use, [3] develops SimGNN. SimGNN inputs a pair of graphs and outputs a similarity score.

The last step is to evaluate the patient similarity computation. We can employ patient similarity analysis that was used for recommendations and evaluated through metrics like AOC, precision, recall and F1.

3   Proposed Method

In this section we describe the method implemented, starting from the chosen dataset to the structure of the inputs, and the selected similarity metrics.

The data is chosen from MIMIC-III (Medical Information Mart for Intensive Care III) [7], which is a free database that contains intensive care unit information of over 40,000 real patients. For our experiments, we select the tables that relate to lab tests taken, which are Lab events and D_Labitems table. The time component of Lab events is used as it is vital in determining if two patients are similar and helps to identify the progression of the disease. The graph structure proposed in Fig. 2 represents the information for one patient, where TW represents different time windows, and LT is the lab test administered. Since all of the lab tests and time points are accounted for in the implementation, weights will not be a part of the computation.

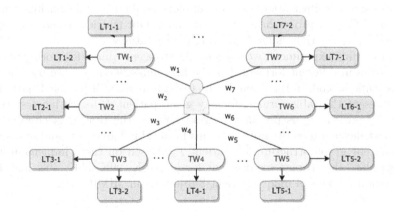

Fig. 2. Graph representation of a patient's lab tests within a time window

For our experimentation on vector representation of Lab events dataset, we propose vector metrics like Jaccard Index, Edit distance [16,18,19] and DTW. Jaccard Index finds the intersection between two given sets and takes the ratio of the intersection with the union of the two sets, while Edit Distance calculates the minimum number of insert, remove, and replace operations that are applied to transform one string to another. The third metric, DTW, is used to measure the similarity between two time-series data. For our experiment, we will use FastDTW [12], which reduces the running time to linear.

For our implementation of Jaccard Index, Edit distance, and DTW, five lab events in a given time window are randomly selected for one patient, P_x, and compared to every five elements per time window randomly selected of the second patient, P_y. This is so that even though two patients may not have a similar history from the beginning, they might gain similarities after some time. After we finish one loop, the maximum score is taken and added to the final result. This final result will be larger for patients with more extended medical histories; therefore, we account for length before assigning ranks.

For the experimentation on graph representation, we will use three metrics. The first metric is GED. Since our graph contains a large number of nodes, our implementation uses Hungarian Algorithm [11]. The second algorithm is Graph2Vec [10] with Cosine Similarity. Unsupervised representation can be learned for an entire graph that can be used for tasks, such as graph classification. After obtaining the embeddings of each graph, we can use distance measurements like cosine similarity to measure the similarity between the two embeddings. Finally, we use SimGNN introduced by [3] that find a representation that takes into account structures and features associated with the graph nodes.

4   Experiments

The experiments illustrated below are implemented on a single machine with 2.7 GHz Intel Core i7 processor of 16 GB memory. Our metrics are not carried out in parallel, and python is used for all implementations.

For the first experiment, we measure the storage required for storing patients' Lab events in different formats. For the first implementation, we use the table Lab events to collect the Item ID of test given as well as the Chart Time. To represent the table format, for any given patient P_x, each lab test that is taken by patient P_x is put in the array next to the time it is administered. The next method uses the same columns from table Lab events and implements an adjacent graph structure for patient P_x. Based on these implementations, we calculate the storage requirement for 5K, 10K, 20K, 30K, 40K and 46252 (maximum number of unique patient Subject ID in MIMIC-III).

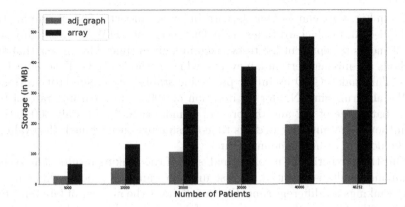

Fig. 3. Amount of storage occupied by different representations of Lab events

From Fig. 3, it can be seen that the storage requirement is reduced by more than 50% when adjacency graph structure is used. This is because the adjacency graph representation removes redundancy by using a single node for common time elements.

For the next experiment, we implement the proposed metrics in Sect. 3. The patients are selected randomly to account for different lengths of medical histories when calculating efficiency. For Jaccard Index, Edit Distance, GED, and DTW, we run our implementations for 10,000 patient lab tests. Graph2Vec is also done for a similar number of patients for 100 iterations with the vector dimension set to 1028 and a learning rate of 0.025. It is important to note that changing the parameters will affect the running time of Graph2Vec.

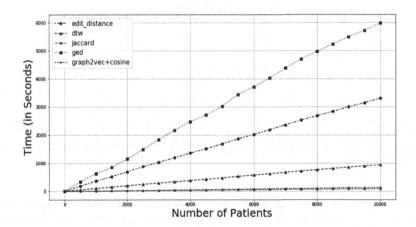

Fig. 4. Running time of different similarity metrics

From Fig. 4, we can see that Jaccard Index produces the fastest running time. In the Sect. 3, for Jaccard Index, Edit Distance, and DTW, we randomly select five elements to represent lab tests taken in a given time. This means that these methods are missing certain lab events, while methods like GED account for all tests. This makes GED useful for producing stable patient similarity results.

We also run SimGNN for an iteration of 100 with a learning rate of 0.025 and a batch size of 512 and the process to more than 2 h for only 300 patients. Even though training the model is time-consuming, once trained, the evaluation can be done in a short amount of time.

Our third experiment involves analysing of the ranking results. The stability of ranking results is essential because unstable rankings produce different similarity rankings for different runs. Figure 5 a shows the number of lab items taken by 100 patients using the time window of 44.89 h found from taking the average interval between patients' lab tests. The time range for each patient starts from 0. As a result, time 0 is where the sum is at its peak. Within the time window, 5 elements are randomly chosen. For this reason, our implementations of Jaccard Index, Edit Distance, DTW are found to give unstable results. Given the number of elements to be selected, we run Jaccard Index for 100 iterations and found the mean on Spearman's correlation. This is done in order to find the correlation between rankings given different number of events selected. As seen in Fig. 5 b, as the number of elements increases the Spearman's correlation also increases, making our ranking more stable.

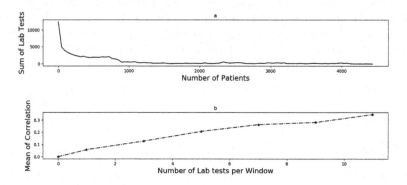

Fig. 5. a) The sum of patients' lab tests performed in a time window of 44.89 h. b) Mean of Spearman's correlation on Jaccard Index rankings for varying number of lab tests per window.

To see how the ranking results are related to each other, inspired by the idea of [22], we build the paragraph for each patient and train a Doc2Vec model for 1,000 iterations on the window size of 5 for 1,000 patients. We then find ranking similarity of Doc2Vec, using Spearman's correlation, to rankings found from Jaccard Index, Edit Distance, and DTW. As discussed earlier, we select 5 elements from each time window. For this reason, we take the correlations for various iterations and take the mean correlations found in every iteration. Although none of them show a high correlation, we find that DTW have a relatively higher correlation to Doc2Vec, as compared to Jaccard Index and Edit Distance, which show a lower correlation.

To summarize our findings, Jaccard Index computes a similarity score in the fastest time. Jaccard Index does not consider order and with the addition of Edit Distance and DTW, Jaccard Index does not take into account all the available information. For stable rankings, GED gives similar ranking for each run although it is not fast. Finally, even though Doc2Vec is not the perfect measure for patient similarity, we implement it so it can be used as a baseline. We find that even if DTW's ranking is not stable, the average of correlation results shows that it is most similar to Doc2Vec.

5 Conclusion

In this paper, we put forth a novel graph structure for tables of hospital databases that can be extended to patients' records that contain a time component. We demonstrate through research that this method accounts for all of the patients' information and occupies a smaller storage. We also discuss the efficiency and ranking results of similarity metrics. In the future, we will integrate knowledge graphs [13] in patient similarity calculations to explain results of similarity scores.

Acknowledgments. This work was supported by National Key R&D Program of China (2018YFB1404401, 2018YFB1402701), NSFC (91646202).

References

1. Ansong, S., Eteffa, K.F., Li, C., Sheng, M., Zhang, Y., Xing, C.: How to empower disease diagnosis in a medical education system using knowledge graph. WISA **2019**, 518–523 (2019). https://doi.org/10.1007/978-3-030-30952-7_52
2. Ao, X., Shi, H., Wang, J., Zuo, L., Li, H., He, Q.: Large-scale frequent episode mining from complex event sequences with hierarchies. ACM TIST **10**(4), 36:1–36:26 (2019). https://doi.org/10.1145/3326163
3. Bai, Y., Ding, H., Bian, S., Chen, T., Sun, Y., Wang, W.: Simgnn: a neural network approach to fast graph similarity computation. In: WSDM, pp. 384–392 (2019). https://doi.org/10.1145/3289600.3290967
4. Barkhordari, M., Niamanesh, M.: Scadipasi: an effective scalable and distributable mapreduce-based method to find patient similarity on huge healthcare networks. Big Data Res. **2**(1), 19–27 (2015). https://doi.org/10.1016/j.bdr.2015.02.004
5. Das, A., Wang, J., Gandhi, S.M., Lee, J., Wang, W., Zaniolo, C.: Learn smart with less: Building better online decision trees with fewer training examples. In: IJCAI, pp. 2209–2215 (2019). https://doi.org/10.24963/ijcai.2019/306
6. Eteffa, K.F., Ansong, S., Li, C., Sheng, M., Zhang, Y., Xing, C.: Application of patient similarity in smart health: a case study in medical education. WISA **2019**, 714–719 (2019). https://doi.org/10.1007/978-3-030-30952-7_72
7. Johnson, A.E., et al.: Mimic-iii, a freely accessible critical care database. Sci. Data **3**, 160035 (2016)

8. Lin, C., Boursier, E., Papakonstantinou, Y.: Approximate analytics system over compressed time series with tight deterministic error guarantees. PVLDB **13**(7), 1105–1118 (2020)

9. Liu, C., Wang, F., Hu, J., Xiong, H.: Temporal phenotyping from longitudinal electronic health records: a graph based framework. In: SIGKDD, pp. 705–714 (2015). https://doi.org/10.1145/2783258.2783352

10. Narayanan, A., Chandramohan, M., Venkatesan, R., Chen, L., Liu, Y., Jaiswal, S.: graph2vec: learning distributed representations of graphs. CoRR abs/1707.05005 (2017). http://arxiv.org/abs/1707.05005

11. Riesen, K., Bunke, H.: Approximate graph edit distance computation by means of bipartite graph matching. Image Vis. Comput. **27**(7), 950–959 (2009). https://doi.org/10.1016/j.imavis.2008.04.004

12. Salvador, S., Chan, P.: Toward accurate dynamic time warping in linear time and space. Intell. Data Anal. **11**(5), 561–580 (2007)

13. Sheng, M., et al.: Clmed: a cross-lingual knowledge graph framework for cardiovascular diseases. WISA **2019**, 512–517 (2019). https://doi.org/10.1007/978-3-030-30952-7_51

14. Tian, B., Zhang, Y., Wang, J., Xing, C.: Hierarchical inter-attention network for document classification with multi-task learning. In: IJCAI, pp. 3569–3575 (2019). https://doi.org/10.24963/ijcai.2019/495

15. Wang, J., Lin, C., Li, M., Zaniolo, C.: Boosting approximate dictionary-based entity extraction with synonyms. Inf. Sci. **530**, 1–21 (2020). https://doi.org/10.1016/j.ins.2020.04.025

16. Wang, J., Lin, C., Zaniolo, C.: Mf-join: Efficient fuzzy string similarity join with multi-level filtering. In: ICDE, pp. 386–397 (2019). https://doi.org/10.1109/ICDE.2019.00042

17. Wang, Y., Chen, W., Li, B., Boots, R.: Learning fine-grained patient similarity with dynamic bayesian network embedded RNNS. In: DASFAA, pp. 587–603 (2019). https://doi.org/10.1007/978-3-030-18576-3_35

18. Wu, J., Zhang, Y., Wang, J., Lin, C., Fu, Y., Xing, C.: Scalable metric similarity join using mapreduce. In: ICDE, pp. 1662–1665 (2019). https://doi.org/10.1109/ICDE.2019.00167

19. Yang, J., Zhang, Y., Zhou, X., Wang, J., Hu, H., Xing, C.: A hierarchical framework for top-k location-aware error-tolerant keyword search. In: ICDE, pp. 986–997 (2019). https://doi.org/10.1109/ICDE.2019.00092

20. Zhao, K., Zhang, Y., Wang, Z., Yin, H., Zhou, X., Wang, J., Xing, C.: Modeling patient visit using electronic medical records for cost profile estimation. In: DASFAA, pp. 20–36 (2018). https://doi.org/10.1007/978-3-319-91458-9_2

21. Zhao, K., et al.: Discovering subsequence patterns for next POI recommendation. In: IJCAI, pp. 3216–3222 (2020). https://doi.org/10.24963/ijcai.2020/445

22. Zhu, Z., Yin, C., Qian, B., Cheng, Y., Wei, J., Wang, F.: Measuring patient similarities via a deep architecture with medical concept embedding. In: ICDM, pp. 749–758 (2016). https://doi.org/10.1109/ICDM.2016.0086

A Twig-Based Algorithm for Top-k Subgraph Matching in Large-Scale Graph Data

Haiwei Zhang, Xiaofang Xie, Yanlong Wen$^{(\boxtimes)}$, and Ying Zhang

College of Computer Science, Nankai University, Tianjin 300350, China
{zhhaiwei,wenyl,yingzhang}@nankai.edu.cn

Abstract. Subgraph matching is considered as a basis query for graph data management, and is used in many domains, such as semantic web and social network analysis. Subgraph isomorphism is an initial solution for the task, which is an NP-complete problem. To speed up the procedure, graph simulation has been presented to match subgraphs with polynomial complexity. Unfortunately, simulation usually loses topology of matched subgraphs. In this paper, we propose an approximation approach for subgraph matching based on twig patterns. First, we transform query graphs into twig patterns and match candidate substructures in graph data. Second, we present an optimized join strategy along with top-k mechanism, including join order selection based on cost evaluation and optimized pruning based on maximum possible score and minimum possible score. Finally, we design experiments on real-life and synthetic graph data to evaluate the performance of our work. The results show that our approach obviously reduces the time complexity and guarantee the correctness for answering the queries of subgraph matching.

Keywords: Subgraph matching · Graph data · Twig patterns · Join optimization · Top-k mechanism

1 Introduction

Graph is widely used as an important data model for various domains in real-world, such as semantic web, social network, bioinformatics network and knowledge base [1]. Several novel graph queries have been used to analyze graph data for important research and usage. Subgraph matching is considered as a basis query for graph data management, which aims to find isomorphic or approximate patterns of a given query graph in the graph data.

With the increasing size of graph data in many domains, subgraph matching in a large-scale graph attracts more attention of researchers. Subgraph isomorphism, which is thought as an NP-complete problem, is the basis and accurate solution for subgraph matching. But it will face to the bottleneck for efficiency while querying in large-scale graph data. Subgraph approximation has been presented to efficiently obtain the matched subgraphs with similar structure to the

© Springer Nature Switzerland AG 2020
G. Wang et al. (Eds.): WISA 2020, LNCS 12432, pp. 475–487, 2020.
https://doi.org/10.1007/978-3-030-60029-7_43

query graphs. It accelerates the query processing, but it relies on data preprocessing and usually changes the original structure compared with the query graphs. In addition, subgraph simulation [2,3] has been used for graph pattern matching in a polynomial complexity of time, but it would get different structure of subgraphs. In recent years, there has been various work to speed up the processing of subgraph matching. Indices have been used to accelerate the query processing, such as RDF-3X [4] and iGraph [5]. However, indices could not be maintained efficiently for large-scale graph data. Especially for subgraph approximation, join optimization [18,19] has been used to pruning the size of intermediate matching results, such as kGPM [6], D-Join [7], CFL [8]. This kind of solutions usually maps the graph data to a simpler schema, such as paths and trees, which match the given query graphs in an efficient way.

In this paper, we propose a novel solution for subgraph matching in large-scale graph data. Our contributions of this paper are as follows. (1) In order to speed up the query processing, query graphs are parsed to twig patterns. We propose an efficient algorithm kSGM for subgraph matching based on twig patterns. (2) We propose a mechanism for join optimization along with top-k strategy to get top k answers without computing the similarities of all the matched subgraphs. (3) We implement various experiments to verify the performance of our approach. Experimental results show that our algorithm for subgraph matching based on twig patterns is advanced to existing work.

The rest of this paper is organized as follows. Section 2 introduces related work on subgraph matching. Preliminaries of subgraph matching problems are mentioned in Sect. 3. Section 4 describes the approach of subgraph matching via twig patterns. Top-k join optimization is presented to speed up the matching processing in Sect. 5. In Sect. 6, experiments are implemented to evaluate the performance of our approaches. Finally, Sect. 7 concludes this paper.

2   Related Work

Subgraph matching is a basic operation for graph data management and is used in many domains, such as semantic web data and social network queries. In the past decades, there are mainly three different directions for subgraph matching, including isomorphism, simulation and approximation.

Subgraph isomorphism has been studied since 1970s, such as Ullman [9], VF2 [10]. Researches on these algorithms focused on good matching orders and effective pruning rules [1]. Unfortunately, subgraph isomorphism can only process small graphs or graph databases, so that it cannot perform efficiently for large-scale graph data. Graph pattern matching is a type of subgraph matching, where graph patterns are special subgraphs constructed by certain rules. Graph simulation changes subgraph matching into relationship matching with bisimulation [11], such as bounded simulation [2] and strong simulation [3].

As is known, graph data can be represented to other simpler models, such as spanning trees [6], sets of paths [13] or nodes [12], which have efficient query processing. Matching by these models is approximated because of the differences between models. kGPM presented by J. Cheng et al. [6] mapped the query graph to a set of probable spanning trees, and matched each spanning tree in graph data by twig-like algorithms. Then, generated matched subgraphs by joining related matched subtrees via optimized processing. Similar to kGPM, Bi F. et al. [8] transformed query graph into Core-Forest-Leaf model(CFL), and matched each substructure by specific algorithms, in which a compressed path index was used to accelerate the join procedure. Unfortunately, kGPM and CFL took more complex procedure for data preprocessing, time occupation might increase with the growing size of query graphs.

Our work is closely related to the approximation matching. The approach we propose in this paper is different from existing works in several aspects. First, we adapt twig patterns to describe query graphs, thus, efficient query algorithms of subtree matching can be used to generate candidates for matched subgraphs. Second, an optimal join algorithm will be performed along with top-k strategy to get the most similar matched subgraphs for the query graph.

3   Preliminaries

In this paper, we focus on the directed labeled graph. We will introduce the notions of our work in this section.

Definition 1 Data graph: A data graph is a directed labeled graph $G = (V, E, L, F_e)$, where: (1) V is a finite set of nodes, (2) $E \subseteq V \times V$ is a set of edges, in which (v, v') denotes an edge from node v to node v', (3) For each node v in V, $L(v)$ is the label of v that specifies the node attribute, (4) F_e is a function defined on E. For each edge (v, v') in E, $F_e(v, v')$ is a integer which denotes the weight of the edge.

Definition 2 Query graph: A query graph is defined as $Q = (V_Q, E_Q, L_Q, f_e)$, where: (1) V_Q is a set of nodes, (2) E_Q is a set of directed edges, as denoted in data graph, (3) L_Q is a function defined on V_Q. For each node u in V_Q, $L_Q(u)$ is the label of u, (4) f_e is a function defined on E_Q. For each edge (u, u') in E_Q, $f_e(u, u')$ is the weight of the edge.

Definition 3 Twig Pattern: Twig pattern of a query graph is defined as a tree pattern $T = (V_T, E_T, L_T, f_{Te})$, where: (1) V_T, E_T, L_T and f_{Te} are defined as those of the query graph, (2) a twig pattern has a unique root node, (3) each non-leaf node q of Q has a set of children, denoted as q.children, (4) each non-root query node q has a unique parent, denoted as q.parent.

Definition 4 Twig Pattern Matching: Given a rooted twig pattern T and a data graph G, a twig pattern matching f is a mapping from V_T to V_G, such that: (1) f preserves the label information: $\forall u \in V_T, l(u) = l(f(u))$, (2) f preserves the structure information: $\forall (u, u') \in E_T$, there is a directed path from $f(u)$ to $f(u')$ in G, (3) the length of shortest path from $f(u)$ to $f(u')$ in G is equal or lesser than the weight of (u, u') in T.

In a matching f, each edge (u, u') in T is mapped to a path from $f(u)$ to $f(u')$ in G. Note that there could be many paths from $f(u)$ to $f(u')$ and we use the length of the shortest paths to characterize the relevance between T and the matching $M_f = \{f(u) | u \in V_T\}$ of T in G.

Definition 5 Score: Given a twig pattern matching $M_f = \{f(u) | u \in V_T\}$ of T in a graph G, we compute its *Score* $S(M_f)$ as,

$$S(M_f) = \sum_{(u,u') \in E_T} \delta_{min}(f(u), f(u')) \tag{1}$$

where $\delta_{min}(f(u), f(u'))$ is the shortest distance from $f(u)$ to $f(u')$ in G.

The less value of $S(M_f)$ means the less distance between $f(u)$ and $f(u')$, and $f(u')$ is more possible to match $f(u)$. Top-k twig pattern matching is to find top-k lowest $S(M_f)$ and further gets all matched substructures in the graph data. The key technique of subgraph matching via twig patterns is join optimization, which is the main problem solved in this paper.

4   Twig Pattern Matching

Twig patterns in Definition 3 are substructures of the query graph. The task of subgraph matching comprises of two stages, substructures matching by twig patterns and top-k join optimization.

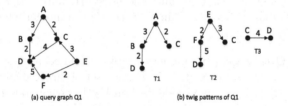

(a) query graph Q1   (b) twig patterns of Q1

Fig. 1. Non-cycle query graph and parsed twig patterns

4.1   Parse the Query Graph

The given query graph always has a complicated structure, which can bring a lot of burden to the matching task. Therefore, we parse the query graph to several twig patterns. Then we conduct the match processing on the data graph for every parsed twig pattern. And we join the matching results of all parsed twig patterns, finally. The joined result is exactly the matching result for the primitive query graph in the data graph.

The query graph Q has complicated structure, just as shown in Fig. 1. Our goal is to parse the Q to a twig set $T = \{T_1, T_2, T_3\}$, such that

(1) $V_{T1} \cup V_{T2} \cup \ldots \cup V_{Tn} = V_Q$,

(2) $E_{T1} \cup E_{T2} \cup \ldots \cup E_{Tn} = E_Q$,

(3) $E_{Ti} \cap E_{Tj} = \emptyset$ $(1 \le i \le n, 1 \le j \le n, i \ne j)$.

We use depth-first-search (DFS) algorithm to implement the decomposition process. For the given query graph Q, we first determine whether there exist the nodes whose in-degrees are zero (eg. A and E) and call them *zero point*. If exist, we can take them as the roots of twigs. We start from the *zero point* A, use DFS algorithm and get $T1\{A, B, C, D\}$. Furthermore, when we start from B and visit D for the first time, we can add the corresponding nodes and edges into $T1$. Unfortunately, when we start from C and visit D for the second time, we can't add node D into $T2$, because D has already existed in $T1$. We also can't add the edge from C to D into $T1$, and keep it in the query graph. The edges that exist in $T1$ are marked *read*. In a similar way, we start from the *zero point* E and get $T2\{E, F, D, C\}$ using DFS. Note that, the edges marked *read* can't be visited again. Finally, only the edge $C \to D$ in the query graph is not *read*, we add this edge and corresponding nodes C and D into $T3\{C, D\}$. If there are no *zero point* in the query graph, cycles exist in Q. In this case, we can use *Tarjan* algorithm [17] to search the strongly connected components (SCC), which can be regarded as a *zero point* as Fig. 2 shows.

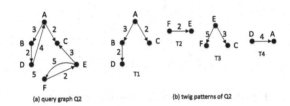

(a) query graph Q2   (b) twig patterns of Q2

Fig. 2. Multi-cycle query graph and parsed twig patterns

4.2 Twig Pattern Matching

We denote T as a query graph which is a rooted twig query graph and $r(T)$ as the root of T. In T, all node labels are distinct.

Transitive Closure. We pre-compute a transitive closure $G_c = (V_c, E_c)$ of $G = (V, E, L, F_e)$ by the algorithm $STACK\_TC$ presented in [14] with the complexity of time $O(min(ns, e_{oct} \log n))$. Here n, e and s is the number of nodes, the number of edges, and the number of strongly connected components in the input data graph G, respectively. e_{oct} is the number of intercomponent tree and cross edges in G. In G_c, $V_c = V$, and an edge (v, v') exists if and only if there is a path in G from v to v'. We also record the length of the shortest path from v to v' in G as the weight of (v, v') in G_c.

Run-Time Graph. A Run-time Graph [15] of Q over G, denoted as $G_R = (V_R, E_R)$, is a subgraph of G_c. An edge (v, v') in G_c is included in G_R if and only if there is an edge (u, u') in T with $L_T(u) = L_T(v)$, $L_T(u') = L_T(v')$ and the length of the shortest paths of (v, v') should not be larger than $f_e(u, u')$. We also store the weight value (length of shortest paths) of an edge in G_R. Clearly, finding the top-k matches of T in G (or G_c) is equivalent to finding the top-k matches in G_R.

Then we can run the top-k twig pattern matching algorithm kTPM proposed in [16] on G_R and get the matching results of T. This time complexity of this algorithm is $O(m_R + k(n_T + \log k))$.

5   Top-k Join Optimization

Twig pattern matching will obtain all matched substructures described by twig patterns for query graph. The matching results for twig patterns shown in Fig. 2 are stored in table-pattern as shown in Fig. 3. Existing top-k processing technique $rank - join$ presented in [18] can be applied to our problem for the join procedure of twig pattern matching. However, the total number of answers in large graph can be enormous. In this section, we focus on the optimization for top-k join.

M1				score
a3	b1	c2	d1	3
a2	b2	c4	d3	3
a1	b4	c1	d2	4
a3	b4	c2	d2	4
a4	b3	c1	d4	4
a1	b2	c3	d1	5
a2	b3	c3	d4	7

M2		score
f1	e2	1
f3	e5	1
f5	e1	1
f3	e2	2
f2	e4	2
f1	e2	3
f4	e3	3

M3			score
e2	f3	c2	3
e1	f2	c1	4
e3	f1	c2	4
e4	f2	c4	5
e5	f1	c4	5
e1	f4	c2	7
e5	f2	c3	8

M4		score
d2	a1	1
d1	a3	2
d2	a4	2
d2	a2	3
d4	a1	3
d3	a5	4
d4	a3	4

Fig. 3. The matching results for twig patterns

5.1   Optimization Based on Cost Evaluation

The $rank - join$ operation can be viewed as the process of spanning the space of Cartesian product of the input tables to get valid join combinations, and the join operation is implemented in a dyadic (two-way) operator. The way that $rank - join$ schedules the next input table can affect the operator response time significantly. We propose an optimization strategy to choose the best join order based on cost evaluation. Frequently used notations are summarized in Table 1.

Table 1. Notations

Notation	Description
$T(R)$	The number of tuples in table R
$R.A$	The column in table R with the label A
$V(R, A)$	The number of distinct values for R.A
$R \cap S = A$	Table R join S on attribute A
$R \infty S$	Table R and S do the joining

Consider the following example to join three ranked inputs M_1, M_2 and M_3. $T(M_1) = 5000$, $T(M_2) = 2000$ and $T(M_3) = 500$. Then there are three join orders and will get different effects:

(1) $T((M_1 \infty M_2) = 1500$, $T((M_1 \infty M_2) \infty M_3) = 120$;
(2) $T((M_1 \infty M_3) = 450$, $T((M_1 \infty M_3) \infty M_2) = 120$;
(3) $T((M_2 \infty M_3) = 200$, $T((M_2 \infty M_3) \infty M_1) = 120$.

The final results are the same. If we get small size of intermediate results, we can reduce the size of data. In the consequent join process, we just need to conduct a small size table, which can improve the overall efficiency. Thus, we will give preference to join order $(M_2 \infty M_3) \infty M_1$ in this example. We aim at finding an optimal join order using a cost-based optimization approach, because the optimal join order has the smallest evaluation cost among all orders that yield different evaluation costs. The evaluation rule is presented as follows.

If $R \cap S = A$, there might be $T(S)/V(S, A)$ tuples in S for every tuple r in R to join. Considering all tuples in R, we may generate $T(R)T(S)/V(S, A)$ tuples in $R \infty S$. In the same way, the $S \infty R$ may generate $T(R)T(S)/V(R, A)$ tuples. Because the $rank - join$ is inner join, $R \infty S$ is equivalent to $S \infty R$. So we evaluate the size of $R \infty S$ is $min(T(R)T(S)/V(S, A), T(R)T(S)/V(R, A))$.

When $n = 3$ (n is the number of input tables), there are three join orders: $T((M_1 \infty M_2) \infty M_3)$, $T((M_1 \infty M_3) \infty M_2)$ and $T((M_2 \infty M_3) \infty M_1)$. And it is easy to evaluate the cost for every order. But when $n = 4$, there are 15 join orders. Along with the increasing n, the number of join orders will grow exponentially. It is relieved that we don't need to give all the join orders in practice. For example, when we consider $M_1 \infty M_2 \infty M_3 \infty M_4 \infty M_5$, we just need to evaluate the 3 join orders of $M_1 \infty M_2 \infty M_3$. Then we use the optimal order to join M_4 and M_5, and only need to evaluate three times again. Now we can get the optimal order for $M_1 \infty M_2 \infty M_3 \infty M_4 \infty M_5$ as well. Using this strategy, we only do six time evaluation for $n = 5$. When $n > 5$, we can do evaluation in the similar way.

5.2   Pruning

In this subsection, we propose a novel top-k pruning strategy which can cancel insignificant tuples as possible. Our technique leads to efficient executions by modeling two sets $S-Source$ and $R-Source$. The joined twig patterns are stored in $S - Source$, and those that have not been joined are stored in $R - Source$.

Assuming that the query graph Q is parsed to n_T twigs. For a twig T_i ($1 \leqslant i \leqslant n_T$), its weight is $W_{T_i} = \sum_{u \in E_{T_i}} W_e(u)$. For the matching result M_{i_j} of T_i, the maximum possible score is $U(T_i) = W_{T_i}$, and the minimum possible score is $L(T_i) = | E_{T_i} |$. That is to say, for the matching results M_{i_j} of T_i, its maximum possible score is the sum of weight value of every edge in T_i. If the score of M_{i_j} is more than W_{T_i}, it is not the matching result of T_i certainly. The minimum possible score of M_{i_j} is the edge number of T_i, that is, every edge can be matched with the shortest path value is 1.

Algorithm 1 shows the procedure of pruning to optimize the top-k join. During the join process, we keep a buffer in which the current top-k results are stored. The object with k^{th} lowest score is denoted as t' which is the object with highest score in buffer. When computing a new result τ during joining, if the number of results in buffer is less than k, we put it into the buffer directly. Otherwise, we perform the following operations: a) if $L(\tau) > U(t')$, we prune the τ and no longer to deal with it in the next operation; b) if $Score(\tau) < Score(t')$, we update buffer and replace t' with τ; c) if a) and b) are both not satisfied, we put τ into candidate set and implement the further joining on it. The significance of case a) is that, if the minimum possible score of τ is less than the maximum possible score of t', then τ will not appear in the final top-k results set certainly. Supposing that, we can get matching result τ' from τ after all the joining operation, and get $\tilde{t}$ from t'. Then, $Score(\tau') \geq L(\tau)$ and $Score(\tilde{t}) \leq U(t')$. Since $L(\tau) > U(t')$, we can know that $Score(\tau') > Score(\tilde{t})$. Moreover, because $U(t') > U(jth)(1 \leq j < k)$ where jth is the object with j^{th} lowest score in buffer, we can know that $Score(\tau') \geq Score(jth')$. Therefore, the final matching result from τ must not appear in the final top-k matching results set, and we can prune it safely. If τ can't satisfy the case a), it illuminate that τ may appear in the final matching results set. Here if τ meets the case b), we need to update the buffer obviously. If τ can't meet case b) as well, we put it into candidate set and perform the join further. The results in buffer and candidate set will both participate in the next round of join operations. When all the twig patterns are joined together, the matching results in buffer are the results we need with the top-k lowest score.

Algorithm 1. Pruning

Require: every twig $T = \{T_1, T_2, \ldots, T_{n_T}\}$, the matching results $M = \{M_1, M_2, \ldots, M_{n_T}\}$ of every twig and the optimal join order.
Ensure: the top-k matching results.
1: set $S - Source = \{T_1, \ldots, T_i\}$, $R - Source = \{T_{i+1}, \ldots, T_{n_T}\}$
2: set the current top-k results are stored in buffer BF
3: $Smax = \sum_{j=i+1}^{n_T} U(T_j)$
4: $Smin = \sum_{j=i+1}^{n_T} L(T_j)$
5: set $\tau \in M_1 \infty M_2 \ldots \infty M_i$
6: $L(\tau) = Score(\tau) + Smin$
7: $U(kth) = Score(kth) + Smax$
8: **if** $L(\tau) > U(kth)$ **then**
9:     prune τ
10: **else if** $Score(\tau) < Score(kth)$ **then**
11:     update current top-k set buffer BF
12: **else**
13:     put τ into candidate
14: **end if**
15: **return** BF

---

6 Experimental Evaluation

We perform various experiments on large-scale real-life and synthetic graph datasets to evaluate the performance of our approach for subgraph matching.

6.1 Experiment Setup

We perform the experiments on a single PC, which has 16 GB DDR3 RAM and 3.4 GHz Intel Core i7-6700 CPU. All algorithms have been coded by C++. Effectiveness and efficiency of our approach will be evaluated in following data sets:

(1) Synthetic dataset. We designed a graph generator to produce synthetic graphs. Graph generation was controlled by three parameters: the number of nodes $|V|$, the number of edges $|E|$ and the number of different labels of nodes $|L|$. The dataset in the experiment is named Synthetic.

(2) Real-life datasets. We selected real-life datasets for comparing the effectiveness and efficiency. **California, Internet, Citation** and **Email** from SNAP[1].
The basic information of each real-life dataset is shown in Table 2.

Table 2. Datasets

Datasets	$|V|$	$|E|$	$|L|$
California	1546	3296	94
Internet	10878	39994	50
Citation	21524	48529	67
Email	265214	420045	1258

(3) Query graphs. We construct three kinds of query graph, non-cycle, multi-cycle and single-cycle query graph (shown in Fig. 1, Fig. 2 and Fig. 4, respectively), to verify the effectiveness of our approach. Furthermore, we unify the query graphs and evaluate the efficiency of our approach compared with existing algorithms.

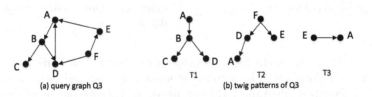

(a) query graph Q3   (b) twig patterns of Q3

Fig. 4. Single-cycle query graph and parsed twig patterns

---
[1] http://snap.stanford.edu/data/index.html.

6.2   Experimental Results

This experiment evaluates the efficiency of our approach kSGM (top-**k** **S**ub**G**raph **M**atching). Firstly, we compare kSGM with existing algorithm VF2, kGPM (top-**k** **G**raph **P**attern **M**atching) and strSim (**str**ong **Sim**ulation). Secondly, we verify the efficiency of our approach with different size of data graphs. Thirdly, we discuss the efficiency of our approach with different size of twig patterns in query graph. Finally, we describe the influence of different k.

VF2 is a traditional algorithm for graph isomorphism. kGPM is proposed as a top-k approximate subgraph matching via spanning tree model. And strSim is the most accurate algorithm of graph simulation. Because of the high complexities of time of running VF2 and strSim, we only use California, Internet and Citation as the datasets in the experiment. The time occupation is shown in Fig. 5. The results show that the complexity of time of kSGM is nearly linear and is more efficient than the counterparts.

We perform our approach on different size of data graph to evaluate the efficiency of the algorithms we proposed. Because of the complexity of real-life datasets, we generate a series of data graph with increasing size for the experiment. The size of labels in query graph is fixed to 60 and the average degree of nodes is set to 3. In addition, k is equal to 3000 for the experiment. Figure 6 shows the time occupation for $rank - join$ (denoted as $Rjoin$) and optimized top-k join(denoted as $Ojoin$). The time occupation is increasing along with the increment of data graph. Fortunately, the optimized join will get more efficient than $rank - join$. And the advantage becomes more evident along with the increasing size of data graph.

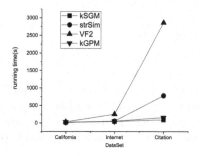
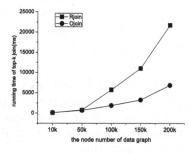

Fig. 5. Comparison to existing work for subgraph matching

Fig. 6. Efficiency of top-k join for subgraph matching in data graphs with different sizes

We evaluate the main impacting factors of top-k join. The main task of subgraph matching via twig patterns is joining all matched twig patterns and obtaining matched subgraphs. On one hand, we execute our algorithm kSGM with different size of twig patterns and different number of twig patterns, respectively. On the other hand, we change the value of k and evaluate the influence to subgraph matching.

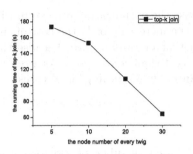 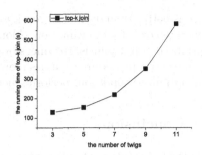

Fig. 7. Efficiency of top-k join with different number of nodes in every twig pattern

Fig. 8. Efficiency of top-k join with different number of twig patterns

First, we perform the experiment on Email dataset. We fix the number of twig patterns $n_T = 3$ in query graphs, and construct different sizes of twig patterns with 5, 10, 20, 30, that is to say, $|V_T| = 5$, 10, 20, 30, respectively. The time occupation of top-k join procedure is shown in Fig. 7. We learn from the figure that for the fixed number of twigs in the given query graph, the time occupation of top-k join will be reduced along with the increasing number of nodes in twig patterns. Then, we construct a query graph with 60 nodes and 299 edges ($|V_Q| = 60$, $|E_Q| = 299$) and change the number of twig patterns in each query processing. The time occupation of top-k join is shown in Fig. 8. For the fixed size of query graph, the more numbers of twig patterns, the more time is occupied for top-k join. Obviously, more twig patterns means more matched twigs to be joined. As a result, less number of twig patterns parsed from query graph will improve the efficiency of subgraph matching.

We compare the efficiency of our approach with different values of k. We use Email as the data graph. $Q1$, $Q2$ and $Q3$ shown in Fig. 1, Fig. 2 and Fig. 4 respectively are used as query graphs. Figure 9 shows the efficiency of matching these query graphs within different k (the x axis in Fig. 9).

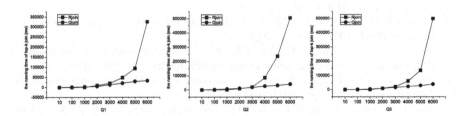

Fig. 9. Efficiency of top-k join within different k

We learn from Fig. 9 that the trends of time occupation of our approach are similar for different query graphs. While k is small, the efficiencies of $rank -$

join is better than optimized top-k join because the optimization procedure will generate cost of join evaluation. Along with the increasing of k, the cost of the optimized top-k join is growing slower than that of $rank - join$, because the optimization procedure occupies more steady cost than $rank - join$. As a result, the optimized top-k join performs better than $rank - join$ in general cases.

7   Conclusion

Traditional solution of subgraph matching is based on subgraph isomorphism, which is considered as an NP-complete problem. Another mechanism based on graph simulation obtains different structures to the given query graphs. In order to address these questions, we propose an approximate approach for subgraph matching in large-scale graph data based on twig patterns along with top-k join optimization. We first parse query graph to twig patterns and match substructures in data graph by twig pattern matching algorithm. We then present an optimized join strategy along with top-k mechanism, including join order selection based on cost evaluation and optimized pruning based on maximum possible score and minimum possible score. We finally perform experiments on real-life and synthetic graph data to evaluate the performance of our work. The results show that our approach obviously speed up the procedure of subgraph matching.

Acknowledgements. This work is supported by "the Fundamental Research Funds for the Central Universities", Nankai University (No. 63201207, No. 63201209 and No. 63201166).

References

1. Kim, J., Shin, H., Han, W.S., et al.: Taming subgraph isomorphism for RDF query processing. PVLDB **8**(11), 1238–1249 (2015)
2. Fan, W., Wang, X., Wu, Y.: Incremental graph pattern matching. ACM Trans. Database Syst. **38**(3), 18 (2013)
3. Ma, S., Cao, Y., Fan, Y., Huai, J., Wo, T.: Capturing topology in graph pattern matching. PVLDB **5**(4), 310–321 (2011)
4. Neumann, T., Weikum, G.: The RDF-3X engine for scalable management of RDF data. VLDB J. **19**(1), 91–113 (2010)
5. Han, W., Lee, J., Pham, M., Yu, J.X.: iGraph: a framework for comparisons of disk based graph indexing techniques. PVLDB **3**(1–2), 449–459 (2010)
6. Cheng, J., Zeng, X., Yu, J.X.: Top-k graph pattern matching over large graphs. In: Proceedings of ICDE, pp. 1033–1044 (2013)
7. Zou, L., Chen, L., Ozsu, M.T., Zhao, D.: Answering pattern match queries in large graph databases via graph embedding. PVLDB **21**(1), 97–120 (2012)
8. Bi, F., Chang, L., Lin, X., et al.: Efficient subgraph matching by postponing cartesian products. In: Proceedings of ACM SIGMOD, pp. 1199–1214 (2016)
9. Ullmann, J.R.: An algorithm for subgraph isomorphism. J. ACM **23**(1), 31–42 (1976)

10. Cordella, L.P., Foggia, P., Sansone, C., Vento, M.: A (sub)graph isomorphism algorithm for matching large graphs. IEEE Trans. Pattern Anal. Mach. Intell. **26**(10), 1367–1372 (2004)
11. Henzinger, M.R., Henzinger, T.A., Kopke, P.W.: Computing simulations on finite and infinite graphs. In: Proceedings of Foundations of Computer Science, pp. 453–462 (1995)
12. Ren, X., Wang, J.: Exploiting vertex relationships in speeding up subgraph isomorphism over large graphs. PVLDB **8**(5), 617–628 (2015)
13. Gupta, M., Gao, J., Yan, X., Cam, H., Han, J.: Top-k interesting subgraph discovery in information networks. In: Proceedings of ICDE, pp. 820–831 (2014)
14. Nuutila, E.: An efficient transitive closure algorithm for cyclic digraphs. Inf. Process. Lett. **52**(4), 207–213 (1999)
15. Gou, G., Chirkova, R.: Efficient algorithms for exact ranked twig-pattern matching over graphs. In: Proceedings of ACM SIGMOD, pp. 581–594 (2008)
16. Chang, L., Lin, X., Zhang, W., Yu, J.X., Zhang, Y., Qin, L.: Optimal enumeration: efficient top-k tree matching. PVLDB **8**(5), 533–544 (2015)
17. Tarjan, R.E.: Depth-first search and linear graph algorithms. SIAM J. Comput. **1**(2), 146–160 (1972)
18. Ilyas, I.F., Aref, W.G., Elmagarmid, A.K.: Supporting top-k join queries in relational databases. VLDB J. **13**(3), 207–221 (2004)
19. Ni, W., Wang, X., Song, W., Li, Y. (eds.): WISA 2019. LNCS, vol. 11817. Springer, Cham (2019). https://doi.org/10.1007/978-3-030-30952-7

Edge Computing and Data Mining

An Advanced Q-Learning Model
for Multi-agent Negotiation
in Real-Time Bidding

Chao Kong$^{(\boxtimes)}$, Baoxiang Chen, Shaoying Li, Jiahui Chen, Yifan Chen,
and Liping Zhang

School of Computer and Information, Anhui Polytechnic University, Wuhu, China
kongchao@ahpu.edu.cn, {bxchen1996,shyli1996,jhchen2000,yfchen1999,
lpzhang1980}@yeah.net

Abstract. This work develops a reinforcement learning method for
multi-agent negotiation. While existing works have developed various
learning methods for multi-agent negotiation, they have primarily focus
on the Temporal-Difference (TD) algorithm (action-value methods) in
general and overlooked the unique properties of parameterized policy.
As such, these methods can be suboptimal for multi-agent negotiation.
In this paper, we study the problem of multi-agent negotiation in real-
time bidding scenario. We propose a new method named EQL, short for
Extended Q-learning, which iteratively assigns the state transition probability and finally converges to a unique optimum effectively. By performing linear approximation of the off-policy critic purposefully, we integrate
Expected Policy Gradients (EPG) into basic Q-learning. Importantly, we
then propose a novel negotiation framework by accounting for both the
EQL and edge computing between mobile devices and cloud servers to
handle the data preprocessing and transmission simultaneously to reduce
the load of cloud servers. We conduct extensive experiments on two real
datasets. Both quantitative results and qualitative analysis verify the
effectiveness and rationality of our EQL method.

Keywords: Q-learning · Multi-agent negotiation · Expected Policy
Gradients · Real-time bidding · Edge computing

1 Introduction

Multi-agent negotiation has attracted significant interest from both academia
and industry due to its wide application in areas like AlphaGo, AlphaZero, real-
time bidding [4], etc. Nowadays, it has become a core component of an ocean of
artificial intelligence services such as asymmetric games, automated auction, etc.
Q-learning is the most prevalent paradigm and methodology in these systems
which visited and updated state-action pairs based on policy it follows. However,

C. Kong and B. Chen—The two authors contributed equally to this work.

© Springer Nature Switzerland AG 2020
G. Wang et al. (Eds.): WISA 2020, LNCS 12432, pp. 491–502, 2020.
https://doi.org/10.1007/978-3-030-60029-7_44

Q-learning often overestimates the action values. Thus, traditional Q-learning algorithms are impeded by the overestimation problem. To solve this problem, a multitude of methods have been proposed to integrate the parameterized policy method into negotiation systems. These methods mainly model the gradient of quiet a few scalar performance measures and show that the improved gradient ascent model has the potential to improve negotiation performance.

Existing policy gradient methods usually assume that there is an army of trajectories needed which often tedious. Thus, the exploration policy assumption may not be suitable for continuous problems since it is not efficient enough. However, most of policy gradient methods usually rely on the sampled trajectory. When the trajectory assumption is incorrect, these methods will suffer from unstable problems.

Internet of Things (IoT) provides a scenario where objects have computational capabilities. The current work is generally focused on security, trust, data mining, data-centric, context-aware computing [3] and so on. As a prevalent computing paradigm, cloud computing offers on-demand services to the end-user. It offers on-demand computing services and dynamic optimization of a shared resource. The current internet-based application leverage cloud services to make serious issues about high latency and mobility-related issues. Edge computing differs from cloud computing for edge computing services located in the edge network and distinguishes characteristics. It brings the utilities of cloud computing closer to the user with fast processing and quick response time [12]. This serves as a great tool for speeding up the EQL algorithm. In summary, our major contributions are as follows:

- We develop an optimization algorithm called EQL to approximate Q-function and minimize the cost function.
- We propose a novel framework for performing the negotiation process efficiently by integrating edge computing into the EQL process.
- Our experimental evaluation on real-world datasets shows that our method outperforms the state-of-the-art multi-agent negotiation methods.

The remainder of the paper is organized as follows. We first review related work in Sect. 2. We formulate the problem in Sect. 3, before delving into details of the proposed method in Sect. 4. We perform extensive empirical studies in Sect. 5 and conclude the paper in Sect. 6.

2   Related Work

Reinforcement Learning (RL) is alongside with supervised learning and unsupervised learning draw much attention in the past decades. Imitate learning process is of much importance. The general problem formulation is the finite Markov Decision Process. However, the existing works of reinforcement are mainly employed in computer games. In this work, we propose EQL, which can approximate value function to solve the problem of the trade-off between exploration and exploitation. This section reviews some relevant literatures [11] from the background knowledge of reinforcement learning to real-time bidding.

2.1   Reinforcement Learning

The early studies of reinforcement can go back to 1927 [15]. Quite a few psychologists extended the idea and the strength and weakness should persist after the reinforcer is withdrawn. Later, the original reinforcement becomes to solve the problem of decision making. It consists of four elements: agent, environment, action, and reward. The goal of reinforcement is to achieve more accumulate rewards.

Temporal-Difference (TD) algorithm is a novel solution to reinforcement learning. It combines the notion of Monte Carlo (MC) and Dynamic Programming (DP). Essentially, the TD target is the estimation by value function and bootstrapping of DP. The existing works to balance the exploration and exploitation can be divided into two categories: on-policy-based and off-policy-based. Sarsa is an anonymous approach [20] to infer the values of state-action pairs:

$$Q(S_t, A_t) \leftarrow Q(S_t, A_t) + \alpha[R_{t+1} + \gamma Q(S_{t+1}, a) - Q(S_t, A_t)]. \tag{1}$$

Watkins et al. [21] develop an off-policy TD control algorithm called Q-learning:

$$Q(S_t, A_t) \leftarrow Q(S_t, A_t) + \alpha[R_{t+1} + \gamma \max_a Q(S_{t+1}, a) - Q(S_t, A_t)], \tag{2}$$

which simplifies the analysis of the algorithm. Then Mnih et al. [14] develop an agent called Deep Q-Network (DQN) that combines Q-learning with a deep convolutional Artificial Neural Network (ANN), and they manifest a reinforcement learning agent can achieve high performance without problem-specific feature. Hasselt ameliorates the DQN algorithm's overestimated problem using Double Deep Q-learning [13], which is a specific adaptation to the DQN algorithm. However, the aforementioned methods are all action-value methods, which need action-value estimation to exist. Policy gradient methods can select actions without consulting a value function. They learn the values of actions and select actions based on their estimated action values: $\theta_{t+1} = \theta_t + \alpha \bigtriangledown J(\theta_t)$.

Konda [17] proposed actor-critic methods which learn approximates to both policy and value. Volodymyr [18] introduced parallel actor-learners with multi-core CPU (A3C). Later, Shi et al. [10] develop method Q-Prop provided an effective way to combine the policy gradient with the efficiency of off-policy. However, it is arduous to converge in some cases. Kamil [6] extended the policy gradient method called Expected Policy Gradients (EPG) to reduce the variance of the gradient estimation.

2.2   Multi-agent System

With the development of computer science and the rise of complexity science, artificial intelligence has gradually penetrated various research communities. The interdisciplinary subject of artificial sociology is proposed and developed. Computational sociology has been proclaimed.

Epstein and Axtell [8] develop the Sugarscape that can simulate the accumulation of artificial social wealth. In this model, the world is composed of cells and there are two areas where resources are concentrated. Each agent can only move in this world. By formulating agent movement rules, environmental rules and replacement rules, we can see the sugar owned by each agent. The number of sugar indicate that there shows a large gap between rich and poor. Use the sugar domain model to verify quite a few social science problems [8]. The shortcoming of the model is that the persuasiveness is flawed and may not be closed to the real economic system.

In the economy community, agent-based modelling [9] published in 2009. It is a breakthrough for it posses traditional statistical algorithms that cannot proceed. It can treat the economy as a complex system and incorporate human adaptation and learning capability into the modeling. If it can avoid arbitrarily, agent-based modeling, it will simulate some trends and behaviors of the economic market.

In 2014, Azoulay et al. [2] advocated an effective agent suitable for Cliff-Edge (CE) and simultaneous Cliff-Edge (SCE) situations. The agent interacts with different human opponents and gains field experience in against unknown opponents. They further compare the performance of the proposed algorithm, and the result indicates that the algorithm is suitable for those scenarios [16]. However, the general applicability of this algorithm is limited.

2.3   Real-Time Bidding

Real-time bidding has been a hot topic in the implementation of AI industry. Wu et al. [22] proposed a mixed model which improve the prediction accuracy. Adikari et al. [1] found a way to select the most pertinent target audience by exploring an auto pricing strategy. Deng et al. [7] presented a novel-preserving real-time bidding protocol to protect user's data privacy. By employing this method, advertisers could not learn the real-time bidding algorithm. Paliwal et al. [19] generalize a winning price model and utilize censored information.

To the best of our knowledge, no one has paid close attention to combine advanced reinforcement learning with real-time resource bidding through multi-agent system. This is the focus and improvement of this work.

3   Problem Formulation

3.1   Preliminaries

Markov Decision Processes (MDP). It is a pivotal approach to simple the process of reinforcement learning model. Let MDP be a tuple (S, A, T, r, γ), where S represents a finite set of state, A is a finite set of action, T is a state transition probability function. A reward function and discount factor are denoted as r and γ respectively. The agents observe a state from the environment and makes a decision based on the state.

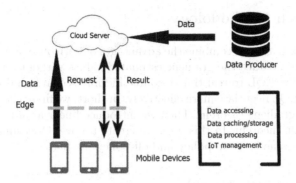

Fig. 1. The framework of edge computing.

Game Theory. Game theory is widely applied for computer science, social science and so on. Game types consist of Cooperative, Symmetric, Zero-sum, Simultaneous and Stochastic outcomes, etc. In this work, we specifically focus on the game theory on Sealed-bid Second Price Auction, $u_i(f_1, ..., f_n) = \sum_v p(v) u_i(f_1(v_1), ..., f_n(v_n), v_1, ..., v_n)$. In general, the second price reward is directly linked to the second high price, which can lead the system to achieve state of equilibrium (every bidding price equals to their ideal value).

Edge Computing. We concentrate on discussing a naive edge computing technique between mobile devices and servers to handle the data preprocessing and transmission simultaneously to reduce the load of cloud servers. The framework of edge computing is shown in Fig. 1. Each edge mobile device can participate in the whole real-time bidding procedure, which contains data accessing, caching, and processing to share calculating tasks. After executing the EQL on the cloud server, the results and requests of the real-time negotiation are sent to mobile devices. In a nutshell, the general naive edge computing framework is to assign computing and storage tasks on mobile devices.

3.2   Problem Definition

In a multi-agent environment, autonomous agents must be capable of adapting their negotiation strategies and tactics to their prevailing circumstances. At first, the agent starts the budget at B. Since the negotiation has an infinite Nash equilibrium, we need to introduce times into account. During t auctions, each budget $b \in B$. In each episode, the agents need to decide the temporarily bid prices according to the information provided by t, b, and x, where x represents a bid request. If the agent bids at price $a \geq m$ (the market price), then the rest budget is $m - a$.

4   Research Methodology

The task of EQL aims to combine the advantages of both on-policy and off-policy methods, where reduce the gradient estimated values without adding variance. We then employ EQL to real-time resource bidding scenario and use edge computing to let EQL negotiate more effectively. In short, we first take the derivation of the policy gradient theorem. Then we introduce linear approximation to the policy gradient theorem. In this way, we create a novel Q-learning called EQL which can combine both on-policy and off-policy methods.

4.1   Proposed EQL Algorithm

Stochastic policy gradients [5] perform gradient ascent on ∇J, where π is parameterised by θ: $\nabla J = \int_s d\rho\pi(a|s) \nabla log\pi(a|s)(Q(a,s) + b(s))$.

When Expected Policy Gradients introduce $I_Q^\pi(s)$, the approximate gradient can be substituted by: $\widehat{\nabla} J = \sum \gamma^t \widehat{I_\pi^Q}(s_t)$.

Based the proposed EPG, we introduce Extended Q-learning which integrates EPG with off-policy Q-learning algorithm. Generally, the policy gradient does not back propagation by error but observation information. By the using of reward, good behavior will increase the probability of the next choice, bad behavior will decrease the next probability of being chosen, which can be formulated as:

$$\nabla_\theta J(\theta) = \mathbb{E}_{\pi_\theta}[\nabla_\theta log\pi_\theta(s,a)Q^{\pi_\theta}(s,a)]. \tag{3}$$

We first introduce a random function $f(x,y)$, where $x = s_t, y = a_t$. Then we utilize linear approximation to present $f(s_t, a_t) = f(s_t, \bar{a}) + f'(s_t, a)|_{a=\bar{a}}(a_t - \bar{a}_t)$. As mentioned above, we can obtain the observation of $\nabla_\theta J(\theta)$ as follows. Due to the page limitation, we omit the derivation.

$$\begin{aligned}
\nabla_\theta J(\theta) &= \mathbb{E}_{\rho_\pi \pi_\theta}[\nabla_\theta log\pi_\theta(s_t, a_t)(Q^{\pi_\theta}(s,a) - f(s_t, a_t)] \\
&+ \mathbb{E}_{\pi_\theta}[\nabla_\theta log\pi_\theta(s_t, a_t)f(s_t, a_t)] \\
&= \mathbb{E}_{\rho_\pi \pi_\theta}[\nabla_\theta log\pi_\theta(s_t, a_t)(Q^{\pi_\theta}(s,a) - f(s_t, a_t)] \\
&+ \mathbb{E}_{\rho_\pi}[\nabla_a f(s_t, a)|_{a=\bar{a}_t} \nabla_\theta \mathbb{E}_\pi[a_t]].
\end{aligned} \tag{4}$$

Here, $\widehat{\mu}$ is the Expected Policy π_θ. Then we can substitute $\widehat{\mu}_\theta(s_t)$ for $\mathbb{E}_\pi[a_t]$:

$$\mathbb{E}_{\rho_\pi}[\nabla_a f(s_t, a)|_{a=\bar{a}_t} \nabla_\theta \mathbb{E}_\pi[a_t]] = \mathbb{E}_{\rho_\pi}[\nabla_a f(s_t, a)|_{a=\bar{a}_t} \nabla_\theta \widehat{\mu}_\theta(s_t)]. \tag{5}$$

Therefore,

$$\begin{aligned}
\nabla_\theta J(\theta) &= \mathbb{E}_{\rho_\pi \pi_\theta}[\nabla_\theta log\pi_\theta(s_t, a_t)(Q^{\pi_\theta}(s,a) - f(s_t, a_t)] \\
&+ \mathbb{E}_{\rho_\pi}[\nabla_a f(s_t, a)|_{a=\bar{a}_t} \nabla_\theta \widehat{\mu}_\theta(s_t)],
\end{aligned} \tag{6}$$

which can combine both off-policy and on-policy methods, and significantly help us to avoid Q-learning method's overestimated action problems. The workflow of our proposed EQL is summarized in Algorithm 1.

Algorithm 1. EQL Algorithm.

Require: critic Q, θ for expected policy gradient π_θ;
Ensure: optimum solution of π_θ: os;
 1: initialize policy π parametrised by θ, initialize a for Q_a
 2: **while** not coverage **do**
 3:     $a_t \sim \pi_\theta(\cdot|s_t), s_{t+1} \sim p(\cdot|s_t, a_t)$
 4:     $\hat{Q}.update(s, a, r)$
 5:     $Compute \bigtriangledown_\theta J(\theta)$
 6: **end while**
 7: **return** os

4.2 Running Example

For example, Fig. 2 gives an example execution of a simple EQL method for
real-time bidding. We represent the state of bidding as a square, where the
black boxes denote the failure bidding state (i.e., sellers and buyers may not
achieve the price consensus); the white ones denote the intermediate bidding
state (i.e., sellers start to consider the price offered by buyers seriously but not
immediately). In terms of intermediate bidding state, EQL executes iteratively
across state 2 to state 3. Specifically, we denote the initial price as S (START)
and the final price as F (FINAL), respectively. When it comes to the black
square, the current state terminates; EQL continues to execute under the white
square state until its convergence (i.e., achieve the final price). It is noteworthy
that all the agents prefer any agreement to none is the worst outcome.

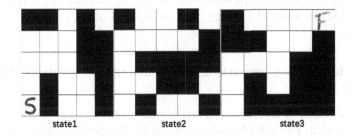

Fig. 2. Running example of EQL.

5 Empirical Study

A promising resource bidding should consist of a satisfying price and less running
time. To achieve the aim for a content result, we consider the resource bidding
system from two perspectives: an effective algorithm that can achieve a satisfying price and our naive edge computing framework that can help to reduce the
running time for price results. Through empirical evaluation, we aim to answer

the following research questions:

RQ1 How does EQL performance compare with state-of-the-art Actor-Critic algorithm and other baselines?

RQ2 Is the EQL helpful to earn a desirable price quickly in the real-time bidding?

RQ3 Can the edge computing techniques help the EQL to train faster?

In what follows, we first introduce experimental settings of EQL, and then answer the above research questions in turn.

5.1 Experimental Settings

Datasets. We use two real datasets in our experiments. One is the lager-scale dataset *Football Manager*[1] which has 159524 records and 89 features. The other is the $FLA\ 18$[2] which has 17995 records and more features up to 184.

Comparative Methods and Evaluation Protocols. We compare EQL with three baselines (Expected Sarsa, Policy Gradient, Actor-Critic)

- Expected Sarsa: Exploit the knowledge by basing the expected value.
- Policy Gradient: Policy is represented by its own function approximator.
- Actor-Critic: This algorithm combines with critic-only and actor-only methods, which can convergence faster.

As we mentioned above, we evaluate our method using speed of convergence. A promising resource bidding should cover a lower final price and faster convergence speed. To evaluate the efficiency of our edge computing component, we also measure the elapsed training time on the top of EQL + edge computing framework.

5.2 Performance Comparison and Convergence Analysis (RQ1 and RQ2)

Football Manager is a real-time bidding game, which means negotiation is an essential issue in football players' trading part. We perform one experiment to compare the EQL with the baselines Expected Sarsa (E-Sarsa), Policy Gradient (PG), and Actor-Critic (AC). Figure 3 plots the bidding results of a football player between two clubs varying the beginning of original price (purple line) to the final price. The original price offered by the seller is 15M CNY. As we can see, the convergence status (i.e., the maximum value) of each method on Football Manager is about 0.76. Moreover, EQL (yellow line with diamond) successfully finds the optima of the bidding results in all three methods. The final price is about 7.0M CNY. This demonstrates EQL's ability to converge to the unique and optimal solution of real-time bidding.

---

[1] https://www.kaggle.com/ajinkyablaze/football-manager-data.

[2] https://www.kaggle.com/thec03u5/fifa-18-demo-player-dataset.

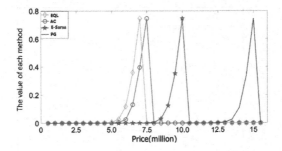

Fig. 3. Performance comparison on Football Manager. (Color figure online)

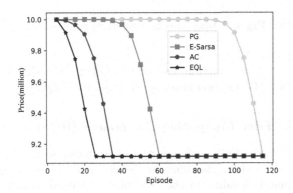

Fig. 4. Convergence analysis on FLA 18.

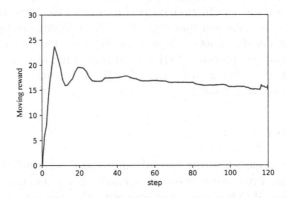

Fig. 5. Reward of each step during training.

Figure 4 demonstrates that EQL outperforms baselines in the real-time bidding scenario by achieving the most satisfy result faster than other methods. The bidding price starts at 10M CNY. After 20 episodes, the EQL is roughly close to 9.08M CNY, which is the price that both buyer and the seller can accept the most content result. Policy Gradient (PG) method converges at 120 episodes.

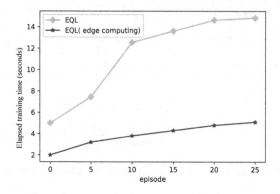

Fig. 6. Utility of edge computing component.

Extended Sarsa (E-Sarsa) method uses one-half of episodes than PG to coverage, and Actor-Critic (AC) method coverage shortly after EQL.

5.3   Utility of Edge Computing Component (RQ3)

As shown in Fig. 5, the EQL+ edge computing framework can obtain each step's reward. When the bidding is going to the end, the reward is general scaling at the same numerical value. In the beginning of several rounds, it is easier to get a reward for each action, but when the bidding continues, the result can not be achieved clearly. We also compare the EQL with EQL+ edge computing framework to perform an ablation study. Fig. 6 plots the elapsed training time of them respectively. We can find that the EQL with edge computing component can almost cut down 400% elapsed training time within 25 episodes. This demonstrates that our proposed EQL+ edge computing framework can achieve the same result much faster than original EQL. The approximate to end-users will benefit our bidding result.

6   Discussion

In this paper, we introduce the task of multi-agent negotiation in a real-time bidding scenario. First, we propose an optimization algorithm called EQL to solve the problem of Q-learning sometimes overestimated action values by combining the stability of policy gradient. Furthermore, we train our EQL model in the cloud. Then, a naive edge computing model is introduced to handle the data prepossessing and transmission simultaneously on mobile devices. Thus the load of cloud servers can be reduced. Finally, the experiment on real data-set manifest that our method can estimate the state transition probability and negotiation result effectively, which considers multiple agents. In the future, we will focus on more case studies and applications in different scenarios.

Acknowledgment. This work was supported by the National Natural Science Foundation of China Youth Fund under Grant No. 61902001 and Initial Scientific Research Fund of Introduced Talents in Anhui Polytechnic University under Grant No. 2017YQQ015.

References

1. Adikari, S., Dutta, K.: A new approach to real-time bidding in online advertisements: auto pricing strategy. INFORMS J. Comput. **31**(1), 66–82 (2019). https://doi.org/10.1287/ijoc.2018.0812
2. Azoulay, R., Katz, R., Kraus, S.: Efficient bidding strategies for cliff-edge problems. Auton. Agents Multi Agent Syst. **28**(2), 290–336 (2014). https://doi.org/10.1007/s10458-013-9227-z
3. Badidi, E., Atif, Y., Sheng, Q.Z., Maheswaran, M.: On personalized cloud service provisioning for mobile users using adaptive and context-aware service composition. Computing **101**(4), 291–318 (2018). https://doi.org/10.1007/s00607-018-0631-8
4. Cai, H., Ren, K., Zhang, W., et al.: Real-time bidding by reinforcement learning in display advertising. In: WSDM 2017, Cambridge, United Kingdom, 6–10 February 2017, pp. 661–670 (2017). https://doi.org/10.1145/3018661.3018702
5. Chou, P., Maturana, D., Scherer, S.A.: Improving stochastic policy gradients in continuous control with deep reinforcement learning using the beta distribution. In: ICML 2017, Sydney, NSW, Australia, 6–11 August 2017, pp. 834–843 (2017). http://proceedings.mlr.press/v70/chou17a.html
6. Ciosek, K., Whiteson, S.: Expected policy gradients. In: AAAI-18, IAAI-18, EAAI-18, New Orleans, Louisiana, USA, 2–7 February 2018, pp. 2868–2875 (2018). https://www.aaai.org/ocs/index.php/AAAI/AAAI18/paper/view/16116
7. Deng, E., Zhang, H., Wu, P., et al.: Pri-RTB: privacy-preserving real-time bidding for securing mobile advertisement in ubiquitous computing. Inf. Sci. **504**, 354–371 (2019). https://doi.org/10.1016/j.ins.2019.07.034
8. Epstein, J.M.: Agent-based computational models and generative social science. Complexity **4**(5), 41–60 (1999). https://doi.org/10.1002/(SICI)1099-0526(199905/06)4:53.0.CO;2-F
9. Farmer, J.D., Foley, D.: The economy needs agent-based modelling. Nature **460**(7256), 685–686 (2009). https://doi.org/10.1119/1.3081304
10. Gu, S., Lillicrap, T.P., Ghahramani, Z., et al.: Q-prop: sample-efficient policy gradient with an off-policy critic. In: ICLR 2017, Toulon, France, 24–26 April 2017, Conference Track Proceedings (2017). https://openreview.net/forum?id=SJ3rcZcxl
11. Han, D., Zhang, J., Zhou, Y., Liu, Q., Yang, N.: Intelligent trader model based on dep reinforcement learning. In: Ni, W., Wang, X., Song, W., Li, Y. (eds.) WISA 2019. LNCS, vol. 11817, pp. 15–21. Springer, Cham (2019). https://doi.org/10.1007/978-3-030-30952-7_2
12. Hassan, N., Gillani, S., Ahmed, E., Yaqoob, I., Imran, M.: The role of edge computing in internet of things. IEEE Commun. Mag. **56**(11), 110–115 (2018). https://doi.org/10.1109/MCOM.2018.1700906
13. van Hasselt, H., Guez, A., Silver, D.: Deep reinforcement learning with double q-learning. In: Proceedings of the Thirtieth AAAI Conference on Artificial Intelligence, 12–17 February 2016, Phoenix, Arizona, USA, pp. 2094–2100 (2016)
14. Hosu, I., Rebedea, T.: Playing Atari games with deep reinforcement learning and human checkpoint replay. CoRR abs/1607.05077 (2016). http://arxiv.org/abs/1607.05077

15. III, L.C.B.: Residual algorithms: reinforcement learning with function approximation. In: Machine Learning, Proceedings of the Twelfth International Conference on Machine Learning, Tahoe City, California, USA, 9–12 July 1995, pp. 30–37 (1995). https://doi.org/10.1016/b978-1-55860-377-6.50013-x

16. Katz, R., Kraus, S.: Modeling human decision making in cliff-edge environments. In: Proceedings, The Twenty-First National Conference on Artificial Intelligence and the Eighteenth Innovative Applications of Artificial Intelligence Conference, Boston, Massachusetts, USA, 16–20 July 2006, pp. 169–174 (2006). http://www.aaai.org/Library/AAAI/2006/aaai06-027.php

17. Konda, V.R., Tsitsiklis, J.N.: Actor-critic algorithms. In: Advances in Neural Information Processing Systems 12, NIPS Conference, Denver, Colorado, USA, 29 November–4 December 1999, pp. 1008–1014 (1999). http://papers.nips.cc/paper/1786-actor-critic-algorithms

18. Mnih, V., Badia, A.P., Mirza, M., et al.: Asynchronous methods for deep reinforcement learning. In: ICML 2016, New York City, NY, USA, 19–24 June 2016, pp. 1928–1937 (2016). http://proceedings.mlr.press/v48/mniha16.html

19. Paliwal, P., Renov, O.: Gradient boosting censored regression for winning price prediction in real-time bidding. In: Li, G., Yang, J., Gama, J., Natwichai, J., Tong, Y. (eds.) DASFAA 2019. LNCS, vol. 11448, pp. 348–352. Springer, Cham (2019). https://doi.org/10.1007/978-3-030-18590-9_43

20. Perkins, T.J., Pendrith, M.D.: On the existence of fixed points for q-learning and Sarsa in partially observable domains. In: ICML 2002, University of New South Wales, Sydney, Australia, 8–12 July 2002, pp. 490–497 (2002). https://doi.org/10.5555/645531.756483

21. Watkins, C.J.C.H., Dayan, P.: Technical note q-learning. Mach. Learn. 8, 279–292 (1992). https://doi.org/10.1007/BF00992698

22. Wu, W.C., Yeh, M., Chen, M.: Predicting winning price in real time bidding with censored data. In: Proceedings of the 21th ACM SIGKDD, Sydney, NSW, Australia, 10–13 August 2015, pp. 1305–1314 (2015). https://doi.org/10.1145/2783258.2783276

Inference Acceleration Model of Branched Neural Network Based on Distributed Deployment in Fog Computing

Weijin Jiang[1,2] and Sijian Lv[1,2(✉)]

[1] School of Computer Science and Technology, Hunan University
of Technology and Business, Changsha 410205, China
me@lvsijian.cn
[2] Key Laboratory of Hunan Province for New Retail Virtual Reality
Technology, Hunan University of Technology and Business,
Changsha 410205, China

Abstract. Research based on deep neural networks (DNN) is becoming more common. In order to solve the problem that DNN needs to consume a lot of performance during the use prediction process and generate unacceptable delays for users, a distributed neural network deployment model based on fog computing is proposed. The distributed deployment of deep neural networks in fog computing scenarios is analyzed. A deployment algorithm based on Solution Space Tree Pruning (SSTP) is designed, and a suitable fog computing node deployment model is selected to reduce the delay of prediction tasks. An algorithm for Maximizing Accuracy based on Guaranteed Latency (MAL) is designed and implemented, and suitable fog computing nodes are selected for different tasks to exit the prediction task. Simulation experiment results show that compared with the method of deploying neural network models in the cloud, the model prediction delay of the distributed neural network model based on fog computing is reduced by an average of 44.79%. Reduced the average computing acceleration framework of similar algorithms by 28.75%.

Keywords: Fog computing · Deep neural network branch definition · Model prediction acceleration · Distributed deployment

1 Introduction

With the development of artificial intelligence technology, the role of deep learning is increasingly significant. The Deep Neural Network (DNN) method has achieved good results on various tasks [1]. More and more intelligent tasks such as intelligent assistants, image recognition, and body recognition require neural networks. Deep neural networks are mainly composed of multiple convolutional layers and fully connected layers. Each of these layers processes the input data, transmits it to the next layer, and outputs the calculation results at the final layer.

As a new computing model, fog computing is one of the core principles of pushing computing power to the edge [2], which has attracted widespread attention from researchers. In the fog computing scenario, the DNN model is deployed on the fog

© Springer Nature Switzerland AG 2020
G. Wang et al. (Eds.): WISA 2020, LNCS 12432, pp. 503–512, 2020.
https://doi.org/10.1007/978-3-030-60029-7_45

computing nodes around the device [3]. Compared to the distance to the cloud service, the fog computing node is much closer to the data source, so low-latency features are easily implemented, and the privacy problem of user data [4] has also been better solved. However, the current fog computing equipment has limited processing capabilities, and a single fog computing node may not be able to complete the prediction task of a complex network model well. Therefore, multiple fog computing nodes are required to jointly deploy a DNN model. The main challenge of deploying a DNN in a fog computing scenario is how to select the appropriate computing node to deploy the model. It is necessary to consider the segmentation of the neural network model [5], the computing requirements of the model, and the network status of the fog computing nodes, in order to optimize the delay when multiple computing nodes run the neural network model cooperatively.

This paper mainly studies how to reduce the model prediction delay of DNN in the fog computing environment. Use the Neurosurgeon deep learning framework to train a DNN model with multiple branches on a cloud server. It is proved that the deployment problem of distributed neural networks is NP-hard. An algorithm based on Solution Space Tree Pruning (SSTP) was designed and implemented to find a fog computing node deployment scheme with minimal running delay for the DNN model. Aiming at the delay requirements and geographical distribution characteristics of data source prediction tasks, an algorithm for Maximizing Accuracy based on Guaranteed Latency (MAL) is proposed. Select appropriate fog computing nodes for different prediction task data sources to return the calculation results. Experimental results show that compared with the method of deploying neural network models in the cloud, the edge-based neural network model prediction method reduces the average delay consumption by 44.79%.

2   Deployment Model and Problem Definition

2.1   Deployment model

(1)   DNN Task Description

There is now an m-layer DNN application, and each layer of the DNN application is regarded as a subtask, that is, each task has m different subtasks. In a task scheduling request, there are n request tasks, and the task set T is denoted as $T = \{T_1, T_2, \ldots, T_n\}$. Among them, each task T_i can be expressed as $T_i = \{t_{i,1}, t_{i,2}, \ldots, t_{i,m}\}$, $t_{i,j}$ represents the j-layer subtask of the i-th task. Each task T_i is generated on the mobile device and reaches the mobile device at a rate of $\lambda_i (i \in [1, a])$.

Since each layer of the DNN is serial, the m subtasks of the same task are also serial. For example, the subtask $t_{i,j}$ must be generated after the predecessor subtask $t_{i,j-1}$ is executed, and the subtask $t_{i,j}$ cannot be generate until the subtask $t_{i,j+1}$ is executed. The execution time of the i-th subtask $t_{x,i}$ of task T_x on node s_j is represented by $time_{i,j}$. The execution time set of subtasks on different nodes is expressed by a two-dimensional matrix $Time$ as:

$$Time = \begin{pmatrix} time_{1,1} & \cdots & time_{1,k} \\ \vdots & & \vdots \\ time_{m,1} & \cdots & time_{m,k} \end{pmatrix}$$

The same subtask is executed on different nodes, and $time_{i,j}$ is small, indicating that the computing power of the node is stronger; For different subtasks executed on the same node, the smaller the $time_{i,j}$, the smaller the task. $D = \{d_1, d_2, \ldots, d_m\}$ represents the data transmission volume between different subtasks of the same task, where d_j represents the data transmission volume from subtask $t_{i,j}$ to subtask $t_{i,j+1}$.

(2) Formal definition

$t_{i,j}.arrival$ represents the moment when the subtask $t_{i,j}$ reaches the node s_y that executes the subtask. The moment when the subtask starts to execute is expressed as $t_{i,j}.begin.t_{i,j}(y)$ represents the queue waiting time of subtask $t_{i,j}$ on node s_y, then:

$$w_{i,j}(y) = t_{i,j}.begin - t_{i,j}.arrival \tag{1}$$

The response time of the task T_i is the time elapsed from the time the task is generated on the mobile device to the completion of all executions, and is equal to the sum of the response times of all subtasks of the task. The response time $r_{i,j}$ of the subtask $t_{i,j}$ is composed of three parts: execution time, transmission time and waiting time:

$$f_{i,j}(y) = time_{j,y} + g_{i,j}(x,y) + w_{i,j}(y) \tag{2}$$

So the response time of task T_i is as follows:

$$f_{resp}(T_i) = \sum_{j=1}^{m} f_{i,j}(y) \tag{3}$$

Define the average response time for all n tasks as follows:

$$f_{ave}(T) = \frac{1}{n} \sum_{i=1}^{n} \sum_{j=1}^{m} f_{i,j}(y) \tag{4}$$

It is necessary to find a scheduling scheme by which n tasks are scheduled to minimize the average response time f_{ave} of n tasks. The scheduling scheme specifies the execution node of each subtask and the order of execution on that node. To solve this problem, the simplest idea is to compare all possible scheduling solutions and find the solution with the shortest average response time. This idea is theoretically feasible, but its complexity is exponential and requires a lot of running time, so a more efficient algorithm is needed to solve this problem.

2.2 Problem Definition

In the fog computing scene, given a fog computing node set $E = \{e_1, e_2, \ldots, e_i, \ldots, e_m\}$ and a set of DNN models with n branches $D = \{d_1, d_2, \ldots, d_z, \ldots, d_n\}$, Where d_z

represents the z-th branch, and the n model branches have a running order requirement. f_{ij} represents the running order of i and j parts of the branch model, m_{ij} represents the delay between fog computing nodes i and j, c_{ij} represents the delay when d_i is deployed on e_j, and x_{ij} represents the i-th model part is deployed to j Node.

Problem 1: Deployment of distributed neural networks. Given the fog computing node set E and the DNN branch model set D, it is required to determine a deployment scheme. Under this deployment scheme, each branch model is deployed to a fog computing node, which minimizes the total delay required for the DNN model to run.

$$\min\left(\sum_{i=1}^{n} \sum_{j=1}^{m} f_{ij}m_{ij}x_{ij} + \sum_{i=1}^{n} \sum_{j=1}^{m} c_{ij}x_{ij}\right) \tag{5}$$

$$\text{s. t.} \quad \sum_{i=1}^{n} x_{ij} = 1; j = 1, 2, \ldots, n$$
$$\sum_{j=1}^{m} x_{ij} = 1; i = 1, 2, \ldots, n \tag{6}$$
$$x_{ij} \in \{0, 1\}$$

Theorem 1. *The deployment problem of distributed neural networks is NP-hard.*

Proof. First introduce the Quadratic Assignment Problem (QAP). QAP can be described as: given n devices and n locations, three $n \times n$ matrices $F = (f_{ij})_{n^*n}$, $D = (d_{ij})_{n^*n}$ and $C = (c_{ij})_{n^*n}$. Among them, f_{ij} represents the traffic between devices i and j, d_{ij} represents the distance between locations i and j, c_{ij} represents the cost of deploying device i at location j, and the product between f_{ij} and d_{ij} is the communication cost. It is required to assign a location to each device and minimize the sum of the communication cost between the facilities and the deployment cost of the facility to the site.

$$min\left(\sum_{i=1}^{n} \sum_{j=1}^{n} f_{ij}d_{p(i)p(j)} + \sum_{i=1}^{n} c_{ip(i)}\right) \tag{7}$$

Where $p(i)$ represents the location where facility i is allocated.

3   Model Deployment Algorithm Design

3.1   Deployment Algorithm Based on Solution Space Tree Pruning

This paper designs a distributed neural network model deployment a algorithm based on Solution Space Tree Pruning. The SSTP algorithm uses the breadth first in the solution space to search for the optimal solution, and uses a single assignment strategy. At each step, an unassigned model is selected and assigned to an idle position. Each assignment calculates the lower bound of the currently assigned problem. If the current lower bound is greater than the known optimal solution lower bound, it means that further exploration of the current solution branch will not result in a better solution lower bound [6]. Then remove the last allocated position and allocate other free positions for the current model; If the current calculation lower bound is less than the known optimal solution lower bound, the algorithm will continue to use the single assignment strategy to continue to assign other objects to the current solution.

Through analysis, the complexity of the algorithm is mainly reflected in the lower bound part of each stage of the calculation problem. The definition method of the lower bound will affect the search efficiency of the algorithm. When n of the problem scale $n \times n$ does not exceed 20, the optimal solution of the problem can be obtained quickly. In the neural network model with n branches considered in the problem, the scale of n will not be very large. This is why the heuristic method is not used and the solution algorithm is used to design the deployment algorithm. The boundary in the algorithm is defined as follows: assuming model i is placed at position j, calculate the lower bound of the current solution $B = C_1 + C_2 + C_3$.

$$C_1 = \sum_{i \in I, j \in J} c_{ij} + \sum_{i,j \in} m_{ij} \tag{8}$$

$$C_2 = min \sum_{i \notin I, j \notin J} \left(c_{ij} + \sum_{t \in I} m_{it} \right) x_{ij} \tag{9}$$

$$C_3 = min \sum_{i \notin I, j \notin J} \left(c_{ij} + m_{ij} \right) x_{ij} \tag{10}$$

Among them: C_1 represents the delay cost of the allocated model, C_2 represents the delay cost of the transmission between the allocated and unallocated models, and C_3 is the delay cost of the model at the unallocated position. For the problem in this paper, the computational strategy of C_3 is to arrange the unassigned compute nodes and the remaining unassigned nodes in ascending order, and then place the model parts that need to be assigned in order.

3.2 Select Node Exit Algorithm

After the deployment model task is completed, a minimal delay deployment scheme for distributed neural network models running on fog computing nodes is obtained. In the actual situation of the prediction task, the distance between the data source and the fog computing node and the network status need to be considered. Predicting the sample to experience different fog computing nodes will lead to changes in the result delay and accuracy.

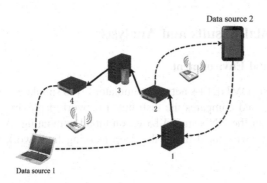

Fig. 1. Running scenario of inference task

As shown in Fig. 1, the edge computing nodes are distributed in the center of the figure, and the nodes marked with numbers are the deployed neural network models. When data source 1 and data source 2 have tasks to infer, first send the data to edge node 1 where the first layer of the neural network model is deployed. Then select edge node 2 that meets the task requirements for data source 2 to return the inference result, and select edge node 4 for data source 1 to return the inference result.

Algorithm 1. *Algorithm for Maximizing Accuracy based on Guaranteed Latency*

Input:

The deployed DNN model network topology;

T: The delay requirement T of the prediction task is required;

B_i: the i-th fog computing node is delayed from the data source B_i;

N: the task exit point N;

$f()$: the predicted delay $f()$;

$\sum_{k=1}^{i} EL_k$: Delay for the prediction task from the 1st to the i-th fog computing node.

Output:

N: that meets the delay requirement of task t.

---

1)   For $k \leftarrow 1$ to n
2) $EL_k \leftarrow f(EL_k)$
3)   For $i \leftarrow 1$ to n
            if $T > B_1 + B_i + \sum_{k=1}^{i} EL_k$
                return i, store i to set I
4)   If $I \neq \emptyset$
5)   return $N = \max I$
6)   else return null

Fog computing applications have soft real-time features, and user requests need to be returned within a certain period of time, and the accuracy of the calculation results also affects the quality of user experience. In order to select a suitable exit point for the prediction task and return the calculation result, this paper designs a selection node exit algorithm. For a prediction task with a given delay requirement, select the target node d from the deployed fog computing node set D to return the calculation result. In the case that the prediction meets the delay requirement, the prediction sample is subjected to more fog computing nodes. Finally, a calculation result that meets the latency requirements and guarantees accuracy is returned for the data source.

4   Experimental Results and Analysis

4.1   Experimental Environment

This article uses the OMNET++ network simulator [7] to simulate the network scenario of fog computing, and compares the distributed network deployment model based on fog computing with the DNN model based on cloud computing. A DNN model with multiple branches was trained under the deep learning framework Neurosurgeon for image classification tasks on the Caltech-101 dataset. The OMNET++ network simulator simulates a network distribution scenario with data sources, fog computing nodes, and cloud servers as the main components.

4.2 Experimental Results

In the current research, there are few distributed neural network deployment frameworks specifically proposed for fog computing, and there is no comparison.

So this article will compare with the existing cloud deployment method of neural network model and the edge computing acceleration framework Edgent proposed by Li et al. [8] in 2019. The trained distributed neural network models are deployed on fog computing nodes and cloud servers, respectively. The comparison experiment will use the traditional cloud server-side deployment model for inference as the baseline. The delay in the experiment in Fig. 2 is the average delay after simulating 15 different network distributions. Most fog computing nodes have low performance and are at the edge of the network, so we set the bandwidth between nodes to 1 Mb/s.

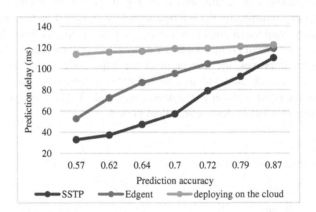

Fig. 2. Comparison of model prediction delays of different methods

The experimental results show that, among the three methods, when the task of the data source exits from the same exit point after being predicted by the model, that is, the accuracy of the task prediction is the same for both methods at this time. Fog-based neural network models use shorter prediction delays than other methods. And considering different prediction accuracy considerations, the prediction[9] delay of the model used in this method is 44.79% lower than the cloud deployment method on average, and 28.75% lower than the Edgent acceleration framework.

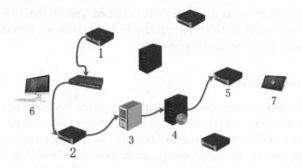

Fig. 3. Network topology

Using the network topology based on Fig. 3, the performance of the neural network model on the edge nodes to infer the tasks of data sources in different geographic locations is analyzed. First, SSTP selected nodes 1, 2, 3, 4, and 5 from a randomly generated edge node distribution set to deploy a branched neural network model.

The experimental results are shown in Fig. 4. It can be found that when the inference accuracy returned by the inferred sample is low, that is, the sample exits the node at the early branch exit point (such as nodes 1, 2, 3). At this time, the data source node 6 is closer to the exit point, so a better delay effect is obtained; As the accuracy of inferred samples improves, the exit point of the edge node at the later stage of the branched neural network model returns the inference result for the data source. At this time, the later exit point is closer to the data source node 7. Therefore, the inferred task of node 7 achieves a better delay effect than node 6.

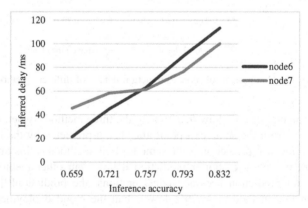

Fig. 4. Task inference delay comparison of different data sources in edge computing scenario

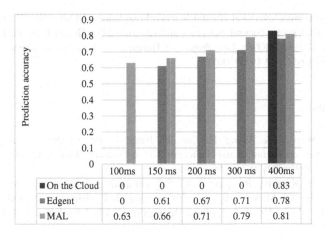

Fig. 5. Comparison of prediction accuracy of algorithms when network bandwidth is limited

The experiment also found that the delay of the prediction task is affected by the network bandwidth. Therefore, consider setting the network bandwidth to 500 kb/s in the experiment. The prediction task at this time is shown in Fig. 5. When the accuracy is 0, it means that the current prediction cannot meet the task delay requirement. When the prediction task delay requirement is 100 ms, only the MAL algorithm in the edge scenario can meet the task's delay requirement; As the prediction task delay requirement decreases, the MAL algorithm performs task prediction by selecting an early branch model and outputs a medium-precision calculation result; When the task delay requirement reaches 400 ms, all three methods can meet the predicted task delay requirement and return the calculation result of the highest precision branch model for the data source.

5 Conclusions

Aiming at the problem that the prediction delay of the existing cloud-based deep learning task model is too high, this paper proposes to deploy a distributed neural network model in the fog computing scenario to reduce the model prediction delay. Designed a deployment algorithm based on Solution Space Tree Pruning to select the appropriate fog computing node deployment for a given DNN model, reducing the delay required for prediction tasks; At the same time, MAL Select Node Exit Algorithm (MAL) is designed to return the calculation results for the prediction tasks with different delay requirements to meet the selection of appropriate nodes. The experimental results show that compared with the traditional method of deploying neural network models in the cloud, the deployment of distributed neural network models in the fog computing scenario presented in this paper reduces the model prediction delay by 44.79% on average.

Acknowledgement(s). National Natural Science Foundation of China (61772196; 61472136); Hunan Provincial Focus Natural Science Fund (2020JJ4249); Hunan Provincial Focus Social Science Fund (2016ZDB006); Key Project of Hunan Provincial Social Science Achievement Review Committee (XSP 19ZD1005).

References

1. Wang, T., Wen, C.-K., Wang, H., Gao, F., Jiang, T., Jin, S.: Deep learning for wireless physical layer: opportunities and challenges. China Commun. **14**(11), 92–111 (2017). https://doi.org/10.1109/cc.2017.8233654
2. Mahmud, R., Kotagiri, R., Buyya, R.: Fog computing: a taxonomy, survey and future directions. In: Di Martino, B., Li, K.-C., Yang, L.T., Esposito, A. (eds.) Internet of Everything. IT, pp. 103–130. Springer, Singapore (2018). https://doi.org/10.1007/978-981-10-5861-5_5
3. Teerapittayanon, S., McDanel, B., Kung, H.-T.: Distributed deep neural networks over the cloud, the edge and end devices. In: 2017 IEEE 37th International Conference on Distributed Computing Systems (ICDCS), pp. 328–339. IEEE (2017). https://doi.org/10.1109/icdcs.2017.226
4. Jiang, W., YuSheng, X., Hong, G., Zhang, L.: Dynamic trust calculation model and credit management mechanism of online trading. SCIENTIA SINICA Informationis **44**(9), 1084–1101 (2014). https://doi.org/10.1360/N112013-00202
5. Liu, W., Wang, Z., Liu, X., Zeng, N., Liu, Y., Alsaadi, F.E.: A survey of deep neural network architectures and their applications. Neurocomputing **234**, 11–26 (2017). https://doi.org/10.1016/j.neucom.2016.12.038
6. Zhang, Y., Li, S., Guo, H.: A type of biased consensus-based distributed neural network for path planning. Nonlinear Dyn. **89**(3), 1803–1815 (2017). https://doi.org/10.1007/s11071-017-3553-7
7. Kanev, A., et al.: Anomaly detection in wireless sensor network of the "smart home" system. In: 2017 20th Conference of Open Innovations Association (FRUCT), pp. 118–124. IEEE (2017). https://doi.org/10.23919/fruct.2017.8071301
8. Li, E., Zeng, L., Zhou, Z., Chen, X.: Edge AI: on-demand accelerating deep neural network inference via edge computing. IEEE Trans. Wirel. Commun. **19**(1), 447–457 (2019). https://doi.org/10.1109/twc.2019.2946140
9. Fan, Y., Shi, Y., Kang, K., Xing, Q.: An inflection point based clustering method for sequence data. In: Ni, W., Wang, X., Song, W., Li, Y. (eds.) WISA 2019. LNCS, vol. 11817, pp. 201–212. Springer, Cham (2019). https://doi.org/10.1007/978-3-030-30952-7_22

Mining the Software Engineering Forums: What's New and What's Left

Wei Yuan[1], Peng Wang[2], Yue Guo[1], Linyang He[1], and Tieke He[1(✉)]

[1] State Key Laboratory for Novel Software Technology,
Nanjing University, Nanjing 210093, China
hetieke@gmail.com
[2] Jiangsu Tongxingbao Intelligent Transportation Technology Co., Ltd.,
Nanjing, China

Abstract. Software maintenance is an important part of the software life cycle. People use bug tracking systems to collect bugs in the system. There are a lot of bug reports in open source software. Researchers conducted a series of studies on these reports, such as automatically determining whether two bug reports were duplicates. This article provides a detailed survey of the researchers' analysis of bug reports. In this paper, we conduct a comprehensive survey of the works concerning the mining of the software engineering forums. Specifically, we formulate these works in a three-dimensional style, i.e., we classify these studies according to the data formats they used, the methodology they are adopting, and most importantly, the questions they are dealing with. With this three dimensional partition, it can be clearly known what has been done, and what is left, along with the question, say, why left? To go further, beyond this three-dimensional partition, we are seeking to add research space through new data, novel techniques, and upcoming research questions.

Keywords: Software maintenance · Data mining · Bug report

1 Introduction

It is difficult for humans to develop software without any problems. So, finding and fixing bugs is an important process in the software life cycle. People use bug tracking systems to collect software system errors discovered by developers, testers, and end-users. The most commonly used bug tracking system is Bugzilla[1]. Many open-source projects use Bugzilla to help them manage their projects. For example, eclipse receives a lot of bug reports every day[2]. But everything has two sides. On the one hand, a lot of reports can help us improve the quality of software, however, on the other hand, handling these reports manually is a very time consuming task. If we can't extract useful value information from these bug reports, then more data doesn't make any sense.

---

[1] https://www.bugzilla.org/.
[2] https://bugs.eclipse.org/bugs/.

© Springer Nature Switzerland AG 2020
G. Wang et al. (Eds.): WISA 2020, LNCS 12432, pp. 513–524, 2020.
https://doi.org/10.1007/978-3-030-60029-7_46

Due to a large number of bug reports, it is unrealistic to rely solely on people to deal with them. People are gradually proposing more and more analytical methods to deal with bug reports. In this article, we will investigate the different problems that researchers have studied in existing erroneous data sets. We want to express what technologies people have used so far to solve the various problems in the bug report.

In order to better review the existing work and look forward to the future work, we sort out the previous work of the researchers from the three dimensions: problem, data and technology. From a problem perspective, some people focus on the detection of bugs, they want to analyze the existing bug reports and extract the characteristics of the bugs. When they receive a new bug report, let the machine automatically determine if this is a real error, whether it is a duplicate of the bug in the existing bug library, and find a bug similar to this bug to solve this problem faster. Some researchers want to automatically identify the severity and priority of new bugs through existing bug reports so that developers can prioritize the most problem-solving issues and do more meaningful things in a limited amount of time. There is a bug in the system, usually because some files are written. Some researchers want to find the relationship between bugs and source files through the existing repair experience, let people locate bugs and fix bugs faster. Some researchers believe that different people have different ability to solve bugs in different fields. They want to learn the distribution of existing bugs and automatically recommend the most suitable developers to solve problems when new bugs occur.

From a data perspective, most of the work already done is based on Bugzilla collecting bug reports. However, the fields that different researchers pay attention to are not the same. Some people only pay attention to the text information in the bug report, such as description, summary field. Some people think that structured information also contains important information. They not only consider text information but also consider structured information such as priorities and components. Still others believe that just analyzing bug reports is not enough. They introduce external data combined with bug reports to solve problems. How to reasonably combine different types of information has always been a difficult point. From a technical point of view, most of the work is based on traditional information retrieval techniques, machine learning and deep learning techniques. In recent years, with the development of big data and the significant improvement of computing power, people think that "big data + complex model" is a better choice. Therefore, neural networks and deep learning have become more and more popular, and they have demonstrated their capabilities in various fields. More and more people are trying to use neural networks instead of traditional information retrieval and basic machine learning algorithms to solve different problems.

2   Bug Report Problem Classification

In this survey, we mainly summarize the problems that others have studied on the bug report from the perspective of the problem and analyze the data and technology they use. We divide the existing research related to the bug report into five categories: bug detection, bug level determination, bug location, bug developer recommendation, and some other issues.

In terms of bug detection, we mainly consider three sub-problems: bug classification, to determine whether a report is a bug; Duplicate bug detection to determine if the two bug reports describe the same problem; Similar bug detection to determine if two bugs are of the same type. The bug level determination is mainly divided into two sub-problems: bug severity prediction and bug priority prediction. The bug location and bug developer recommendations are two separate and widely studied issues. By dividing the bug report problem into different categories, it is easy to know the hot point research area in recent years and conclude future tendencies.

3 Bug Detection

There are three main problems with bug detection. The first question is whether the problem described in the bug report is about the bug. Users not only submit a bug to the tracking system, they also mention other requirements, such as how to make the system more convenient. Thus, we need to identify the real bug and then solve it. The second question is about duplication. Many reports submitted by users refer to the same bug but with different descriptions, we need to be able to determine whether the two bug reports are about the same problem. The third problem is to identify similar bugs. We know that if the two errors are very similar, then the reasons for them may be very similar. By recommending similar bug reports to developers, it may be faster to locate errors and fix bugs.

3.1 Bug Reports Classification

There are a large number of reports that are actually misclassified. Manually classifying bug reports is a very time consuming task. [1] used 90 days to manually sort over 7,000 error reports. Therefore, it is necessary to classify reports automatically.

Researchers firstly used the text field in the bug report to extract features and use this to determine if the new report was a bug. [2] applied topic modeling to the corpus of pre-processed bug reports, and then classified bug reports using decision trees, naive Bayes classifiers, and logistic regression. Experiments implicate that the topic-based model outperforms than the word-based model, and the naive Bayesian model is better than the other two in classification.

Some researchers believe that structured information in the bug report can help judge whether a report is a bug. [3] used a hierarchical Dirichlet process (HDP) and clustering to classify bug reports. The bug report is projected into the topic vector space, then clustering method is utilized to aggregate the bug reports and tag the categories. How to better combine structured and unstructured data has always been an important issue. Some people [4] proposed a hybrid approach to classify error reports. They used the text mining method to extract features from the report summary, and then used the data mining method to combine the structured information features in the error report with the previous stage text features, finally used the Bayesian classifier to predict.

3.2   Duplicate Bug Report

In 2018, [5] found that the proportion of duplicate bugs reached 20% on the bug repositories of Mozilla Core, Firefox and so on. These duplicated reports may be solved by developers for many times, resulting in waste of human resources. So far, the bug tracking system can not detect duplicate bugs automatically when collecting bugs [6]. The work of detecting duplicate bugs can be divided into two categories: One is to apply natural language processing (NLP) techniques to unstructured textual information, such as bug title and bug summary; Another approach focuses on execution information in bug reports.

Duplicate Bug Detection Based on NLP. [7] used BM25 for term weighting to transform bug reports into vector space. They found that the right term weighting is critical for detecting duplicate bug reports.

There are many challenges in using text information. As each person has different speaking habits, different words may be used to express the same concept, so only use text matching is not enough. Hence, it is necessary to analyze the semantic of bug report. [8] treated each bug report as a text document and used it to train word embedding models [9]. Using the trained word embedding model, They converted the error report into a vector and further trained the deep neural network above these bug report vectors to understand the distribution of duplicate bug reports and non-repeated bug reports. In addition to the above method, there are some other attempts. [10] proposed a combination of Latent Dirichlet Allocation (LDA) and word embedding method to determine whether it is a duplicate bug report. The idea of this approach is to use LDA's higher recall rate to first exclude the most dissimilar bug reports, and then use the word embedding model in the remaining reports to calculate the similarity among reports.

Duplicate Bug Detection Based on Execution Information. The bug execution information describes the context in which the bug occurred. The context of repeated bugs is the same. [11] proposed a repetitive error detection involving both execution information and natural language information. [12] considered to utilize domain knowledge and context of software. In the Android bug tracking system experiment, they found that the detection of bug reports can be improved by considering keywords in the Android domain. Some people think that the more features of a bug report you have, the more detailed you can portray a bug. [13] defined 25 features in the bug report as the basis for the classification, most of which were generated by the TakeLab system, and then used the SVM training model to classify the bugs. The topic model enables efficient semantic analysis and text mining. [14] proposed a novel duplication detection method based on the topic model. They combined the similarity of each report in the topic space with the similarity of the classified information of each report to predict duplication.

Deep learning methods also be involved in this area. [15] proposed a search and classification model combining CNN and LSTM [16]to solve the problem of repeated bugs. They used the vanilla single layer neural network to handle structured information in bug reports, used LSTM to handle short descriptions, used CNN to process long descriptions, and finally combined them to learn bug reporting features. Some people [5] proposed to use word embedding and Convolution Neural Networks to calculate the similarity between bug reports, because this not only pays attention to textual similarities but also achieves semantic similarity. This method not only considers the textual information in the report, but also combines domain-specific features (i.e., Component, Create time and Priority, etc.) to better detect duplicate bug reports. Over time, more and more new models have been proposed for specific problems. [17] used stack traces and hidden Markov models to automatically detect duplicate bug reports. Based on their research, they recognized the obvious benefits of using stack trace information. They believe that this information can improve the accuracy of the detection of repeated bug reports. They used recall rate and Mean Average Precision (MAP) to evaluate their models on Firefox and GNOME datasets and found better results than baseline models.

3.3   Similar Bug Report

Similar errors mean that bugs in several bug reports are related to common code files. Unlike duplicate bug reports, we generally think that two reports are similar reports when they have more than 50% modified common files. By recommending similar bug reports to developers, we can help them locate the cause of the error faster and solve new bugs efficiently.

[18] combined traditional information retrieval technology and word embedding technology, and considered the title, description, and other component information in the bug report to recommend similar reports to developers. They experimented with similar bug recommendations, and their approach has better performance than NextBug [19]. Inspired by this, [20] proposed a new method for using document embedding models in order to further improve the performance of the method. They added a new document embedding vector component to the existing three components. This component focuses on mining the potential relationships between the two bug reports at the document level for better results.

4   Bug Level Determination

4.1   Bug Reports Severity

The bug report generally includes a severity, which helps the developer to resolve the serious error first. The severity is divided into crashes, errors, low efficiency and minors. Automatically detecting the degree of severity can benefit bug report processing, letting high severity bug reports be processed preferentially. [21] extracted the concept word in the bug report to construct the concept profile (CP). When a new bug report is encountered, they calculate the degree of similarity between the report and the CP severity concept to determine the severity of the bug. How to determine the severity of the profile requires people to do it manually. [22] use unsupervised methods to determine whether the severity of the bug report is correctly assigned. They used a Gaussian mixture model to group similar bug reports, then assigned severity labels to grouped bug reports, and finally used supervised algorithms to predict the severity of unmarked bug reports.

Just thinking about textual information is not enough. [23] proposed a nearest neighbor method based on information retrieval to predict the severity of bugs. They used the extended BM25 document similarity function to select the k reports that are most similar to the new bug report, and then predicted the severity of the new bug based on the severity of the k reports. In addition to considering the text information in the bug report, they also considered structured information like product and component.

4.2 Bug Reports Prioritization

[24] proposed to use different machine learning algorithms such as Naïve Bayes, decision trees, and random forests to predict the priority of reported bugs. They experimented on two feature sets. The first feature set is based on the textual description of the error report. The second feature set is based on the predefined metadata of the bug report. Experiments showed that the classification results of random forests and decision trees are better than Naïve Bayes, and the results of the second feature set are better than the first feature set. [25] thought that previous researchers did not take into account the reporter's sentiment when predicting the priority of bug reports. In the bug report, if the submitter's description is very anxious, the severity and priority of the bug may be high. Therefore, they extracted features from the bug report, then used the sentiment words involved in the bug report summary to calculate the sentiment value of each bug report, and then combined the two to train and predict the priority of the bug report. Rich, unstructured information in bug reports was also involved. [26] extracted the temporal, textual, author, related-report, severity, and product in the bug report as features, and then used the linear regression model to determine the priority of the report. They defined the priority of five bug reports and then used the thresholding approach to solve the problem of data imbalance.

5 Bug Localization

To solve the bug, the system developer needs to locate the source file that caused the bug which would be difficult especially in a huge system. The biggest challenge in auto-locating bugs is the mismatch between the terms used in the bug report and the terms used in the source file. There are lots of methods for solving this issue. One of the directions is analyzing the source code file. [27] believed that structured information based on code structure such as class name and a method name can help us locate bugs better. Their method only required source code and error reporting. However, the terms used in the error report used to describe the error may not be the same as the terms used in the source file. [28] combined DNN and information retrieval (IR) techniques to locate error files associated with bugs. They used information retrieval techniques to calculate textual similarities between error reports and source files. DNN is used to link the specific terms in the bug report to the terms in the source file. [29] found that if the error report lacks rich and structured information, the information retrieval technology often does not work well, and too much stack tracking information does not help the positioning.

Some researchers believed that considering the version history can better locate potential error locations. [30] proposed AmaLgam, a model that combines historical data, similar reports, and structural information to locate files related to bugs and achieved good performance. To help better understand existing code, researchers use information retrieval techniques to map bug reports to associated code units. [31] proposed a variant of 15 vector space models based on tf-idf to form a new composite model. They used the VSM model and AmaLgam [30] to calculate the weighted sum of suspicious files to locate bug files.

6 Bug Developer Recommendation

When we encounter a new bug, which developer should I assign to fix it? In general, we can randomly assign bugs to the developer, but it is deficient. Generally, developers have their own expertise area, so it is better to recommend bug reports that belongs to this area for them to fix up. Researchers attempt to address this issue by analyzing different kinds of data, such as text information, structured information and developer profile, and so on.

[32] believed that most of the previous work focused only on open source projects. They used convolutional neural networks and word embedding to build auto-recommended developers to fix bugs and apply the technology to industrial projects and open source projects. They believed that there are two main challenges. The first challenge is that in multinational companies whose native language is not English, bug reports often appear in their native language and English. The second challenge is that industrial projects are different from open source projects, and there may be many specific terms in specific fields. They mainly extracted the two text fields of description and summary in the bug report as features. They also proposed the idea of manual and automated classification cooperation and introduced the experience used in the industrial development environment.

Textual information in bug reports alone often does not yield satisfactory results. [33] introduced a highly scalable recommendation system for bug reporting assignments. In addition to considering the textual information of the report, they also used structured information such as component id, product id, and bug severity as feature data. They used convolutional neural networks and recurrent neural networks as deep learning classifiers. At the same time, they also made certain restrictions on developers. They believe that only developers who are still active in the project should be assigned bug fixes. They believe that only developers who have been fixing bugs for a while can be considered an active developer. They opened up further research directions in optimizing training speed and predictive performance.

Summarizing the bugs that each developer has fixed in the past is a good way to portray developers. [34] proposed to create an activity profile for the history of all activities of all users in the bug tracking system, then model it according to this file, and then recommend the appropriate developer to solve the bug through this model. Through this file, we can probably know the role of the developer in this system and the areas of expertise. Although we can better distribute bugs to developers through configuration files, it takes time to mine the configuration files for each developer's historical data. [35] proposed a new method called DevRec, which consists of two types of analysis, bug reports based analysis and developer based analysis. The bug-based analysis is mainly to find bugs similar to the newly collected bugs from the bug repository. By analyzing the bug fixers, he can found potential fix developers for new bugs. They converted the features in the bug report into vectors to calculate the similarity between the two reports. The developer-based analysis measures the distance between the bug report and the developer, correlating the developer with the characteristics of the bug report. They combined these two analyses for optimal performance. [36] proposed a unified model based on learning ranking technology, which combines activity-based technology to find out which developers have solved similar bugs and location-based techniques to find the right developer for the bug location.

7 Other Problems

7.1 Generating Bug Fixes

Although the bug report tells the developer that there is a defect in the system, it may not be able to fix the defect due to the lack of a development environment, and the defect remains in the system. [37] proposed a method called R2Fix, which automatically generates bug fixes from bug reports. They chose buffer overflow, null pointer error and memory leak to evaluate the proposed method, because the repair methods of these three types of errors are relatively simple, and the repair mode can be found. When R2Fix received a new bug report, it analyzed the bug report, determined which error belongs to the above three types of errors, and finally generates a possible patch to fix the bug. In the verification experiment, R2Fix automatically generated the correct patch for 57 errors with an accuracy of 71.3%, and it also found potential errors that the tester did not find. Due to the difficulty of automatically generating bug fix, there is still a long way to go to generalize automatic patch generation.

7.2   Automatic Vulnerability Recognition

Automatically identify potential bugs will greatly improve the efficiency of software maintenance. In order to find unrecognized vulnerabilities in the open-source library and fix it, Some people [38] proposed an automatic vulnerability identification system. They used a variety of machine learning classifiers as basic classifiers to extract features from bug report submissions and reports themselves and automatically discovered unrecognized errors in submitted reports through natural language processing and machine learning techniques. However, because of the complexity of this issue, existing methods perform not well. Future work will continue exploring new methods to solve this problem.

7.3   Bug Report Summarization

In order to help developers quickly understand the information in the bug report, [39] proposed a two-layer semantic model (TSM) to extract important information from the report. They first used the extended NR (ENR) model to preserve the sentences with important semantics in the report, then used BRC (Bug Report Classifier) to extract the text features from these sentences, and finally used the logistic regression training model to select the sentences with high scores to generate the abstract of the article. [40] explored the use of deep neural networks to generate a summary of bug reports. They used bug report preprocessing, unsupervised network training and summary generation to assign scores to sentences in bug reports, and then dynamically selected sentences with high scores to generate summaries.

8   Conclusion - What's the Outlook?

We have presented a comprehensive survey of bug report, categorizing current bug report tasks based on problem, data and technology and summarizing the current situation for each tasks. What's the next for bug report? We end with future potential directions by applying past insights to the current situation. Firstly, different types of data will be used to better analyze bug reports; Then, advanced models will be invented to better address specific problems; Last but not least, the efficiency and practicability of methods will be considered.

Acknowledgment. This work was partially supported by the Fundamental Research Funds for the Central Universities (14380023), and the Equipment Development Department Pre-Research Foundation of China (31511110310).

References

1. Herzig, K., Just, S., Zeller, A.: It's not a bug, it's a feature: how misclassification impacts bug prediction. In: Proceedings of the 2013 International Conference on Software Engineering, pp. 392–401. IEEE Press (2013)

2. Pingclasai, N., Hata, H., Matsumoto, K.I.: Classifying bug reports to bugs and other requests using topic modeling. In: 2013 20th Asia-Pacific Software Engineering Conference (APSEC), vol. 2, pp. 13–18. IEEE (2013). https://doi.org/10.1109/APSEC.2013.105

3. Limsettho, N., Hata, H., Monden, A., Matsumoto, K.: Automatic unsupervised bug report categorization. In: 2014 6th International Workshop on Empirical Software Engineering in Practice, pp. 7–12. IEEE (2014). https://doi.org/10.1109/IWESEP.2014.8

4. Zhou, Y., Tong, Y., Gu, R., Gall, H.: Combining text mining and data mining for bug report classification. J. Softw.: Evol. Process **28**(3), 150–176 (2016). https://doi.org/10.1109/ICSME.2014.53

5. Xie, Q., Wen, Z., Zhu, J., Gao, C., Zheng, Z.: Detecting duplicate bug reports with convolutional neural networks. In: 2018 25th Asia-Pacific Software Engineering Conference (APSEC), pp. 416–425. IEEE (2018). https://doi.org/10.1109/APSEC.2018.00056

6. Sun, C., Lo, D., Wang, X., Jiang, J., Khoo, S.C.: A discriminative model approach for accurate duplicate bug report retrieval. In: Proceedings of the 32nd ACM/IEEE International Conference on Software Engineering-Volume 1, pp. 45–54. ACM (2010). https://doi.org/10.1145/1806799.1806811

7. Yang, C.Z., Du, H.H., Wu, S.S., Chen, X.: Duplication detection for software bug reports based on BM25 term weighting. In: 2012 Conference on Technologies and Applications of Artificial Intelligence, pp. 33–38. IEEE (2012). https://doi.org/10.1109/TAAI.2012.20

8. Budhiraja, A., Dutta, K., Reddy, R., Shrivastava, M.: DWEN: deep word embedding network for duplicate bug report detection in software repositories. In: Proceedings of the 40th International Conference on Software Engineering: Companion Proceedings, pp. 193–194. ACM (2018). https://doi.org/10.1145/3183440.3195092

9. Xia, C., He, T., Wan, J., Wang, H.: Ensemble methods for word embedding model based on judicial text. In: Ni, W., Wang, X., Song, W., Li, Y. (eds.) WISA 2019. LNCS, vol. 11817, pp. 309–318. Springer, Cham (2019). https://doi.org/10.1007/978-3-030-30952-7_31

10. Budhiraja, A., Reddy, R., Shrivastava, M.: Poster: LWE: LDA refined word embeddings for duplicate bug report detection. In: 2018 IEEE/ACM 40th International Conference on Software Engineering: Companion (ICSE-Companion), pp. 165–166. IEEE (2018)

11. Wang, X., Zhang, L., Xie, T., Anvik, J., Sun, J.: An approach to detecting duplicate bug reports using natural language and execution information. In: Proceedings of the 30th International Conference on Software Engineering, pp. 461–470. ACM (2008). https://doi.org/10.1145/1368088.1368151

12. Alipour, A., Hindle, A., Stroulia, E.: A contextual approach towards more accurate duplicate bug report detection. In: 2013 10th Working Conference on Mining Software Repositories (MSR), pp. 183–192. IEEE (2013). https://doi.org/10.1109/MSR.2013.6624026

13. Lazar, A., Ritchey, S., Sharif, B.: Improving the accuracy of duplicate bug report detection using textual similarity measures. In: Proceedings of the 11th Working Conference on Mining Software Repositories, pp. 308–311. ACM (2014). https://doi.org/10.1145/2597073.2597088

14. Zou, J., Xu, L., Yang, M., Yan, M., Yang, D., Zhang, X.: Duplication detection for software bug reports based on topic model. In: 2016 9th International Conference on Service Science (ICSS), pp. 60–65. IEEE (2016)

15. Deshmukh, J., Podder, S., Sengupta, S., Dubash, N., et al.: Towards accurate duplicate bug retrieval using deep learning techniques. In: 2017 IEEE International Conference on Software Maintenance and Evolution (ICSME), pp. 115–124. IEEE (2017). https://doi.org/10.1109/ICSS.2016.16
16. Hochreiter, S., Schmidhuber, J.: Long short-term memory. Neural Comput. 9(8), 1735–1780 (1997). https://doi.org/10.1162/neco.1997.9.8.1735
17. Ebrahimi, N., Trabelsi, A., Islam, M.S., Hamou-Lhadj, A., Khanmohammadi, K.: An HMM-based approach for automatic detection and classification of duplicate bug reports. Inf. Softw. Technol. 113, 98–109 (2019). https://doi.org/10.1016/j.infsof.2019.05.007
18. Yang, X., Lo, D., Xia, X., Bao, L., Sun, J.: Combining word embedding with information retrieval to recommend similar bug reports. In: 2016 IEEE 27th International Symposium on Software Reliability Engineering (ISSRE), pp. 127–137. IEEE (2016). https://doi.org/10.1109/ISSRE.2016.33
19. Rocha, H., Valente, M.T., Marques-Neto, H., Murphy, G.C.: An empirical study on recommendations of similar bugs. In: 2016 IEEE 23rd International Conference on Software Analysis, Evolution, and Reengineering (SANER), vol. 1, pp. 46–56. IEEE (2016). https://doi.org/10.1109/SANER.2016.87
20. Hu, D., et al.: Recommending similar bug reports: a novel approach using document embedding model. In: 2018 25th Asia-Pacific Software Engineering Conference (APSEC), pp. 725–726. IEEE (2018). https://doi.org/10.1109/APSEC.2018.00108
21. Zhang, T., Yang, G., Lee, B., Chan, A.T.: Predicting severity of bug report by mining bug repository with concept profile. In: Proceedings of the 30th Annual ACM Symposium on Applied Computing, pp. 1553–1558. ACM (2015). https://doi.org/10.1145/2695664.2695872
22. Pushpalatha, M., Mrunalini, M.: Predicting the severity of open source bug reports using unsupervised and supervised techniques. Int. J. Open Source Softw. Process. (IJOSSP) 10(1), 1–15 (2019). https://doi.org/10.4018/IJOSSP.2019010101
23. Tian, Y., Lo, D., Sun, C.: Information retrieval based nearest neighbor classification for fine-grained bug severity prediction. In: 2012 19th Working Conference on Reverse Engineering, pp. 215–224. IEEE (2012). https://doi.org/10.1109/WCRE.2012.31
24. Alenezi, M., Banitaan, S.: Bug reports prioritization: which features and classifier to use? In: 2013 12th International Conference on Machine Learning and Applications, vol. 2, pp. 112–116. IEEE (2013). https://doi.org/10.1109/ICMLA.2013.114
25. Umer, Q., Liu, H., Sultan, Y.: Emotion based automated priority prediction for bug reports. IEEE Access 6, 35743–35752 (2018). https://doi.org/10.1109/ACCESS.2018.2850910
26. Tian, Y., Lo, D., Sun, C.: Drone: predicting priority of reported bugs by multi-factor analysis. In: 2013 IEEE International Conference on Software Maintenance, pp. 200–209. IEEE (2013). https://doi.org/10.1109/ICSM.2013.31
27. Saha, R.K., Lease, M., Khurshid, S., Perry, D.E.: Improving bug localization using structured information retrieval. In: 2013 28th IEEE/ACM International Conference on Automated Software Engineering (ASE), pp. 345–355. IEEE (2013). https://doi.org/10.1109/ASE.2013.6693093
28. Lam, A.N., Nguyen, A.T., Nguyen, H.A., Nguyen, T.N.: Combining deep learning with information retrieval to localize buggy files for bug reports (n). In: 2015 30th IEEE/ACM International Conference on Automated Software Engineering (ASE), pp. 476–481. IEEE (2015). https://doi.org/10.1109/ASE.2015.73

29. Rahman, M.M., Roy, C.: Poster: improving bug localization with report quality dynamics and query reformulation. In: 2018 IEEE/ACM 40th International Conference on Software Engineering: Companion (ICSE-Companion), pp. 348–349. IEEE (2018)

30. Wang, S., Lo, D.: Version history, similar report, and structure: putting them together for improved bug localization. In: Proceedings of the 22nd International Conference on Program Comprehension, pp. 53–63. ACM (2014). https://doi.org/10.1145/2597008.2597148

31. Wang, S., Lo, D., Lawall, J.: Compositional vector space models for improved bug localization. In: 2014 IEEE International Conference on Software Maintenance and Evolution, pp. 171–180. IEEE (2014). https://doi.org/10.1109/ICSME.2014.39

32. Lee, S.R., Heo, M.J., Lee, C.G., Kim, M., Jeong, G.: Applying deep learning based automatic bug triager to industrial projects. In: Proceedings of the 2017 11th Joint Meeting on Foundations of Software Engineering, pp. 926–931. ACM (2017). https://doi.org/10.1145/3106237.3117776

33. Florea, A.-C., Anvik, J., Andonie, R.: Parallel implementation of a bug report assignment recommender using deep learning. In: Lintas, A., Rovetta, S., Verschure, P.F.M.J., Villa, A.E.P. (eds.) ICANN 2017. LNCS, vol. 10614, pp. 64–71. Springer, Cham (2017). https://doi.org/10.1007/978-3-319-68612-7_8

34. Naguib, H., Narayan, N., Brügge, B., Helal, D.: Bug report assignee recommendation using activity profiles. In: Proceedings of the 10th Working Conference on Mining Software Repositories, pp. 22–30. IEEE Press (2013). https://doi.org/10.1109/MSR.2013.6623999

35. Xia, X., Lo, D., Wang, X., Zhou, B.: Accurate developer recommendation for bug resolution. In: 2013 20th Working Conference on Reverse Engineering (WCRE), pp. 72–81. IEEE (2013). https://doi.org/10.1109/WCRE.2013.6671282

36. Tian, Y., Wijedasa, D., Lo, D., Le Goues, C.: Learning to rank for bug report assignee recommendation. In: 2016 IEEE 24th International Conference on Program Comprehension (ICPC), pp. 1–10. IEEE (2016). https://doi.org/10.1109/ICPC.2016.7503715

37. Liu, C., Yang, J., Tan, L., Hafiz, M.: R2Fix: automatically generating bug fixes from bug reports. In: 2013 IEEE Sixth International Conference on Software Testing, Verification and Validation, pp. 282–291. IEEE (2013). https://doi.org/10.1109/ICST.2013.24

38. Zhou, Y., Sharma, A.: Automated identification of security issues from commit messages and bug reports. In: Proceedings of the 2017 11th Joint Meeting on Foundations of Software Engineering, pp. 914–919. ACM (2017). https://doi.org/10.1145/3106237.3117771

39. Yang, C.Z., Ao, C.M., Chung, Y.H.: Towards an improvement of bug report summarization using two-layer semantic information. IEICE Trans. Inf. Syst. **101**(7), 1743–1750 (2018). https://doi.org/10.1587/transinf.2017KBP0016

40. Li, X., Jiang, H., Liu, D., Ren, Z., Li, G.: Unsupervised deep bug report summarization. In: Proceedings of the 26th Conference on Program Comprehension, pp. 144–155. ACM (2018). https://doi.org/10.1145/3196321.3196326

Fast Trading and Price Discovery in the Financial Crisis: Evidence from the Taiwan Futures Market

William T. Lin[1], Zi-Huang Huang[1], and Shih-Chuan Tsai[2(✉)]

[1] Tamkang University, New Taipei City, Taiwan
yungshuncn@hotmail.com, huangwelcomeyou@hotmail.com
[2] National Taiwan Normal University, Taipei City, Taiwan
chuant@ntnu.edu.tw

Abstract. This paper used millisecond-level intraday data from the Taiwan futures market during the financial crisis to propose an effective data processing method using the program and a non-SQL database. Fast traders were classified based on the investors' trading volume and position size. First, the state space model was used to decompose the prices. It was discovered that fast trading (FT) can cause permanent price increments, which are independent of temporary prices. FT during the financial crisis helped improve price efficiency and liquidity. Second, the activity of FT is based on public information, which makes price discovery during a high-Volatility Index (VIX) period possible and causes an increase in the adverse selection cost of non-fast traders (non-FT).

Keywords: Fast trading · Price discovery · Futures market · Financial crisis · Data mining

1 Introduction

Recently, the world economy has been hit by the effects COVID-19, and the circuit breaker mechanism for important indexes in the financial markets of various countries has been triggered many times. Investor panic in the market is running high, and the possibility of a repeat of the 2008 financial crisis is great. However, the use of technology to process information as noted by Brogaard et al. [5] and the advantages of high-frequency traders are more obvious, the information during the financial crisis has been obtained earlier than in other investors, according to Foucault et al. [11]. Additionally, parsing high-frequency trading (HFT) and information from intraday data is a complex process. It is difficult to acquire the data as well as manage the analysis of the large amount of intraday data. The purpose of this study is to establish a set of data processing methods based on the intraday data provided by the Taiwan futures exchange during financial crisis, using a program and database, in order to analyze investors and price discovery and to provide suggestions for dealing with similar data in the future.

© Springer Nature Switzerland AG 2020
G. Wang et al. (Eds.): WISA 2020, LNCS 12432, pp. 525–536, 2020.
https://doi.org/10.1007/978-3-030-60029-7_47

Data mining integrates theories and technologies in artificial intelligence, statistics, database and machine learning, among other fields, as noted by Wei Yu and Shijun Li [29]. Data mining in the financial market is mainly used to analyze the financial statements or major announcements of stocks and sometimes used to simulate optimal portfolio returns. While HFT is an important aspect of the international financial market, with its market share growing rapidly in the past decade by Van Kervel and Menkveld [28], most of its information comes from the macroeconomic news bulletin, as noted by Andersen et al. [1], and the imbalance of the order book, as noted by Cao et al. [7]. Carrion [8], Conrad et al. [9], Hasbrouck and Saar [15] and Hirschey [16] found that HFT can predict short-term prices and plays an important role in liquidity supply.

Currently, however, the discussion about HFT focuses more on the European and American markets and less on emerging markets. The Taiwan futures market is an emerging market that is dominated by individual investors, according to Lin et al. [22], and no real HFT exists due to laws, regulations, machinery and equipment. However, this research defines fast traders in the Taiwan future market because they are important members in the international financial market who exhibit behavior similar to that of HFT, mostly in market operations. The FT are defined using the state space model to analyze their information ability and impact on prices, to explain their contribution to price discovery, and to interpret price discovery using the least square method.

2   Data

2.1   Data Source

The main source of the data in this paper is the millisecond-level intraday data provided by the Taiwan Futures Exchange (TAIFEX), which contains the details of each investor's orders and trades and the best five bid and ask prices and volume of commodities. There are more than 30 data fields in total. In 2018, the Taiwan futures exchange was ranked 15th in the world in terms of trading volume, which means that it is not only widely favored by international capital but also climbs the regional representative indexes or intraday Taiwan Stock Exchange Capitalization Weighted Stock Index (TAIEX) in seconds from the Taiwan Stock Exchange (TESE) website. The sample interval that was used is from January 15, 2008, to November 19, 2008, including a total of 209 trading days. The total size of the dataset is more than 200 Gb, and the average number of daily observations is more than 200,000. These data collected during the financial crisis have a variety of market conditions and are of great research significance and value.

2.2   Extraction and Processing

Brogaard et al. [5] found that HFT like to trade commodities with a large volume and that are highly liquid. Lin et al. [21] and Lin et al. [23] also pointed out that a contract with a large trading volume has better liquidity and can therefore more accurately capture the real situation and characteristics of the market and reflect the transaction price accordingly. Therefore, TAIEX Futures (TXF), the most popular commodity traded in the Taiwan futures exchange, was selected as the research sample to represent the Taiwan futures market. TXF has a total of five contracts to trade every day. The contract value is calculated by multiplying its points by 200 NT$; therefore, to buy a bite, the actual manipulation position is approximately 9,000 * 1 * 200 NT$ = 1,800,000 NT$. The trading time is from 8:45:00–13:45:00, and the minimum rise and fall unit is 1 point, i.e., 200 NT$. Lin et al. [23] discuss the contract expiration effect in the futures market, where information is converted with the expiration of the contract in the futures market maturity. To remove the expiry effect, every nearby futures contract, with the days to maturity less than 5 days, is replaced by the next nearby futures contract, as its daily trading volume is less than that of the next nearby futures contract as done in Lin et al. [21].

3 Methodology

3.1 Framework

The research process and structure, as well as the processing of the data, are described here because of the large amount of millisecond-level intraday data studied. First, the transaction data that were extracted from the two exchanges were sorted out. Since the type of data that were provided by the different exchanges were inconsistent, all the data were converted into JSON format after removing the exchange identification fields. Second, the sorted data were added to the local server's MongoDB database. Then, Python is also a good choice for developers who want to improve their coding skills to Zhang et al. [31]. JupyterLab was built, and Python was used to introduce first-level data and to conduct data processing and collation, according to the study by Gai et al. [12] that discussed how to use the nanosecond-level intraday data method, perform multithreaded calculation, and view millisecond day trading records. Finally, R was used to perform the subsequent model analysis of the sorted secondary data. The state space model both used toolkits from R, the details of which are explained below.

3.2 Fast Trading (FT)

As mentioned above, programs similar to HFT have appeared in the Taiwan futures market; therefore, this section discusses how to define FT in the Taiwan futures market. Brogaard et al. [5] pointed out that HFT do not leave positions until the next day. Antoine et al. [2] and the SEC [27] believe that HFT are characterized by a large daily trading volume and close to closing or neutral positions at the end of the trading day. This study uses Brogaard et al. [6] and Kirilenko et al. [19] to define the HFT method, beginning with the trading

volume and position size of each investor. Since this study focuses on futures, the proportion of the trading time of investor contracts in the total trading time is added. The three screening conditions are as follows:
(1) days total more than 30%;
(2) make up more than 0.02% of the trading volume;
(3) never hold more than 30% of the daily trading volume at one time within the trading day.

The details of the investors are arranged in Table 1 in order to provide a more intuitive distinction between FT and non-FT. Table 1 shows that the average activity of FT is higher than that of non-FT. Additionally, the ratio of positions held at the end of the day for FT is smaller than that the maximum positions held during the day. This means that FT will concentrate on positions held during trading and will leave or remain neutral at the end of the trading. Overall, the number and the proportion of total trading volume of FT are much smaller than those of non-FT, but this also indicates that FT have more capital. It was determined that the selected FT has a small trading scale, a large trading market value and other characteristics that are generally accepted for HFT. This also proves that this classification method is effective and lays a solid foundation for future research.

Table 1. Trader type statistics.

	FT (1)	Non-FT (2)
Volume	2.39	1.36
Dollar volume ($million)	3.53	1.99
Max position/volume	5.80%	57.44%
Position/volume	0.95%	39.78%
Volume/total volume	24.86%	75.14%
People	24	8486

In addition, in the market transaction, both the orders and the trades contain information, but the information contents are different. To explore the price discovery ability of FT, the message frequency of previously classified investors was sorted in Table 2 for comparison. First, Table 2 shows that the proportion of order price change for non-FT is relatively high, which means that it is more likely to be affected by sliding cost. Second, because each order represents the information issued by the investor, the deletion and cancel of the order can be understood as changes in the information. In Table 2, the proportion of the FT in the deletion and cancel orders and the best first bid and ask prices are higher than those of the non-FT. This demonstrates that the information is concentrated in the best first bid and ask prices and also reflects the technical advantage of the FT, which is reflected not only in the computer equipment but also in the trading skills. Antonine et al. [2] discuss the order to trade ratio (OTR), which observes whether investors will make a large number of orders, delete and modify orders, that is, measure the message flow. Generally, the value of HFT is relatively high, and this is also reflected in Table 2.

Table 2. Message frequency.

	FT (1)	Non-FT (2)
Trade-change price	8.97%	10.71%
Worsening cancel	65.01%	42.82%
Order placement at NBBO	36.60%	38.17%
Order cancel at NBBO	6.96%	3.40%
Order to trade ratio	3.23%	2.43%
Total number of orders obs	35,084	87,728
Total number of trades obs	15,918	62,216

3.3   The Optimal Time Interval

A reasonable fixed time interval should be selected for this part of the study because of the use of intraday data. Brogaard et al. [4] and Brogaard et al. [6] used 10 s as the time interval in their research on HFT and price discovery. Lin et al. [23] chose a 5 s interval to study the information advantage in the Taiwan option market. Diebold and Strasser [10] pointed out that a too-short time interval would lead to too much noise when conducting market microstructure research. Therefore, in order to strengthen the rigor of the research, this study sorted out the TXF of the Taiwan futures market data and obtained the results shown in Table 3.

Panel A is the time ratio of order to price movements, where the larger the time interval is, the larger the proportion of price movements caused by the order. It was found that at 10 s, on average, nearly 90% of orders cause price movements. If the time interval is relaxed to 30 or 60 s, the difference is not significant. Panel B is the time ratio of order to trade, which means a longer order time with a higher trade ratio. These statistics show that most orders can be executed in a very short time. Within 60 s, at least 90% of the orders are traded, which means that 90% of the information is reflected. The magnitude of price movements and the number of observations also depend on the time interval. Longer time intervals can reduce the number of observations and stabilize price movements; therefore, a balance must exist between the number of observations and the size of price movements. Therefore, this study used 60 s as the time interval.

Table 3. Order and trade in time interval.

1 s	5 s	10 s	30 s	60 s
Panel A: Time Ratio of Order to Price Movements				
57.06%	91.85%	97.29%	99.44%	99.66%
Panel B: Time Ratio of Order to Trade				
54.56%	70.63%	76.74%	84.70%	88.73%

3.4 State Space Model

Hasbrouck [14] pointed out that the trade price can be divided into random-walk and stationary components. The random part can be used to monitor the effectiveness of the price, and the stationary part is the difference between the effective price and the actual price, which is called mispricing. The previous information share model was intended to explore the relationship between price lead and lag and the information content; therefore, this section will use the state space model to disassemble prices and disaggregate investors' activities according to liquidity. According to the research methods of Brogaard et al. [5] and Menkveld et al. [25], the state space model is used to divide the price of the TXF into a permanent price and a transitory price:

$$p_t = m_t + s_t \tag{1}$$

where p_t is the $ln(midprice_t)$ at time interval t for TXF and is composed of a permanent price m_t and a transitory price s_t. The permanent price is modeled as a martingale:

$$m_t = m_{t-1} + w_t \tag{2}$$

The permanent process characterizes information arrivals, where w_t represents the permanent price increments. Then, the following regression discussion is established for transitory price and permanent price increments:

$$s_t = \alpha s_{t-1} + \gamma^{All} FT_t^T + \xi_t \tag{3}$$

$$w_t = \beta^T FT_t^T + \beta Vol_t + \varepsilon_t \tag{4}$$

where, FT_t^T is the logarithmic of the FT volume in the interval, and Vol_t is the logarithmic of the trading volume in the interval. This regression formula differs from that of Brogaard et al. [6] in that the Taiwan futures market is still an emerging market. The trading volume before and after FT was not found to be highly autocorrelated after the correlation analysis.

Brogaard et al. [5] and Carrion [8] noted that the trade prices of the two types of investors are divided into liquidity demand and liquidity supply according to the relationship between buying and selling, driving and leading and lagging, and the regression is as follows:

$$s_t = \alpha s_{t-1} + \gamma_{FT}^{D/S} FT_t^{D/S} + \gamma_{Non-FT}^{D/S} Non - FT_t^{D/S} + \xi_t \qquad (5)$$

$$w_t = \beta_{FT}^{D/S} FT_t^{D/S} + \beta_{Non-FT}^{D/S} Non - FT_t^{D/S} + \varepsilon_t \qquad (6)$$

Among them, the volume of FT is divided into liquidity demand (FT_t^D) and liquidity supply (FT_t^S). The trading volume of non-FT is divided into liquidity demand $(Non\text{-}FT_t^D)$ and liquidity supply $(Non\text{-}FT_t^S)$. In R, there are several databases that can be used for the state space model. In this study, Petris and Petrone [26] used the sspir's package for the calculation. KFAS can also be used, and the instructions for the specific use method can be found on GitHub.

3.5   FT Activity and Return

Previous studies aimed to determine the price discovery capability of FT activity to analyze the influence of two types of investors on the transitory price and the permanent price increment by using the state space model. This can be viewed as the robustness test of these previous studies that was used to further understand the role of FT in the price discovery process from order to trade in order to further strengthen and expand the available information.

It is well known that public information is complicated and multifarious and that the interpretation of the information and market prices is often ambiguous. During financial crises, information tends to become more complex and richer. However, the order to trade is the behavior of various investors after the interpretation of the information; therefore, the information in the order book is extracted with multiple variables. Cao et al. [7] found an imbalance between the amount of liquidity available for trading and predicted short-term price changes. Therefore, the limit order book imbalance (LOBI) variables proposed in Brogaard et al. [5] were used as the agent variables of the information in the limit order book.

In this paper, ordinary least squares (OLS) regression was used to explore two relationships. During financial crisis, the uncertainty of information and the drying up of liquidity will increase the VIX, and the excessive VIX will also increase the uncertainty of the market, therefore affecting the FT. This study defined a VIX greater than 30 as a dummy variable and included it in the regression:

$$Ret_t = \alpha_1^3 + \beta_1^3 FT_{t-1} + \beta_2^3 VIX_{t-1} + \beta_3^3 VIX_{t-1} * FT_{t-1} + \beta_3^4 LOBI_{t-1} + \beta_4^5 Ret_{t-1} + \varepsilon_t^3 \qquad (7)$$

In the above regression, Ret_t is the absolute value of the return as the price explanatory variable because the Taiwan futures exchange can short trade, and it can also represent the price change. FT_t respectively represent the amount of order volume, trading volume, liquidity demand and supply of FT and take the logarithmic. VIX_t is a dummy variable, represented by 0 or 1. $OrderSize_t$ is the logarithmic of all order volumes in interval, which represents the collection of all the delegate information. $LOBI_t$ s a summary of the information from the price depth in the interval. If this value is negative, it represents a sell message. The Taiwan futures market provides only the best five bid and ask prices order volume, so the value is calculated as the difference in the best five bid and ask prices order volume divided by the sum.

4 Result Analysis and Discussion

4.1 Descriptive Statistics

The main explanation of the variables is discussed in this section, and will further use descriptive statistics in the analysis. As shown in Table 4, FT during financial crisis, it is still continuing to order and trade, and the liquidity supply to demand ratio is higher with the FT. Carrion [8] believes that HFT will provide liquidity and vice versa when the market is in a state of abundant liquidity. Of course, in extreme cases, the HFT will withdraw from the liquidity supply as reported by Kirilenko et al. [19]. Grossman and Miller [13] showed that they would choose to supply liquidity during the imbalance of orders and put those investors who hold opposite views into a passive position. In addition, during the crisis, the LOBI in the market was negative, which indicates that the overall investor was selling information. Since the minimum variable associated with trading volume appears to be 0, this means that there will be a brief vacuum during trading, which is also in line with the high degree of panic in the market during times of crisis.

Table 4. Descriptive statistics.

	FT^O	FT^T	FT^D	FT^S	non-FT^D	non-FT^S	Ret	LOBI	OrderSize	Vol
Mean	5.23	4.24	1.08	2.87	2.18	4.08	0.0006	−0.03	6.75	5.78
Std	1.21	1.18	1.05	1.21	1.08	1.03	0.0007	0.38	0.74	1.03
Min	0.00	0.00	0.00	0.00	0.00	0.00	0.0000	−1.00	0.00	0.00
Max	7.98	7.34	4.56	6.48	5.59	6.99	0.0152	1.00	8.79	8.82

4.2 State Space Model of FT and Prices

This section mainly describes the results of the state space model and the disassembly of prices by using the Kalman filter, which is divided into the permanent price and transitory price, where the difference in permanent price is the permanent price increment. In this study, the activities of the FT and the non-FT and liquidity dismantling of the two parts of the regressive transitory price and the permanent price increment are explained, and the results are shown in Table 5.

Table 5. State space model of FT and prices.

	Transitory component (1)	Permanent price increments (2)
Panel A: By FT^{All} and Prices		
FT^{All}	−0.00**	0.00*
(t-stat.)	(−2.00)	(1.74)
Panel B: By Liquidity Demand, FT^D and Non-FT^D, and Prices		
FT^D	−0.00	−0.00
(t-stat.)	(−0.76)	(−0.48)
Non-FT^D	−0.00	0.00
(t-stat.)	(−0.39)	(1.27)
Panel C: By Liquidity Supply, FT^S and Non-FT^S, and Prices		
FT^S	0.00	0.00***
(t-stat.)	(0.85)	(2.84)
Non-FT^S	−0.00*	0.00***
(t-stat.)	(−1.71)	(2.08)

Robust t-statistics in parentheses *** p<0.01, ** p<0.05, * p<0.1

Panel A shows the transitory price and permanent price increment with the FT activities. By column (1), as can be seen, the FT is significantly negative, and this represents that a transitory price is unrelated to FT activities, and in column (2), the permanent price increment and FT have a positive significant relationship. The representative FT actually have the information, cause permanent price increment, and take this information advantage in and out of the futures in the financial crisis. While Panel B is the liquidity demand analysis of FT and non-FT, the two do not have much of a relationship with the transitory price part and the permanent price increment, and the liquidity supply of its activities has certain research significance.

In the Panel C column (2), it is shown that the liquidity supply of both types of investors has a positive and significant relationship with the permanent price increment, and the t-test statistic of the FT is larger, which can be interpreted as providing liquidity for the FT when the market dries up. In addition, previous statistics showed that non-FT accounts for a large proportion of the total trading volume. Kaniel et al. [17] and Kelley and Tetlock [18] found that individual investors in emerging markets can obtain positive remuneration in options markets with higher thresholds of knowledge and technology. Therefore, it is believed that there are some sophisticated retail investors that have a certain capacity for non-FT.

4.3 FT Activity and Return

The purpose of this section is to further explore the relationship between FT and the return in order to further explain its effect on price discovery and market efficiency, as well as its performance in times of a high-VIX index. Malinova et al. [24] believe that the information reduction of HFT leads to an increase in the spread of the transaction cost for other investors. Lee [20] found that HFT mainly consumes liquidity in the Korean market, contrary to the findings in Western markets, and that it has no effect on price discovery and market efficiency. Using commodity future data of the DaLian Commodity Exchange, Zhao and Wan [30] also found that high-frequency traders mainly consume liquidity. The regression results are shown in Table 6.

Table 6. FT activity with high-VIX.

	Order (1)	Trade (2)	Demand (3)	Supply (4)
Constant	0.00***	0.00***	0.00***	0.00***
(t-stat.)	(19.35)	(23.86)	(78.49)	(43.97)
Lag1 FT	0.00***	0.00***	0.00***	0.00***
(t-stat.)	(3.93)	(7.04)	(10.07)	(10.37)
Lag1 VIX	−0.00***	−0.00***	0.00***	−0.00***
(t-stat.)	(−6.47)	(−10.42)	(17.39)	(−3.89)
Lag1 (VIX*FT)	0.00***	0.00***	0.00***	0.00***
(t-stat.)	(16.90)	(21.86)	(12.12)	(20.00)
Lag1 LOBI	−0.00**	−0.00	−0.00***	−0.00**
(t-stat.)	(−2.49)	(−1.45)	(−3.47)	(−2.43)
Lag1 Ret	0.25***	0.23***	0.24***	0.23***
(t-stat.)	(31.63)	(28.29)	(29.32)	(27.91)

Robust t-statistics in parentheses *** $p<0.01$, ** $p<0.05$, * $p<0.1$

Table 6 shows the results when the VIX in the market is included in the regression and uses the lag period to explore the impact of FT activities on the return. The FT order volume in (1) shows that during the high-VIX period, the results are all significantly positive, but smaller t-test values. This means that the tendency for the FT during high-VIX periods to entrust the impact on the return is slow, but the FT still has the capacity for price discovery and price efficiency, which is associated with FT capital being relatively abundant. The results of (2) for FT activity are similar to those for (1). The results from (3) and (4) strengthen the previous analysis. In addition, since a high-VIX inevitably leads to the exhaustion of market liquidity, the liquidity supply t-test value of FT is large, which also conforms to the previous idea that FT provides liquidity when market liquidity is low and at the same time aggravates the adverse selection cost of non-FT.

5   Conclusion

This study collected and sorted the millisecond-level intraday data of the Taiwan futures exchange and the Taiwan stock exchange and used Python, R and MongoDB to process the data. The TXF, the most traded commodity in the Taiwan futures market, was the research object, and the investors who were similar to the HFT in the international market were defined by their trading volume and position size and called FT. Using the state space model, it was found that the FT in the Taiwan futures market is informative and has the capacity for price discovery, which also enhances market efficiency. The subsequent regression model also indicated that FT activities lead to the use of bad prices for non-FT and increase the adverse selection cost of non-FT, which is similar to previous HFT research results.

In addition, recent research on HFT was also explored from the limit order book of Brogaard et al. [6]. Baldauf et al. [3] thinks can be strengthened by different models or information connotations. They also provide a direction for future research on FT. The research also discovered effective observation values for exploring price discovery and improves price prediction. It provides a good learning sample for deep learning in the future and machine learning of FT behavior to improve the success rate.

References

1. Andersen, T.G., Bollerslev, T., Diebold, F.X., Vega, C.: Micro effects of macro announcements: real-time price discovery in foreign exchange. Am. Econ. Rev. **93**(1), 38–62 (2003)
2. Antoine, B., Cyrille, G., Carlos, A.R., Christian, W., Steffen, N.: High-frequency trading activity in EU equity markets. Econ. Rep. **1**, 1–31 (2014)
3. Baldauf, M., Mollner, J.: High-frequency trading and market performance. J. Finance **75**, 1495–1526 (2020)
4. Brogaard, J., Carrion, A., Moyaert, T., Riordan, R., Shkilko, A., Sokolov, K.: High frequency trading and extreme price movements. J. Financ. Econ. **128**(2), 253–265 (2018)
5. Brogaard, J., Hendershott, T., Riordan, R.: High-frequency trading and price discovery. Rev. Financ. Stud. **27**(8), 2267–2306 (2014)
6. Brogaard, J., Hendershott, T., Riordan, R.: Price discovery without trading: evidence from limit orders. J. Finance **74**(4), 1621–1658 (2019)
7. Cao, C., Hansch, O., Wang, X.: The information content of an open limit-order book. J. Futures Mark.: Futures Options Deriv. Prod. **29**(1), 16–41 (2009)
8. Carrion, A.: Very fast money high-frequency trading on the NASDAQ. J. Financ. Mark. **16**(4), 680–711 (2013)
9. Conrad, J., Wahal, S., Xiang, J.: High-frequency quoting, trading, and the efficiency of prices. J. Financ. Econ. **116**(2), 271–291 (2015)
10. Diebold, F.X., Strasser, G.: On the correlation structure of microstructure noise: a financial economic approach. Rev. Econ. Stud. **80**(4), 1304–1337 (2013)
11. Foucault, T., Hombert, J. Roşu, I.: News trading and speed. J. Finance **71**(1), 335–382 (2016)

12. Gai, J., Choi, D. J., O'Neal, D., Ye, M., Sinkovits, R.S.: Fast construction of nanosecond level snapshots of financial markets. Concurr. Comput. Pract. Exp. **26**(13), 2149–2156 (2014)
13. Grossman, S.J., Miller, M.H.: Liquidity and market structure. J. Finance **43**(3), 617–633 (1988)
14. Hasbrouck, J.: Assessing the quality of a security market: a new approach to transaction-cost measurement. Rev. Financ. Stud. **6**(1), 191–212 (1993)
15. Hasbrouck, J., Saar, G.: Low-latency trading. J. Financ. Mark. **16**(4), 646–679 (2013)
16. Hirschey, N.: Do high-frequency traders anticipate buying and selling pressure?. Available at SSRN 2238516 (2019)
17. Kaniel, R., Saar, G., Titman, S.: Individual investor trading and stock returns. J. Finance **63**(1), 273–310 (2008)
18. Kelley, E.K., Tetlock, P.C.: How wise are crowds? Insights from retail orders and stock returns. J. Finance **68**(3), 1229–1265 (2013)
19. Kirilenko, A., Kyle, A.S., Samadi, M., Tuzun, T.: The flash crash high-frequency trading in an electronic market. J. Finance **72**(3), 967–998 (2017)
20. Lee, E.J.: High frequency trading in the Korean index futures market. J. Futures Mark. **35**(1), 31–51 (2015)
21. Lin, W.T., Tsai, S.C., Chiu, P.: Do foreign institutions outperform in the Taiwan options market? North Am. J. Econ. Finance **35**, 101–115 (2016)
22. Lin, W.T., Tsai, S.C., Zheng, Z., Qiao, S.: Does options trading convey information on futures prices? North Am. J. Econ. Finance **39**, 182–196 (2017)
23. Lin, W.T., Tsai, S.C., Zheng, Z., Qiao, S.: Retrieving aggregate information from option volume. Int. Rev. Econ. Finance **55**, 220–232 (2018)
24. Malinova, K., Park, A., Riordan, R. Do retail investors suffer from high frequency traders?. Available at SSRN 2183806 (2018)
25. Menkveld, A., Koopman, S.J., Lucas, A.: Modeling around-the-clock price discovery for cross-listed stocks using state space models. J. Bus. Econ. Stat. **25**, 213–225 (2007)
26. Petris, G., Petrone, S.: State space models in R. J. Stat. Softw. **41**(4), 1–25 (2011)
27. U. S. Securities and Exchange Commission: Equity market structure literature review Part II High frequency trading. Staff of the Division of Trading and Markets (2014)
28. Van Kervel, V., Menkveld, A.J.: High-frequency trading around large institutional orders. J. Finance **74**(3), 1091–1137 (2019)
29. Yu, W., Li, S.: Research on financial data analysis based on data mining algorithm. Concurr. Comput.: Pract. Exp. **31**(10), e4780 (2019)
30. Zhao, Y., Wan, D.: Institutional high frequency trading and price discovery evidence from an emerging commodity futures market. J. Futures Mark. **38**(2), 243–270 (2018)
31. Zhang, J., Xu, L., Li, Y.: Classifying Python code comments based on supervised learning. In: Meng, X., Li, R., Wang, K., Niu, B., Wang, X., Zhao, G. (eds.) WISA 2018. LNCS, vol. 11242, pp. 39–47. Springer, Cham (2018). https://doi.org/10.1007/978-3-030-02934-0_4

Data Privacy and Security

A Risk Analysis of Android Children's Apps

Min Li, Haroon Elahi, and Shuhong Chen[✉]

School of Computer Science, Guangzhou University,
Guangzhou 510006, People's Republic of China
shuhongchen@gzhu.edu.cn

Abstract. The Android security mechanism, building on the privacy self-management model, treats all users equally and expects them to evaluate and manage Apps risks, which can put the privacy and security of vulnerable users at risk. In this paper, we perform a risk analysis of 90 Android Apps meant for children and downloaded from an official App store in China. We consider factors, including the App rating, the number of downloads, developer reputation, and requested Permission to assess the risks of a Children App. We use an analytic hierarchy process-based method for this purpose. Our analysis identifies that about 16% of these Apps pose high and medium risks to their users, and the rest of the Apps are not risk-free either. We propose that there is a need to follow privacy as a shared responsibility model, and the Android security mechanism should also be re-engineered, while considering the privacy protection needs of its users, like children.

Keywords: Android security · Children Apps · Risk analysis · Privacy self-management

1 Introduction

The Android security mechanism has a downside – it ignores the differences among users regarding their abilities and protection needs and builds around a privacy self-management model [10,13]. For example, the European Union's General Data Protection Regulation acknowledges that children may be less aware of risks associated with the processing of their data [14]. Yet, the Android security model does not make any exceptions.

Android App stores distribute many Apps designed and developed for children, and children, widely use Android smartphones [1]. Previous research suggests that children Apps can pose different threats to their users [9,12]. However, the scope of these studies is narrow. The purpose of conducting this research is to explore the privacy and security risks of children's Apps available in China. In this regard, the main contributions of this research are as follows.

1. This is the first study that evaluates the risks of children Apps in China. Overall, this is one of the very few studies published in this domain.

© Springer Nature Switzerland AG 2020
G. Wang et al. (Eds.): WISA 2020, LNCS 12432, pp. 539–546, 2020.
https://doi.org/10.1007/978-3-030-60029-7_48

2. We download 90 children Apps from one of the most popular Official App stores in China, and perform their risk analysis by adapting a state-of-the-art risk evaluation method [2]. However, contrary to the original model, we do not consider the market place factor in our analysis. Because all of apps are downloaded from the same marketplace, this means all of apps don't exist different value at this factor.
3. We discover that 36% of the evaluated Apps pose high risks to the privacy of their users and security of the devices. We also find that rest of the Apps cannot be treated as safe either.

The rest of the paper is organized as follows. Section 2 contains the related work. Section 3 explains the methods and procedure used to perform a risk analysis of the selected Apps. Section 4 presents the results. Section 5 discusses different implications of the results of this study, and finally, Sect. 6 concludes this paper.

2   Related Work

Different researchers have focussed on assessing the risks of Android Apps [2,4,7]. However, there are only a handful of studies that study their potential risks for children. The use of smartphones has introduced new privacy and security concerns. Liccardi et al. [8] assessed 38,842 Apps claiming to be suitable for young users for Children's Online Privacy Protection Act (COPPA) compliance. They found that half of the Apps had the ability to collect private data, and only 6% presented privacy policy. They also discovered that even the parents lacked understanding and knowledge of what data was accessed. Hu et al. [6] pointed out that Children Apps could contain adult content or the content not appropriate for underage mobile users. Etaher and Weir [5] conducted a school-based survey to understand the use of smartphones by children. They learned that children were keen on installing new Apps and updated the existing ones. They discovered that boys were more cautious than girls in installing new Apps. Their results also suggest that girls participating in their study received more upsetting content than boys and were subject to frequent online attacks and bullying. Liu et al. [9] analyzed 67,778 Children Apps collected from Google Play Store and discovered that 53% of these Apps included targeted Ads, 19.6% were connected to social networks, and 22.5% had in-App purchases. These findings indicated threats such as data collection for targeted marketing, data sharing on social networks, and potential financial losses.

Although these studies identify different risks of Children Apps, they have their limitations. For example, Hu et al. [6] focus on the presence of adult content in Children's Apps and Etaher and Weir [5] study on the consequences of careless App installations by School-goers. Likewise, Liccardi et al. [8] consider COPPA compliance, and Liu et al. [9] analyze Apps only for risks of targetted Ads, social media APIs, and in-App purchases. This study focusses on risk evaluation of popular Children Apps downloaded from three App stores from a privacy self-management point of view.

3   Method

In this paper, we adapt a method proposed by Dini et al. [2]. Contrary to the original method, we do exclude the marketplace factor from our risk analysis because we deal with only the Apps downloaded from trusted sources. We categorize Apps as a *Trusted, Medium-Risk*, or *High-Risk*. These categories are based on the abilities of Apps to threaten the privacy of users and the security of their devices. In our analysis, we treat assessing the trustworthiness of an App as a multi-criteria decision-making problem and use the Analytic Hierarchy Process (AHP) to solve it. AHP finds the best alternative among several given options to reach a goal [11]. We also review App permissions to learn specific Apps risks. In this section, we explain the steps performed for conducting this research.

3.1   Data Collection

We downloaded 90 Android children's Apps from Huawei App store. We also collected the Apps' information (marketplace, developer, the number of downloads, and the user rating) from homepage of respective Apps in the marketplace. Overall, these Apps have been downloaded for about 1.1 billion times. We used Androguard[1], an open-source application programming interface (API), for reverse engineering of Android Apps and extracted Android App Permissions from APK files.

3.2   Risk Assessment

Risk assessment can reveal different threats of Apps [3]. We performed the following steps for assessing the risks of 90 Apps evaluated in this research.

Determining Threat Level of Permissions. We extract the Permissions of the Apps from their manifest files and determine their threat levels. While determining the threat-levels, we consider the impact of a Permission on confidentiality and integrity. Since each Permission affects the confidentiality of data and integrity of a smartphone device in a different way, it is possible to assign them threat levels accordingly. These scores can help in calculating global threat scores by combining, weighting and normalizing. If x is the threat score in $[0,1]$, and every 0.2 as a interval, we treat the threats as no-threat, negligible, limited, significant, relevant and maximum from 0 to 1.

Pairwise Comparisons. We use four criteria to evaluate the risk of children's Apps. These include, the global threat score σ derived from App Permissions, developer δ, and user rating ρ, which users gave after using the App. We build a comparison matrix for pairwise comparison. Each value in the matrix determines the relative importance of two criteria, which is between zero and 9. If the value

---

[1] Androguard API, Available: https://github.com/androguard/androguard.

is higher, it means the one is more important than the other and vice versa. Comparison matrix is a reciprocal matrix, therefore $a_{ji} = \frac{1}{a_{ij}}$, and if $i = j$, then $a_{ii} = 1$.

Global Threat Score. Global threat score (σ) is calculated using the following formula.

$$\sigma = \frac{\sum_{i=1}^{n} w_p.pt_i + w_s.st_i + w_f.ft_i}{1 + \lceil \lg(1 + n) \rceil} \tag{1}$$

In Eq. 1, n is the number of Permissions in a children App. pt_i, st_i, and ft_i are the privacy, system, and financial threat scores of the ith Permission. w_p, w_s, and w_f are the weights assigned to a given Permission from privacy, system integrity and financial threat point of view. In this paper, since our focus is mainly on the privacy threats, we assign higher weight to privacy threat. We assign the weights as following. $w_p = 0.4$, $w_s = 0.3$, and $w_f = 0.3$. For every criterion, we set the three alternatives. They are trusted, medium-risk, and high-risk. Then for every criterion, we compare the relevance of three alternatives and get the score between 0 and 9. This score was further divided into three categories. We use an App's global threat score for this purpose. The Apps with a threat score [0–4] are trusted. The Apps with a threat score [4–7] are middle-risk, and those with a threat score [7 to+] are high-risk.

Consistency Analysis. We obtain thirteen complete comparison matrices. In order to make sure that all of the matrices are consistent, we perform a consistency analysis. The steps for our consistency analysis are as follows.

1. Consistency index CI is calculated using Eq. 2

$$CI = \frac{\lambda_{max} - n}{n - 1} \tag{2}$$

2. We calculated the consistency ratio (CR) using Eq. 3.

$$CR = \frac{CI}{RI} \tag{3}$$

In Eq. 3, RI is a random index and it can be obtained using a search table. If the CR is equal to or less than 0.1, the comparison matrix is considered accepted. Otherwise, there exists some inconsistency in these matrices.

Calculating Local Priorities. We need to normalize these matrices for calculating local priorities. We use asymptotic normalization coefficient (ANC) for this purpose. ANC has the following steps.

1. Get a new matrix A by dividing b_{ij} by sum of each matrix's column B_j. Thus every new element a_{ij} will be achieved using Eq. 4.

$$a_{ij} = \frac{b_{ij}}{B_j} = \frac{b_{ij}}{\sum_{i=1}^{n} b_{ij}} \tag{4}$$

2. Finally, the criterion's local priority of every alternative is calculated (Eq. 5).

$$P_{c_k}^a = W_i = \frac{A_i}{n} = \frac{\sum_{i=1}^n a_{ij}}{n} \tag{5}$$

For each matrix, we can get the local priority of four criteria is [0.36 0.28 0.12 0.23], the local priority of Permission from low threat to high threat is [0.69 0.23 0.08], [0.11 0.26 0.63], [0.08 0.69 0.23]. The local priority of developer is [0.11 0.31 0.58]. The local priority of user rating is [0.68 0.13 0.19], [0.11 0.26 0.63], [0.13 0.66 0.21]. With the decreasing number of downloads, the local priority of downloads is [0.65 0.12 0.23], [0.54 0.11 0.35], [0.26 0.41 0.33], [0.10 0.53 0.37], [0.08 0.66 0.26].

Computing Global Priorities. $P_g^{a_i}$ is the global priority of the alternative of a_i, i.e. trusted, high-risk, and medium-risk. It is the sum of the products of global priority of relevant criterion c_j, $P_g^{c_j}$ multiplied by the local priority $P_{c_j}^{a_i}$ of each criterion's alternative (Eq. 6). In Eq. 6, k is the number of alternatives. After that, we get the global priorities of the three alternatives. By comparing these three values, it's becomes obvious that the App belongs to which alternative.

$$P_g^{a_i} = \sum_{j=1}^k P_g^{c_j} P_{c_j}^{a_i} \tag{6}$$

4 Results

In this paper, we assessed 90 children's Apps for privacy and security risks. As given in Table 1, our analysis reveals that five of them are high-risk Apps. These are the Apps that pose threats to user privacy and device security. Twenty-eight more Apps are medium-risk. These are the Apps that pose threats to device security. We also reviewed the Permissions for individual Apps and found that all the 89 Apps could collect information including phone number, current cellular network information, the status of any ongoing calls, and a list of any phone accounts registered on the device. We also found that 88 and 72 Apps could write and read external storage, respectively. Moreover, 82 Apps could access the coarse location (with a 300-m precision), and 25 Apps had access to the fine or exact location. 59% of these Apps collect information about the neighboring devices. Another 94% can install new Apps and 11% user system alert windows that can be used to trick a user into believing that a message is from the Android operating system.

5 Discussion

Children do not have the abilities and skills needed for privacy self-management [14]. Therefore, App providers should participate in protecting the privacy of children. Yet, the results of our risk assessment of 90 children's Apps find that App providers are ignoring their corresponding responsibility. And sometime users' evaluations in appstore will mislead fresh user in security. In Fig. 1, it's the ratio of every factor's average value. We can see that there is just a little different between three risk levels.

Table 1. Mid-risk and high-risk apps

App name	Global threat	User rating	Number of download	Result
com.babigongzhushengbao.bao.apk	3.22	3.8	360,000	High risk
com.crazystore.android$_g$litterslime.mi.apk	2.12	2.9	290,000	High risk
com.fatcat.FashionNailArtSalonGame.huawei.apk	3.08	3.4	500,000	High risk
com.fatcat.injectiondoctor.huawei.apk	1.58	3.2	460,000	High risk
com.hello.SchoolDentistDoctor.huawei.apk	1.58	3.9	170,000	High risk
com.mengyinps.pizzashop.apk	2.26	3	330,000	High risk
com.newbee.cooking.apk	1.9	3.2	780,000	High risk
com.yiqizuoye.jzt.apk	7.07	2.9	100,000,000	High risk
com.ynwareapp.meirenyumakeup.apk	2.74	2.6	730,000	High risk
cn.mama.pregnant.apk	6.15	4.5	100,000,000	Mid-risk
com.dw.btime.apk	5.14	3.9	100,000,000	Mid-risk
com.hhdd.kada.apk	6.31	4.8	17,310,000	Mid-risk
com.mampod.ergedd.apk	4.14	4.7	58,460,000	Mid-risk
com.mobbanana.bbmhwmx.huawei.apk	4.23	3.5	2,110,000	Mid-risk

In the past, Liccardi et al. [8] found that half of the Apps that they assessed could collect private data. They also discovered that even the parents lacked understanding and knowledge of what data were accessed. However, since Android follows a privacy self-management based security model that requires users to grant Apps Permissions at the runtime, we argue that even the knowledge of parents cannot guarantee the privacy protection of children. Contrary to their study, we also found that although only 16% of the Apps fell in high and mid-risk categories, most of the evaluated Apps collected sensitive information. This is particularly concerning when previous studies suggest that while children are keen on installing new Apps, they can be careless about the risks of these Apps [5]. In such instances, the distribution of Apps that collect private information of their users who are unable to make such complex decisions as understanding App requirements to release their data is a serious concern. Since children Apps may include ad libraries, social media connections, and in-App purchases feature [9,15], the private data collected by these Apps can affect the wellbeing of their users and security of the devices.

We note that the problem is at two ends; first, the Android security assumes that children can perform privacy self-management, and second, the App providers collect excessive data indiscriminately. The use of models like privacy as a shared responsibility and making privacy and security by design principles mandatory can save children from such risks.

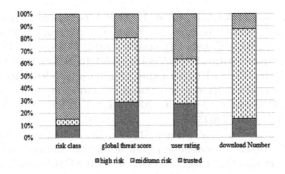

Fig. 1. The proportion of every threat level in each factor.

6   Conclusion

In this paper, we conducted a risk analysis of 90 children's Apps downloaded from an official App store in China. We found that about 36% of these Apps posed significant risks to user privacy and device security. We also discovered that all the evaluated Apps collected private user information and information that may affect the security of the devices. Keeping in view the privacy self-management requirement and the abilities of children to manage Apps, the findings of this study suggest that there is a need to follow privacy and security by design principles strictly. Further, the Android security mechanism should also be re-engineered to serve the privacy protection needs of its users, like children.

Acknowledgments. This work is supported in part by the National Natural Science Foundation of China under 61632009, in part by the Guangdong Provincial Natural Science Foundation under Grant 2017A030308006 and High-Level Talents Program of Higher Education in Guangdong Province under Grant 2016ZJ01.

References

1. Andone, I., Blaszkiewicz, K., Eibes, M., Trendafilov, B., Montag, C., Markowetz, A.: How age and gender affect smartphone usage. In: Lukowicz, P., Krüger, A., Bulling, A., Lim, Y., Patel, S.N. (eds.) Proceedings of the 2016 ACM International Joint Conference on Pervasive and Ubiquitous Computing, UbiComp Adjunct 2016, Heidelberg, Germany, 12–16 September 2016, pp. 9–12. ACM (2016)
2. Dini, G., Martinelli, F., Matteucci, I., Petrocchi, M., Saracino, A., Sgandurra, D.: Risk analysis of android applications: a user-centric solution. Future Gener. Comput. Syst. **80**, 505–518 (2018). https://doi.org/10.1016/J.FUTURE.2016.05.035
3. Elahi, H., Wang, G., Chen, J.: Pleasure or pain? An evaluation of the costs and utilities of bloatware applications in android smartphones. J. Netw. Comput. Appl. **157**, 102578 (2020). https://doi.org/10.1016/j.jnca.2020.102578

4. Elahi, H., Wang, G., Peng, T., Chen, J.: AI and its risks in android smartphones: a case of google smart assistant. In: Wang, G., Bhuiyan, M.Z.A., De Capitani di Vimercati, S., Ren, Y. (eds.) DependSys 2019. CCIS, vol. 1123, pp. 341–355. Springer, Singapore (2019). https://doi.org/10.1007/978-981-15-1304-6_27

5. Etaher, N., Weir, G.R.S.: Understanding children's mobile device usage. In: 2016 IEEE International Conference on Cybercrime and Computer Forensic (ICCCF), Vancouver, BC, Canada, 12–14 June 2016, pp. 1–7. IEEE (2016). https://doi.org/10.1109/ICCCF.2016.7740437

6. Hu, B., Liu, B., Gong, N.Z., Kong, D., Jin, H.: Protecting your children from inappropriate content in mobile apps: an automatic maturity rating framework. In: Bailey, J., et al. (eds.) Proceedings of the 24th ACM International Conference on Information and Knowledge Management, CIKM 2015, Melbourne, VIC, Australia, 19–23 October 2015, pp. 1111–1120. ACM (2015). https://doi.org/10.1145/2806416.2806579

7. Li, B., Wang, G., Elahi, H., Duan, G.: A light-weight framework for pre-submission vetting of android applications in app stores. In: Wang, G., Bhuiyan, M.Z.A., De Capitani di Vimercati, S., Ren, Y. (eds.) DependSys 2019. CCIS, vol. 1123, pp. 356–368. Springer, Singapore (2019). https://doi.org/10.1007/978-981-15-1304-6_28

8. Liccardi, I., Bulger, M., Abelson, H., Weitzner, D.J., Mackay, W.E.: Can apps play by the COPPA rules? In: Miri, A., Hengartner, U., Huang, N., Jøsang, A., García-Alfaro, J. (eds.) 2014 Twelfth Annual International Conference on Privacy, Security and Trust, Toronto, ON, Canada, 23–24 July 2014, pp. 1–9. IEEE Computer Society (2014). https://doi.org/10.1109/PST.2014.6890917

9. Liu, M., Wang, H., Guo, Y., Hong, J.I.: Identifying and analyzing the privacy of apps for kids. In: Chu, D., Dutta, P. (eds.) Proceedings of the 17th International Workshop on Mobile Computing Systems and Applications, HotMobile 2016, St. Augustine, FL, USA, 23–24 February 2016, pp. 105–110. ACM (2016). https://doi.org/10.1145/2873587.2873597

10. Liu, X., Chen, J., Xia, X., Zong, C., Zhu, R., Li, J.: Dummy-based trajectory privacy protection against exposure location attacks. In: Ni, W., Wang, X., Song, W., Li, Y. (eds.) WISA 2019. LNCS, vol. 11817, pp. 368–381. Springer, Cham (2019). https://doi.org/10.1007/978-3-030-30952-7_37

11. Saaty, T.L.: Decision making with the analytic hierarchy process. Int. J. Serv. Sci. 1(1), 83–98 (2008). https://doi.org/10.1504/IJSSCI.2008.017590

12. Scott, K.M., Richards, D., Londos, G.: Assessment criteria for parents to determine the trustworthiness of maternal and child health apps: a pilot study. Health Technol. 8(1–2), 63–70 (2018). https://doi.org/10.1007/S12553-018-0216-8

13. Solove, D.J.: Privacy self-management and the consent dilemma. Harv. Law Rev. 126, 1880–1903 (2013)

14. The European Parliament and the Council of the European Union: Regulation (EU) 2016/679 (GDPR). Off. J. Eur. Union L119, 1–88 (2016)

15. Zhang, S., Wang, G., Bhuiyan, M.Z.A., Liu, Q.: A dual privacy preserving scheme in continuous location-based services. IEEE Internet Things J. 5(5), 4191–4200 (2018). https://doi.org/10.1109/JIOT.2018.2842470

Top-k Frequent Itemsets Publication of Uncertain Data Based on Differential Privacy

Yunfeng Zou[1], Xiaohan Bao[2], Chao Xu[1], and Weiwei Ni[2(✉)]

[1] State Grid Jiangsu Marketing Service Center Metrology Center,
Nanjing 210036, China
[2] School of Computer Science and Engineering, Southeast University,
Nanjing 211189, China
wni@seu.edu.cn

Abstract. Privacy preserving frequent itemset mining (PPFIM) on uncertain data is booming with the increasing attention to data privacy. Existing methods use a filter function satisfies exponential differential privacy to obtain top-k frequent itemsets, and add Laplace noise to their supports to achieve the privacy protection release of top-k frequent itemsets. The privacy protection mechanism is independent of the mining process, resulting in the accuracy of the top-k frequent itemsets being affected by the k value. When the algorithm is applied to large-scale frequent itemsets, balance the data availability and privacy security is difficult. In view of the above deficiencies, the privacy protection mechanism and the mining process are integrated to design a PPFIM strategy, achieve the separation of noise addition and top-k filtering, and avoid the dependence of the algorithm accuracy on the k value. A candidate level information extraction strategy is designed to reduce the search space and effectively reduce the privacy budget by utilizing the feature of upper limit threshold in uncertain data sets. On this basis, a novel algorithm Uncertain difference privacy level frequent itemset mining (UDP-LFIM) is proposed. Theoretical analysis and experimental demonstrate that the top-k itemsets published by the algorithm can guarantee the accuracy on the premise of satisfying differential privacy.

Keywords: Frequent itemsets mining · Uncertain data · Differential privacy · Top-k itemset publication

1 Introduction

Uncertain data refers to data described by probability or measured by likelihood [1]. Uncertain data may contain sensitive information of data owners, and mining uncertain data inevitably brings privacy leakage [2, 3]. The released Top-K frequent itemsets [13, 15, 16] of uncertain data need to protect its sensitive information from leaking while ensuring availability. At present, the research on PPFIM of uncertain data is still in its infancy. Ref. [3] proposed a PPFIM algorithm Frequent itemsets mining for uncertain data based on differential privacy (FIMUDDP) [9–11]. Although FIMUDDP can satisfy differential privacy, the privacy protection mechanism is independent of the mining

© Springer Nature Switzerland AG 2020
G. Wang et al. (Eds.): WISA 2020, LNCS 12432, pp. 547–558, 2020.
https://doi.org/10.1007/978-3-030-60029-7_49

process and the algorithm uses the exponential mechanism which makes it difficult to take into accounts both data utility and privacy security [12, 14].

In response to the above problems, UDP-LFIM is proposed for uncertain data leveraging differential privacy. To realize the integration of the mining process and privacy protection processing process, a Laplace mechanism is introduced to disturb the support and minimum support threshold of itemsets. It ensures that the published frequent itemsets and their support satisfy differential privacy constraints. Besides, privacy protection frequent itemsets and their support are generated simultaneously, which avoids the influence of k value on the accuracy. Aiming at the problem that the exponential mechanism is not suitable for large-scale frequent itemsets, a strategy of extracting level information of candidate items is designed by virtue of the upper threshold characteristic of uncertain data sets. Level information is used to screen and reduce the search space, which effectively reduces the privacy budget and improves the efficiency of the algorithm.

The main contributions of this paper are as follows:

- The integration of the mining process and the privacy protection process is realized, and the addition of differential noise is separated from the top-k itemset screening. The sensitivity of the differential noise data is independent of the k value, and the accuracy of the algorithm is independent of the k value.
- The level information can be used to reduce the search space of frequent itemsets, to effectively reduce the privacy budget and improve the time efficiency.
- The UDP-LFIM algorithm is proposed to obtain the frequent itemset that satisfies the differential privacy and its support degree, to ensure that the output top-k itemset satisfies the differential privacy.

The organization structure of the paper is as follows: Sect. 2 introduces relevant work, Sect. 3 describes problems and relevant definitions, Sect. 4 introduces UDP-LFIM algorithm, and analyzes the privacy protection effect of the algorithm, Sect. 5 designs experiments to verify the effectiveness of the algorithm, and finally summarizes the whole paper and looks forward to the following work.

2 Related Work

Differential privacy has been widely applied in privacy-preserving data mining. However, the research on frequent itemset mining for uncertain data based on differential privacy is still in its infancy. Ref. [3] proposed FIMUDDP, a frequent itemset mining algorithm for uncertain data based on differential privacy. It designs a filtering function satisfying exponential differential privacy to filter top-k frequent itemsets from all frequent itemsets. Subsequently, Laplace noise is added to the support of the itemset to realize its privacy-preserving publication. However, FIMUDDP algorithm still has the following shortcomings: (1) The privacy protection mechanism is independent of the mining process, resulting in the accuracy of the published top-k frequent itemsets being affected by the k value. (2) The algorithm uses an exponential mechanism. When applied to a large and frequent itemset, the privacy budget is large, and it is difficult to balance data availability and privacy security. In summary, all existing methods have the defect that the privacy protection mechanism is independent of the mining process, and balance privacy protection security and data availability is difficult.

3   Problem Description and Basic Idea

3.1   Problem Description

Uncertain data privacy protection of frequent itemsets mining under the condition of ensuring the safety of individual privacy, release Top - K frequent itemsets and accurately as possible support, specific need to satisfy the following constraints: (1) K value has a little influence on the accuracy of the algorithm (2) Take into account the security and availability of data privacy, and ensure a lower privacy budget.

To achieve the above goals, the main ideas are as follows: using the characteristics of uncertain data with upper threshold, designing hierarchical information extraction strategy, effectively reducing search space. To avoid the influence of k value on mining results, the idea of incorporating privacy protection mechanism into mining process is adopted, and differential noise adding operation is embedded into mining process, which is separated from top-k frequent itemsets selection, so that the accuracy of mining results is not directly related to k. The proposed method can simultaneously obtain frequent itemsets and their supports satisfying differential privacy.

3.2   Related Concepts

Let Item $= \{x_1, x_2 \ldots x_m\}$ be the set of all items, $X = \{x_1, x_2 \ldots x_k\}$, is a subset of Item, called k-itemset, and k is the number of items in X. Let D represent an uncertain data set, $D = \{t_1, t_2 \ldots t_n\}$, where t_j is called a transaction, each transaction is a subset of Item. The data exist in uncertain datasets exists in the form of probability, like $P(x_q, t_j)$, presenting the existence probability of item x_q in transaction t_j, $0 < P(x_q, t_j) \leq 1$. The symbol TID is the identification number of the transaction t_j in the database, and $|T|$ indicates the number of transactions in the database T [4–8].

Definition 1 [5]. Itemset support. Let X denotes an itemset, the itemset support $P(X, t_j)$ is defined as the cumulative multiplication of each item x's existence probabilities in t_j, where x is an item in X (Table 1).

Table 1. Sample of the uncertain dataset

TID	Transaction record
t_1	a: 0.5 b: 0.6 c:0.7 d:0.3 e:0.9
t_2	a: 0.7 b:0.4 d:0.8
t_3	b: 0.3 c:0.9 f:0.7
t_4	a: 0.5 b:0.7 c:0.6

$$P(X, j) = \prod_{x \in X} P(x, t_j) \tag{1}$$

Definition 2 [5]. Expected supports. Expected supports of X is denoted as expSup(X). It is defined as the accumulation of the itemset supports of X related to all records in the uncertain dataset D.

$$expSupX = \sum_{j=1}^{n} P(X, t_j) = \sum_{j=1}^{n} \prod_{x \in X} P(x, t_j) \tag{2}$$

Definition 3 [5]. The upper limit threshold of item x on record t_j. For given t_j in D, the upper limit threshold item x on t_j is denoted as $P^{cap}x, t_j$. It is defined as the product of the existence probability of item x and the maximum probability of items other than x in the record t_j.

$$P^{cap}x, t_j = \begin{cases} P(x_i, t_j) \times M_1, h > 1 \\ P(x_i, t_j), h = 1 \end{cases} \tag{3}$$

Let $h = |t_j|$ be the length of t_j, namely the number of items in t_j, $M_1 = max_{q \in [1,h]}$, $q \neq xP(x_q, t_j)$.

Theorem 1 [5]. The probability of recording any k-item set X (k ≥ 2) containing x in t_j is always less than or equal to the upper threshold of the record t_j, that is $P(X, t_j) \leq P^{cap}x, t_j$, X ∈ t_j.

Definition 4 [5]. Upper limit threshold of expected support for item x. The upper limit threshold of expected support for item x is denoted as $expSup^{Cap}x$. It is defined as the sum of the upper limit bound probability thresholds for all transaction records t_j containing x.

$$expSup^{Cap}x = \sum_{j=1}^{n} \left(P^{cap}x, t_j | x \in t_j \right), n = |DB| \tag{4}$$

Theorem 2 [5]. Given any k-itemset X containing item x in DB, $expSupX \leq expSup^{Cap}x$.

Differential Privacy based Top-k Frequent Itemsets Mining Algorithm for Uncertain Data

3.3   Algorithm Ideas

As shown in Fig. 1, the algorithm can be divided into three stages, namely preprocessing stage, frequent itemset mining stage and top-k frequent itemsets filtering stage. The preprocessing stage includes data preprocessing and level information extraction. Data preprocessing operations can obtain candidate sets satisfying differential privacy and corresponding noise support. In frequent itemset mining stage, Laplace noise is added to the support of itemsets, and the searching space of frequent itemset is reduced by level information and downward closure property of Apriori, thus the privacy budget is effectively reduced. Finally, the k most frequent items can be filtered from the obtained frequent itemsets.

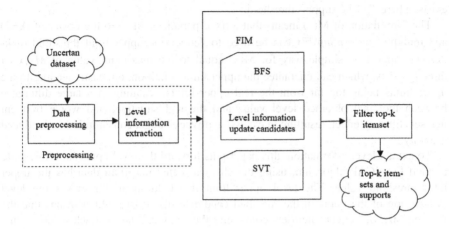

Fig. 1. Algorithm idea

3.4   Preprocessing

Due to space limitations, the data preprocessing process will not be repeated here. The following describes the level information processing strategy.

Definition 5. $M_2(x, t_j)$. $M_2(x, t_j)$ is the second-largest probability in the transaction record t_j except for x itself. The value is

$$M_2(x, t_j) = max_{r \in t_j - q, r \neq x}(P(x_r, t_j)), \tag{5}$$

Where q satisfies $max_{q \in t_j, q \neq x} P(x_q, t_j)$

Definition 6. Level iteration parameter of item x, denoted as M(x). M(x) is defined as an overall estimate of $M_2(x, t_j)$ for each record, the value of M(x) is

$$M(x) = max_{1 \leq l \leq L} \sqrt[l]{\frac{\sum_{j=1}^{n} P^{cap}x, t_j \times M_2(x, t_j)^l}{expSup^{Cap}x}} \tag{6}$$

Where L is the length limit of data set D records.

Theorem 3. Given in the DB any k-item set containing x, where k $\geq$ 2, M satisfies the Definition 6.

$$expSupX \leq expSup^{Cap}x \times M^{k-2} \tag{7}$$

Due to space limitation, prove omission.

Definition 7. Level information, denoted as level(x). For a given item x, the maximum value of k (k > 1) that k-itemset containing this item is frequent is level(x), means that level(x) guarantees $expSup^{Cap}x \times M^{\text{level(x)}-3} \leq T$ and $T \leq expSup^{Cap}x \times M^{\text{level(x)}-2}$ exists, where T is the support threshold.

The formulation of M(x) means that a fixed parameter M(x) to the power of (k−2) and multiply the $expSup^{Cap}x$ can be used to denote *the* upper limit threshold of k-itemset support. A simple way for M(x) equal to the maximum value of $M_2(x, t_j)$ directly exists, which can guarantee the upper limit condition, but easily leading to the upper limits being too far from the real upper limits, resulting in a large difference between the final obtained level value and the real situation, and worse frequent itemsets filtering effect, resulting in excessive privacy budget consumption and reduced usability.

To extract level information, this paper puts forward the level upper limit threshold method LevelCAP algorithm, using the characters that uncertain data has the upper limit threshold value. The level upper limit threshold method retrieves the level information of the items in the specified candidate list (IList), which means that the itemsets size of frequent itemsets containing the candidate items. Each record in the original method corresponds to an upper limit, the LevelCAP algorithm estimates the M_2 values of all records, acting as M. Using M value as the upper limit of supports can approximate the real value as much as possible and guarantee the upper limit condition constraint. Such an M value not only avoids the space loss consumed by saving M_2 for all recorded values, but also approximates the real upper limit as far as possible. The closer the calculated upper limit is to the real upper limit, the more accurate the obtained level value is, the more the search prune effect is, the smaller the privacy budget is consumed, the more accurate the mining results are when privacy is guaranteed.

3.5   Frequent Itemset Mining for Privacy Protection Based on Differential Privacy

The main idea of Frequent itemset mining for privacy protection based on differential privacy, denoted as LevelPrivSearch, using breadth-first search idea and building a search tree that contains all frequent itemsets which can obtain frequent itemsets and supports at the same time and ensure the correctness constraint and descending order constraint of the subsequent filtering top-k itemsets, using level information to reduce the search space in the search process and a variant of SVT to determine whether the itemsets are frequent or not, to ensure that the frequent itemsets output and supports meet differential privacy and effectively reduce the privacy budget. The concrete implementation is to use the level information of candidate items first, arrange the candidate items in descending order, and use the level information to continuously update the list of candidate items of each node of each layer. Suppose the candidate item list is x_Set, the candidate item is x, and the number of items in the current node itemset is num. Update conditions for x_Set is that (1) The resulting itemset that is merging x with any (num − 1) - item subset of the current node itemset is frequent (2) the level information level(x) is greater than or equal to num + 1. If item x satisfies both conditions (1) and (2), x_Set retains this item. In the updated x_Set, each item is merged with the current node itemset to form a new itemset. SVT algorithm is used to determine whether it is frequent or not. If frequent, add it into the (num + 1) - level node list, otherwise, it cost no privacy budget. When the candidate items of all nodes are empty, the establishment of the search tree is completed. Each node of the search tree stores frequent itemsets and their corresponding supports, and all frequent itemsets and their supports can be obtained by traversing the search tree.

3.6 Privacy Protection Effect Analysis

The privacy protection effect of the algorithm needs to measure two indicators, privacy protection degree and algorithm availability. Experiments prove that UDP-LFIM algorithm guarantees the accurate availability of the released top-k itemsets. This section will prove that UDP-LFIM algorithm meets the differential privacy requirements.

Theorem 4. In the condition the data scale is large enough, level information acquisition will not disclose user privacy.

Prove. For two neighbor data sets D and D' that differ by one record, proving that the level information values obtained are not different is necessary. Level information is related to the upper limit threshold value of item x, denoted as $expSup^{Cap}x$ and the level iteration parameter M according to definition 8. When the size of the dataset is large enough, the value of a record has little effect on the calculation of the upper limit threshold from Eq. (8), the $expSup^{Cap}x$ of D and D' is basically the same. For the level iteration parameters M values, the values are linked to the cumulative sum of $P^{cap}x, t_j \times M_2(x, t_j)$, $expSup^{Cap}x$ and 1. When the dataset is large enough, the $expSup^{Cap}x$ of D and D' and the cumulative sum of $P^{cap}x, t_j \times M_2(x, t_j)$ are not affected by a single record, and the number of items 1 in a transaction dataset is given and does not change with a single record. In summary, there are almost no difference in the level iterative parameter M values calculated for two data sets that differ only by one record, thereby ensuring that there is no difference in the level information calculated, that is, under the condition that the data scale is large enough, level information acquisition will not disclose user privacy.

Firstly, whether the UDP-LFIM algorithm satisfies ε-difference privacy will be analyzed. The UDP-LFIM algorithm consists of four specific steps, namely, data preprocessing, level information acquisition, using level information to obtain frequent itemsets and supports that satisfy differential privacy, and obtaining top-k itemsets. Step 1 use privacy budget $ε_1$, where $\frac{ε_1}{2}$ is used to gain the $\tilde{\lambda}$ and the other $\frac{ε_1}{2}$ is used to gain $\widetilde{IList}$ both using the exponential mechanism. Theorem 4 guarantees that step 2 level information extraction will not leak privacy, so it does not need to be processed. Step 3, the algorithm LevelPrivSearch involved in the original data access is the SVT algorithm called in the getFI function used to obtain whether the itemsets are frequent or not. This article uses the sparse vector technology (SVT) variation, to ensure that this count query meets the differential privacy, and infrequent itemsets of the query consume no privacy budget, thus effectively reducing the privacy budget. Step 4, obtaining top-k item sets, which are directly selected from frequent itemsets that satisfy differential privacy, without access to original data and privacy concerns. UDP-LFIM algorithm satisfies differential privacy according to the sequential combinatorial nature of differential privacy. Differential privacy is built on a solid mathematical basis, and the ε directly reflects the protection level of data privacy. Experiments show that the UDP-LFIM algorithm can guarantee the accuracy of mining results while consuming a small privacy budget ε, and the privacy protection effect is better.

4   Experiment Results

At present, there are many researches on FIM of certain data based on differential privacy, but few researches on FIM of uncertain data. To verify the effectiveness of UDP-LFIM method, this paper uses real datasets for experimental verification and compare it with FIMUDDP algorithm under the same test conditions.

4.1   Evaluation

Precision and relative error RE were used to evaluate the experimental results. Let F be the first k most frequent itemsets mined from uncertain datasets by algorithm; $\tilde{F}$ is the true top-k frequent itemsets. Precision can be obtained from Eq. (8). $S_e(X)$ represented itemsets expected supports, and $\tilde{S}_e(X)$ is the noisy expected support of the said item set X, RE can be obtained from Eq. (9).

$$\text{precision} = \frac{F \cap \tilde{F}}{K} \tag{8}$$

$$\text{RE} = \text{median}_{X \in \tilde{F}} \frac{\left|\tilde{S}_e(X) - S_e(X)\right|}{S_e(X)} \tag{9}$$

4.2   Experimental Data and Environment

The experiment uses six datasets downloaded from http://fimi.ua.ac.be/data/ and introduces probability values that follow a gaussian distribution with a mean value of 0. 5 and a variance of 0.125 for each data. Experimental environment: Windows 10 operating system, Intel Core i5-4200 M CPU @ 2.5 GHZ, memory 8 GB.

4.3   Analysis of Experimental Results

UDP-LFIM algorithm can mine frequent itemsets from uncertain datasets, and the published top-k itemsets satisfy differential privacy. Due to the randomness of the differential privacy algorithm, this paper carries out 10 repeated tests on each group of data and takes the average value. Figure 2 shows the trend of the accuracy precision of UDP-LFIM and FIMUDDP algorithms on the datasets changing with the privacy budget, where K = 30, Fig. 3 shows the trend of the accuracy precision of UDP-LFIM and FIMUDDP algorithms changing with the value of K on the datasets, where the privacy budget is 1.6, and Fig. 4 shows the trend of the relative error (RE) of UDP-LFIM and FIMUDDP algorithms changing with the privacy budget on the datasets, where K = 30.

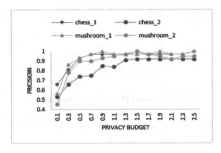 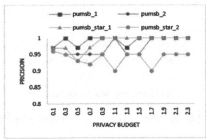

(a) Chess and Mushroom Dataset   (b) Pumsb and Pumsb_starDataSet

Fig. 2. The precision of UDF-LFIM and FIMUDDP in dataset changing with privacy

Figure 2(a) shows the trend of UDP-LFIM and FIMUDDP algorithms' accuracy in datasets chess and mushroom as the privacy budget increases, where the horizontal coordinate is privacy budget, and the change interval set is 0.1–2.5. In the figure, 1 represents UDP-LFIM algorithm, and 2 represents FIMUDDP algorithm. Figure 2(a) shows that with the increase of the privacy budget, the accuracy of UDP-LFIM and FIMUDDP gradually improves and tends to 1, which is consistent with the presupposition. The larger the privacy budget is, the less noise is added, the less availability of data is lost, and the higher the accuracy of mining results are. Secondly, the accuracy of UDP-LFIM algorithm under the same privacy budget is generally higher than that of FIMUDDP algorithm, indicating that UDP-LFIM algorithm has a better privacy protection effect and can more effectively guarantee the availability of data under the condition of data security protection.

BudgeTheoretical analysis indicates that the accuracy of UDF-LFIM algorithm is independent of K value, while the accuracy of FIMUDDP algorithm is affected by k value. To prove this point, design experiments compare the accuracy of UDF-LFIM and FIMUDDP algorithms under the same privacy budget and different K value conditions. In Fig. 3(a), the accuracy of UDF-LFIM algorithm on chess dataset remains basically stable, while the accuracy of FIMUDDP algorithm shows a significant decline trend with the increase of K value. The experimental results of multiple datasets prove that the accuracy of UDF-LFIM algorithm in different distributed datasets is independent of K value, which overcomes the problem of FIMUDDP.

Figure 4 compares the relative error trend of UDF-LFIM and FIMUDDP algorithms on datasets with the increase of the privacy budget. Figure 4(a) is the experimental results of the algorithms on chess, mushroom and pumsb datasets, and Fig. 4(b) is the experimental results of the algorithms on pumsb_star, T40I0D100K and accidents datasets. Experimental results show that the relative errors of the algorithms decrease with the increase of the privacy budget and tend to 0. From a theoretical

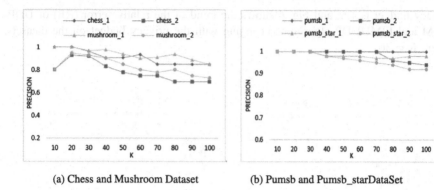

(a) Chess and Mushroom Dataset   (b) Pumsb and Pumsb_starDataSet

Fig. 3. The precision of UDF-LFIM and FIMUDDP in dataset changing with K value.

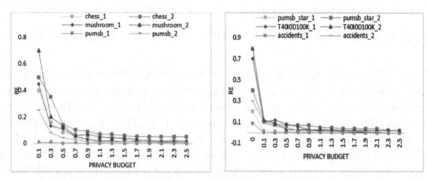

(a) Chess,Mushroom and Pumsb Datasets   (b) Pumsb_star,T40I0D100K and Accidents

Fig. 4. The RE of UDF-LFIM and FIMUDDP in dataset changing with privacy budge

perspective, the larger the privacy budget is, the smaller the scale parameter of the Laplace distribution subject to the noise will be. The noise value keeps decreasing with the increase of the privacy parameter, and the relative errors reflect the value of the added noise. It can also be seen in Fig. 4 that the smaller the datasets supports are, the slower the convergence rate of RE increases with the increased privacy parameters. For example, chess has the smallest supports and the slowest convergence rate, and accidents has the largest supports and the fastest convergence rate.

5 Conclusion

To release the more accurate uncertain dataset top-k frequent itemsets and protect the user's private data from being leaked to meet differential privacy, this paper proposes the UDP-LFIM algorithm. The UDP-LFIM algorithm mainly has two contributions. One is to use the uncertainty of the uncertain dataset to extract the level information and use the level information to narrow the search space of the frequent itemsets, effectively reduce the privacy budget and improve the time efficiency. Secondly, the frequent itemsets and the corresponding noise supports are mined under the condition of satisfying the differential privacy, and the top-k itemsets of the output satisfy the differential privacy and are arranged in descending order of support to satisfy the query semantics. The accuracy of the algorithm is independent of the k value. The validity of UDP-LFIM algorithm is verified by theoretical analysis and experimental comparison.

Acknowledgement. This work is supported by the National Natural Science Foundation of China (No. 61772131), the State Grid Corporation of China Project (5700-202018268A-0-0-00).

References

1. Tong, Y., et al.: Mining frequent itemsets over uncertain databases. Proc. VLDB Endow. **5** (11), 1650–1661 (2012). https://doi.org/10.14778/2350229.2350277
2. Leung, C.K.-S., et al.: Privacy-preserving frequent pattern mining from big uncertain data. In: 2018 IEEE International Conference on Big Data (Big Data) (2018). https://doi.org/10.1109/BigData.2018.8622260
3. Ding, Z., et al.: Frequent itemsets mining for uncertain data based on differential privacy. Appl. Res. Comput. **35**(321.07), 28–32 (2018). (Chinese). https://doi.org/10.3969/j.issn.1001-3695.2018.07.004
4. Chui, C.-K., Kao, B., Hung, E.: Mining frequent itemsets from uncertain data. In: Zhou, Z.-H., Li, H., Yang, Q. (eds.) PAKDD 2007. LNCS (LNAI), vol. 4426, pp. 47–58. Springer, Heidelberg (2007). https://doi.org/10.1007/978-3-540-71701-0_8
5. Leung, C.K.S., Mateo, M.A.F., Brajczuk, D.A.: A tree-based approach for frequent pattern mining from uncertain data. In: Washio, T., Suzuki, E., Ting, K.M., Inokuchi, A. (eds.) PAKDD 2008. LNCS (LNAI), vol. 5012, pp. 653–661. Springer, Heidelberg (2008). https://doi.org/10.1007/978-3-540-68125-0_61
6. Leung, C.K.S., Tanbeer, S.K.: Fast tree-based mining of frequent itemsets from uncertain data. In: Lee, S., Peng, Z., Zhou, X., Moon, Y.S., Unland, R., Yoo, J. (eds.) DASFAA 2012. LNCS, vol. 7238, pp. 272–287. Springer, Heidelberg (2012). https://doi.org/10.1007/978-3-642-29038-1_21
7. Leung, C.K.S., Tanbeer, S.K.: PUF-tree: a compact tree structure for frequent pattern mining of uncertain data. In: Pei, J., Tseng, V.S., Cao, L., Motoda, H., Xu, G. (eds.) PAKDD 2013. LNCS (LNAI), vol. 7818, pp. 13–25. Springer, Heidelberg (2013). https://doi.org/10.1007/978-3-642-37453-1_2
8. Dwork, C.: A firm foundation for private data analysis. Commun. ACM **54**(1), 86–95 (2011). https://doi.org/10.1145/1866739.1866758
9. Dwork, C., McSherry, F., Nissim, K., Smith, A.: Calibrating noise to sensitivity in private data analysis. In: Halevi, S., Rabin, T. (eds.) TCC 2006. LNCS, vol. 3876, pp. 265–284. Springer, Heidelberg (2006). https://doi.org/10.1007/11681878_14
10. Dwork, C.: Differential privacy: a survey of results. In: Agrawal, M., Du, D., Duan, Z., Li, A. (eds.) TAMC 2008. LNCS, vol. 4978, pp. 1–19. Springer, Heidelberg (2008). https://doi.org/10.1007/978-3-540-79228-4_1
11. Wu, J., Ni, W., Zhang, S.: Generalization based privacy-preserving provenance publishing. In: Meng, X., Li, R., Wang, K., Niu, B., Wang, X., Zhao, G. (eds.) WISA 2018. LNCS, vol. 11242, pp. 287–299. Springer, Cham (2018). https://doi.org/10.1007/978-3-030-02934-0_27
12. Lyu, M., Su, D., Li, N.: Understanding the sparse vector technique for differential privacy. Proc. VLDB Endow. **10**(6), 637–648 (2017). https://doi.org/10.14778/3055330.3055331

13. Zhang, X., et al.: An accurate method for mining top-k frequent pattern under differential privacy. J. Integr. Plant Biol. **51**(1), 104–114 (2014). (Chinese). https://doi.org/10.7544/issn1000-1239.2014.20130685
14. Gan, W., et al.: Frequent pattern mining with differential privacy based on transaction truncation. J. Chin. Comput. Syst. **36**(11), 2583–2587 (2015). (Chinese)
15. Liang, W., Chen, H., Zhang, J., Zhao, D., Li, C.: An effective scheme for top-k frequent itemset mining under differential privacy conditions. Sci. China Inf. Sci. **63**(5), 1–3 (2020). https://doi.org/10.1007/s11432-018-9849-y
16. Xiong, X., et al.: Frequent itemsets mining with differential privacy over large-scale data. IEEE Access **6**, 2887 (2018). https://doi.org/10.1109/ACCESS.2018.2839752

A Method for Resisting Adversarial Attack on Time Series Classification Model in IoT System

Zhongguo Yang[1,2(✉)], Han Li[1,2], Mingzhu Zhang[1,2], Jingbin Wang[3], and Chen Liu[1,2]

[1] School of Information Science and Technology, North China University of Technology, Beijing, China
{yangzhongguo,lihan,liuchen}@ncut.edu.cn,
18710202239@163.com
[2] Beijing Key Laboratory on Integration and Analysis of Large-Scale Stream Data, Beijing, China
[3] Changqing Oilfield Company General Manager Office, Xian, China
wangjb2_cq@petrochina.com.cn

Abstract. IoT device is often associated with corresponding datasets, algorithms, and infrastructure. However, many potential threats exist in IoT basic infrastructure when deep learning algorithms are applied in these devices. Typically, a deep learning method is widely applied as the basic decision algorithm to classify the time series data, which is an important task in IoT data application. Nevertheless, they are vulnerable to adversarial examples, which bring potential risks to some fields such medical and security in which, a minor disturbing in the time series data could lead to wrong decision. In this paper, we show some white-box attack and random noise attack against time series data. Moreover, we show an adversarial example generated method which only changes one value of the original time series. To resist the adversarial attack, we train an adversarial examples detector to differentiate the adversarial examples from normal examples based on deep features. The adversarial examples detector could filter the adversarial examples before future impair happening. Experiments on UCR data sets show 97% of adversarial examples could be successfully detected generated by two common attack methods: FGSM and BIM.

Keywords: Time series classification (TSC) · Adversarial attack · Random noise attack · IoT algorithm security

1 Introduction

Internet of Things (IoT) systems rely on various detection, classification, and prediction tasks [1–3] to learn from and adapt to the underlying spectrum environment. Deep learning technology has emerged as a powerful tool to perform these tasks in an automated way. However, due to the adversarial attacks in these algorithms, the application security problem is becoming serious day by day.

© Springer Nature Switzerland AG 2020
G. Wang et al. (Eds.): WISA 2020, LNCS 12432, pp. 559–566, 2020.
https://doi.org/10.1007/978-3-030-60029-7_50

Typically, as is shown in Fig. 1, time-series data classification models are popular application in some crucial and sensitive system.

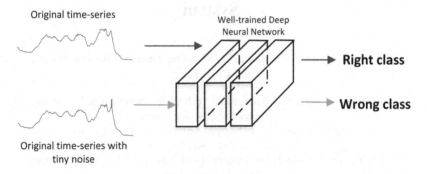

Fig. 1. Adversarial attack in IoT data with a deep neural network.

In this paper, we focus on the adversarial attack and defense on time series data classification task which owns a wide application scenario in IoT system. The state-of-art method in time series data classification model is based on deep learning methods [4].

With aims to resist the adversarial attack, we design an extra adversarial examples detector based on the deep feature extracted from deep neural network. The detector trained on one CNN model could detect adversarial examples generated by other CNN models. The experiments show the detector could successfully distinguish most of the adversarial examples generated by different attack method in different attack ratio ε setting.

2 Related Work

Time Series Classification (TSC) problems are encountered in various real-world data mining tasks ranging from health care [5–7] and food safety [8] power consumption monitoring [9]. As deep learning models have revolutionized many machine learning fields, researchers recently started to adopt these models for TSC tasks [4].

The classification of time series data (IoT) is the core problem in many application fields. Following the advent of deep learning, researchers started to study the vulnerability of deep networks to adversarial attacks [10]. In the context of image recognition, an adversarial attack involves modifying an original image so that the changes are almost undetectable by a human.

There are much defense work for image processing, including data compression [11], data randomization [12], depth compression network [13], gradient regularization [14], defensive distillation [15] etc.

However, there are few studies on attack defense against temporal data. Fawaz in his work [5] showed us the serious problem exists in the time-series classification deep learning model. Buckman et al. [16] have shown the input discretized method could resist adversarial attacks.

The deep learning method will learn important feature automatically from original time series which also could be used to differentiate adversarial examples from normal examples. Inspired by this idea, we trained an adversarial examples detector based on these high level features.

3   Adversarial Attacks in IoT Time Series Data

We choose ResNet [17] as the baseline neural network architecture and Coffee [18] data set as typical time series data to illustrate the adversarial attack phenomena in IoT fields.

3.1   TSC Adversarial Attacks and Fast Gradient Sign Method

Fawaz [4] first introduced some definition of the TSC problem and the adversarial examples. They argue that a time series could be represented as $X = [x_1, x_2, \ldots, x_T]$. The length of X is a natural number T. Moreover, there is a well-trained deep learning model $f(\cdot) \in F : \mathcal{R}^T \to Y$. Where $\mathcal{R}$ is a real number space and Y is the label space of time-series. The goal of adversarial example is to find another example $X\prime$ to be a perturbed version of X with the constraint that $\|X - X\prime\| < \varepsilon$ and $Y \neq Y\prime$. A good illustration of these definitions is shown in Fig. 2 (a).

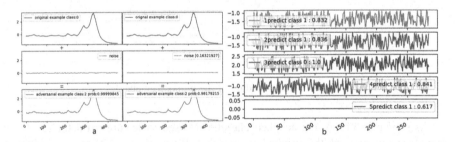

Fig. 2. (a) The adversarial examples for perturbation in whole data (left) and one point (right). (b) The noise data could be predicted as any one class with high confidence.

As shown in Fig. 2 (a), a very tiny perturbation to original example will lead to a huge mistake for the classifier which illustrate the potential insecurity in IoT system if this model is deployed in some decision making devices.

3.2   Random Noise Attack

Nguyen [19] et al. discovered a new type of attack, named False Positive attack, where adversarial examples are classified by deep neural networks with high confidence (99%), which is unrecognizable to human. We trained a ResNet model to classify the time series and some randomly generated noise data. Obviously, the random time series should be rejected by the classifier in low confidence. However, we found these random noise data classified into class one with high confidence, which indicates there is a potential risk in the model. Some samples of random noise time series and their predicted label are shown in Fig. 2 (b).

As shown in Fig. 2 (b), we find that the random noise or even zero values could also lead to a high confidence output. Consequently, the model could not be deployed directly to intelligence device.

4   Proposed Method and Experiment

Through the analysis in Sect. 3, it is obvious that there are some potential risks existing in deep learning models even they achieved best performance in time series classification. We design an adversarial example detection method based on deep featuring extraction. The process is shown in Fig. 3.

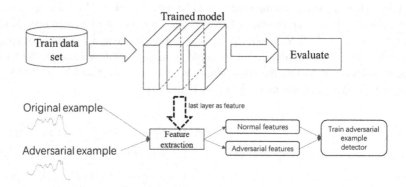

Fig. 3. The procedure of the proposed method.

As shown in Fig. 3, the attacked classifier is used as feature extractor to extract deep representation of the original time series. When the well trained classifier is deployed to IoT systems, our detector will be an extra algorithm to distinguish the input time series to resist the adversarial attack.

4.1   Data Sets and Comparison Method

The experiments are conducted the datasets of the UCR archive [20] which consists of different types of time series data from many fields including food security, electricity industry, image, sensors. The data set consists of 375 univariate time series length of 720 and its classes are Kettle, Microwave, and Toaster. The data set is a typical representation of IoT time series and the application example is an important task in the intelligence device.

For comparison, we set our ResNet architecture the same as [4] have adopted. The ResNet takes time series as input and the K possible classes as output. In each block of the ResNet the convolution kernel size is 8,5,3 which means it will take the neighbor 8,5,3 points into consideration to learn some useful features. There are three blocks in the ResNet we adopted and they comprised respectively of 64,128 and 128 filters.

4.2 Adversarial Example Detector and Defense Result and Their Analysis

We apply XGBoost as the basis classifier to distinguish the normal examples and adversarial examples. The training adversarial examples are generated by Fast Gradient Sign Method (FGSM) and Basic Iterative Method (BIM) [21].

Some data sets are chosen as a case study to show the effectiveness of our method. First, we set a different attack ratio ε to generate enough adversarial examples. The ten-fold cross-validate method is used to train the XGBoost classifier. The deep feature for adversarial example and normal example are shown in Fig. 4.

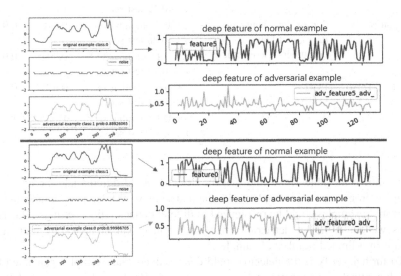

Fig. 4. The difference of deep feature for original example and adversarial example.

From Fig. 4, we could see that the difference in deep feature between adversarial examples with the original example is prominent. At the same time, we find the deep feature of the adversarial example is different from deep feature of the attacked target examples. The experience analysis illustrates the practicability of training a detector to distinguish adversarial examples from normal examples based on deep features.

Here, given the huge difference existing in deep features, we apply the XGBoost algorithm as the base classifier to classify adversarial example and norm examples.

We generate multiple adversarial examples by method of FGSM, BIM with different attack ratio settings ε. The accuracy of adversarial examples detector is evaluated by using ten-fold cross-validation and one hundred repeated process. The evaluated result is shown in Table 1.

Table 1. The accuracy of adversarial example detector on UCR data sets.

Data set	Data description	Adversarial examples	FGSM	BIM
Coffee	Food spectrographs	$\varepsilon \leq 0.08$ training set $\varepsilon > 0.08$ test set	0.744	0.6488
		$\varepsilon \leq 0.12$ training set $\varepsilon > 0.12$ test set	0.9973	0.9821
Meat	Food spectrographs	$\varepsilon \leq 0.08$ training set $\varepsilon > 0.08$ test set	0.9694	1
		$\varepsilon \leq 0.12$ training set $\varepsilon > 0.12$ test set	0.9667	0.9667
GunPoint	Motion with hand	$\varepsilon \leq 0.08$ training set $\varepsilon > 0.08$ test set	0.9422	0.7267
		$\varepsilon \leq 0.12$ training set $\varepsilon > 0.12$ test set	1	0.9985
Computers	Sensor data: behavioral data about how consumers use electricity	$\varepsilon \leq 0.08$ training set $\varepsilon > 0.08$ test set	0.9462	0.9407
		$\varepsilon \leq 0.12$ training set $\varepsilon > 0.12$ test set	0.994	0.9563
CinCECGTorso	ECG	$\varepsilon \leq 0.08$ training set $\varepsilon > 0.08$ test set	0.6667	0.9972
		$\varepsilon \leq 0.12$ training set $\varepsilon > 0.12$ test set	1	1

As shown in Table 1, we find that the detector could differentiate the adversarial examples from normal examples with high accuracy. While, it is interesting, the adversarial examples could be recognized in different attack ratio ε settings. This result means we could train a detector to resist different extent of attack by training the detector in a limited number of samples.

To further verify if the detector could detect adversarial examples generated by a black-box attack, we generate some adversarial examples by adding some random tiny noise to the original examples. The adversarial examples detector is trained on examples generated by FGSM and BIM correspondingly. The detection accuracy is illustrated in Table 2.

Table 2. Adversarial examples recognition accuracy for Black-box attack.

Data set	Train set come from FGSM	Train set come from BIM
Coffee	0.1558	0.1818
Meat	0.03	0.1133
GunPoint	0.4062	0.5938
Computers	0.2666	0.9779
CinCECGTorso	0.005013	0.2995

As shown in Table 2, the performance of detector depends on the train data set and differs for different data sets. Compared with FGSM, the detector shows better result based on data set come from BIM attack. It is noteworthy that data sets such as black-box attack examples for Computers are easier to detect compared with other data sets. We found that the examples in Computers are not similar to each other which exhibit big differences in their own adversarial examples. However, the original examples in Coffee and other data sets are very similar to each other, and their corresponding adversarial examples are also like each other. The detector fails to recognize them from normal examples.

In conclusion, the recognition result of adversarial examples shows it is possible to distinguish adversarial examples from normal examples. More research on applying deep learning models in IoT systems is needed considering the existence of black-box attacks.

5   Conclusions

In this paper, we analyze some typical adversarial attack examples in IoT system from the perspective of applying deep learning method in time series classification. Furthermore, the sensitive points attack examples and the method of constructing them are shown based on susceptible field extraction.

We proposed a method to distinguish adversarial examples from normal examples based on deep feature extraction from deep neural network. However, the white box attack launched by FGSM is not easy to eliminate. Experiments show the detector could detect most of white-box attack examples but fails to recognize the black-box attack examples.

Acknowledgment. This work is supported by National Key R&D Plan (No: 2018YFB14 02500), the Research Start-up Fund of North China University of technology (110051360002), General Project of Science and Technology Plan of Beijing Education Commission (No. KM201810009004).

References

1. Mohammadi, M., Al-Fuqaha, A., Sorour, S., Guizani, M.: Deep learning for IoT big data and streaming analytics: a survey. IEEE Commun. Surv. Tutor. 20(4), 2923–2960 (2018)
2. Li, H., Ota, K., Dong, M.: Learning IoT in edge: deep learning for the internet of things with edge computing. IEEE Netw. 32(1), 96–101 (2018)
3. Li, H., Yu, B., Zhao, T.: An anomaly pattern detection method for sensor data. In: Ni, W., Wang, X., Song, W., Li, Y. (eds.) WISA 2019. LNCS, vol. 11817, pp. 270–281. Springer, Cham (2019). https://doi.org/10.1007/978-3-030-30952-7_28
4. Fawaz, H.I., Forestier, G., Weber, J., Idoumghar, L., Muller, P.-A.: Adversarial attacks on deep neural networks for time series classification (2019)
5. Ismail Fawaz, H., Forestier, G., Weber, J., Idoumghar, L., Muller, P.-A.: Deep learning for time series classification: a review. Data Min. Knowl. Discov. 33(4), 917–963 (2019). https://doi.org/10.1007/s10618-019-00619-1

6. Abdelfattah, S.M., Abdelrahman, G.M., Wang, M.: Augmenting the size of EEG datasets using generative adversarial networks. In: Proceedings of the International Joint Conference on Neural Networks, vol. 2018-July (2018)

7. Wang, Z., Yan, W., Oates, T.: Time series classification from scratch with deep neural networks: a strong baseline. In: Proceedings of the International Joint Conference on Neural Networks, vol. 2017-May, pp. 1578–1585 (2017)

8. Nawrocka, A., Lamorsk, J.: Determination of food quality by using spectroscopic methods. In: Advances in Agrophysical Research (2013)

9. Zheng, Z., Yang, Y., Niu, X., Dai, H.N., Zhou, Y.: Wide and deep convolutional neural networks for electricity-theft detection to secure smart grids. IEEE Trans. Ind. Inf. **14**(4), 1606–1615 (2018)

10. Yuan, X., He, P., Zhu, Q., Li, X.: Adversarial examples: attacks and defenses for deep learning. IEEE Trans. Neural Netw. Learn. Syst. 1–20 (2019)

11. Dziugaite, G.K., Ghahramani, Z., Roy, D.M.: A study of the effect of JPG compression on adversarial images (2016)

12. Xie, C., Wang, J., Zhang, Z., Zhou, Y., Xie, L., Yuille, A.: Adversarial examples for semantic segmentation and object detection. In: Proceedings of the IEEE International Conference on Computer Vision, vol. 2017-October, pp. 1378–1387 (2017)

13. Gu, S., Rigazio, L.: Towards deep neural network architectures robust to adversarial examples (2014)

14. Ros, A.S., Doshi-Velez, F.: Improving the adversarial robustness and interpretability of deep neural networks by regularizing their input gradients. In: 32nd AAAI Conference on Artificial Intelligence, AAAI 2018, pp. 1660–1669 (2018)

15. Distillation as a Defense to Adversarial Perturbations Against Deep Neural Networks (2018)

16. Buckman, J., Roy, A., Raffel, C., Goodfellow, I.: Thermometer encoding: one hot way to resist adversarial examples. ICLR **19**(1), 92–97 (2018)

17. He, K., Zhang, X., Ren, S., Sun, J.: Deep residual learning for image recognition. Comput. Vis. Pattern Recognit. 770–778 (2016)

18. Briandet, R., Kemsley, E.K., Wilson, R.H.: Discrimination of Arabica and Robusta in instant coffee by Fourier transform infrared spectroscopy and chemometrics. J. Agric. Food Chem. **44**(1), 170–174 (1996)

19. Nguyen, A., Yosinski, J., Clune, J.: Deep neural networks are easily fooled: high confidence predictions for unrecognizable images. In: Proceedings of the IEEE Computer Society Conference on Computer Vision and Pattern Recognition, 07–12-June, pp. 427–436 (2015)

20. Dau, H.A., et al.: The UCR time series classification archive. arXiv (2018)

21. Kurakin, A., Goodfellow, I., Bengio, S.: Adversarial examples in the physical world (2016)

Variable-Length Indistinguishable Binary Tree for Keyword Searching Over Encrypted Data

Heyu Wang, Qin Jiang$^{(\boxtimes)}$, Hui He, Yong Qi, and Xu Yang

School of Electrical and Information Engineering, Xi'an Jiaotong University,
Xi'an 710071, China
{why615347842,qjiang16}@stu.xjtu.edu.cn, {huihe,qiy}@mail.xjtu.edu.cn,
sunshine561@163.com

Abstract. We consider the problem of privacy-preserving keyword search on the clouds. In this paper, we propose a new searchable encryption scheme that satisfies efficient and secure query processing. In particular, our scheme utilizes a novel data structure called *Indistinguishable Bloom Filter (IBF)* to construct the first *variable-length Indistinguishable Binary Tree (vIBtree)*. Prior IBF based searchable encryption scheme takes tremendous time to construct a huge secure index. Our scheme overcomes these two shortcomings by constructing an inconsiderable secure index within an insignificant time. Meanwhile, we propose a corresponding trapdoor algorithm for vIBtree. Furthermore, we analyze that our scheme is secure under the adaptive $IND - CKA$ security model. We implemented our scheme in C language. Our experiments show that vIBtree can speedup the secure index construction time and reduce the storage overhead for the data structure over known, state of the art alternatives.

Keywords: Searchable Encryption (SE) · Indistinguishable Bloom Filter (IBF) · Indistinguishable Binary Tree (IBtree) · Cloud computing

1 Introduction

1.1 Motivation

The cloud has significantly matured in the last several years. It centralizes unlimited resources and delivers elastic services to end-users. For reasons of cost and convenience, users often store their data on remote cloud servers that may offer better connectivity. However, privacy has become the key concern as data owners may not fully trust public clouds, including concerns about data security and user's privacy. The untrusted server tends to contribute to privacy and security issues.

As a solution, the data is encrypted before outsourcing to a cloud service [1–4]. However, data encryption makes it hard to retrieve data selectively from the server. The usual searching analogy could not get through the encrypted data.

© Springer Nature Switzerland AG 2020
G. Wang et al. (Eds.): WISA 2020, LNCS 12432, pp. 567–578, 2020.
https://doi.org/10.1007/978-3-030-60029-7_51

To address this issue, many Searchable Encryption (SE) schemes have been proposed, which usually consists of the following steps: A data owner generates its encrypted document and the encrypted index. The encrypted document and the encrypted index are then outsourced to the cloud. To search a keyword, a data user generates a so-called trapdoor. With the trapdoor, the cloud can search on the encrypted index and return related documents.

Unfortunately, it is insufficient to protect the data using encryption, there are side channels for the adversary to get valid information. For instance, by selecting the queries sent to the server adaptively, the attacker can infer the content of the queries. Moreover, it is unwise to store the data structures without considering the storage as a large space can increase the attack suffice for the server and maintenance cost for the clients.

1.2   Threat Model

We assume that the server in our paper is semi-honest. The attackers on the system can observe the interaction between the client and the server and can not prevent the schemes executing correctly but attempt to obtain valid information such as the content of the user queries, and the data items. We also assume that the user communicates with the server through a secure channel and the user is not compromised. In our implementation, the security of cryptographic primitives is also assumed.

1.3   Security Goal

Let $VBT = (VBT\_Setup, VBT\_Enc, VBT\_Tok, VBT\_Query, VBT\_Dec)$ be searchable scheme, k be the security parameter, $\zeta = \{\mathbf{Gen}^{sym}, \mathbf{Enc}^{sym}, \mathbf{Dec}^{sym}\}$ be the $IND - CPA$ encryption scheme. We now describe our VBT scheme as follows:

- $(SK) \leftarrow \mathbf{VBT\_Setup}(1^k, S)$. Take k and $S = (D_1, D_2,..., D_n)$ as input. Use input k to execute $\mathbf{Gen}^{sym}$ two times and output SK_I, SK_v. SK_v is used for encrypt the data items. SK_I is used to encrypt the variable bloom filer tree. In addition, we set $SK = (SK_I, SK_v)$
- $(E\_IV, E\_S) \leftarrow \mathbf{VBT\_Enc}(SK, IV, S)$. Take SK, inverted index IV ,document 29 No collection $S = (D_1, D_2,..., D_n)$ as input. Use input SK_I, SK_v to execute Enc^{sym} to encrypt the IV, S and output E_IV, E_S respectively.
- $t \leftarrow \mathbf{VBT\_Tok}(SK_I, R)$. Use input SK_I and $R = (key_1$ and key_2 and key_3 and ... and $key_m)$ to calculate t using $\mathbf{Enc}^{sym}$.
- $C\_Result \leftarrow \mathbf{VBT\_Query}(t, E\_IV)$. Use input search token t, and encrypted inverted index E_IV to perform query operations and output the encrypted ID of the documents satisfying the search token t.
- $result \leftarrow \mathbf{VBT\_Dec}(SK_I, C\_Result)$. Use input SK_I and C_Result to execute $\mathbf{Dec}^{sym}$ to calculate the search result $result$.

Similar to [5], we assume that the level security of VBT is adaptive IND-CKA security model. More specifically, to achieve this level of security, the main challenge for VBT is to construct an index I which is adaptive IND-CKA.

1.4 Our Contributions

In this paper, we study the fundamental problem of processing keyword queries in a privacy-preserving manner. Our main contributions are summarized as follows.

First, we propose an SSE based keyword query scheme that satisfies requirements of efficient query processing and adaptive security. We utilize a data structure IBF [5] to construct the first variable-length IBtree as shown in Fig. 4.

Second, we extend the trapdoor computation algorithm for fixed-length IBtree to supporting searching over our variable-length IBtree. Data users who share the same secret keys with data owner can generate a trapdoor according to our algorithm, by sending the trapdoor to the cloud they can perform a keyword search over the encrypted data items in $O(logn)$ time, where n is the total number of data items inserted into our variable-length IBtree. The index reveals no information about its contents without valid trapdoors.

Third, we implement both our scheme and prior fixed-length IBtree scheme [5] in C language.[1] We evaluate our scheme with the fixed-length scheme on three aspects: IBtree construction time, IBtree memory usage and keyword searching time. Experimental results show that our scheme significantly speeds up IBtee construction time and reduces IBtree memory overhead compared with the state of the art.

2 Related Work

Song et al. developed a Searchable Symmetric Key Encryption (SSKE) scheme [6] that allows a user, given a trapdoor for a word, to test if a ciphertext block contains the word. Boneh et al. developed Searchable Public Key Encryption (SPKE) [7], which is the public key equivalent of SSKE. But their schme is inefficient because the server performs a linear scan through all data on every query. In [8], Goh presented an efficient scheme for keyword search over encrypted data using Bloom filters. Determining whether a document contains a keyword can be done securely in constant time. However, the scheme does not support secure conjunctive search. Curtmola et al. [9] proposed the first sub-linear scheme, which is more suitable for a static document set than a dynamic one [10].

A significant amount of work has been done in privacy-preserving keyword queries [11]–[16]. In [11], Qi et al. proposed a compressed and private data sharing framework that provides efficient and private data management. In [12], S. Kamara et al. proposed a SSE scheme that satisfies sublinear search time, security against adaptive chosen keyword attacks, compact indexes and the ability to add and delete files efficiently. In [13] and [14], they used deterministic encryption to enable search on encrypted data with existing database and search techniques. Jiang et al. [17] proposed an oblivious data structure to support efficient boolean queries using Intel SGX. However, it focuses on access pattern leakage and does not consider the adaptive security for the data structure which

---

[1] https://github.com/HeyuWang/Indistinguishable-bloom-filters.

is important for the private search. Li et al. [5] proposed a data structure *Indistinguishable Bloom Filter* for storing index elements, and utilized it to construct a fixed-length *Indistinguishable Binary Tree (IBtree)*. However, the cost of a fixed-length IBtree scheme in terms of build tree time and memory usage increases quickly with the number of leaf nodes.

3   Background

3.1   Notations and Terminology

This section explains the notation and terminology used in this paper.

$D = \{d_1, d_2, ..., d_n\}$—A collection of n data items.

$W_i = \{\omega_1, \omega_2, \cdots, \omega_n\}$—A keyword set of a given data item d_i.

$HMAC_{K_i}(.)$ —A keyed hash function which is used to hash a keyword ω_i, $SHA - 256$ is used in our $HMAC$. The outcome are 256 bits.

$SHA1(.)$—A hash function that is used to determine the "chosen" cell position in an IBF, the outcome are 160 bits.

B: An array B of m twins which stands for the IBF.

3.2   System Model

In this paper, we consider a cloud-data-sharing system consisting of three entities, i.e., data owners, cloud server and data users. In particular, we consider Searchable Symmetric Encryption (SSE) schemes as symmetric encryption based privacy-preserving schemes are significantly more efficient than asymmetric ones. In SSE, a data owner builds a *secure index I* for a data set D and encrypts each data item $d_i \in D$ into $((d_i)_K)$ using a secret key K that is shared between the data owner and data users. Then the data owner outsources the secure index I along with the set of encrypted data $\{(d_1)_K, (d_2)_K, \cdots, (d_n)_K\}$ to the cloud. Given a keyword query q, the data user generates a *trapdoor t_q* for q and sends t_q to the cloud. Base on t_q and the secure index I, the cloud can determine which encrypted data items satisfy q without knowing the content of data and query. In this process, the cloud should not be able to infer privacy information about data items and queries such as data content, query content, and the statistical properties of attribute values. The index I should leak no information about the data items in D.

3.3   Indistinguishable Bloom Filter (IBF)

To achieve adaptive security and privacy-preserving query scheme, we adopt a new data structure called IBF. Li *et al.* [5] proposed Indistinguishable Bloom Filter (IBF) in 2017. As Fig. 1 shows, an IBF is an array B of m twins. Each twin consists of two cells where each cell stores one bit, and the two bits remain opposite. Figure 1 shows an initialized IBF with 11 twins (22 bits) without any encrypted keyword added. Two cells of each twin are not equivalent; one of them

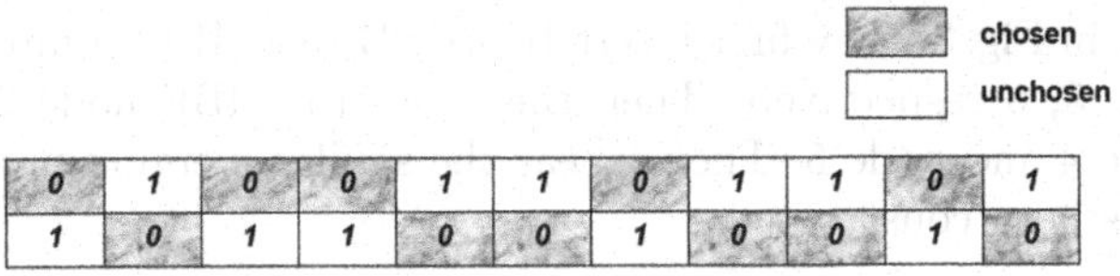

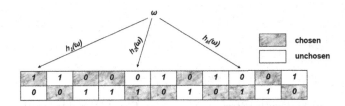

Fig. 1. An initialized IBF without any keyword inside with 11 twins.

is labelled with "chosen", which is the bit used to store information. The other is the "unchosen" cell that is used to mask information inside this twin.

As shown in Fig. 2, after inserting the keyword inside the IBF, the bit in "chosen" cell of each twin is one, and the bit in "unchosen" cell of each twin is 0, the other unmapped position keeps the "chosen" cell 0 and "unchosen" cell 1. Compared with BF, IBF chooses a random cell to store the information that the position is mapped by associating a random number with it, which also eliminates the correlation among different IBFs. IBF construction algorithm can be found in [5].

Fig. 2. An IBF which has been inserted.

False Positive Rate of Indistinguishable Bloom Filter: According to [8], after using k hash functions to insert n distinct elements into an array of size m, the probability that bit i in the array is 0 is $(1-(1/m))^{kn} \approx e^{-kn/m}$. Therefore, the probability of a false positive is $(1-(1-(1/m))^{kn})^k \approx (1-e^{-kn/m})^k$. Since m and n are typically fixed parameters, we compute the value of k that minimizes the false positive rate. The derivative of the right hand side of the equation with respect to k gives a global minimum of $k = (ln2)(m/n)$, with a false positive rate of $(1/2)^k$. We can round k to the nearest integer.

3.4 Fixed-Length Indistinguishable Binary Tree

In a fixed-length indistinguishable binary tree, each terminal and nonterminal node is an IBF of the identical size. We briefly described their IBtree construction algorithm here. Details about corresponding trapdoor algorithm and query process can be found in [5]. Li et al. construct an IBtree in two stages. First, they insert each keyword into a separate leaf IBF node. Second, they construct each nonterminal IBF node by merging its left and right children.

As shown in Fig. 3, they first insert $W(d_1)$, $W(d_2)$, $W(d_3)$ into the terminal IBF nodes 4, 5, 6 respectively. Then they construct IBF node 2 through the union of node 4 and node 5. They repeat the similar merging process until the root node has been constructed.

4   Variable-Length IBtree Scheme

4.1   Variable-Length IBtree

A variable-length IBtree is a binary tree. Each node is a constructed IBF. A children IBF node is only half the size of its parent node and stores approximately half the keywords of its parent. Figure 4 shows a simple variable-length IBtree a root node size of 12 twins.

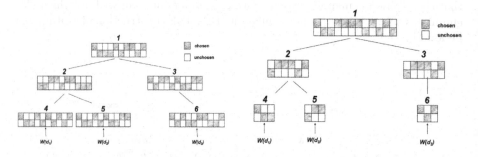

Fig. 3. A Fixed-length IBtree.   **Fig. 4.** Variable-length IBtree.

Variable-Length IBtree Construction: In a fixed-length IBtree, each IBF node occupies the same memory as the root IBF node. Thus the memory required at each level increases exponentially, which is extremely a waste of storage because each child IBF node represents a keyword set that is approximately half of its parent. Besides, twins need to be initialized also increases exponentially. The initialization stage requires three hash operations for each twin. Therefore the fixed-length IBtree spend a tremendous amount of time on the initializing stage. Considering these two severe disadvantages, we propose to construct a variable-length IBtree to overcome these two shortcomings. Our variable-length IBtree construction steps are listed below.

Algorithm 1: VBT_Setup

Input: n encrypted data documents $E_k(d_i)(1 \le i \le n)$ and
corresponding data indices (W_i), where $W_i = \{\omega_1, \omega_2, \cdots, \omega_n\}$
Output: a constructed variable-length IBtree
initialization stage :
for *each node in the variable-length IBtree* **do**
  | 1.*allocate memory.*
  | 2. *initialize it.*
terminal nodes insertion stage :
for *each terminal node* **do**
  | *insert keywords set individually into it.*

nonterminal nodes insertion stage:;
for *each nonterminal node* **do**
  | *insert all keywords added to its left children and right children*
  | *into it directly.*

4.2 Trapdoor Algorithm for Variable-Length IBtree

We need a trapdoor algorithm to enable encrypted searching over the variable-length IBtree. In a fixed-length IBtree, $t_q = \{(h_1(\omega_i), h_{k+1}(h_1(\omega_i))), (h_2(\omega_i), h_{k+1}(h_2(\omega_i))), \cdots, (h_k(\omega_i), h_{k+1}(h_k(\omega_i)))\}$ where $h_j(\omega_i) = \text{HMAC}_{K_j}(\omega_i)\%$ $m(1 \le j \le k + 1)$. We are not allowed to use the trapdoor algorithm for fixed-length IBtree here. Because in our variable-length IBtree, the different level has different twins, which means m is different. We extend original fixed-length IBtree trapdoor algorithm, supposing the variable-length IBtree has h levels, then data users need to calculate h trapdoors for each level. Denoting n is the number of data items stored in the leaf node of variable-length IBtree, $above(n)$ is the number of all nonterminal nodes above terminal nodes. Then the height of the tree is $h = (log_2(n + above(n)) + 1)$(if we label the height of root node with 1). Thus, our variable-length trapdoor t_q^j consists of $h * k$ pairs. For each level j we calculate corresponding k pairs trapdoor $t_q^j(\omega_i) = \{(h_1^j(\omega_i), h_{k+1}^j(h_1^j(\omega_i))), (h_2^j(\omega_i), h_{k+1}^j(h_2^j(\omega_i))), \cdots, (h_k^j(\omega_i), h_{k+1}^j(h_k^j(\omega_i)))\}(1 \le i \le n, 1 \le j \le h)$, i is used to denote different keyword ω and j is used to denote different level in the variable-length IBtree. We show our variable-length trapdoor algorithm below.

Algorithm 2: VBT_Tok

Input: a keyword ω_i
Output: a trapdoor t_q^j of $h * k$ pairs
Get the height of the variable-length IBtree h by
communicating with cloud server;
for *each height $j(1 \le j \le h)$* **do**
  | $t_q^j(\omega_i) = \{(h_1^j(\omega_i), h_{k+1}^j(h_1(\omega_i))), (h_2^j(\omega_i), h_{k+1}^j(h_2(\omega_i))), \cdots,$
  | $(h_k^j(\omega_i), h_{k+1}^j(h_k^j(\omega_i)))\}(1 \le i \le n)$

4.3   Query Processing

After receiving a trapdoor of $h * k$ pairs for a specified keyword from data user, cloud server can search over the variable-length IBtree with algorithm listed below. Cloud server also perform searching in a top-down manner, in each level $j(1 \leq j \leq h)$, cloud server using corresponding trapdoor $t_q^j(\omega_i) = \{(h_1^j(\omega_i), h_{k+1}^j(h_1^j(\omega_i))), (h_2^j(\omega_i), h_{k+1}^j(h_2^j(\omega_i))), \cdots, (h_k^j(\omega_i), h_{k+1}^j(h_k^j(\omega_i)))\}(1 \leq i \leq n)$ to match the specified keyword ω_i, if a match is found at root node, then it's child can be searched according to the corresponding trapdoor. For each ordered pair of hashes in t_q^j, we use the first hash value to locate the twin that the corresponding keyword is hashed to, the second hash value locates the "chosen" cell in the twin. If the cloud determines that t_q matches no data items, the query processing terminates. Otherwise, the cloud processes t_q^j in each level in a top-down manner.

Algorithm 3: VBT_Query

Input: h trapdoors for a specified keyword ω_i
Output: if a match is found, then return relevant encrypted document.
　　　　　otherwise return nothing
for *each height* $j(1 \leq j \leq h)$ **do**
> **if** *match is found in either a node* **then**
>> **if** *is leaf* **then**
>>> | return encrypted document
>>
>> **else**
>>> └ keep searching
>
> **else**
>> └ break

4.4   Security Analysis

To give the security analysis, we firstly define two leakage functions based on the threat model in Sect. 1.

$L_{enc}(\text{D})$: Given the document data set D, this function outputs the number of n documents, the max keywords m contained in the document in D, the document identifiers $ID = \{id_1, id_2, ..., id_n\}$, the number of nodes in the created index and the size of each node of the index.

$L_{que}(\text{D,I,R,t})$: Given the document set D, the index tree I, the search query R and given point in time t, this function outputs the search pattern at time t which contains the information about whether R is the same as queries performed before time t or not and the access pattern at time t which contains the information about which documents contain R.

Theorem 2 (Security): The variable IBtree is IND-CAK (L_{enc}, L_{que})-security against an adaptive adversary.

Proof (Sketch). Similar to the proof in [5], we can construct a simulator S using L_{enc} and L_{que}. More specifically, using L_{enc}, S can simulate variable IBtree index I. Using L_{que}, S can simulate queries on I. And S can simulate the search token and the query results which are indistinguishable to the adversary as the pseudo-random function and CPA-secure encryption algorithm. Thus, our scheme is IND-CKA security against an adaptive adversary.

4.5   Advantages and Disadvantages of Our Scheme

Disadvantages of Variable-Length IBtree. A fixed-length indeed has some advantages over our variable-length IBtree scheme. In the fixed-length scheme, data users only need to generate one trapdoor for a specified keyword w_i, while in our scheme, data users need to generate h trapdoors, each trapdoor for each level where $h = \lceil log(n) \rceil + 1$. The overhead brought by increased trapdoors is negligible in nowadays's network bandwidth. For example, when we insert 10000 data documents into the leaf nodes of our variable-length IBtree, we only need to generate 12 more trapdoors for the other IBF node to perform the matching operation.

Advantages of Variable-Length IBtree. Our variable-length IBtree scheme significantly reduces IBtree construction time and IBtree index size. Supposing there are n data items inserted into the IBtree of both scheme. In our scheme, each level occupies the same memory as a root node. Thus, index size is proportional to the height($logn$) of a tree. In a fixed-length IBtree scheme, index size is proportional to the number of nodes(n). Therefore, our scheme reduces index size logarithmically. Meanwhile, the number of twins initialized is proportional to the index size. Consequently, our scheme saves index construction time logarithmically.

5   Experiment Evaluation

We realized both fixed-length IBtree scheme and variable-length IBtree scheme in C language. We evaluate both schemes on three aspects of IBtree construction time, IBtree index size and single keyword searching time. We first use a simple simulation data set to observe the tendency of both schemes. Then we experiment on the real-world data set "America NSF Research Award Abstract set [18]".

5.1   Data Sets

We use a simulation data set to see the tendency of both schemes before using a real-world data set. We randomly generate several (3 to 11) fixed-length words for each leaf node in our simulation data set. We choose a real-world data set that is the America NSF Research Award Abstract set, which consists of 129 K abstracts describing NSF awards from 1990–2003. We extract keywords from each abstract. The number of keywords in a document ranges from 15 to 454, and the average number of keywords in a document is 99.27. We use the real word data set to generate 1 data set, which contains 10 document sets with a size range of 500 to 5000.

5.2   Implementation Details

Our experiment platform is equipped with 3.20 GHz Intel Core i7-8700 CPU and 16 GB RAM. The operating system running on the server is Ubuntu(version:18.03), and the kernel is Linux Ubuntu 5.3.0-51-generic. In our experiments, we use $SHA256$ as the hash function of $HMAC$. We fix the root node size to 50000 bytes when we use the simulation data set. To minimize the false positive rate when we use the real-world data set, we set the root node size m according to $k = ln2(m/n)$ for different collections of data items. We use 3 hash functions in both data sets.

5.3   Experiment Results

Simulation Data Set Experiment Results: In Fig. 5 (a) and (b), with leaf nodes growing from 250 to 1500, construction time and index size of a fixed-length scheme increase more rapidly than that of our scheme. In Fig. 5 (c), we can observe that single keyword searching time of both schemes is in the millisecond scale. The simulation data set experiment result demonstrate that our scheme significantly reduces the construction time and index size.

Real-world Data Set Experiment Results: The real-world data set experiment results are demonstrated in Fig. 6. We analyse each experiment result in detail below.

Experiment 1 (IBtree construction time): With data set sizes growing from 500 to 5000, the index construction time of fixed scheme and variable scheme grows from 18 min to 29 h, and from 42 s to 10 min respectively in Fig. 6 (a). Experimental results show that our variable-length scheme significantly reduces IBtree construction time.

Experiment 2 (IBtree memory usage): The evaluated results of IBtree memory usage are shown in Fig. 6 (b). With data items ranging from 500 to 5K, the memory usage of fixed-length scheme and variable-length scheme grows from 102 MB to 10 GB, and from 1 MB to 14 MB, respectively. One can observe clearly that the memory occupied by the fixed-length IBtree increases exponentially. However, the memory usage of the variable-length IBtree is proportional to the levels of the IBtree, as we described in Sect. 4.4 Experimental evaluation shows that the more the number of terminal nodes we put, the larger the memory of our scheme saved compared with the fixed-length scheme.

Experiment 3 (keyword searching time): Experimental results show that the keyword query processing time is in the millisecond scale. The time consumed in keyword searching in a fixed-length IBtree scheme is an order of magnitude the same as our variable-length IBtree scheme. Because both schemes locate relevant position of a keyword in an IBF by hash operation. Figure 6 (c) shows the single keyword searching time in two schemes. We can see that with data items ranging from 500 to 5K, searching time of fixed-length scheme and variable-length scheme varies from 2 ms to 6 ms, and from 2 ms to 10 ms respectively.

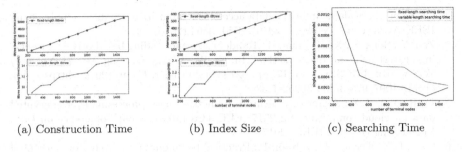

(a) Construction Time   (b) Index Size   (c) Searching Time

Fig. 5. Simulation data set results

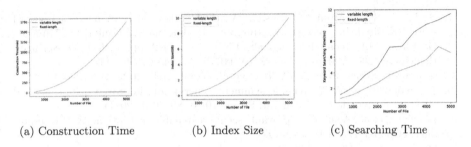

(a) Construction Time   (b) Index Size   (c) Searching Time

Fig. 6. Real data set results

6   Conclusion and Future Work

In this paper, we first propose a variable-length IBtree scheme to enable the keyword searching over the encrypted data in a popular client and cloud paradigm based on the data structure IBF. We extend the original trapdoor algorithm for a fixed-length IBtree to get our trapdoor algorithm. In IBtree construction algorithm, much time is consumed by initializing IBFs. Therefore, our scheme theoretically cost less time and occupy less memory. Experimental results show our scheme has significantly reduced the IBtree construction time and memory usage. Keyword searching time of our scheme is an order of magnitude the same as the fixed-length IBtree scheme. Future work will be done to enable range queries through a variable-length IBtree over encrypted data.

Acknowledgement. This work is in part supported by the National Natural Science Foundation of China (No.61672421).

References

1. Pasupuleti, S.K., Subramanian, R., Rajkumar, B.: An efficient and secure privacy-preserving approach for outsourced data of resource constrained mobile devices in cloud computing. J. Netw. Comput. Appl. **64**, 12–22 (2016)
2. Qi, S., Zheng, Y., Li, M., Liu, Y., Qiu, J.: Scalable industry data access control in RFID-enabled supply chain. IEEE/ACM Trans. Network. **24**(6), 3551–3564 (2016). https://doi.org/10.1109/TNET.2016.2536626

3. Qi, S., et al.: Scalable industry data access control in RFID-enabled supply chain. IEEE/ACM Trans. Network. **24**(6), 3551–3564 (2016)

4. Zou, Y., Song, S., Xu, C., Luo, H.: Random sequence coding based privacy preserving nearest neighbor query method. In: Ni, W., Wang, X., Song, W., Li, Y. (eds.) WISA 2019. LNCS, vol. 11817, pp. 327–339. Springer, Cham (2019). https://doi.org/10.1007/978-3-030-30952-7_33

5. Li, R., Liu, A.X.: Adaptively secure conjunctive query processing over encrypted data for cloud computing. In: 2017 IEEE 33rd International Conference on Data Engineering (ICDE), IEEE (2017)

6. Song, D.X., David, W., Adrian, P.: Practical techniques for searches on encrypted data. In: Proceeding 2000 IEEE Symposium on Security and Privacy. S&P 2000. IEEE (2000)

7. Baek, J., Safavi-Naini, R., Susilo, W.: Public key encryption with keyword search revisited. In: Gervasi, O., Murgante, B., Laganà, A., Taniar, D., Mun, Y., Gavrilova, M.L. (eds.) ICCSA 2008. LNCS, vol. 5072, pp. 1249–1259. Springer, Heidelberg (2008). https://doi.org/10.1007/978-3-540-69839-5_96

8. Goh, E.-J.: Secure indexes. IACR Cryptol. ePrint Arch. **2003**, 216 (2003)

9. Curtmola, R., et al.: Searchable symmetric encryption: improved definitions and efficient constructions. J. Comput. Secur. **19**(5), 895–934 (2011)

10. Bösch, C., et al.: A survey of provably secure searchable encryption. ACM Comput. Surv. (CSUR) **47**(2), 1–51 (2014)

11. Qi, S., et al.: CPDS: Enabling Compressed and Private Data Sharing for Industrial IoT over Blockchain. IEEE Transactions on Industrial Informatics (2020)

12. Kamara, S., Charalampos, P., Tom, R.: Dynamic searchable symmetric encryption. In: Proceedings of the 2012 ACM Conference on Computer and Communications Security (2012)

13. Bellare, M., Boldyreva, A., O'Neill, A.: Deterministic and efficiently searchable encryption. In: Menezes, A. (ed.) CRYPTO 2007. LNCS, vol. 4622, pp. 535–552. Springer, Heidelberg (2007). https://doi.org/10.1007/978-3-540-74143-5_30

14. Amanatidis, G., Boldyreva, A., O'Neill, A.: Provably-secure schemes for basic query support in outsourced databases. In: Barker, S., Ahn, G.-J. (eds.) DBSec 2007. LNCS, vol. 4602, pp. 14–30. Springer, Heidelberg (2007). https://doi.org/10.1007/978-3-540-73538-0_2

15. Ballard, L., Kamara, S., Monrose, F.: Achieving Efficient conjunctive keyword searches over encrypted data. In: Qing, S., Mao, W., López, J., Wang, G. (eds.) ICICS 2005. LNCS, vol. 3783, pp. 414–426. Springer, Heidelberg (2005). https://doi.org/10.1007/11602897_35

16. Damiani, E., et al.: Balancing confidentiality and efficiency in untrusted relational DBMSs. In: Proceedings of the 10th ACM Conference on Computer and Communications Security (2003)

17. Jiang, Q., et al.: PBSX: a practical private boolean search using Intel SGX. Inf. Sci. **521**, 174–194 (2020)

18. Bache, K., Lichman, M.: UCI machine learning repository (2013). http://archive.ics.uci.edu/ml

WLTDroid: Repackaging Detection Approach for Android Applications

Junxia Guo, Dongdong Liu, Rilian Zhao, and Zheng Li(✉)

College of Information Science and Technology,
Beijing University of Chemical Technology, Beijing, China
{gjxia,lizheng}@mail.buct.edu.cn

Abstract. Huge number of mobile applications are downloaded every year. These benefits promote the rapid development of the mobile application industry, especially Android applications for its openness. But, because of the low cost and good profit, Android application repackaging also developed quickly, which can make a kind of malware applications and publish them to the Android market. Therefore, in order to defend against this dangerous technology, repackaging detection technology has been continuously studied in recent years. Contrary to the repackaging detection technique, obfuscation techniques are used in the application repacking to avoid detection. This makes the effectiveness of many existing methods be affected. In this paper, we propose a novel approach based on Dynamic Whole Layout Tree extraction, that we call WLT-Droid, which can avoid the interference of the layout file obfuscation. The experimental results show that the approach proposed in this paper can resist the obfuscation affect better than other repackaging detection methods. In addition, our approach is more accurate than the existing method RepDroid.

Keywords: Repackaging detection · Android application · Obfuscation · Layout view transfer

1 Introduction

With the rapid development of global smart phones market, smart phone applications (apps for short) have also been rapidly developed. In the year 2018, global App downloads exceeded 194 billion, up 35% from 2016, and consumer spending reached $101billion, up 75% from 2016 [1]. However, application repackaging is becoming a serious threat to the Android ecosystem, both at the security of app users and the revenue of app developers [11]. Zhou et al. [22] found out that among 1260 malicious apps, 1083 (86%) were spread by repackaging other apps. Another study [5] showed that 14% of apps' revenue and 10% of apps' users are illegally stolen by repackaged apps.

The work described in this paper is supported by the National Natural Science Foundation of China under Grant No. 61702029, No. 61872026 and No. 61672085.

G. Wang et al. (Eds.): WISA 2020, LNCS 12432, pp. 579–591, 2020.
https://doi.org/10.1007/978-3-030-60029-7_52

In recent years, many Android app repackaging detection methods have been proposed. They are divided into three kinds, which are code-based detection, resource-based detection, and UI-based detection. For the code-based methods, large amount of app codes make time consuming. Meanwhile, the code obfuscation methods can cause low code similarity. Researchers [4] found that original app and repackaged app have high similarity at the resource level. Thus, resource-based detection methods are proposed, which detect repackaged apps by analyzing the type and amount of the decompiled resource files (such as: xml files, pictures and audios, etc.). However, this kind of methods still have problems with resource obfuscation and redundant resources in repackaged app. Layout-based detection methods have been proposed because of the high similarity between the original app and the repackaged app at the layout level. However, layout files can be easily interfered with or confused during repackaging, resulting in a decrease in layout similarity.

Considering that the repackaging app need to maintain the same user interface and user interaction, so we assume that the purpose-based measures may be more effective. Because that no matter how the features change, the high similarity of user interface and user interaction never change. Therefore, in this paper, we designed a new approach called whole layout tree (WLT), which is a purpose-based measure. We use interface transfers to represent user interactions. By analyzing user interfaces and interface transfers of running apps, we can completely extract the external view layout information and the inner view transfer information. According to these two kinds of information, we can build an WLT for app repackaging detection. The evaluation criterion is that WLT similarity and repackaging possibility have a positive correlation. The primary contributions of this paper are summarized as follows:

- We design the Whole Layout Tree based on the Android application's interface layouts and transfers. It represents the application's external display information and internal operation logic information. Even with the obfuscation of code, resources and layout, it can still accurately express the application.
- We proposed a new repackaging detection method named WLTDroid for Android applications based on the WLT. WLTDroid is a purpose-based method which has strong resistance.

The rest of this paper is organized as follows: Sect. 2 describes design principles, definitions and features of WLT in detail. Section 3 shows the architecture and implementation of our repackaging detection method. In Sect. 4 we set up the experiments and evaluate the method proposed in this paper. Related work is briefly introduced in Sect. 5. Section 6 concludes our work.

2   Design of Whole Layout Tree

Basically, repackaging detection is using metric methods to find similar applications. To address the problems that existing methods are not resistant to obfuscation technology and easily avoided by attackers, we propose a new measurement method based on Whole Layout Tree (WLT). It is a kind of expression for dynamic user interface layout of Android applications. We explain the detail of WLT, including the design principle, definition and features.

2.1 Design Principles of WLT

As we mentioned, repackaging detection methods need to have strong resistance and good accuracy. Thus, we design WLT with two design principles: strong resistance and high accuracy.

When using applications, users mainly interact with applications' interface instead of directly access applications' source code, resources or XML layout files. Therefore, users are more sensitive to applications' changes in interface than the changes in code, resources, XML layout files. To cheat users, repackaging attackers can modify everything except user interface and interface transfer. Based on above facts, WLT should be designed from the user interface and interface to get strong resistant.

The way how to denote and compare the similarity of the user interface and interface transfer will determine the accuracy of this kind of repackaging detection methods. In an Android application, the user interface is composed of widgets. Each widget has many attributes. Those widget attributes jointly determine the final display of the widgets in the user interface. Therefore, those widget information and widget attribute information should be properly recorded and used for ensure the accuracy.

2.2 Definition of WLT

The WLT is defined as follows[1].

Definition 1. *Whole Layout Tree (WLT)*

The Whole Layout Tree (WLT) is a multi-fork tree with a hierarchical structure which is consists of a limited number of nodes and edges, denoted as T (W, R), where W is a finite set of widgets, R is a finite set of widgets' relationship.

The node set W is consisted of 3 types of nodes, which are "Container Node", "Display Node", and "Transfer Node". Container Nodes are those who have sub-widgets, such as FrameLayout and so on. Display Nodes are those who do not have any sub-widgets, such as TextView and so on. Transfer Nodes are those who can trigger interface transfer, such as Button and so on.

The edge set R is consisted of 2 kinds of relationships. (1) $iR <a, b>$, represents that the widget a and widget b are inclusion relationship, in which the widget a and b belong to the same interface view, and widget a is a layout widget which contains widget b. (2) $tR <a, b>$, represents that the widget a and widget

---

[1] An example figure is given at https://github.com/pro-resrc/figure. The app has four interface views, with 3 view transfers, 79 container nodes, 54 display nodes, 3 transfer nodes, 132 inclusion edges and 3 transfer edges.

Algorithm 1. Whole Layout Tree Generation

Input: HomeActivity(HomeView) from the testing APK
Output: a whole layout tree,WLT
 1: dump view layout,HomeLayout from HomeActivity(HomeView)
 2: extract core view layout,cLayout from HomeLayout by Filter
 3: InitializationWLT(cLayout)
 4: extract visual components,taskComponent from HomeLayout
 5: **while** WLT is changing AND taskComponent is not empty **do**
 6:     todoComponent=taskComponent.pop()
 7:     transfer,newView=todoComponent.simulate()
 8:     add transfer into WLT
 9:     dump view layout,newLayout from newView
10:     extract core view layout,cNewLayout from newLayout by Filter
11:     updateWLT(cNewLayout)
12:     update taskComponent according to newLayout
13: return WLT

---

b are transfer relationship, in which the widget a and b do belong to two different interface views, widget a can trigger the response event and leads the transfer to the new view which widget b belongs to.

2.3 Features of WLT

According to the definition, a WLT should have following features:

1. WLT has only one root node.
2. The parent nodes of a WLT should be container nodes or transfer nodes.
3. The leaf nodes of a WLT should be display nodes.
4. A container node connects to one or more different types of child-nodes, those nodes are connected by inclusion edges.
5. A transfer node connects to a node which belongs to another interface view.
6. The order of the interface view that appears during a user operating the app is the order of the view layout in the WLT.

2.4 Generation of WLT

According to the extracted interface view layout xml file and the view transfer information, we start to generate WLT. Firstly, we initialize WLT with the home view's layout content. Then query the transfer node's information and connect the root node of the transferred new interface view as a child node to the transfer node with a transfer edge. Next, a transfer attribute is added to the transfer node for recording the related information. The above operation is repeated until the WLT no longer changes or there is no more transfer information. In addition, we make several optimization in WLT generation to make the process more efficient. Algorithm 1 shows the pseudocode of generating a WLT.

(a) When transfer to a interface view in the blacklist, only record the transfer information without the interface view information.

(b) If a interface view is already appeared in the WLT, it will not be added to the WLT again. We only keep the transfer information.

(c) When the task stack is empty or WLT does not change in 3 min, the WLT generation process ends.

3   Repackaging Detection Method Based-on WLT

Based on the definition and features of WLT, we design a similarity measurement method for the repackaging detection. In addition, we implement a repackaging detection method, which is named WLTDroid. We explain the details in this section.

3.1   Similarity Measurement of WLT

Hashing algorithms are widely used in file similarity comparison for its good performance. For comparing the similarity of WLT, we design a method based on the Context Triggered Piecewise Hashing (CTPH) [9]. However, the tree-structured texts of WLT can not be used for the CTPH algorithm. In addition, the information of the inclusion and transfer edges need to be calculated for the similarity measurement.

Therefore, we transform the WLT into a kind of sequential text information from two levels, which are interface layout and interface transition. First, we convert each interface layout individually, then convert the processed interface layout as a whole according to the order of interface transition by using a method similar to the binary tree preorder traversal algorithm.

For the inclusion edge information in WLT, we convert the inclusion information to the sequential information as text directly. For the transfer edge information in WLT, in addition to converting the transfer information into sequence information, a special weight needs to be given because of the special.

We have added the transfer attributes to all the transfer nodes as an identifier in the WLT generation step, so those two types of edges have been distinguished in the preprocessed WLT.

The CTPH algorithm first calculates the hash value of the testing text, then get the similarity score by calculating the minimum edit distance between the two hashes. The preprocessed WLT already contains the view layout information and the view transfer information, so the similarity score calculated by the CTPH algorithm represent the similarity of the apps. The formula for calculating the SimilarityScore is show in Formula 1, where pp means preprocess.

$$SimilarityScore = CTPH(WLT_1{}^{pp}, WLT_2{}^{pp}) \qquad (1)$$

The similarity score ranged from 0 to 100. A score of 0 means that two WLTs are completely dissimilar and the possibility of repackaging for these two apps is extremely low. A score of 100 means that two WLTs are exactly the same, and those two apps most likely to be repackaged. We set a threshold θ, when the similarity score of two apps exceeds the threshold, it represents that the two apps have high similarity which need to do the signature verification for the final confirmation.

3.2 Framework of WLTDroid

The framework of repackaging detection method based on WLT is showed in the Fig. 1, which is consisted of four parts. They are data extraction, birthmark generation, similarity measurement and signature verification.

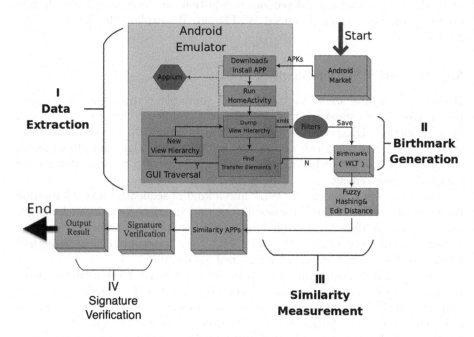

Fig. 1. The framework of WLTDroid

Firstly, WLTDroid dynamically extracts the view layout information of a testing app, and records the view transfer information. Testing app may jump to other apps (such as sharing button, etc.), whose view information does not belong to the original app, so we only save the view transfer information without further data extraction. In addition, for a special widget who names *WebView* we only retain the information of the native widgets.

However, the original layout contents contain redundant information, which do not have any effect on the structure of the interface layout and transfer, but will seriously affect the results of similarity measurement. Because that the attacker can easily add this kind of redundant information in the xml file for avoiding the detection. Therefore we need to filter such redundant information.

(a) Redundant UI widgets: Layout widgets that contain only one or no child are redundant. (b) Redundant attributes of widgets: There are some redundant attributes of the widgets. We found six redundancy attributes which are *text, package, bounds, resource-id, index, and content-desc*. For example, although the bounds attributes directly reflect the position of the widget, a subtle changes will not lead to a huge change in the view which may affect the results of our detection.

Next, generation module will construct WLT. Then, the similarity score between WLTs is calculated when the WLTs no longer change. Finally, if the score exceeds the setted threshold, signature verification will starts.

4   Experimental and Evaluation

Here we design two research questions to verify the effectiveness of our repackaging detection method.

RQ1: Does the WLTDroid has strong resistant when facing the obfuscated apps?

RQ2: How is the accuracy of the WLTDroid?

4.1   Dataset

The data sets used for the experiments are all from the Wandoujia app store (the largest third-party app store in China) [17] and F-Droid (an open source Android app store) [6]. We used three data sets named SR, S_1 and S_2. SR contains 30 Android apps from Wandoujia, which is used for resistance experiment. It includes the latest version of 30 apps randomly selected from Wandoujia's top 50 apps. We leverage Shengtao Yue's work (RepDroid) [12] to create S_1 and S_2 for Accuracy experiment. Unfortunately, we can only download 45 (41 of 45 apps can run) of 58 apps for S_1, and 115 (103 of 115 apps can run) of 125 apps for S_2 because that they are out of stock. The apps cover different fields, such as News, Reading, Education, etc. Totally, 190 apps are used in the experiments.

4.2   Resistance Testing

The resistance testing experiment adopts the current four major Android app packers in China (360, Tencent, Aliyun, Bangcle) to obfuscate or encrypt data set SR. Through code obfuscation, resource file encryption, and Dex file packing, we can simulate the real world Android repackaging environment. Then set the obfuscated app and the original app as a pair. In addition, we use XML files for obfuscation by adding redundant XML files to the decompiled 30 original apps to generate obfuscated apps.

We perform detection using WLTDroid, SUIDroid [11], AndroGuard [8], and FSquaDRA [20] for comparison with the existing methods. AndroGuard is a code-based repackaging detection method. FSquaDRA is a code-and-resource-based repackaging detection method. SUIDroid is a layout-based repackaging detection method.

Using the 30 original apps and the apps obfuscated by above five packers, totally 150 comparison pairs are prepared for the experiment as shown in the upper part of Table 1. The results are shown in Fig. 2.

Table 1. Obfuscated or encrypted apps

Repackaging type	Data set	Original app count	Repackaged app count
360	$5 * SR$	$5 * 30$(Wandoujia$_{top50}$)	30
Tencent			30
Aliyun			30
Bangcle			30
RedundantXML			30
IJiami	$4 * S_1$	30(Wandoujia)	30
AndroCrypt		$3 * 11$(F-droid)	11
FakeActivity			11
NestedLayout			11

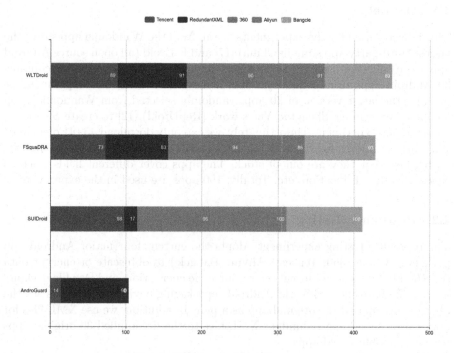

Fig. 2. Results of resistance testing

The horizontal axis represents similarity scores, also represents resistance of methods. The four stacked bars represent four repackaging detection methods. The stacks of five colors represent five types of app packers. The total similarity score of detections is 500 points. The scores from high to low are: 450 points for WLTDroid, 428 points for FSquaDRA, 411 points for SUIDroid, and 104 points for AndroGuard.

We can found that WLTDroid performs much better than AndroGuard. AndroGuard detects repackaged apps based on the decompiled source code. Once the source code was confused, it will be affected much. Except for the XML packer, AndroGuard almost cannot work for the other four packers. In contrast, SUIDroid performs poorly when detecting redundant XML packer. Only scores 17 points (WLTDroid scores 91 points). But, it performs well when detecting the other four types of app packers. That is because SUIDroid is based on analyzing the decompiled XML layout files. Once the attacker adds a large number of redundant XML files to the repackaged app, it will cause the repackaged app's schema layout to be completely different from the original app's schema layout, which makes SUIDroid get the wrong result. FSquaDRA has a better performance than AndroGuard and SUIDroid. FSquaDRA uses a combination of source code and resource files for detecting repackaged apps. Code obfuscation still affects FSquaDRA and may lead to instability during detection.

4.3   Accuracy Testing

We use S_1 and S_2 for Accuracy experiment. We compare the accuracy results of WLTDroid with RepDroid, because that RepDroid is also a repackaging detection technology based on interface layout and transition. The experiment is divided into two steps. Step one, we use S_1 to determine the thresholds of WLTDroid and RepDroid. Step two, we use S_2 to evaluate the counts of false positive(FP) and false positive rates(FPR) of WLTDroid and RepDroid, besides time consumption.

S_1 contains 41 apps that can run, of which 30 apps are from Wandoujia, 11 apps are from F-Droid. We use Ijiami to encrypt 30 Wandoujia's apps, and use three encryption tools: AndroCrypt, FakeActivity and NestedLayout, to encrypt 11 11 F-Droid apps. Totally 33 encrypted apps are used here. The encrypted apps and the original apps are combined as a pair. There are a total of 63 comparison pairs as shown in lower part of Table 1. S_2 contains 103 apps that can run from Wandoujia. The apps are compared with each other, and formed 5253 comparison pairs.

1) Determine the threshold

By comparing all similar scores of 63 comparison pairs, we select the lowest score as the best threshold. Because that 63 comparison pairs all belong to repackaged pairs, similarity scores represent the effectiveness of repackaging detection method, where the lowest similarity score represents the lower limit of the method. The FNR and FPR of the method with this threshold are both to be at a lowest level. For WLTDroid, 59 similar scores are not less than 80 points, 34 similar scores are not less than 90 points, 4 similar scores are between 75 and 80 points, and the lowest similarity score is 75 points. For RepDroid, 59 similar scores are not less than 80, 45 similar scores are not less than 90, 4 similar scores are between 76 and 80, and the lowest similarity score is 76 points.

2) Evaluate FP, FPR, and time consumptions
We use WLTDroid and RepDroid to detect all the 5253 comparison pairs of apps. The detection results are shown in Fig. 3. We use logarithmic coordinates to make them more intuitive. The horizontal axis represents similar score sections. The vertical axis represents the number of comparison pairs belong to the section.

According to statistics, the distribution of similarity score is almost the same. More than half of the comparison pairs score 0. In detail, there are 6 comparison pairs with a value equal to or greater than 75 points for WLTDroid. Through manually check, we find that 3 pairs are actually repackaged apps, and the other 3 pairs belong to FP. The FPR of WLTDroid is 0.057%. There are 7 comparison pairs with a value equal to or greater than 76 points for RepDroid. Through manually check, there are 3 comparison pairs are repackaged apps, and the other 4 pairs belong to FP. The FPR of RepDroid is 0.076%.

In addition, we recorded the time consuming of two methods. The results are shown in Table 2. It shows that the most time consuming part is to generate the birthmark. However, there is 36% shorter for our method.

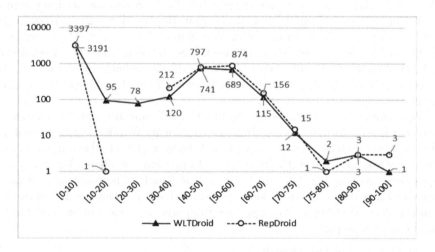

Fig. 3. Distribution of the similarity for WLTDroid and RepDroid

Table 2. Accuracy and Time-consuming for WLTDroid and RepDroid

		WLTDroid	RepDroid
3 * S1	θ	75	76
	FN	0	0
	FNR	0%	0%
2 * S2	FP	3	4
	FPR	0.057%	0.076%
2 * Time consumptions	Birthmark generation	10.02 min	15.74 min
	Similarity calculation	0.029 s	0.034 s

5   Related Work

Most of the studies in earlier years were based on decompiled source code. For example, DroidMOSS [21] handle the Dalvik bytecode to get the features of code by using the fuzzy hashing, then calculate the similarity of the two apps by using the edit distance algorithm [10]. DNADroid [3] generate program dependency graph as the program features based on the source code, then use the subgraph isomorphism algorithm to calculate the similarity of two dependency graphs. Steve Hanna et al. designed a repackaging detection system named Juxtapp, hashing the extracted features of testing apps' code for detection [7]. Jonathan Crussell et al. proposed AnDarwin, which combines the code information with other related information, for example the store information of testing app, for detection [2].

Later, many resource-based methods are proposed. FSquaDRA [20] decompile the testing app to extract all the code and resource files, then encode them with a hashing algorithm for calculating the similarity with Jaccard distance. ViewDroid [19] builds the view graph of the testing app by counting keywords in the app's source code for detection. ResDroid [14] calculates all the resource files contained in the apk file and transforms them into vectors for detection. Research [15] extracts the resource files, such as pictures, videos and so on to detecting repackage applications by using machine learning methods.

Recent years, the layout-based methods for repackaging detection are often used. DroidEagle [16] gets the layout files in the testing apk file, then uses the edit distance to calculate the similarity between individual layouts. SUIDroid [11] also gets the layout files from the testing apk file, then combine multiple layout files to get the app's schema layout for detection. RepDroid [12] proposes a layout group graph (LGG) based repackaging detection method, which built LGG from UI behaviors. Research [13] extracts user interfaces from mobile apps and analyzes the extracted screenshots to detect repackaging apps. RegionDroid [18] proposed an approach based on the app UI regions extracted from app's runtime UI traces.

The method in this paper is different from the above methods. This method dynamically extracts the interface layout and view transfer information of the testing apps, then combines those two types of information to form a whole layout tree (WLT) for similarity detection.

6 Conclusion

We proposed a novel repackaging detection approach, WLTDroid, for Android applications by dynamically extracting the interface layout and transfer information of the testing apps. WLTDroid firstly builds the whole layout tree (WLT) for the apps, then calculate the similarity score. If the score exceeds the threshold which is set according to the experiment, signature verification will start to give the final result. The experimental results show that WLTDroid has strong resistance than the other three existing methods. Meanwhile, the accuracy of WLTDroid is good.

References

1. State of Mobile 2019. https://www.appannie.com/en/go/state-of-mobile-2019/?utm_source=AppStats2019&utm_medium=appdata&utm_campaign=AppAnnie
2. Crussell, J., Gibler, C., Chen, H.: AnDarwin: scalable detection of semantically similar android applications. In: Crampton, J., Jajodia, S., Mayes, K. (eds.) ESORICS 2013. LNCS, vol. 8134, pp. 182–199. Springer, Heidelberg (2013). https://doi.org/10.1007/978-3-642-40203-6_11
3. Crussell, J., Gibler, C., Chen, H.: Attack of the clones: detecting cloned applications on android markets 81(13), 2454–2456 (2012)
4. Gadyatskaya, O., Lezza, A.-L., Zhauniarovich, Y.: Evaluation of resource-based app repackaging detection in android. In: Brumley, B.B., Röning, J. (eds.) NordSec 2016. LNCS, vol. 10014, pp. 135–151. Springer, Cham (2016). https://doi.org/10.1007/978-3-319-47560-8_9
5. Gibler, C., Stevens, R., Crussell, J., Chen, H., Zang, H., Choi, H.: AdRob: examining the landscape and impact of android application plagiarism. In: Proceedings of the International Conference on Mobile Systems, Applications, and Services, pp. 431–444 (2013)
6. Gultnieks, C.: f-droid.org/packages/ (2018)
7. Hanna, S., Huang, L., Wu, E., Li, S., Chen, C., Song, D.: Juxtapp: a scalable system for detecting code reuse among android applications. In: Flegel, U., Markatos, E., Robertson, W. (eds.) DIMVA 2012. LNCS, vol. 7591, pp. 62–81. Springer, Heidelberg (2013). https://doi.org/10.1007/978-3-642-37300-8_4
8. Kim, J.-H., Im, E.-.G.: AndroGuard: similarity analysis for android application binaries. J. Korean Inf. Sci. Soc. (2014)
9. Kornblum, J.: Identifying almost identical files using context triggered piecewise hashing. Digit. Invest. 3, 91–97 (2006)
10. Levenshtein, V.I.: Binary codes capable of correcting deletions, insertions, and reversals. Soviet Phys. Doklady 10, 707–710 (1966)
11. Lyu, F., Lin, Y., Yang, J., Zhou, J.: SUIDroid: an efficient hardening-resilient approach to android app clone detection. In: 2016 IEEE Trustcom/BigDataSE/ISPA, pp. 511–518. IEEE (2016)
12. Ma, J.: RepDroid: an automated tool for android application repackaging detection. In: ICPC (2017)

13. Malisa, L., Kostiainen, K., Och, M., Capkun, S.: Mobile application impersonation detection using dynamic user interface extraction. In: Askoxylakis, I., Ioannidis, S., Katsikas, S., Meadows, C. (eds.) ESORICS 2016. LNCS, vol. 9878, pp. 217–237. Springer, Cham (2016). https://doi.org/10.1007/978-3-319-45744-4_11

14. Shao, Y., Luo, X., Qian, C., Zhu, P., Zhang, L.: Towards a scalable resource-driven approach for detecting repackaged android applications. In: Proceedings of the 30th Annual Computer Security Applications Conference, pp. 56–65. ACM (2014)

15. Sibei, J., Lingyun, Y., Yi, Y., Yao, C., Purui, S., Dengguo, F.: An anti-obfuscation method for detecting similarity among android applications in large scale. J. Comput. Res. Dev. **51**(7), 1446 (2014)

16. Sun, M., Li, M., Lui, J.: DroidEagle: seamless detection of visually similar android apps. In: Proceedings of the 8th ACM Conference on Security & Privacy in Wireless and Mobile Networks, p. 9. ACM (2015)

17. Wang. www.wandoujia.com/apps (2018)

18. Yue, S., Sun, Q., Ma, J., Tao, X., Xu, C., Lu, J.: RegionDroid: a tool for detecting android application repackaging based on runtime UI region features. In: IEEE International Conference on Software Maintenance & Evolution (2018)

19. Zhang, F., Huang, H., Zhu, S., Wu, D., Liu, P.: ViewDroid: towards obfuscation-resilient mobile application repackaging detection (2014)

20. Zhauniarovich, Y., Gadyatskaya, O., Crispo, B., La Spina, F., Moser, E.: FSquaDRA: fast detection of repackaged applications. In: Atluri, V., Pernul, G. (eds.) DBSec 2014. LNCS, vol. 8566, pp. 130–145. Springer, Heidelberg (2014). https://doi.org/10.1007/978-3-662-43936-4_9

21. Zhou, W., Zhou, Y., Jiang, X., Ning, P.: Detecting repackaged smartphone applications in third-party android marketplaces. In: ACM Conference on Data and Application Security and Privacy, pp. 317–326 (2012)

22. Zhou, Y., Jiang, X.: Dissecting android malware: characterization and evolution. In: 2012 IEEE Symposium on Security and Privacy (SP), pp. 95–109. IEEE (2012)

Evolutionary-Based Image Encryption with DNA Coding and Chaotic Systems

Shiyue Qin[1], Zhenhua Tan[2(⊠)], Bin Zhang[2], and Fucai Zhou[2]

[1] School of Computer Science and Engineering, Northeastern University,
Shenyang, China
qinsy556@stumail.neu.edu.cn
[2] Software College, Northeastern University, Shenyang, China
tanzh@mail.neu.edu.cn

Abstract. In order to obtain encrypted image with optimal correlation coefficient, we propose a new evolutionary-based image encryption scheme in this paper, based on DNA coding, chaotic systems, and a genetic algorithm (GA). Firstly, the DNA encoding operation is performed over the original plain image. Then logistic mapping and hyperchaotic systems are used to create the number of encrypted images over the DNA matrices. We fix these encrypted images as the initial population for the GA, and the GA is applied to them to determine the best one based on the fitness function. The correlation coefficient is used to define the fitness function in this paper. Finally, after the evolution, the encrypted image with the lowest correlation coefficient will be obtained as the final result. The novelty of this research is in using the logistic map over the DNA matrix of the original image directly, which will make the initial group of encrypted images more secure and with lower initial fitness for the GA. Therefore, the scheme will achieve high fitness in fewer iterations and retain the efficiency of the encryption. Experimental results confirm that the proposed scheme not only has a good encryption effect, but also has the ability of resisting statistical analysis attacks, brute force attacks, and differential attacks.

Keywords: Image encryption · DNA coding · Genetic algorithm · Logistic mapping · Chen's hyper-chaotic

1 Introduction

Image encryption is one of the most efficient and common methods for protecting the contents of an image. The outsourcing storage of encrypted images can effectively alleviate the problem of privacy disclosure. However, the image has various inherent features, such as a high correlation coefficient among adjacent pixels, conventional encryption techniques apply to the text structure data are not very effective for image encryption [3]. In order to propose effective image encryption methods with the abilities of resisting different attacks, several image encryption schemes have been proposed based on different methods. Such as the

G. Wang et al. (Eds.): WISA 2020, LNCS 12432, pp. 592–604, 2020.
https://doi.org/10.1007/978-3-030-60029-7_53

chaotic mapping systems [2,8], DNA encoding operations [17,19], and even with the help of evolutionary algorithms [1,5].

The chaotic system has various properties, such as the high sensitivity to initial values, pseudorandom, and so on. Based on these properties, the chaotic system can improve the security of the image encryption according to Shannon's information theory [13]. The chaos-based image encryption methods are all composed of two steps of permutation and diffusion [6]. The permutation refers to reallocate the positions of image pixels, and the diffusion refers to change the grey value of each pixel one by one. During the diffusion, the changing of any pixel' s grey value will affect the other pixel value. The chaotic system can be divided into a one-dimensional chaotic system and a hyper-chaotic system (with multiple dimensions) with different advantages. Applying the hyper-chaos into image encryption or combine the one-dimensional chaos system with it can achieve higher security for the image encryption [2,14].

Recently, some of the chaotic based image encryption schemes are combined with the DNA encoding operations [9,15], which improve more confusion and diffusion capabilities during the encryption process. Generally, for a plain image, it will be converted into a DNA matrix firstly, then the DNA matrix will be encrypted with different methods. Wei et al. [17] encoded the original color image into three DNA matrices, and used Chen's hyper-chaotic maps to scramble each matrix, and finally performed DNA sequence addition operation over itself. Zhang et al. [19] encoded the original image and another one key image firstly and scramble, and finally performed DNA sequence addition operation over these two DNA matrices. Mehdi et al. [18] encrypted the plain image based on DNA XOR operation principle and chaotic map function, additionally used RNA rule for further encryption.

Moreover, evolutionary algorithms have also been implemented in some image encryption schemes to further improve the quality of encrypted images [1,5,11]. In evolutionary-based schemes, the initial population of encrypted images are generated under different encryption conditions, and the encrypted image with the best fitness value will be outputted as the final result after the evolution. Enayatifar et al. [5] used a weighted discrete imperialist competitive algorithm (WDICA) combined with the chaotic map for image encryption, where calculating the fitness function from the two opposite-trend objectives, the entropy and the correlation coefficient. Then, they [4] proposed a GA-based encryption scheme with different DNA masks. Manjit et al. [10] used the adaptive differential evolution algorithm for helping to optimize the parameters of the chaotic system.

This paper proposed a GA-based method for image encryption using DNA encoding and chaotic systems. The fitness function is defined based on the correlation coefficient so that the encrypted image with the optimal correlation coefficient will be obtained finally. First of all, the original plain image is converted into DNA matrices. Different logistic map generated based on the original plain image is used to diffuse the DNA matrices into specified numbers of intermediate matrices. Then, different hyper-chaotic maps are used for permutation to create an initial population of GA, where the initial values of each hyper-chaotic map are generated based on each intermediate matrix. Finally, a GA is used to optimize the outputs. The novelty of this research is in using the logistic map over the DNA matrix of the original image directly, which will make the initial group of encrypted images more secure and with lower initial fitness for the GA. Therefore, the GA will achieve high fitness in fewer iterations and retain the efficiency of the encryption.

The rest of the paper is organized as follows: Sect. 2 gives basic concepts of chaotic mapping, hyper-chaotic mapping, and DNA coding. Section 3 concentrates on the proposed method in detail. Section 4 presents experimental results, including statistical analysis, Brute-force attack analysis, and differential attack analysis. Meanwhile, it also concludes with comparisons. Finally, Sect. 5 discusses the concluding of this work.

2   Preliminaries

2.1   Chaotic Systems

Logistic mapping is a one-dimensional chaotic mapping [1]. It generates keys with only one parameter r, as shown in Eq. (1), where X_0 is the primary value and $X_n \in [0, 1]$. Consider the case of $r = 3.999$ in this paper.

$$X_{n+1} = rX_n(1 - X_n)   \tag{1}$$

Additionally, 4D hyper-chaotic Chen system [3] is used in this paper to generate four chaotic sequences, which can be described by the following equation Eq. (2). The system is in the chaotic state when $a = 36, b = 3, c = 28, d = 16, -0.7 \leq k \leq 0.7$. Consider the case of $k = 0.2$ in this paper.

$$\begin{cases} \dot{x} = a(y - x); \\ \dot{y} = -xz + dx + cy - q; \\ \dot{z} = xy - bz; \\ \dot{q} = x + k; \end{cases}   \tag{2}$$

2.2   DNA Coding

According to base-pairing rules of DNA, the nucleic acid bases A (adenine) and T (thymine), G (guanine) and C (cytosine) are complementary pairs. Since 00 and 11, 01, and 10 are also complementary pairs in the binary system, 00, 01, 10, and 11 can be encoded by using four bases A, C, G, and T, respectively. There are eight binary coding rules for DNA coding and decoding [16], as shown in Table 1. Moreover, the addition and subtraction operations based on DNA codes are shown in Table 2. For instance, a pixel value 219 with the binary format $(10110111)_2$ can be converted into eight different DNA codes as follows: Rule 1 (CTGT), Rule 2 (GTCT), Rule 3 (CAGA), Rule 4 (GACA), Rule 5 (AGTG), Rule 6 (TGAG), Rule 7 (ACTC) and Rule 8 (TCAC).

2.3   Genetic Algorithm

A genetic algorithm (GA) is a type of evolutionary algorithm, which aims at searching for final solutions among potential solutions through several rounds of evolution. Basically, GA firstly generates the numbers of solutions as the initial population. Then, GA performs the iteration to produce new solutions. In this paper, the iteration is mainly categorized into two steps: (1) Selection: Chosen parts of the solutions by using the fitness function for creating the offspring. (2) Crossover: Create new solutions based on the current generation by swapping pixels with the crossover probability. In this manner, the optimal solutions in the current generation will be selected into the offspring, and the optimum result will output after the iterations.

Table 1. Coding rules for DNA.

	A	T	C	G
Rule 1	00	11	10	01
Rule 2	00	11	01	10
Rule 3	11	00	10	01
Rule 4	11	00	01	10
Rule 5	10	01	00	11
Rule 6	01	10	00	11
Rule 7	10	01	11	00
Rule 8	01	10	11	00

Table 2. Add and Sub rules.

	A	T	C	G
+A	C	G	A	T
+T	G	C	T	A
+C	A	T	C	G
+G	T	A	G	C
−A	C	A	G	T
−T	G	C	T	A
−C	A	T	C	G
−G	T	G	A	C

3   Proposed Scheme

Step1 (Setup). For an $M \times N$ plain-image, divide it into four equal parts with the block number $B = 0, 1, 2, 3$. Convert each part into a matrix P_B with the size $(M' \times N')$, where $M' = M/2$ and $N' = N/2$. The original plain-image can be represented as $\begin{bmatrix} P_{B=0} & P_{B=1} \\ P_{B=2} & P_{B=3} \end{bmatrix}$.

Step2 (Logistic Maps Generation). For each matrix P_B, generate a specified numbers of logistic maps. Let $P[a, b]$ denotes the value on the a_{th} row and the b_{th} column in a matrix P. The primary value of each logistic map can be calculated by using the first fifteen pixels of each column in P_B by $X_0^K = \frac{1}{2^{12}}\{P_B[0, K] \oplus \ldots \oplus P_B[7, K] + \sum_{j=0}^{14} P_B[j, K]\}$, and the corresponding logistic mapping $X^K = \{X_0^K, X_1^K, X_2^K, \ldots\}$, where $X_{n+1}^K = 3.999 * X_n^K * (1 - X_n^K)$ for $n = 0, 1, 2, \ldots$. In conclusion, for each matrix P_B, N' different logistic map are generated in this step.

Step3 (DNA Encoding and Diffusion). For each matrix P_B, firstly encode it into N' different DNA matrices $P_{K,B}$ with the size $(4 * M', N')$ using Table 1. The DNA encoding rule number for $P_{K,B}$ is determined by $[(B \oplus K) \bmod 8] + 1$. Then, diffuse each DNA matrix $P_{K,B}$ with the corresponding logistic map X^K to be an intermediate matrix $Pb_{K,B}$ as in Eq. (3). Therefore, given a matrix $Pb_{K,B}$, we can recover the primary value and the corresponding logistic map as long as we know the value K of this matrix. Finally, each matrix P_B are generated into N' different intermediate matrixes $Pb_{K,B}$.

$$
\begin{aligned}
&n = 0; \\
&\textbf{For } a = 0; a{<}4 * M'; a{+}{+} \\
&\quad \textbf{For } b = 0; b{<}N'; b{+}{+} \\
&\qquad \textbf{If } b \neq K \text{ and } (a \neq 0 \text{ or } a \neq 1 \text{ or } \ldots \text{ or } a \neq 59) \\
&\qquad\quad Pb_{K,B}[a,b] = P_{K,B}[a,b] \oplus (X_n^K * 3); \\
&\qquad\quad n = n + 1; \\
&\qquad \textbf{Else} \\
&\qquad\quad Pb_{K,B}[a,b] = P_{K,B}[a,b];
\end{aligned}
\tag{3}
$$

Step4 (Permutation with Hyper-chaotic). For each intermediate matrix $Pb_{K,B}$, compute the initial values $CK_{K,B} = \{c_0, c_1, c_2, c_3\}$ for the hyper-chaotic system based on pixel values of $Pb_{K,B}$ firstly, where $c_i = (\sum_{j=0}^{M'-1} \sum_{k=0}^{N'-1} Pb_{K,B}[i * M' + j, k])/(3 * M' * N')$. Then, a Chen's hyper-chaotic system can be constructed as in Sect. 2.1 with $CK_{K,B}$. From the hyper-chaotic system, two pseudo-random sequences $Y_0 = (y_1, y_2, ..., y_{4*M'})$ and $Z_0 = (z_1, z_2, ..., z_{N'})$ can be obtained. Perform ascending to Y_0 and Z_0 , new sequences Y and Z can be obtained, where Y is the new index of Y_0 and Z is the new index of Z_0. In order to hide the initial values from the matrix, add each elements of $Pb_{K,B}$ in a row-by-row way before permutation according to DNA addition operations based on Table 2 as follow.

$$
\begin{aligned}
&temp = 0; \\
&\textbf{For } a = 0; a{<}4 * M'; a{+}{+} \\
&\quad \textbf{For } b = 0; b{<}N'; b{+}{+} \\
&\qquad Pb_{K,B}[a,b] = Pb_{K,B}[a,b] + temp; \\
&\qquad temp = Pb_{K,B}[a,b];
\end{aligned}
\tag{4}
$$

Then, perform the permutation for $Pb_{K,B}$ based on pseudo-random sequences Y and Z as $P'_{K,B}[a,b] \leftarrow Pb_{K,B}[Y[a], Z[b]]$, where $a = 0, 1, ..., 4 * M' - 1$ and $b = 0, 1, ..., N'-1$. Finally, after decoding with the corresponding DNA rules, N' different intermediate matrixes $P'_{K,B}$ will be obtained based on each original matrix P_B where $B = 0, 1, 2, 3$.

Step5 (Generating the Initial Population). Merge intermediate matrices $P'_{K \in [0, N'-1], B \in [0,3]}$ with the same K together into an encrypted image. After merging, the first group of N' encrypted images are generated as the initial population. These encrypted images are the first group of potential solutions.

Each encrypted image I can be represented as Eq. (5). In the initial population, Eq. (5) satisfies $K0 = K1 = K2 = K3$.

$$I = \begin{bmatrix} P'_{K0,B=0}, P'_{K1,B=1} \\ P'_{K2,B=2}, P'_{K3,B=3} \end{bmatrix} \quad (5)$$

Meanwhile, the secret key of an encrypted image I is integrated into a six-tuple key $\{CK_{K0,0}, CK_{K1,1}, CK_{K2,2}, CK_{K3,3}, K0, K1\}$. According to our crossover method, $K0$ will be kept the same as $K2$ and $K1$ will be kept the same as $K3$ in the future evolution.

Step6 (Measuring the Correlation Coefficient of Encrypted Images). The fitness function F is defined based on the correlation coefficient as $F = 1 - |cr_V + cr_H + cr_D|$. cr_V, cr_H, cr_D denotes the vertical direction, the horizontal direction, and the diagonal direction of the correlation coefficient, respectively.

Step7 (Generating the Offspring). Roulette-wheel [7] is the selection method in the GA used in this paper. After the selection, the crossover will be performed based on the crossover probability. Consider crossover probability is 0.70 in this paper. For each pair of encrypted images as a parent, the matrixes with $B = 1, 3$ are swapped between the two parents.

For instance, there are two parents as follow.

$$Parent1 = \begin{bmatrix} P'_{K0=8,0}, P'_{K1=8,1} \\ P'_{K2=8,2}, P'_{K3=8,3} \end{bmatrix} \quad Parent2 = \begin{bmatrix} P'_{K0=5,0}, P'_{K1=5,1} \\ P'_{K2=5,2}, P'_{K3=5,3} \end{bmatrix} \quad (6)$$

The two children will be created as the new generation as follow.

$$Child1 = \begin{bmatrix} P'_{K0=8,0}, P'_{K1=5,1} \\ P'_{K2=5,2}, P'_{K3=8,3} \end{bmatrix} \quad Child2 = \begin{bmatrix} P'_{K0=5,0}, P'_{K1=8,1} \\ P'_{K2=8,2}, P'_{K3=5,3} \end{bmatrix} \quad (7)$$

Furthermore, the secret key of each children image is also should be changed. The secret key of $Parent1$ is $\{CK_{8,0}, CK_{8,1}, CK_{8,2}, CK_{8,3}, 8, 8\}$ and the secret key of $Parent2$ is $\{CK_{5,0}, CK_{5,1}, CK_{5,2}, CK_{5,3}, 5, 5\}$, while the secret key of $Child1$ is changed into $\{CK_{8,0}, CK_{5,1}, CK_{8,2}, CK_{5,3}, 8, 5\}$ and the secret key of $Child2$ is changed into $\{CK_{5,0}, CK_{8,1}, CK_{5,2}, CK_{8,3}, 5, 8\}$.

Step8 (Performing the Convergence). Keep the iteration of Step 6 and Step 7 until the stopping condition is met. The stopping condition can be set as the case of that the best correlation coefficient does not significantly change in two successive iterations or can be set as a fixed number of iterations.

Finally, the solution with the lowest correlation coefficient is chosen as the result cipher image. The corresponding secret key is needed to decrypt it. The process of the proposed method is depicted in Fig. 1.

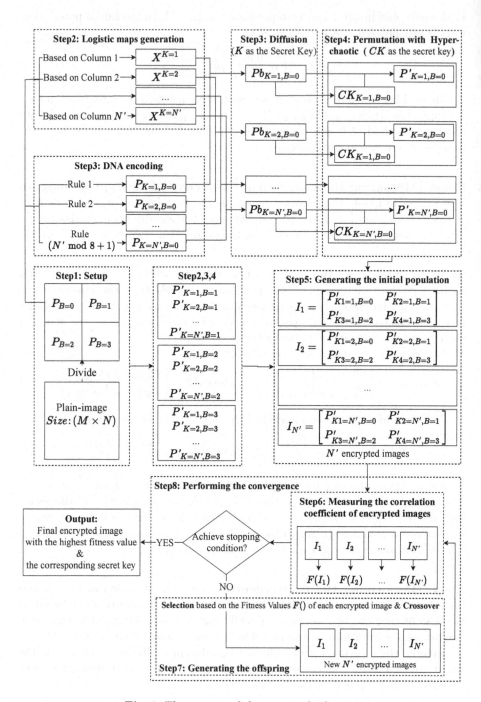

Fig. 1. The process of the proposed scheme.

4   Experimental Results and Performance Analyses

Visual Studio 2017 on a laptop with an Intel Core i7, 1.80 GHz CPU, 8 GB memory with a Windows 10 operating system was utilized to perform the proposed scheme. In this research, standard grey level images "Lena" with the dimensions of 256 * 256 and 512 * 512 are used. The GA crossover probability is set to 0.7.

4.1   Statistical Analysis

Histogram Analysis. The histogram can directly reflect the distribution of all the grey level values throughout the image. The histogram for an ideal encrypted image should be uniform. The histogram of the Lena's image before and after encryption is shown in Fig. 2. The histogram of the cipher image is fair, and totally different from the plain image.

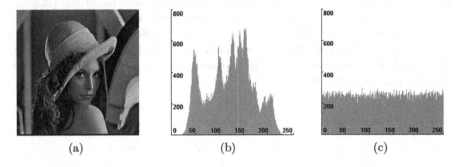

(a)   (b)   (c)

Fig. 2. For (a) Original Lena image, the histogram of (a) before encryption (b) and after encryption (c).

Correlation Analysis. Correlation of adjacent pixels in horizontal, diagonal, and vertical directions of a meaningful image will have a strong regularity. It should be approached to 0 of an ideal encrypted image based on the standard formula. 4000 pairs of adjacent pixels in Lena's image are randomly selected, and the correlation coefficients for each direction are shown in Fig. 3.

In addition, execute 30 times of the proposed scheme on the standard Lena's image with different size with different iterations, and the results are listed in Tables 3 and 4. Corr(Best) denotes the best performance of the correlation coefficient, Corr(Avg) denotes the average performance. Note that the proposed GA-based scheme is not a deterministic process. Hence it cannot ensure that every result is the best case. Corr(Avg) is good enough, that it is close to 0 in each direction of the encrypted image, which means the proposed scheme has a strong statistical attack ability.

Information Entropy Analysis. The entropy of an image can reflect the degree of uncertainty of this image. The ideal entropy calculated by the standard formula should be approached to 8. The entropy of the encrypted image is listed in Table 5. It can be seen that the average entropy is 7.9978 and 7.9993, which confirms the effectiveness of the proposed encryption method.

4.2   Brute-Force Attack Analysis

Key Space Analysis. In the proposed method, the initial values $CK_{K,B}$ of Chen's system, and the column number $K \in [0, 127]$ of each block are used as the secret keys. Consider the precision of Chen's system is 10^{-5}, the keyspace is $10^{5*4*4} + 2^{7*2} \cong 2^{3.32*80+14} = 2^{279}$. Therefore, the secret key space seems to be large enough to avoid brute-force attacks.

Key Sensitivity. To investigate the key sensitivity analysis of the proposed method, perform the encryption with the same secret key by alteration in one arbitrary bit during the **Step 2** (the secret key used in **Step 4** is generated based on the result of **Step 2**). The corresponding encrypted images are depicted in Fig. 4, which clearly shows that the proposed scheme is sensitive against any alteration of the initial key.

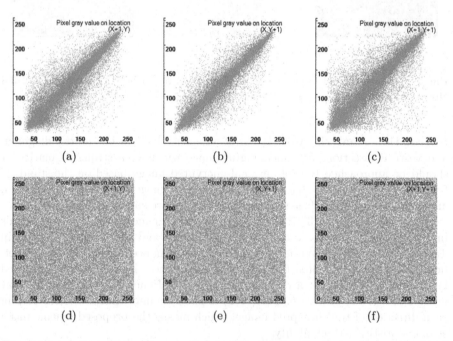

(a)   (b)   (c)

(d)   (e)   (f)

Fig. 3. Correlation of two adjacent pixels in different direction before (top section) and after (bottom section) encryption.

Table 3. Correlation coefficient analysis in different iterations for 256 * 256 Lena.

Iteration		5	30	50	70	100
Corr(Best)	Vertical	0.0003	−0.0005	0.0003	0.0003	0.0003
	Horizontal	−0.0017	0.0007	−0.0007	−0.0007	0.0009
	Diagonal	0.0014	−0.0014	0.0015	0.0015	0.0003
Fitness		0.0034	0.0026	0.0025	0.0025	0.0015
Entropy		7.9969	7.9970	7.9978	7.9978	7.9972
Corr(Avg)	Vertical	0.0012	0.0022	0.0012	0.0019	0.0006
	Horizontal	0.0020	0.0008	0.0015	0.0009	0.0014
	Diagonal	0.0017	0.0010	0.0013	0.0011	0.0011
Fitness		0.0049	0.0040	0.0040	0.0039	0.0031
Entropy		7.9972	7.9969	7.9971	7.9973	7.9974

Table 4. Correlation coefficient analysis in different iterations for 512 * 512 Lena.

Iteration		5	30	50	70	100
Corr(Best)	Vertical	−0.0006	0.0001	−5.17E−05	−0.0005	−0.0002
	Horizontal	0.0001	0.0005	0.0006	8.82E−07	0.0002
	Diagonal	0.0023	0.0006	0.0004	−0.0003	0.0004
Fitness		0.0030	0.0012	0.0010	0.0008	0.0008
Entropy		7.9989	7.9990	7.9993	7.9990	7.9993
Corr(Avg)	Vertical	0.0008	0.0003	0.0005	0.0008	0.0007
	Horizontal	0.0009	0.0011	0.0013	0.0004	0.0006
	Diagonal	0.0025	0.0011	0.0002	0.0008	0.0006
Fitness		0.0042	0.0025	0.0020	0.0020	0.0019
Entropy		7.9990	7.9990	7.9989	7.9993	7.9993

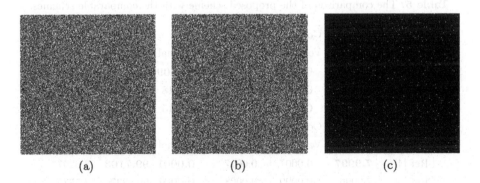

(a)   (b)   (c)

Fig. 4. (a) cipher-image with the original secret key, (b) cipher-image with the same key as (a) with 1 bit changed and (c) differences between (a) and (b).

4.3   Differential Attack Analysis

Differential attack analysis attempts to evaluate if a minor alteration of the plain image can make a considerable difference in the related encrypted image or not. The evaluation index of this kind of attack is UPCR(Number of pixels change rate) and UACI(Unified average changing intensity). Based on the standard calculation formulas for NPCR and UACI, the ideal value for NPCR is 100%, and for UACI is 33% when against differential attack.

Table 5 shows the experimental results. For a 256 * 256 Lena's image, the UPCR value of the proposed scheme is 99.6353%, and the value of UACI is 33.2881%, and 99.6302% and 33.4128% for a 512 * 512 Lena's image. The result is very close to the ideal value of resisting differential attack. Therefore, the proposed method is sensitive to the minor change of the plain image.

4.4   Comparison

Compare the proposed scheme with some schemes based on an evolutionary algorithm. The comparison is listed in Table 5. The best result in each column is shown with a bold font. For the 256 * 256 Lena's image, the proposed scheme outperforms the other algorithms in most of the indices, whereas that in Ref [1] yields a better outcome when considering the diagonal correlation and UCAI. Moreover, from the Tables 3 and 4, the correlation coefficient is achieved higher even in early 5 iterations. Hence, the proposed scheme can achieve a better correlation only in a handful of iterations.

In addition, for the 512 * 512 Lena's image, the proposed scheme also achieves the best correlation coefficient in the vertical and horizontal direction. Meanwhile, the UCAI is the best in comparing. Note that the GA algorithm in the proposed scheme performs the selection only on the respect of the correlation coefficient, it cannot take care of the entropy simultaneously. Therefore, it is possible that, for example during the experiment for 512 * 512 Lena's image,

Table 5. The comparison of the proposed scheme with the comparable schemes.

256-size	Entropy	Correlation coefficient			NPCR	UCAI
		Vertical	Horizontal	Diagonal		
Ref [1]	**7.9978**	0.0093	−0.0054	**−0.0009**	97.1394	**33.1084**
Ref [12]	**7.9978**	0.0018	0.0011	0.0018	99.6006	33.3479
Proposed	**7.9978**	**0.0003**	**−0.0007**	0.0015	**99.6353**	33.2881
512-size	Entropy	Correlation coefficient			NPCR	UCAI
		Vertical	Horizontal	Diagonal		
Ref [4]	**7.9997**	0.0007	0.0017	**0.0001**	**99.7103**	33.6297
Ref [5]	7.9996	−0.0009	0.0008	**0.0001**	99.6837	33.5735
Proposed	7.9993	**0.0002**	**−0.0002**	0.0004	99.6302	**33.4128**

there exist some intermediate results that have both the better entropy and correlation coefficient than Ref [5] and Ref [4], but the fitness of these results is not the best. Consequently, these intermediate results cannot be outputted from the GA as the final results.

5   Conclusions

This paper proposed an evolutionary-based image encryption scheme based on DNA coding and chaotic systems, which aims at generating an encrypted image with the best correlation coefficient. In the proposed scheme, the plain image is converted into a DNA matrix firstly. Then, several encrypted images are generated as the initial populations by using a logistic map and a hyper-chaotic scheme. A genetic algorithm is used for evolution to decide the best solution with the lower correlation coefficient. The numerical experiments demonstrate that the proposed method can effectively reduce the correlation of the encrypted image in all of the three directions. Meanwhile, the indices of entropy, UCAI, and NPCR are also close to the ideal value. The comparisons with the existing evolutionary-based image encryption methods support the superior security and effectiveness of our proposed method. In future research work, it is needed to study evaluating the fitness function based on the two or more objectives to optimize the performance of the encryption.

Acknowledgments. This research was funded by the National Key Research and Development Program of China under Grant No. 2019YFB1405803; the National Natural Science Foundation of China under Grants No. 61772125, No. 61872069, No. 61772127, and No. 61402097; and CERNET Innovation Project under Grant No. NGII20190609.

References

1. Abdullah, A.H., Enayatifar, R., Lee, M.: A hybrid genetic algorithm and chaotic function model for image encryption. AEU Int. J. Electron. Commun. **66**(10), 806–816 (2012)
2. Alghafis, A., Firdousi, F., Khan, M., Batool, S.I., Amin, M.: An efficient image encryption scheme based on chaotic and deoxyribonucleic acid sequencing. Math. Comput. Simul. **177**, 441–466 (2020)
3. Chen, J.X., Zhu, Z.L., Fu, C., Yu, H., Zhang, L.B.: A fast chaos-based image encryption scheme with a dynamic state variables selection mechanism. Commun. Nonlinear Sci. Numer. Simul. **20**(3), 846–860 (2015)
4. Enayatifar, R., Abdullah, A.H., Isnin, I.F.: Chaos-based image encryption using a hybrid genetic algorithm and a DNA sequence. Opt. Lasers Eng. **56**, 83–93 (2014)
5. Enayatifar, R., Abdullah, A.H., Lee, M.: A weighted discrete imperialist competitive algorithm (WDICA) combined with chaotic map for image encryption. Opt. Lasers Eng. **51**(9), 1066–1077 (2013)
6. Fouda, A., Effa, J., Sabat, S., Ali, M.: A fast chaotic block cipher for image encryption. Commun. Nonlinear Sci. Numer. Simul. **19**, 578–588 (2014). https://doi.org/10.1016/j.cnsns.2013.07.016

7. Gupta, N., Shekhar, R., Kalra, P.K.: Congestion management based roulette wheel simulation for optimal capacity selection: probabilistic transmission expansion planning. Int. J. Electr. Power Energy Syst. **43**(1), 1259–1266 (2012)
8. Han, C.: An image encryption algorithm based on modified logistic chaotic map. Optik **181**, 779–785 (2018). https://doi.org/10.1016/j.ijleo.2018.12.178
9. Hu, T., Liu, Y., Gong, L.-H., Ouyang, C.-J.: An image encryption scheme combining chaos with cycle operation for DNA sequences. Nonlinear Dyn. **87**(1), 51–66 (2016). https://doi.org/10.1007/s11071-016-3024-6
10. Kaur, M., Kumar, V.: Adaptive differential evolution-based lorenz chaotic system for image encryption. Arab. J. Sci. Eng. **43**(12), 8127–8144 (2018). https://doi.org/10.1007/s13369-018-3355-3
11. Kaur, M., Chahar, V.: Color image encryption technique using differential evolution in nonsubsampled contourlet transform domain. IET Image Proc. **12**, 1273–1283 (2018)
12. Mahmud, M., Rahman, A.U., Lee, M., Choi, J.Y.: Evolutionary-based image encryption using RNA codons truth table. Opt. Laser Technol. **121**, 105818 (2020)
13. Shannon, C.: Communication theory of secrecy systems. **28**, 656–715 (1948). https://doi.org/10.1002/j.1538-7305.1949.tb00928.x
14. Wang, X., Zhang, H.: A novel image encryption algorithm based on genetic recombination and hyper-chaotic systems. Nonlinear Dyn. **83**(1), 333–346 (2015). https://doi.org/10.1007/s11071-015-2330-8
15. Wang, X., Zhang, Y.Q., Bao, X.M.: A novel chaotic image encryption scheme using DNA sequence operations. Opt. Lasers Eng. **73**, 53–61 (2015). https://doi.org/10.1016/j.optlaseng.2015.03.022
16. Watson, J., Crick, F.: A structure for deoxyribose nucleic acid. Nature **171**, 737–738 (1953). https://doi.org/10.1007/BF02834980
17. Wei, X., Guo, L., Zhang, Q., Zhang, J., Lian, S.: A novel color image encryption algorithm based on DNA sequence operation and hyper-chaotic system. J. Syst. Softw. **85**, 290–299 (2012). https://doi.org/10.1016/j.jss.2011.08.017
18. Yadollahi, M., Enayatifar, R., Nematzadeh, H., Lee, M., Choi, J.Y.: A novel image security technique based on nucleic acid concepts. J. Inf. Secur. Appl. **53**, 102505 (2020). https://doi.org/10.1016/j.jisa.2020.102505
19. Zhang, Q., Guo, L., Wei, X.: A novel image fusion encryption algorithm based on dna sequence operation and hyper-chaotic system. Optik Int. J. Light Electron Opt. **124**(18), 3596–3600 (2013)

Blockchain

Cross-Lingual Public Opinion Tracing Based on Blockchain Technology

Ye Liang[(⊠)] and Ying Qin

Artificial Intelligence and Human Languages Lab, School of Information Science and Technology, Beijing Foreign Studies University, Beijing, China
liangye@bfsu.edu.cn

Abstract. One of the important issues of public opinion governance in the internet age is to study the characteristics and laws of the generation, development and spread of public opinion on complex social networks and to set up an effective tracing model based on blockchain technology. As a decentralized shared database, blockchain technology is tamper-resistant and traceable. Because of this, it is an important tool for containing the spread of fabricated public opinion, maintaining national security and stability, and protecting corporate image and personal interests. Given the current cross-lingual tracing technology and the fact that blockchain is decentralized and immutable, this project aims for establishing a blockchain-based cross-lingual public opinion tracing system targeting at the characteristics and laws of the spread of public opinion in social network. This study focuses on four aspects: the architecture that is based on complex chain structure; the consensus-based safety and liveness of the system; the management of the system based on smart contracts; standards and rules of this blockchain-based tracing technology. Through the study, a commonly used multilingual tracing system is expected to be built by using the sophisticated blockchain technology.

Keywords: Cross-lingual · Public opinion · Tracing · Blockchain

1 Introduction

The rise of China in multiple fields raised the concern from the existing major powers. For Western conservatives, taking advantage of the anonymity on the internet and secretly creating unfavorable international public opinion has become a new choice to curb China's development [1]. With the deepening of internationalization, the cross-lingual frequent and real-time flow of information has become the norm on the internet. Especially on social networks, multilingual netizens are often the key force in the cross-lingual information dissemination. Under such background, when public opinion arises in one language, it will immediately be spread to social networks of other languages. Since the information dissemination is difficult to supervise, the authenticity of information is difficult to distinguish, and the number of multilingual netizens is relatively limited, in cross-lingual transmission, information is extremely susceptible to interference from individual consciousness of netizen at key communication nodes. In this way, key communication nodes may guide public opinion tendentiously. Because

© Springer Nature Switzerland AG 2020
G. Wang et al. (Eds.): WISA 2020, LNCS 12432, pp. 607–617, 2020.
https://doi.org/10.1007/978-3-030-60029-7_54

of this, studying the characteristics and laws of the generation, development and spread of public opinion on complex social networks and setting up an effective tracing model based on blockchain technology has become an important issue of public opinion governance in the internet age. It is an important tool to contain the spread of fabricated public opinion, to maintain national security and stability, and to protect corporate image and personal interests [2, 3].

Blockchain is a decentralized shared database that involves fields like mathematics, cryptography, and the internet. Putting public opinion data on the chain makes it tamper-resistant, fully recorded, transparent and collectively maintained [4–6]. Because of this, exploring cross-lingual public opinion tracing system in social network to analyze the dissemination route of information among netizens using different languages is one of the most promising areas where blockchain and natural language processing (NLP) can be combined [7–9].

2 Related Work

The so-called public opinion refers to the social attitudes, opinions, and emotions the public holds to a social event in certain social space. In today's society where the internet is highly developed, the anonymity, freedom of expression, and the rapid spread and large scale of information on the internet gradually makes voicing opinions and ideas through the internet a way for the mass to express themselves. Since the internet has become the main social space and form of public opinion dissemination and expression, in many cases, public opinion refers specifically to online public opinion that holds the internet as the communication medium, and various events as its center. Online public opinion is a set containing the views, attitudes, emotions, expression, and the opinion dissemination and interaction of netizens, and the subsequent influences. Public opinion spread through the internet influence all aspects of the society. In recent years, the more frequent occurrence of public events caused by the spread of public opinion or rumors impacted the country's economic development and social stability. More and more public opinions originate and thrive on the internet. Eventually they attract social attention and cause sensation. Therefore, tracing the source of online public opinion has become a new problem to be solved.

2.1 Overseas Development Trends

Linguistics focuses on communication, while computer science focuses on algorithms and complexity. The combination of the two will solve many key problems in natural language processing. Therefore, it has become the main investment direction of the "big five", namely Facebook, Apple, Microsoft, Google, and Amazon, which are also referred to as "FAMGA" [10–12]. The interdisciplinary research of blockchain and natural language processing technology involves deep integration of informatics, linguistics and management. Despite the rapid development in blockchain technology, the large-scale applications of natural language processing are still challenging, it is necessary to build new application modes in computer technologies such as distributed data storage, point-to-point transmission, consensus mechanisms, and data encryption algorithms [13, 14]. At present, the performance and practicality of blockchain technology cannot support the application building of large-scale natural language processing. Scalability, privacy, and interoperability are still the main fields FAMGA companies are working on [15, 16]. Research and exploration on both fields will certainly promote the developments of linguistics and management.

In the related research of online public opinion management, public opinion is divided into positive, neutral and negative [17, 18]. Among them, the unrestricted spread of negative public opinion on the Internet will be harmful to the organizations, societies or countries involved [19, 20]. Researchers have had some findings on tracing the source of negative public opinion caused by the spread of fabricated information in one language [21]. As for cross-lingual public opinion transmission and tracing, the existing research findings are quite limited and little research applied blockchain technology to address to the problem [22].

2.2 Domestic Development Trends

In China, blockchain technology and natural language processing have risen to the national strategic level in science and technology and was deemed as key advanced technologies together with quantum communications, artificial intelligence, virtual reality, big data cognitive analysis, unmanned transportation. China articulated that it is necessary to strengthen innovation, experimentation and application of related technologies to be the leader of the new generation of information technology for a new round of technological innovation and industrial transformation in the world. At the same time, the speeding up of laying down related industry standards, national standards and international standards will provide key technical standards to blockchain and natural language processing and will also promote the ecological development of related industries.

At the same time, the combination of blockchain and natural language processing technology can be applied to public opinion analysis and tracing. It also has broad application prospects in epidemiological tracking, product anti-counterfeiting tracing, and learning experience certification. Blockchain and natural language processing technology can also play an important role in this covid-19 outbreak. We can apply them in 5 directions: (1) Early warning of the outbreak. Although SARS has become a thing of the past and the novel coronavirus has also been effectively controlled, there is still possibility that another epidemic might appear a few years later. Comprehensively look into the past epidemic data can be effective in the early warning of the next epidemic. One method is to analyze the dissemination of information using various information on the Internet. By identifying the characteristic patterns of public opinion during the outbreak from various information dissemination networks, early warning can be made. (2) Epidemic data management. The high infection rate of the epidemic makes it important to reduce the spread of the virus. Blockchain and natural language processing technology can be used to manage the epidemic-related information reported by the disease control centers, and to establish a visual map of population flow. In this way, the critical path of flow can be found to advice the government on epidemic management. (3) Medical case sharing. The spreading nature of the novel coronavirus was first misunderstood by many people. People later discovered the important feature that the virus infects people during incubation period. The sharing of medical cases is therefore very important. Only when experts have seen many cases can we quickly diagnose the unknown new diseases and speed up the treatment. (4) Supply management. To avoid fraud in supply and make proper and full use of supply, it is necessary to monitor and manage the data and process of supply for correctness. (5) Public opinion control. Blockchain and natural language processing technology can be used to collect relevant content on internet during the outbreak, extract the top-ranking issues and track them. These would be beneficial to refuting rumor and guiding public sentiment.

2.3   Problems in Existing Public Opinion Tracing Research

At present, all cross-lingual tracing researches using blockchain technology store tracing information in the blockchain and utilize the decentralized nature of the blockchain to achieve a decentralized tracing system. In these tracing systems, the security of tracing information is guaranteed because the blockchain that stores the information is tamper-resistant [23, 24]. However, the security of the blockchain is not achieved without cost. The security of the blockchain is usually related to the number of nodes in the blockchain network, and factors like processing power. The number of nodes or the processing power in the blockchain used by common companies is hardly comparable to the widely used block chains like Bitcoin or Ethereum. This means that it can be costly for these technology companies to ensure the security of the blockchain they build.

In addition, the existing domestic public opinion research analyses mainly focus on English and Chinese. Few studies focus on non-English languages, and the distribution of research in different languages is extremely uneven. Therefore, considering the cross-lingual characteristics of the spread of public opinion, there is critical need to study the basic theories and key technology of cross-lingual public opinion analysis. In this way, non-English languages will shore up their weak spots in public opinion analysis through referring to languages like English and Chinese that are abundant in corpus resources [25].

3   The Architecture of Cross-Lingual Public Opinion Tracing

Cross-lingual public opinion tracing architecture based on blockchain technology contains layers of data source, data, network, consensus, contract and application from bottom to top (Fig. 1).

The data source refers to opinions and information from various social media in forms of text, image, and audio. Blockchain can track and manage the information on the blockchain by integrating data and information from different sources. The Internet of Things is used when collecting the source data.

The data layer constitutes data blocks, chain structure, timestamp, data encryption such as hash algorithm, Merkle tree and asymmetric encryption and other technologies. This layer is to design block's data structure, embed the timestamp in the block header, encrypt by using the hash algorithm. The block content is also encrypted through

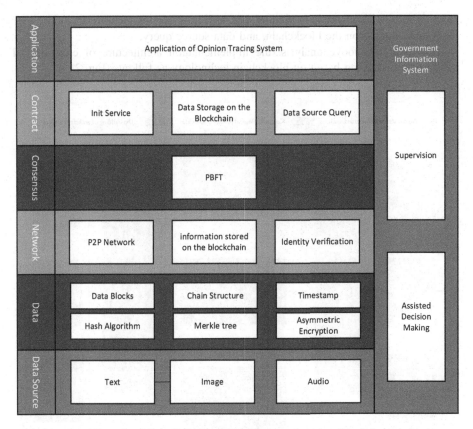

Fig. 1. Cross-lingual public opinion tracing architecture based on blockchain technology.

asymmetric encryption for the convenience of verifying contents on different nodes. Thus, a chain structure can be formed once all blocks are connected.

The network layer defines the rule for adding and accessing relevant nodes and circulating information stored on the blockchain on the internet. It also contains the identity verification.

The consensus layer is to design the verification algorithm used in every node after the node receives the information so that consistency of information stored on different nodes can be achieved. The Byzantine Fault Tolerance algorithm is mainly used to improve overall safety and liveness of the system.

The contract layer includes script codes or smart contracts that are self-executing once they are triggered. Smart contracts can diversify the function of blockchain and support more applications in the upper layer. The chaincode is deployed for the implementation of smart contracts. Functions such as recording, self-verifying and tracing information can be called through the chain code.

The application layer is the layer that the blockchain-based cross-lingual opinion tracing system is deployed in the network. It implements functions such as modularization, automatic data binding, and user interface interactions, including user login, data storage on the blockchain, and data source query.

Based on the above analysis, the implementation architecture of cross-lingual public opinion tracing based on blockchain technology as follows (Fig. 2).

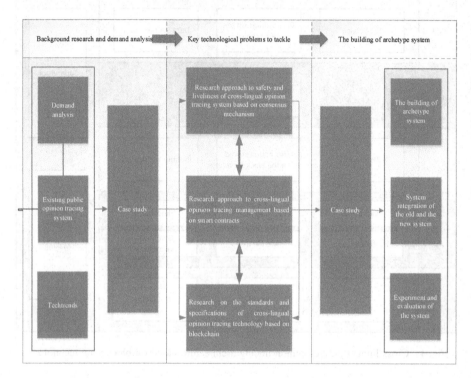

Fig. 2. Cross-lingual public opinion tracing implementation architecture.

4   The Critical Problems of Cross-Lingual Public Opinion Tracing

In order to establish a blockchain-based cross-lingual public opinion tracing system targeting at the characteristics and laws of the spread of public opinion in social network. This study focuses on four aspects: the architecture that is based on complex chain structure; the consensus-based safety and liveness of the system; the management of the system based on smart contracts; standards and rules of this blockchain-based tracing technology. Through the study, a commonly used multilingual tracing system is expected to be built by using the sophisticated blockchain technology.

4.1   Architecture Based on Complex Chain Structure

Most of the current tracing systems adopt centralized methods to track the information. That means the interaction between new and old systems must be properly handled if we want to decentralize the system with distributed ledger technology as used in blockchain. So how to upgrade the old system with blockchain technology but without changing the original tracing system is a major challenge facing the R&D.

This section studies the overall architecture of the cross-lingual opinion tracing system based on the complex chain structure. First, by studying the overall framework of the existing opinion tracing system, this section sorts out its loopholes and limitations, and develops solutions based on blockchain technology. The overall architecture includes public opinion media analysis, public opinion data collection and preprocessing layer, consensus algorithm layer, network communication layer, application decision layer and application component layer. Since the public blockchain under the traditional architecture cannot be edited or modified, and that private blockchain is independent of each other and individual problem cannot be linked up, this project takes consortium blockchain as the major architecture and combine it with the public blockchain to form the complex blockchain structure in order to enable various public opinion media and regulatory institutions to run an individual node on the consortium blockchain, ensuring controllability while improving decentralization of the system. In addition, considering the compatibility between the existing opinion tracing system and the new one, this section will discuss how to upgrade the chain, optimize and merge the system without changing the existing tracing logic, and, in the end, to construct architecture schemes for public opinion tracing system based on blockchain technology.

4.2 Consensus-Based Safety and Liveness of the System

To ensure that data cannot be corrupted, the liveness of information on the blockchain is usually lost during multiple links from editing, publishing to becoming a sensation. Thus, it is necessary to maintain the data's liveness while ensuring the non-repudiation. The safety of system is another key area in this study because opinions may involve information about national security and privacy.

Blockchain is a decentralized system with distributed ledger technology, which means all nodes within the system have to be consistent. This involves duplication and update of status through consensus mechanism. Consensus mechanism realized the consistency of all nodes within the system and ensured the effectiveness and orderliness of data in the same time. This section will be focusing on the consensus-based safety and liveness of the system and will analyze on the 3 main consensus algorithms namely Proof-of-Work algorithm, Delegated Proof of Stake algorithm and Proof of Authority algorithm to construct a consensus algorithm targeted at opinion tracing scenarios' application. Moreover, taking security requirement and the possible complex situations in different environments between media into consideration, this research will be researching on system fault tolerance under consensus mechanism and will be focusing on making sure the safety and liveliness of nodes without having more than $(n-1)/3$ failure nodes. Ultimately, both the safety of the cross-lingual opinion tracing system and the liveliness of nodes within the system have been guaranteed.

The entire link of public opinion information ranges from editing, publishing to dissemination and involves many different media. Moreover, the addition, deletion and modification of information content means that the information will be continuously updated. What comes with it is the increase of the number of tasks carried by the entire system. Data exchange always occurs within systems, and different system nodes might be at different states. For the application of blockchain with distributed ledger technology, the consistency between different nodes is the most basic and critical problem to be solved. Because it is related to the scalability and fault tolerance of the system. The existing consensus mechanism includes Proof-of-work algorithm (PoW), Proof of Stake algorithm (PoS) and Delegated Proof of Stake algorithm (DPoS). Among them, DPoS uses a certain number of representatives to authenticate the proxy nodes, which can greatly reduce the number of nodes participating in the verification. DPoS is the core of this project. This project realizes the management of public opinion information based on the alliance chain, and builds a corresponding task-based consensus mechanism, which can not only effectively enable communication between nodes, but also improve the interactive efficiency of system. In addition, considering factors like the change of media, the overall system requires high fault-tolerance. The study uses the Byzantine Fault-Tolerance algorithm (PBFT) to ensure the safety and liveliness of nodes without having more than $(n-1)/3$ failure nodes. Finally, on the premise of ensuring the stability and high availability of system, this project achieves efficient management of opinion tracing system.

4.3   Management Based on Smart Contracts

Traditionally, the self-verification is conducted by media that release the information, so misinformation often occurs. Under such a background, it is essential to develop a safety automation system that can trace and self-verify the information.

Smart contract are digitally specified commitments, including the protocol on which contract participants can execute the commitments. For blockchain technology, a smart contract is code on the chain that could run independently to realize automatic initiation and automatic business handling of information interaction among blockchain nodes. Blockchain can encapsulate the complex behavior of distributed nodes through the programmability of smart contracts, and smart contracts can be implemented in a trustless and executable environment thanks to the decentralization of blockchain. This section studies the approach to the cross-lingual opinion tracing management based on smart contracts. The approach is to write corresponding smart contract algorithm program according to the tracing requirements and process, so as to add, delete, modify, and check the public opinion information from different media. In addition, considering the complexity of the information and the huge amount of data, a cross-lingual public opinion self-check program should be developed to realize the automatic content review of the public opinion on the chain and report the problematic opinion. At the same time, with the location access on intelligent terminal, it could authenticate the operator and therefore restrict the information review and modification by various personnel. When the operation goes out of the scope, the system would automatically alarm and record the behavior. Finally, based on smart contracts, programs and algorithms for cross-lingual public opinion tracing are formed, which enables automatic processing of public opinion blockchain, self-checking, tracing, and other operations.

Most of the traditional public opinion inspection, information tracing and other businesses are manually operated. Throughout the process, time stamps and electronic signatures are added to the data to achieve management and traceability. However, many manual operations make the system extremely vulnerable to threats, such as information changes caused by user negligence or internal tampering, while the system has no record of such a situation. Blockchain-based smart contracts is the solution to this problem. Smart contract is a computerized transaction protocol that is intermediary-free, self-verifying, and self-executing. First, the triggering scenarios and response rules of the terms are preset in the smart contract, negotiated and signed by the node parties, and then submitted along with the transactions. Then through the P2P network, the contract is transmitted, verified, and stored in a specific block on the chain. The user, after obtaining the returned contract address and contract interface, could invoke the contract by initiating a transaction. Compared with the traditional information system, smart contracts make the transfer system more intelligent, automated, and efficient, while saving human resources. In addition, blockchain data of the system based on smart contracts is tamper-resistant and traceable even for the anonymous information, which can ensure the reliability of the opinion recorded during the circulation. With the information recorded on the blockchain, it is possible to achieve reliable and fast information traceability. What's more, corresponding smart contract programs could solve problems regardless of the media, information attributes or specific problems, making it more convenient for data browsing, querying, and management.

4.4   Standards and Rules of This Blockchain-Based Tracing Technology

The government needs to formulate requirements on disclosing information, provide a whole-chain supervision and examine the information. Yet the blockchain is about decentralization so a whole-chain supervision is impossible. This study, therefore, seeks to find a solution to the problem so that the reliability and privacy of data can be secured while the government can have a whole-chain supervision.

Based on the needs of opinion tracing of relevant stakeholders, namely, public opinion media, government regulatory agencies, and users that releases public opinion, this research builds the front-end and back-end system of opinion tracing through the establishment of blockchain application development and quality assurance specifications, the lower layer platform's technical architecture and quality assurance specifications, and platform deployment and operations specifications of information. Barcode, QR code or RFID are used in the log in and access of the front-end construction based on open source architecture of smart contracts and the back-end construction consists of the smart contract and basic blockchain services that includes node management, member services, sequencing and account maintenance of the opinion tracing system. The research also suggests working with relevant government departments and local governments to promote the construction of local standards and industry standards based on the blockchain-based opinion tracing system.

Public opinion information is generated on the Internet and is spread on the Internet. The whole process involves many different languages and media. The traditional opinion tracing system is limited by language and media so that it cannot effectively manage the entire opinion network in a unified manner or further achieve a complete tracing of public opinions. By summarizing different traceability systems, the project forms technical standards and specifications of cross-lingual opinion tracing system by combining the building process of complete archetype system. Presumably, the project will provide technical foundation and new ideas for related works in the future.

5 Conclusion

The project takes opinion tracing as the main social setting and conducts relevant research on key technology based on the decentralized and immutable characteristics of blockchain. In cross-lingual transmission, information is extremely susceptible to interference from key communication nodes. In this way, key communication nodes may guide public opinions tendentiously. To solve this problem, this project studies existing opinion tracing systems, generalizes a universal cross-lingual opinion tracing system, and forms the corresponding standards and specifications. Compound chain structure is used to build practical public opinion system for different media and objects. In addition, to deal with the complicated and overloaded information that the system may encounter, the project applies public opinion self-verifying algorithms based on smart contracts, and researches on the safety and liveliness of the system based on the consensus mechanism. Finally, the project constructs a cross-lingual opinion tracing system by tackling every key technological problems and solves them in practical blockchain application scenarios. The project also provides new research ideas in opinion tracing and blockchain.

Acknowledgement. This work was supported in part by the Social Science Foundation of China (No. 15CTQ028), and the First-class Disciplines Construction Foundation of Beijing Foreign Studies University (No. YY19SSK02, No. 2020SYLZDXM040).

References

1. Liang, Y., Xu, L., Huang, T: Sentiment tendency analysis of NPC & CPPCC event in German news. In: 2019 16th International Conference on Web Information Systems and Applications, pp. 298–308 (2019). https://doi.org/10.1145/2976749.2978341
2. Gervais, A., Karame, G., Wüst, K., et al.: On the security and performance of proof of work blockchains. In: ACM SIGSAC Conference on Computer and Communications Security, vol. 2016, pp. 3–16 (2016). https://doi.org/10.1145/2976749.2978341
3. Tosh, D., Shetty, S., Foytik, P., et al.: CloudPoS: a proof-of-stake consensus design for blockchain integrated cloud. In: 2018 IEEE 11th International Conference on Cloud Computing (CLOUD) (2018). https://doi.org/10.13140/rg.2.2.20169.44640
4. Sousa, J., Bessani, A.: Separating the WHEAT from the chaff: an empirical design for geo-replicated state machines. In: 34th IEEE Symposium on Reliable Distributed Systems (SRDS), pp. 146–155 (2015). https://doi.org/10.1109/srds.2015.40
5. Wood, G.: Ethereum: a secure decentralised generalised transaction ledger. Ethereum Project Yellow Paper (2014). https://doi.org/10.1017/CBO9781107415324.004
6. Prusty, N.: Building Blockchain Projects. Packt Publishing Ltd., Birmingham (2017)

7. Poon, J., Dryja, T.: The bitcoin lightning network: scalable off-chain instant payments (2016). https://lightning.network/lightning-network-paper.pdf

8. Zhu, Y.: SEBDB: semantics empowered blockchain database. In: IEEE 35th International Conference on Data Engineering (ICDE). IEEE (2019). https://doi.org/10.1109/icde.2019.00198

9. Li, X., Jiang, P., Chen, T., et al.: A survey on the security of blockchain systems. Future Gener. Comput. Syst. (2019). https://doi.org/10.1016/j.future.2017.08.020

10. Microsoft: The Coco framework technical overview. White paper (2017)

11. Ben-Sasson, E., Chiesa, A., Genkin, D., et al.: SNARKs for C: verifying program executions succinctly and in zero knowledge. In: Canetti, R., Garay, J.A. (eds.) CRYPTO 2013. LNCS, vol. 8043, pp. 90–108. Springer, Heidelberg (2013). https://doi.org/10.1007/978-3-642-40084-1_6

12. Ben-Sasson, E., Chiesa, A., Tromer, E., et al.: Succinct non-interactive zero knowledge for a von neumann architecture. In: 23rd USENIX Security Symposium, pp.781–796 (2014)

13. Nguyen, B., Cachin, C., Yellick, J., et al: Multichannel consensus (2016). https://docs.google.com/document/d/1eRNxxQ0P8yp4Wh__Vi6ddaN_vhN2RQHP-IruHNUwyhc/edit

14. Carlyle, J.: CORDA performance: to infinity & beyond (2018). https://www.r3.com/wp-content/uploads/2018/04/Corda-Performance-ENG.pdf

15. Croman, K., Decker, C., Eyal, I., et al.: On scaling decentralized blockchains. In: Clark, J., Meiklejohn, S., Ryan, P.Y.A., Wallach, D., Brenner, M., Rohloff, K. (eds.) FC 2016. LNCS, vol. 9604, pp. 106–125. Springer, Heidelberg (2016). https://doi.org/10.1007/978-3-662-53357-4_8

16. Angelis, S., Aniello, L., Baldoni, R., et al: PBFT vs proof-of-authority: applying the CAP theorem to permissioned blockchain. In: 2nd Italian Conference on Cyber Security, pp. 1–11 (2018)

17. Su, Y., Zhang, C., Li, J.: Cross-lingual entity query from large-scale knowledge graphs. In: Cai, R., Chen, K., Hong, L., Yang, X., Zhang, R., Zou, L. (eds.) APWeb 2015. LNCS, vol. 9461, pp. 139–150. Springer, Cham (2015). https://doi.org/10.1007/978-3-319-28121-6_13

18. Jin, H., Li, C., Zhang, J., et al.: XLORE2: large-scale cross-lingual knowledge graph construction and application. Data Intell. 1(1), 77–98 (2019). https://doi.org/10.1162/dint_a_00003

19. Wang, Z., Li, Z., Li, J., et al.: Transfer learning based cross-lingual knowledge extraction for Wikipedia. In: ACL, pp. 641–650 (2013)

20. Milan, D., Julio, H., Markus, A., et al.: DBpedia NIF: open, large-scale and multilingual knowledge extraction corpus. Computing Research Repository abs/1812.10315 (2018)

21. Gottschalk, I., Demidova, E.: EventKG: a multilingual event-centric temporal knowledge graph. In: Gangemi, A. (ed.) ESWC 2018. LNCS, vol. 10843, pp. 272–287. Springer, Cham (2018). https://doi.org/10.1007/978-3-319-93417-4_18

22. Dagmar, G., Maria, M.: Body-Mind-Language: multilingual knowledge extraction based on embodied cognition. In: AIC, pp. 20–33 (2017)

23. Cabral, B.S., Glauber, R., Souza, M.: CrossOIE: cross-lingual classifier for open information extraction. In: Quaresma, P., Vieira, R., Aluísio, S., Moniz, H., Batista, F., Gonçalves, T. (eds.) PROPOR 2020. LNCS (LNAI), vol. 12037, pp. 368–378. Springer, Cham (2020). https://doi.org/10.1007/978-3-030-41505-1_35

24. Wang, Z., Li, J., Tang, J.: Boosting cross-lingual knowledge linking via concept annotation. In: International Joint Conference on Artificial Intelligence, pp. 2733–2739 (2013)

25. Chen, M., Tian, Y., Yang, M., et al: Multilingual knowledge graph embeddings for cross-lingual knowledge alignment. In: International Joint Conference on Artificial Intelligence, pp. 1511–1517 (2017). https://doi.org/10.24963/ijcai.2017/209

Proof of Activity Consensus Algorithm Based on Credit Reward Mechanism

Dong Wang[1], Chenguang Jin[1(✉)], Han Li[1], and Marek Perkowski[2]

[1] School of Software, Henan University, Kaifeng 475001, China
905335851@qq.com
[2] Portland State University, Portland, OR 97201, USA

Abstract. Proof of Activity (PoA) is a key algorithm to reach consensus among nodes. In current PoA, N online representative nodes are only used to create one transaction block, and the probability of creating a block by malicious nodes cannot be controlled, which leads to a serious waste of computing power. An improved algorithm proposed in this paper introduces credit reward mechanism to replenish the missing trust in current PoA. It also can control the probability that the node obtains the right of creating block head and trading block according to the credit value and set up the reward and punishment scheme according to the proportion of credit value, which decrease the cost of good node generating block with increasing credit value and increase the cost of malicious node creating block significantly. The algorithm uses Byzantine fault tolerant idea and follow-the-Satoshi mechanism to select representative nodes through multi-level selection and set different workload for the nodes at different levels, by which the probability of malicious nodes creating transaction blocks is reduced effectively. The experimental results show that the number of transaction blocks in a block header in CPoA is 1.75 times increase than PoA. The reward and punishment scheme can achieve the purpose of dealing with malicious nodes quickly. When the proportion of malicious nodes increased from 30% to 70%, the average decline rate of their overall credit value increased about 1.7 times, which reduces the probability of malicious nodes creating blocks, increases the cost of malicious nodes creating blocks, and enhances the stability of the system.

Keywords: Consensus algorithm · Credit value · Byzantine fault tolerant · Follow-the-Satoshi mechanism

1 Introduction

The essence of blockchain is a kind of distributed database, which has the characteristics of decentralization, non-tampering, traceability and multi-party common maintenance [1]. By integrating P2P protocol, asymmetric encryption, consensus algorithm, blockchain structure and other technologies, the problem of data credibility is solved [2]. In the distributed storage structure of blockchain, the consensus algorithm encapsulated in the consensus layer mainly solves various consensus problems, such as, which node records information in the system and how to achieve the consistency of recording information among nodes [3]. At present, there are many researches on the

© Springer Nature Switzerland AG 2020
G. Wang et al. (Eds.): WISA 2020, LNCS 12432, pp. 618–628, 2020.
https://doi.org/10.1007/978-3-030-60029-7_55

application of blockchain technology. such as Xue studied the trustworthy management framework of blockchain in electronic records [4]. Zhang studied the security performance of blockchain in E-commerce [5]. Different consensus algorithms are used in different application scenarios. Appropriate consensus algorithm is the key to enhance data query efficiency, storage efficiency and system stability [6].

Proof of work (PoW) [7] is used to achieve consensus among nodes in bitcoin. However, it needs too much more resources to find effective random numbers despite that it use computing power to ensure system security [8]. Longer time needed by achieving data consistency directly leads to low block rate and high delay of PoW [9]. Moreover, it will lead to centralization [10] if several polls control computing power. In order to solve the problems of resource waste and low block rate in PoW, King proposed proof of stake (PoS) consensus algorithm in 2012 [11], in which coinage was introduced to adjust the probability of node obtaining the accounting right effectively [12]. Although PoS does not need high hardware and power costs, it still uses computing power to create blocks, and it is possible to launch attacks to fork the blockchain and achieve double spend when malicious nodes have accumulated considerable coinage [13]. Proof of activity (PoA) consensus algorithm [14] combines the characteristics of PoW and PoS, which controls the speed of creating block head by difficulty target of the hash problem, then uses follow-the-Satoshi algorithm to select representative nodes to create block body. This step-by-step way of creating block effectively improves efficiency of creating blocks and ensures the security of blocks. But there are still some problems in PoA, such as, the serious waste of computing power, the inability to effectively control the malicious node building block, and the long time to reach consensus.

Poof of Activity Consensus Algorithm Based on Credit Reward Mechanism (CPoA) is proposed in this paper, which introduces reward mechanism according to credit, improves the way of building block head and controls the cost of building block by using credit. It draws on Byzantine fault tolerance idea [15, 16] to select N representative nodes to create multiple transaction blocks. By which the number of transaction blocks in one block head is effectively increased, the number of verification nodes is reduced, the efficiency of reaching consensus is improved furtherly and the stability of the system is enhanced.

2   Related Research

In order to solve a series of problems, such as, long node consensus time, low efficiency in creating blocks and system security, there appeared some methods, such as, reducing the block interval, increasing the size of the block and classifying nodes by marking it. According to these methods, many extended algorithms based on traditional consensus algorithms have been developed. Eyal proposed the Bitcoin-NG scheme in order to expand the capacity and improve the efficiency of block generation [17]. Its important idea is to subdivide traditionally created blocks into key blocks and microblocks so that bookkeeping and competing bookkeeping rights can be executed in parallel, the long-term system freeze between the selection of two leader nodes is thereby eliminated in bitcoin. The subordination of key blocks and micro-blocks is established through public-private key pair encryption, which enables a leader node to create multiple micro-blocks and increase the transaction capacity of the blocks. However, due to the lack of node trust, the malicious leader node can repeatedly send different states to different nodes through the generated micro-blocks, which will damage the stability of the system.

The trustworthiness is introduced for nodes in literature [18], which is dynamically measured by logistic regression model. At the same time, the block creation process is subdivided into two process of creating empty blocks and packaged transactions, which effectively avoids bribery attacks and equity crushing attacks. However, it does not increase the transaction volums held by the blocks.

3   Algorithm Design

The different difficulty of building block head is given to different nodes according to its credit to avoid the disadvantage of computing power centralization; The selection scheme of representative nodes is improved by referring to Byzantine fault tolerance and follow-the-Satoshi, which effectively prevent nodes from doing evil through alliance; It no longer needs to keep every representative nodes online in the stage of transaction block generation, which greatly avoids the waste of computing power.

3.1   Block Creation Process

Block creation process is divided into two steps. Firstly nodes create block head by PoW and its credit; Then N representative nodes are selected by Byzantine fault tolerance and follow-the-Satoshi to create transaction block. The specific process is as follows:

(1) **Generating block head.** Nodes use the PoW algorithm integrating its credit to contest to generate a block head with requirements. In order to control the difficulty of generating block heads, *Credit* parameter is added in the hash algorithm of the original PoW (as shown in Formula 1). The higher the credit value of the node, the lower the difficulty of creating the block heads. It is broadcast to the whole network for verification when a legal block header is generated.

$$hash(pre\_hash, node\_id, index, nonce, timestamp) < D \times Credit \qquad (1)$$

Where D is the target domain value, pre_hash is the hash value of the predecessor block, $node\_id$ is the node address that generated the block header, $index$ is the index of the block in the blockchain, $nonce$ is the random number to be found.

(2) **Selecting participant nodes.** T participant nodes were selected by Byzantine fault tolerance and follow-the-Satoshi mechanism. The K nodes that want to participate in the selection are sorted according to the product ($credit\_coins$) of credit value($credit$) and unspent output($coins$) (as shown in Formula 2) in order to reduce the cost of participation of high-credit nodes. Then use the follow-the-Satoshi mechanism to select T participant nodes from the K applicant nodes, and if the ordinal number of a

participant node is denoted by T_id, there is $T\_id \leq T$ (there are cases where a node is selected multiple times). The detailed process is shown by the example of selecting a participant: random number X between 1 and $\sum_{n=1}^{K} credit\_coins$ is generated by the node that generates the block header. then W node which is the first one to satisfy the formula $\sum_{n=1}^{W} credit\_coins \geq X$ is found in the ordered K nodes, where W is the selected partic-ipant. In order to prevent a single node from being selected multiple times to make the signature result too centralized, the Byzantine fault tolerance is used to determine the ratio which is the times a single node is selected (T_id_count) to the total number of participants (T_ratio). If it is more than *1/3* (Byzantine fault tolerance can accommo-date *1/3* of the malicious nodes), T participants must be re-selected, as shown in Formula 3.

$$credit\_coins = coins \times \frac{credit}{credit\_max} \tag{2}$$

$$T\_ratio = \frac{T\_id\_count}{T} < \frac{1}{3} \tag{3}$$

(3) Selecting representative. N representative nodes were randomly selected from T participants by referring to Byzantine fault tolerance ideas. In order to prevent the power centralization of representative nodes, it is necessary to re-select representative nodes when the ratio of single representative node selected is too high, as shown in Formula 4. Considering that the transaction blocks created by N representative nodes are verified by T participants, so the proportion of N representative nodes in T partic-ipants should not be too large. N cannot exceed *1/3* of T, which is shown in Formula 5.

$$N\_ratio = \frac{N\_id\_count}{N} < \frac{1}{3} \tag{4}$$

$$N \leq \frac{T}{3} - 1 \tag{5}$$

(4) Create multiple transaction blocks in one block header. The first *N-1* repre-sentative nodes of the selected N representative nodes are responsible for verifying and signing block header. The *Nth* representative node is responsible for verifying block header and packing the transaction information and the signatures of the first *N-1* representative nodes into a transaction block. After generating a transaction block, N representative nodes need to be randomly sorted. The generated block is validated by T participants. It is also linked to the block header if more than two-thirds participants consider think it is valid, otherwise the transaction block will be abandoned. After that, *N-1* operation is still performed according to the process of creating the transaction block. And the multiple transaction blocks that pass the validation together with the block header form a complete block.

(5) **Distribution of rewards.** When a complete block is generated, the rewards in this cycle will be distributed uniformly according to the reward and punishment scheme. The schematic diagram of creating a block is shown in Fig. 1.

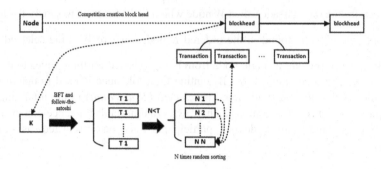

Fig. 1. Schematic diagram of creating blocks.

3.2   The Reward and Punishment Scheme Introducing Credit Value

By introducing the credit value into the reward and punishment scheme, nodes can get the credit value and coins if they honestly participate in block creation. The node with higher credit value is more likely to create the block head. All the credit values of Nodes that sign with malicious intent will deducted besides their all coins. The ease of creating a block header is directly proportional to the credit value, which make it more expensive for nodes with low credit values to create a block header.

Reward Scheme. If the block head created by a node is verified to be valid by most nodes, this node can obtain corresponding rewards, as shown in Formula 6:

$$reward\_result = \left(1 + \frac{pre\_credit}{credit\_max}\right) \times reward\_credit \tag{6}$$

pre_credit is the original credit value of the node, *credit_max* is the maximum credit value of the node, *reward_credit* is the credit value of this reward, and *reward_result* is the reward result.

When a valid transaction block is created, if the representative node's signature on this transaction block is finally correct, this representative node will be rewarded with a certain credit value issued by the system. However, how much reward it obtains is determined by the ratio of the credit value to the maximum credit value and the number of correct signatures (*right_num*), as shown in Formula 7, the objective of which is to motivate the node to maintain the stability of the system.

$$reward\_result = \left(1 + \frac{pre\_credit}{credit\_max}\right) \times reward\_credit \times right\_num \tag{7}$$

Punishment Scheme. The newly created block head needs to be verified by the whole network nodes. It is signed with the private key [19] if the node considers it is valid. The credit value of a properly signed node remains unchanged at this stage. While the credit value of a node with a malicious signature or a node that does not participate in the signature is deducted according to Formula 8.

$$punish\_result = \left(1 + \frac{pre\_credit}{credit\_max}\right) \times punish\_credit \times \beta \qquad (8)$$

Where β is the penalty coefficient, which is used to control that the penalty value is β times ($\beta \geq 1$) the reward value, $punish\_credit$ represents the penalty value. It can be seen from the formula that the penalty of malicious nodes increases with the increase of their credit value, which make it more difficult to create block headers in the future. This is to prevent the node from doing evil when credit values accumulate at a high level, and it is also a major improvement on the existing PoA penalty scheme (it just deducts the amount of money owned by the node, and it doesn't make it more difficult for malicious nodes to create block heads in the future).

A transaction block needs T participants to verify and sign it after it is created. Nodes with malicious signatures or no signatures are punished by combining the proportion of credit values and the number of malicious acts (*error_num*) during the block validation phase. The penalty to the participant increases with the number of malicious signatures, as shown in Formula 9:

$$punish\_result = \left(1 + \frac{pre\_credit}{credit\_max}\right) \times punish\_credit \times \beta \times error\_num \qquad (9)$$

This kind of punishment will urge participants to do evils as little as possible during the process of transaction block creation, otherwise the deducted credit value will increase linearly and the cost of subsequent participation in creating blocks will also increase significantly.

4   Experimental Verification and Analysis

Thirty nodes were simulated using Python for experimental validation of CPoA. Initially, the number of participant nodes $T = 10$, the number of representative nodes $N = 3$, the initial credit value of the nodes is set to 0.5, the maximum credit value is 1, and the credit value reward is set to 0.05. When the credit value reaches the maximum credit value 1, it is reset to the initial credit value.

4.1   Impact of Different Penalty Factors on Malicious Nodes

It can be seen from Fig. 2, 3, and 4 that the credit value of a malicious node decreases faster with a higher penalty factor when the proportion of malicious nodes is the same. Compared Fig. 2, 3, 4, it is clear that with the increasing proportion of malicious nodes in the system, the overall credit value of malicious nodes decreases faster. For example, when the penalty factor $\beta = 1$ and the proportion of malicious nodes increases from 30% to 70%, the average decline rate of their overall credit value increased about 1.7 times. Therefore, when some malicious nodes try to do evil through alliances, the credit value decreases faster with more malicious nodes, and the more cost it need to create subsequent blocks.

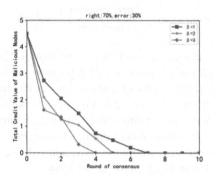

Fig. 2. The influence of 30% malicious nodes under different penalty coefficients.

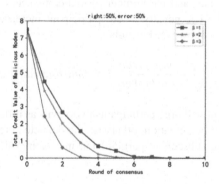

Fig. 3. The influence of 50% malicious nodes under different penalty coefficients.

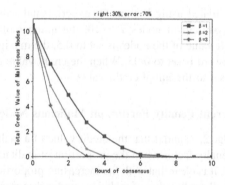

Fig. 4. The influence of 70% malicious nodes under different penalty coefficients.

4.2 Impact of Malicious Nodes on Overall Credit Value

The common feature of Fig. 5, 6 and 7 is that the overall credit value of the system fluctuates from time to time when there are 30% malicious nodes in the system, but the malicious nodes have no effect on the stability of the system. When there are 50% malicious nodes in the system, the overall credit value will decrease continuously at the beginning and then tend to be flat, and the overall credit value cannot be restored to the state at the beginning of the system, which has a certain impact on the stability of the system. The overall credit value drops even faster when there are 70% malicious nodes, which seriously destroys the stability of the system. why the overall credit value of the system does not change significantly when there are 30% malicious nodes is due to integrate Byzantine fault tolerance into CPoA, which ensures the stability of the system.

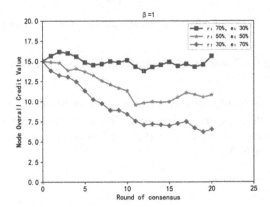

Fig. 5. $\beta = 1$: impact of different ratios of malicious nodes on overall credit value.

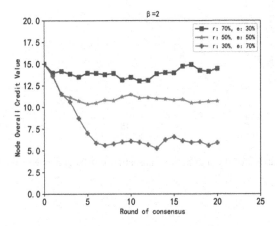

Fig. 6. $\beta = 2$: impact of different ratios of malicious nodes on overall credit value.

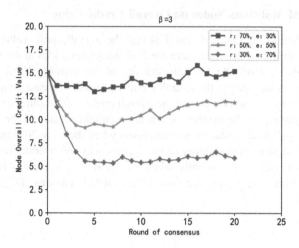

Fig. 7. $\beta = 3$: impact of different ratios of malicious nodes on overall credit value.

4.3 Comparing the Number of Transaction Blocks Generated by CPoA and PoA

It is learn from Fig. 8 that the number of transaction blocks generated by CpoA is significantly more than PoA at the same time. This is due to the fact that a block header can hold up to N transaction blocks in CPoA, while just one transaction block holded in one block header in PoA.

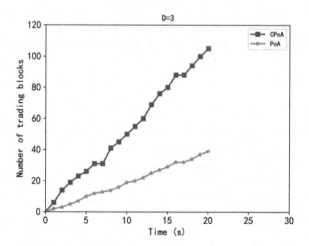

Fig. 8. Comparison of the number of transaction blocks generated by CPoA and PoA.

4.4   Comprehensive Analysis

From the changes of the overall credit value of all nodes and the malicious nodes, we learn that the overall credit value of malicious nodes decreases at a faster rate as the penalty factor increases When the proportion of malicious nodes reaches 30% or even 70%, while the overall credit value of all nodes can remain stable. It shows that the reward and punishment scheme introduced in the system can quickly deal with malicious nodes and reduce the probability of creating blocks by malicious nodes, which ensures the stability of the system.

The credit value is introduced into reward and punishment scheme so that the probability of a node participating in the creation of a block is closely related to its credit value, which ensures the security and stability of the system during the step-by-step block creation process.

In the process of creating blocks, each representative nodes selection in PoA just create one transaction block, while CPoA can create more transaction blocks. So the number of transaction blocks created in CPoA increases about 1.75 times than PoA in the same time, which saves the time wasted in the selection process of the representative nodes and increases the number of transaction blocks held by a block.

5   Conclusion

As the core of block chain technology, consensus algorithm solves the distributed consistency problem of data [20]. Different consensus algorithms are chosen for different scenarios and businesses. Proof of activity consensus algorithm based on credit reward mechanism(CPoA) proposed by us introduces credit reward mechanism to replenish the missing trust in the current PoA, which decrease the cost of good node generating block with the increasing of its credit value and increase the cost of malicious node creating block significantly. And the stability of the system is guaranteed by referring to Byzantine fault tolerant idea.

Further work is to study better credit reward and punishment scheme and the representative nodes selection scheme to make the blockchain system more secure and stable.

References

1. Yuan, Y., Wang, F.Y.: Blockchain: the state of the art and future trends. Acta Autom. Sin. **42** (4), 481–494 (2016). https://doi.org/10.16383/j.aas.2016.c160158. (in Chinese)
2. Qian, W.N., Shao, Q.F., Zhu, Y.C., et al.: Research problems and methods in blockchain and trusted data management. J. Softw. **29**(1), 150–159 (2018). https://doi.org/10.13328/j.cnki. jos.005434. (in Chinese)
3. Han, X., Yuan, Y., Wang, F.Y.: Security problems on blockchain: the state of the art and future trends. Acta Automatica Sin. **45**(1), 206–225 (2019). https://doi.org/10.16383/j.aas. c180710. (in Chinese)

4. Shao, Q.F., Jin, C.Q., Zhang, Z., et al.: Blockchain: architecture and research progress. J. Softw. **41**(5), 969–988 (2018). https://doi.org/10.11897/SP.J.1016.2018.00969. (in Chinese)

5. Xue, S., Zhao, X., Li, X., Zhang, G., Xing, C.: A trusted system framework for electronic records management based on blockchain. In: Ni, W., Wang, X., Song, W., Li, Y. (eds.) WISA 2019. LNCS, vol. 11817, pp. 548–559. Springer, Cham (2019). https://doi.org/10.1007/978-3-030-30952-7_55

6. Zhang, Y., Li, X., Fan, J., Nie, T., Yu, G.: A blockchain based secure e-commerce transaction system. In: Ni, W., Wang, X., Song, W., Li, Y. (eds.) WISA 2019. LNCS, vol. 11817, pp. 560–566. Springer, Cham (2019). https://doi.org/10.1007/978-3-030-30952-7_56

7. Nakamoto, S.: Bitcoin: a peer-to-peer electronic Cash system. White Paper (2008). https://bitcoin.org/bitcoin.pdf

8. Vries, A.D.: Bitcoin's growing energy problem. Joule **2**(5), 801–805 (2018). https://doi.org/10.1016/j.joule.2018.04.016

9. Tschorsch, F., Scheuermann, B.: Bitcoin and beyond: a technical surveys on decentralized digital currencies. IEEE Commun. Surv. Tutor. **18**(3), 2084–2123 (2016). https://doi.org/10.1109/COMST.2016.2535718

10. Shen, X., Pei, Q.Q., Liu, X.F.: Survey of block chain. Chin. J. Network Inf. Secur. **2**(11), 11–20 (2016). https://doi.org/10.11959/j.issn.2096-109x.2016.00107. (in Chinese)

11. King, S., Nadal, S.: PPCoin: peer-to-peer crypto-currency with proof-of-stake. White Paper (2012)

12. He, P., Yu, G., Zhang, Y.F., et al.: Survey on blockchain technology and its application prospect. Comput. Sci. **44**(4), 1–7 (2017). (in Chinese)

13. Houy, N.: It will cost you nothing to 'kill' a proof-of-stake crypto-currency. Soc. Sci. Electr. Publ. **34**(2), 1038–1044 (2014). https://doi.org/10.2139/ssrn.2393940

14. Bentov, I., Lee, C., Mizrahi, A., et al.: Proof of activity: extending bitcoin's proof of work via proof of stake. ACM SIGMETRICS Perform. Eval. Rev. **42**(3), 34–37 (2014)

15. Fan, J., Yi, L.T., Shu, J.W.: Research on the technologies of Byzantine system. J. Softw. **24**(6), 1346–1360 (2013). https://doi.org/10.3724/SP.J.1001.2013.04395. (in Chinese)

16. Castro, M., Liskov, B.: Practical byzantine fault tolerance and proactive recovery. Acm Trans. Comput. Syst. **20**(4), 398–461 (2002). https://doi.org/10.1145/571637.571640

17. Eyal, I., Gencer, A. E., Sirer, E.G., et al.: Bitcoin-NG: A scalable blockchain protocol. Computer Science. In: 13th USENIX Symposium on Networked Systems Design and Implementation, pp. 45–59. USENIX Association (2016)

18. Huang, J.H., Xia, X., Li, Z.C., et al.: Proof of trust: mechanism of trust degree based on dynamic authorization. J. Softw. **30**(9), 2593–2607 (2019). https://doi.org/10.13328/j.cnki.jos.005772. (in Chinese)

19. Lin, I.C., Liao, T.C.: A survey of blockchain security issues and challenges. Int. J. Network Secur. **19**(5), 653–659 (2017). https://doi.org/10.6633/IJNS.201709.19(5).01

20. Zhu, L.H., Gao, F., Shen, M., et al.: Survey on privacy preserving techniques for blockchain technology. J. Comput. Res. Dev. **54**(10), 2170–2186 (2017). https://doi.org/10.7544/issn1000-1239.2017.20170471. (in Chinese)

Blockchain and Distributed System

Xu Zhao[1], Zhiwei Lei[2], Guigang Zhang[3], Yong Zhang[1], and Chunxiao Xing[1(✉)]

[1] Beijing National Research Center for Information Science and Technology
(BNRist), Department of Computer Science and Technology,
Research Institute of Information Technology, Institute of Internet Industry,
Tsinghua University, Beijing 100084, China
msyhzhaoxu@163.com
{zhangyong05,xingcx}@tsinghua.edu.cn
[2] Joint Research Center for Industry Trust Blockchain Application Technology,
Beijing 100084, China
thefirst@126.com
[3] Institute of Automation, Chinese Academy of Sciences, Beijing 100190, China
zhangguigang@163.com

Abstract. In recent years, blockchain technology has received more and
more attention. Blockchain is a storage technology for public decentralized databases. The emergence of blockchain technology makes it possible
to solve the trust problem of distributed system nodes within the wide
area network. This article elaborated on the current advantages and disadvantages of blockchain technology and the main problems facing the
development of wide-area distributed systems. This paper attempts to
combine the ideas of blockchain and distributed system, and gives a
complete design scheme of blockchain for building a wide area network
distributed system. The designed blockchain not only retains the tamper-
proof and traceable characteristics of the existing blockchain technology, but also can overcome the node trust problem of the wide-area
distributed system.

Keywords: Blockchain · Distributed system

1 Introduction

Massive data calculation is not uncommon in today's common Internet scene.
There are many massive computing scenarios in the industrial world. For these
application scenarios to function properly, they must rely on powerful servers
[14]. For these massive computing scenarios, the traditional solution is usually
to use a centralized system, such as the IBM mainframe. This system has the
disadvantages of complex structure, poor scalability, low fault tolerance, difficult to improve performance, and difficult to maintain. Therefore, distributed
systems More and more attention has been paid by industry and academia. A distributed system is a system composed of multiple computer nodes. Its software

This work was supported by National Key R&D Program of China (2018YFB1404401).

G. Wang et al. (Eds.): WISA 2020, LNCS 12432, pp. 629–641, 2020.
https://doi.org/10.1007/978-3-030-60029-7_56

or hardware components are distributed on different computers, and message transmission and work coordination are performed through network communication. The distributed system has the characteristics of easy expansion and high fault tolerance. To improve the performance of the distributed system, only horizontal expansion is needed, and the number of machines can be increased. Therefore, the distributed system is very suitable for use in the Internet industry. At the same time, machine learning, data mining and other technologies are developing rapidly and their application fields are becoming more and more extensive. These technologies are extremely popular in scientific research and are currently in many traditional industries such as catering, tourism, and manufacturing. Ordinary people can also use These technologies efficiently produce applications. The application programs that accompany these new computer technologies often generate large amounts of data. To process these data, it is usually necessary to build a distributed cluster, but the price of building a distributed cluster is very expensive and it is difficult for ordinary people to afford it. In addition, distributed system clusters have high requirements for node reliability. There must be complete trust between the nodes. No malicious nodes can appear in the cluster. Generally, distributed systems can only use local area networks for communication. If a distributed system can be used in a wide area network, a large number of users can voluntarily form computers into clusters, eliminating the consumption of building distributed clusters, so that ordinary people can also enjoy the convenience brought by distributed systems at any time. At this point, it is necessary to solve the problem of trust between computer nodes of the WAN.

The emergence of blockchain technology makes it possible to solve the trust problem between computer nodes in the WAN. For this situation, the blockchain technology uses "distributed ledger" as a solution, which is essentially a decentralized data storage technology. In blockchain technology, there is no central node, each node is completely equal, and all nodes jointly maintain a public database through an encryption protocol. Blockchain technology applies cryptography technology to the decentralized network. The data stored in it is nontamperable and traceable. It is a very effective solution to the trust problem of the decentralized network.

Therefore, this article combines the ideas of blockchain technology and distributed systems, and proposes a blockchain for building a wide area network distributed system. This blockchain can solve the trust problem generated by the wide area network computer nodes to form a distributed system problem. This kind of blockchain makes it unnecessary for any user with massive computing needs to build a distributed cluster, but only needs to join the blockchain network at a lower threshold to enjoy the convenience brought by distributed computing. The designed blockchain has improved and redesigned the existing blockchain technology based on distributed computing scenarios.

2   Foundation of Blockchain

The development history of blockchain is relatively short. The rise of this technology is mainly due to the popularity of a series of virtual digital currencies represented by Bitcoin. These digital currencies are the initial applications of blockchain. 9 Bitcoin originated in 2008 during the financial crisis. [10] describes the underlying architecture of the Bitcoin system, which is the original appearance of the blockchain architecture. Blockchain technology was only realized by financial institutions around 2014. Ledger technology has a potential impact on finance and various industries no less than the invention of the retest bookkeeping method.

With the continuous development of blockchain technology, blockchain technology is divided into 1.0, 2.0 and 3.0 stages. In the 1.0 stage, a series of virtual currencies represented by Bitcoin were used as application scenarios. Blockchain 2.0 mainly applied blockchain technology to the financial field to enable it to serve the financial Internet industry and make value conversion more convenient, safe, and To save costs, two of the most critical technical features are asset on-chain and smart contracts; Blockchain 3.0 technology is to expand the application scope of blockchain to various fields to realize the "blockchain +" era. At present, blockchain technology is still in the process of development, and the specifications and standards of blockchain technology are still being researched and formulated worldwide.

Blockchain [13] belongs to a distributed decentralized computing and storage paradigm in technical architecture, specifically It is a continuously growing distributed ledger technology jointly maintained by multiple parties. Its core technology includes distributed network Cryptography that cannot be tampered with network and timing. Ledger and distributed consensus mechanism, this blockchain can build an autonomous letter Any environment. From the perspective of data structure, Fig. 1 shows the blockchain data structure of Bitcoin. You can see the blockchain The system packages a series of transactions into blocks, except for the genesis bloc. The other blocks except the first block above are linked to the previous block through the block hash value, and each block contains a mathematical certificate that verifies the previous block on its valid link. Combined with the chain structure and cryptographic algorithm, the blockchain can easily verify whether the data input has been tampered with, can be easily traced, and can ensure data security.

The blockchain architecture [3] is very robust. The platform it builds includes public chains, alliance chains, and private chains. Public chains are an environment that assumes no trust as a premise, while alliance chains and private chains are relative to The system built by multi-center and weak center mode, so the blockchain technology is not necessarily a complete decentralized architecture. In different application environments, the network architecture of the blockchain can be designed according to needs. For the unfeasible network architecture, the blockchain system needs to make the nodes in the network reach the consistency of the state. Therefore, they all rely on their respective consensus mechanisms, that is, the consensus protocol. There are more than 10 popular consensus mech-

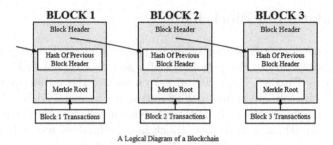

Fig. 1. Data Structure of Bitcoin Blockchain

anisms, including Proof of Work (Pow), Proof of Stake Algorithm (Pos), Byzantine Fault Tolerance Algorithm (PBFT), etc. The consensus algorithm is the core of blockchain technology, which determines who To account, which in turn affects the safety and reliability of the system.

The overall architecture of the blockchain can be summarized as a hierarchical model. As shown in Fig. 2, The blockchain hierarchy model can be roughly divided into 6 layers: data layer, network layer, consensus layer, incentive layer, contract layer and application layer, which can be flexibly selected according to different scenarios The elements are used to build the system, and the data layer, network layer, and consensus layer are necessary elements of the blockchain.

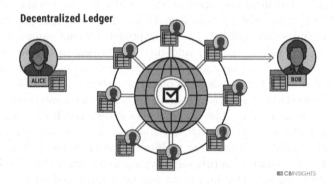

Fig. 2. Overall Architecture of Blockchain

The functional modules corresponding to different levels in the blockchain hierarchy model are different.

- Data layer: This layer is an abstraction of the underlying chain data structure of the blockchain, which combines non-paired encryption, Merkle tree data structure and timestamp technology. It is the lowest data structure of the entire model.

– Network layer: The middle layer is the network layer, and the technologies involved include point-to-point networking communication mechanisms, data verification, and data dissemination.
– Consensus layer: It is the encapsulation of the consensus mechanism algorithm adopted by the system. The consensus layer is the key to the system reaching consensus. This layer needs to resolve inconsistent behaviors caused by errors and malicious behaviors in the real environment.

The blockchain uses asymmetric encryption technology [9,15], which is the technical basis for ensuring the security of the blockchain. As shown in Fig. 3, Asymmetric encryption contains two keys. The system generates keys according to a certain method (for example: Hash, SHA-256, etc.), Obtain the private key by generating a random number, and then use elliptic curve encryption to generate its corresponding public key, and because the private key of SHA-256 can reach 2256. Under the existing conditions, the possibility of launching the private key is extremely slim. Therefore, this process is considered irreversible, so as to ensure the security of its key. The information is encrypted and decrypted.

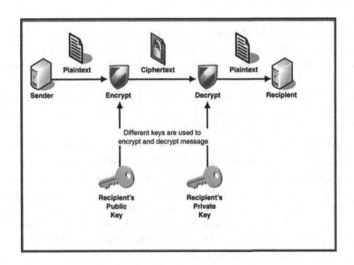

Fig. 3. Overall Architecture of Blockchain

In summary, the technical characteristics of the blockchain include the following characteristics: distributed structure, trust mechanism, openness and transparency, and timing cannot be tampered with. Blockchain is an emerging technology with "disruptive", these features can bring some new solutions for commercial networks, industry business and traditional cloud environment, bring higher operational efficiency and Lower operating costs.

3   Distributed System Theory

3.1   CAP and BASE

Distributed systems, especially distributed storage CAP is the most basic theory. CAP is Eric Brewer [2]'s conjecture on data consistency (Consistency), service availability (Availability) and partition fault tolerance (Partition-tolerance) at the 2000 PODC conference. In 2002, Seth Gilbert and Nancy Lync [8] h proved CAP theoretically. After that, the CAP theory officially became the basic theorem in the field of distributed systems. The theorem pointed out that for distributed systems, the following three points cannot be achieved at the same time:

– Consistency: the data of all nodes are consistent at all times.
– Availability: Read and write requests to the system can always be completed successfully.
– Partition tolerance: When a node is down or a network partition causes a message loss, the system can still provide services that meet consistency and availability

According to the CAP theory, the three points cannot be satisfied at the same time. In a distributed system, P is inevitable. If you select CP, the write operation is performed in one partition, and the other partition is unreachable, which will inevitably cause data inconsistency, so you must reduce availability; if you choose AP, after the system partition, but still provide external services There must be data inconsistency. The three CAPs are not equal, P is the basis, and there is a trade-off between C and A.

C (consistency) in CAP theory refers to strong consistency, that is, all nodes in the system contain the same data. In a real environment, strong consistency guarantees often reduce system availability. The BASE theory was proposed by Dan Pritchett, an architect at eBay, and was published as a paper on ACM in 2008 [11]. In a real system, it is impossible to give up the availability for strong consistency, all of which are to balance the consistency and availability, reduce the strong consistency to improve the system availability, BASE theory is derived from the practice summary of the distributed system, will be strong consistency It is the practical evolution of the CAP theory that the sex is reduced to the final consistency, thereby improving the usability as much as possible. BASE theory refers to Basically Available, Soft State and Eventually Consistent.

– Basically Available: In the case of node downtime or network partitioning in the distributed system, it is allowed to lose part of the consistency and ensure that the core is available.
– Soft State: Relative to strong consistency, it must be ensured that the data copies of multiple nodes are consistent. Soft state allows the data copies to be synchronized between different nodes of the system with a certain delay, allows the data copies to be in an inconsistent state, and does not require every moment.

– Eventually Consistent: Although there may be inconsistent moments, after a period of data synchronization, the data copies in the system can eventually remain consistent.

The core idea of BASE theory is to improve availability by reducing consistency, allowing data to have inconsistent time windows, but after a period of time, the final consistency can be achieved.

3.2 Distributed Consensus Algorithm

In consideration of improving reliability, accelerating user access and enhancing fault tolerance in distributed systems [6,7], there is generally a copy mechanism, that is, multiple copies of data, and the common copy strategy is three copies. Since the data will have multiple copies, then there must be how to solve the problem of consistency of multiple copies. In distributed systems, because machines are distributed in different regions and communicate with each other through message transmission, network abnormalities between nodes, such as message duplication, loss, delay, out-of-order, and network partition, must exist, plus nodes may be down Machine, which requires that the distributed consensus algorithm must resolve the data consistency in the case of the above failures.

Paxos and Raft are very well-known consensus algorithms in the field of distributed systems to solve consistency problems. Compared with Paxos' obscure and difficult to engineering, Raft's implementation is relatively concise and follows human intuition. It can be said that Raft is for the purpose of Invented to solve Paxos is difficult to understand and difficult to achieve.

3.3 Data Partition

The main goal of partitioning is to make full use of multiple storage nodes and distribute the data and query load evenly across all nodes. If the nodes share the load equally, then theoretically 10 nodes should be able to handle 10 times the data volume and 10 times the read and write throughput of a single node. If the partitions are not uniform, some partition nodes will bear more data volume and query load than other partitions, and data skew will occur. Tilt will cause a serious drop in partition efficiency. In extreme cases, all the load may be concentrated on one partition node, which means that 10 nodes and 9 are idle, and the bottleneck of the system is on the busiest node. This kind of load is not serious. The proportional partition becomes the hotspot of the system.

Only considering key-value data which means that the data is accessed through keywords. Like an encyclopedia. Throughout the book, you can find an item by title. The arrangement of entries is similar to an index, and how to organize entries is particularly important. The current mainstream data distribution strategy, that is, how to organize entries, can be basically divided into three types: keyword-based interval distribution, keyword-based hash value distribution, and consistent hashing.

The partitioning method based on the key interval is to allocate a continuous range of key intervals to each partition. If you know the upper and lower limits of the keyword interval, you can directly send a request to the node, such as an encyclopedia, which is to get the desired book directly from the shelf. The interval segments of keywords do not have to be evenly distributed, mainly because the data itself may not be uniform. In order to distribute the data more evenly, the partition boundaries should be adapted to the distribution characteristics of the data itself. Keyword-based interval partitioning systems include BigTable, HBase and MongoDB. Each partition is stored in order according to keywords, which can easily support interval query.

Keyword-based hash distribution will cause a large amount of data migration in the face of changes in the number of underlying storage nodes. The solution is consistent hashing [5]. Consistent hashing [4] was first proposed by Karger et al., and is a method of evenly allocating load. It was originally used in Internet caching systems such as content distribution networks. Consistent hashing maps the hash value of the data to a hash ring connected end to end, and also maps the storage nodes to the hash ring. After the key is hashed, it is stored on the first storage node found clockwise.

Data partitioning is an important part of a distributed database. Briefly, keyword-based interval partitioning has good interval query characteristics, but some access modes will cause hot spots and need to be supplemented by a load balancing mechanism; based on keyword hash value Partitioning is simple to implement, but loses the good interval query characteristics and changes in the number of underlying nodes will cause a large amount of data migration; consistent hashing solves the problem of data migration when the underlying storage node changes, but the underlying layer uses consistent hashing as the architecture All storage systems use final consistency, and scenarios that require strong data consistency require careful consideration.

4   Distributed System on Blockchain

The previous chapters introduced the cryptography, network related knowledge involved in the blockchain, and the core issues that the distributed system needs to solve. The existing blockchain technology mainly has the following disadvantages:

1. Unfair property distribution: In the current blockchain technology, the mainstream consensus [1] mechanisms are PoW, PoS, DPoS, etc. The PoW mechanism uses violent search for random numbers as the working mode. The higher the computing power, the higher the probability of a computer getting a reward. This mechanism has led to the emergence of a large number of Bitcoin "mining pools". Mastering makes it difficult for ordinary users to get rewards. PoS and DPoS are consensus algorithms based on the shareholding system, that is, nodes with more assets are more likely to receive rewards, which will also lead to unfair distribution of wealth in the network, and the gap between the rich and the poor is increasing;

2. Waste of computing resources: This situation is very significant in blockchain model of Bitcoin. A large number of computers compete for new blocks through a large number of meaningless operations, resulting in a lot of computing resources being wasted;

3. Blockchain data occupies too much storage space: due to the distributed storage feature of the blockchain technology, any computer node that joins the blockchain network needs to synchronize the full amount of blockchain data. Taking Bitcoin as an example, the current joining The Bitcoin network needs to synchronize more than 100 GB of blockchain data, which is difficult for ordinary users to accept;

4. The rules of the blockchain cannot be adjusted according to user needs: the operating rules, parameters, and algorithms of the existing mainstream blockchain models cannot be changed, and users cannot affect these rules. The parameters in the network cannot be adjusted according to users Adjust the overall demand of Bitcoin. For example, Bitcoin's existing new block generation mechanism generates a new block in about 10 min, but this also causes the entire network to conduct a new transaction too slowly. If the parameter can be based on Customized by user needs, it can be flexibly applied to different scenarios.

This chapter will propose and design a blockchain model that can effectively improve and overcome the shortcomings of the above-mentioned blockchain. This model combines blockchain technology with distributed computing and distributed data processing technology to improve and overcome the above While having the disadvantages of blockchain technology, it also supports users to effectively use the computing power in the network for customized distributed computing or big data processing [12].

4.1 Network Topology Design

There are two types of computer nodes in the network: computing nodes and storage nodes. The nodes in the network are divided into two layers according to different functions: the outer layer is a computing network composed of computing nodes, responsible for performing distributed computing; the inner layer is a storage network composed of storage nodes, responsible for the storage of blockchain data and maintain. Nodes in the same network are connected to each other, and the computing network can only send messages to the storage network in one direction.

There are three main functional modules in the network:

1. The computing network performs distributed computing through task distribution and aggregation. The role of the computing node is to participate in distributed computing tasks or distributed data processing tasks in the network. Each computing node has two states: distribution state and computing state. When a computing node actively initiates a distributed computing or distributed data processing task, it will enter the distribution state and is

responsible for splitting a larger computing task into multiple subtasks. And distributed, at this time the idle computing nodes around it will enter the computing state, responsible for receiving these subtasks and computing.

2. The storage network is responsible for the storage and maintenance of blockchain data, and constructs new blocks by receiving the calculation result data from the calculation network.

3. The open source algorithm community is a distributed code base maintained by the computing network and the storage network. Users can upload the source code of the custom algorithm to the open source algorithm community, and all nodes will synchronize the custom algorithm and update the local program through the open source algorithm community, which allows users to easily customize in the blockchain network Own distributed computing tasks.

4.2   Computing Network Design

Distributed computing systems usually perform distributed computing or big data processing through task distribution and aggregation. The computing network in the blockchain described in this article also works in this way. In the idle state of the computing network, each node is completely equal, and each node can initiate a distributed computing or big data processing task in the idle state. Among them, the node that initiated the task will enter the distribution state, responsible for splitting a larger computing task into multiple subtasks and distributing, at this time the surrounding computing nodes will enter the computing state, responsible for receiving these subtasks and processing. When a complete distributed computing or big data processing task is completed, the nodes participating in this task will return to a completely equal idle state.

Nodes in the distribution state are called distribution nodes, and nodes in the calculation state are called computing nodes. A block in the blockchain is composed of block units with multiple tasks. Note that when the node is in the distribution state, the Mapper module and Collector module inside the node are turned on at the same time.

The steps for the distribution node and the computing node to work together are:

1. The Mapper module splits the task into multiple sub-tasks according to the task selected by the user, and distributes it to the nodes involved in the calculation;

2. The computing node executes the task immediately after receiving the subtask;

3. After the computing node completes the subtask, it immediately sends the processing result of the subtask to the Collector module of the distribution node. The Collector module organizes the results of the subtasks of the same distributed task into the same block unit. The subtask is not processed, the Mapper module will distribute the subtask to the node that just completed the task;

4. If all the sub-tasks have been processed and the result is obtained, the block unit of the calculation network in the calculation will be sent to the storage network. After the calculation is completed, the nodes participating in the calculation return to the idle state.

4.3   Storage Network Design

Each storage node is composed of four areas: block unit temporary storage area, complete calculation result block temporary storage area, calculation execution area and blockchain storage area.

- The block unit temporary storage area is used to receive and temporarily store the calculation result block unit from the calculation network and the transaction block unit generated by the user initiated transaction. These block units will be added to the new block when the next new block is generated.
- The complete calculation result block temporary storage area is used to receive and temporarily store complete blocks sent by other storage nodes. When a storage node generates a new calculation result block, it will broadcast the block to all nodes in the storage network, and at the same time generate a corresponding transaction block according to the new calculation result block. After the transaction block generation is completed, it will be broadcast again to All nodes in the storage network. Therefore, after receiving a complete calculation result block, the storage node will not directly add it to the blockchain, but will temporarily store it and wait for the corresponding transaction block to arrive, only when the calculation result block corresponds to its corresponding After all the transaction blocks are received, the blocks in temporary storage area of the complete calculation result block will be deleted by the node and added to the blockchain.
- The calculation execution area is the core area where the storage node executes the proof of work. The proof of work is a series of algorithms defined by the user. Generally, the higher the frequency of the computer's central processing unit (CPU), the more efficient the storage node performs proof of work. The higher, the more rewards you get.
- For storage nodes, the operation of ending the calculation of proof of work and updating the blockchain is after receiving the transaction block and receiving the calculation result block corresponding to the transaction block. This is because if the node that completes the proof of work is The downtime occurs after broadcasting the new block. At this time, if other nodes receive the new calculation result block and add it directly to the blockchain, the transaction block corresponding to the calculation result block will never be generated because the transaction area Blocks will only be broadcast by nodes that have completed proof of work. So when receiving a calculation result block, the storage node will only temporarily store the block and record the current proof of work completion progress, and continue to calculate the proof of work

will not stop until the corresponding transaction block is received This time the block is calculated and updated, which also ensures that even if the first node that completes the proof-of-work is down, the node that subsequently completes the proof-of-work will ensure the correct propagation of the block.

5   Conclusion

This article details the current development trend of blockchain technology and distributed systems, and points out the main problems faced by distributed systems. According to the distributed characteristics of the blockchain technology itself, we propose a solution that combines the blockchain and the distributed system to solve the problem of node trust and consistency under the current wide-area distributed system. Through this design, more users can join the distributed system at low cost, which is conducive to improving the scalability of the distributed system and promoting its development.

References

1. Baliga, A.: Understanding blockchain consensus models. Persistent **4**, 1–14 (2017)
2. Brewer, E.A.: Towards robust distributed systems. In: PODC, vol. 7. Portland, OR (2000)
3. Crosby, M., Pattanayak, P., Verma, S., Kalyanaraman, V., et al.: Blockchain technology: beyond bitcoin. Appl. Innov. **2**(6–10), 71 (2016)
4. Karger, D., Lehman, E., Leighton, T., Panigrahy, R., Levine, M., Lewin, D.: Consistent hashing and random trees: distributed caching protocols for relieving hot spots on the world wide web. In: Proceedings of the Twenty-Ninth Annual ACM Symposium on Theory of Computing, pp. 654–663 (1997)
5. Kleppmann, M.: Designing data-intensive applications: the big ideas behind reliable, scalable, and maintainable systems. "O'Reilly Media, Inc." (2017)
6. Lakshman, A., Malik, P.: Cassandra: a decentralized structured storage system. ACM SIGOPS Operat. Syst. Rev. **44**(2), 35–40 (2010)
7. Lakshman, A., Malik, P., Ellis, J.: Facebook's cassandra paper, annotated and compared to apache cassandra 2.0 (2014)
8. Lynch, N., Conjecture, G.S.B.: the feasibility of consistent, available, partition-tolerant web services (2002)
9. Enhanced attribute based encryption for cloud computing: Na, S.K., GV, R.L., Ba, B. Procedia Computer Science **46**, 689–696 (2015)
10. Nakamoto, S.: Bitcoin: A peer-to-peer electronic cash system. Technical report, Manubot (2019)
11. Pritchett, D.: Base: an acid alternative. Queue **6**(3), 48–55 (2008)
12. Sharples, M., Domingue, J.: The blockchain and kudos: a distributed system for educational record, reputation and reward. In: Verbert, K., Sharples, M., Klobučar, T. (eds.) EC-TEL 2016. LNCS, vol. 9891, pp. 490–496. Springer, Cham (2016). https://doi.org/10.1007/978-3-319-45153-4_48
13. Wattenhofer, R.: Distributed Ledger Technology: The Science of the Blockchain. CreateSpace Independent Publishing Platform (2017)

14. Xue, S., Zhao, X., Li, X., Zhang, G., Xing, C.: A trusted system framework for electronic records management based on blockchain. In: Web Information Systems and Applications - 16th International Conference, WISA 2019, Qingdao, China, 20–22 September 2019 (2019), Proceedings. pp. 548–559 (2019). https://doi.org/10.1007/978-3-030-30952-7_55

15. Zhou, X., Tang, X.: Research and implementation of RSA algorithm for encryption and decryption. In: Proceedings of 2011 6th International Forum on Strategic Technology, vol. 2, pp. 1118–1121. IEEE (2011)

Efficient Patient-Friendly Medical Blockchain System Based on Attribute-Based Encryption

Yan Sun, Wei Song[(⊠)], and Yuan Shen

School of Computer Science, Wuhan University, Wuhan 430072, China
songwei@whu.edu.com

Abstract. In recent years, there have been rapid advances in electronic medical record (EMR) sharing with the fast development of blockchain technology. Using blockchain technology to build the electronic medical record sharing system can effectively solve the problem that medical data is hard to share between different hospitals' databases. However, existing EMR sharing system using blockchain technology is mainly for medical institutions, and the patients as the data owner even can't easily share their own personal EMR when they need to use it. The patient's demand for management of personal EMRs is largely ignored. At the same time, the doctor's longitudinal access to personal EMRs requires a fairly large overhead, which severely causes treatment delay. This paper proposes a patient-friendly blockchain electronic medical data sharing system based on main-sub structured blockchain, which enables patients to manage their personal medical data directly and enhances the authorized users' efficiency of access to personal EMRs significantly. Experimental results demonstrate that our scheme is efficient enough to support the real medical record sharing applications.

Keywords: Electronic medical record · Main-sub blockchain · Medical data sharing · Attribute-based encryption

1 Introduction

EMR system can facilitate the medical diagnosis process, but it is limited by the traditional management mechanism. In particular, EMR data are stored and managed by the hospital which generated it. However, the real data owner, the patient, cannot conveniently manage their own medical records. When they visit other hospitals, they are not available to provide the doctor their past medical records, because their past records were stored in the hospital's private database. Interoperability challenges between different hospital systems give tough problems to data sharing. It is difficult for patients to share the medical data because of lack of personal data management and sharing. At the same time, due to the centralized management of hospitals, EMR cannot resist the threats of personal information leakage as well.

To meet the requirements on medical data sharing, some researchers [1–3] have proposed some relative schemes about cloud storage technologies to provide suitable solutions to share medical data. However, cloud service providers (CSP) still face some significant problems in persuading hospitals to use centralized cloud services due to the

G. Wang et al. (Eds.): WISA 2020, LNCS 12432, pp. 642–653, 2020.
https://doi.org/10.1007/978-3-030-60029-7_57

risks of personal privacy disclosure. After all, a whole lot of data stored in third parties is not reassuring.

Blockchain is a distributed and shared database, characterized by decentralization, non-tampering, traceability and openness. In recent years, blockchain has been proposed to be a promising solution to achieve EMR sharing with security and privacy preservation due to its advantages of non-tampering and traceability. Therefore, compared with the cloud-based electronic medical record (EMR) sharing system, blockchain has the advantages of decentralization and traceability, so it has higher security level. But most existing systems are mainly for medical institutions, which ignores the patient's requirements of personal data management and control. When the patients want to share their personal EMR to someone, they cannot provide their own medical records.

Generally speaking, there are still three major challenges that need to be over-come:

- The EMRs are distributed in different hospitals. Therefore, it can only be visited through the hospital where the patient was treated. When the patient is going to another hospital, his or her EMR can't be shared between hospitals, which can cause misdiagnosis and unnecessary repeated physical examination. How to deal with interoperability problems between different hospital systems is the first challenge.
- Besides the interoperability issue, the patient pays more attention to personal privacy as the EMR contains many sensitive information [4]. The patient needs a more personalized method to share their EMR, which means the authorized attributes of attribute set can be decided by the patient. Therefore, a fine-grained access control mechanism is highly desirable for the real EMR sharing applications.
- Furthermore, when the data user accesses to the patient's EMR which they need, the overhead of query personal EMR is quite big. As the blockchain length increases, the query overhead becomes larger.

Motivated by the above issues, we deign a medical blockchain system using attribute-based encryption which meets all the practical requirements mentioned above with strong security guarantee. The main contributions of this paper can be summarized as:

- First, we design a main-sub medical blockchain system, all the patients' medical records are stored in the blockchain, the authorized medical institutions can share these data for treatment and research. Therefore, interoperability between different hospital systems is well addressed.
- Second, we introduce the attribute-based encryption method [5]. By it, the patients are allowed to directly manage and share their own EMR with strong security guarantee. And the data users can only access to authorized personal EMR without leaking out other sensitive information, which effectively avoids personal information disclosure.
- Third, we store the patient's EMR according to unique personal ID in the sub blockchain, so that various data users can efficiently access the medical data by comparing the hash value of attribute set which reduce the overhead of query and computation greatly.

- Finally, we formally prove the security of the proposed scheme and conduct experiments which can also demonstrate that our scheme is efficient.

The rest of this paper is organized as follows. In Sect. 2, we discuss related work about medical data sharing based-on blockchain, and present the necessary background to understand the proposed scheme. In Sect. 3, we present our scheme. Our experimental analysis is outlined in Sect. 4, and the conclusion is presented in Sect. 5.

2 Related Work

Blockchain is widely used in the field of data sharing, due to its characteristics of decentralization, openness, anti-tampering and traceability.

Prior work by Zyskind et al. [6] has demonstrated the use of blockchain protocols for permission management. They implement a trusted blind escrow service, storing encrypted data in trusted third party hosting services while logging pointers on the blockchain, but the third party brings the risk of data disclosure. Kish proposed the blockchain for hypothetical key management in a medical context [7].

Ariel et al. build on these ideas and develop MedRec: a decentralized record management system to handle EMRs, using blockchain technology [8]. But the cost of querying personal EMR is relatively large.

Esposito et al. [9] detailed the drawbacks of using cloud storage technology to establish a data sharing system in the medical field. They also raised the possible challenges of using blockchain technology in medical data sharing (such as privacy protection). However, the article does not propose practical schemes to address these challenges. These studies [10] only proposed a framework, but did not implement specific security data access authorization scheme.

Kai Fan et al. [11] proposed a blockchain-based information management system, MedBlock, to handle patients' information. But it used the bread crumbs which will bring additional amount of data, leading to excessive storage consumption.

Therefore, we need an encryption scheme that provides a more efficient way to control data access based on data user's attributes [12]. The concept of attribute-based encryption (ABE) was first proposed by Sahai and Waters [13], and in this context can be employed to encrypt the EMRs in the medical blockchain system. The ABE scheme allows users to access the data when their attributes satisfy the data owner's requirements. And we have designed the retrieval mechanism on the sub blockchain scheme that provides better performance of accessing the patient's personal medical records.

In this paper, we propose an efficient patient-friendly system based on main-sub blockchain (PSBM) to share electronic medical records among authorized users. The sub blockchain allows the patient define who can access their EMR, while the main blockchain makes the users find the information they want in an efficient way. Moreover, the medical blockchain system is suitable for multi-task oriented applications and can also achieve load balance.

3 The Proposed PSBM Scheme

Blockchain provides an important support for the sharing of medical data. But there are still problems that the patient as the data owner lacks control over their own medical data, so an efficient patient-friendly medical blockchain scheme is needed. As for individual data users, only the data user who is authorized by the patient can access the data through the sub blockchain. But for those who are not in the attribute collection, the patient cannot authorize the data to them. And for medical institutions (including research, education and statistics), this system model allows medical staff to access to anonymous medical data through the main blockchain.

In order to ensure the data owner's management of personal medical data, patients also need to set up corresponding attributes collection and compute the hash value of each attributes collection. In advance, the patients update personal EMR adding authorized attributes collection in the sub blockchain, so in case of individual query access, the data user in the attributes collection can access the EMR at a rapid speed.

3.1 System Model

The system model includes six entities: key generation center, medical institutions, data owner, main blockchain, sub blockchain and data user. The proposed model is illustrated in Fig. 1.

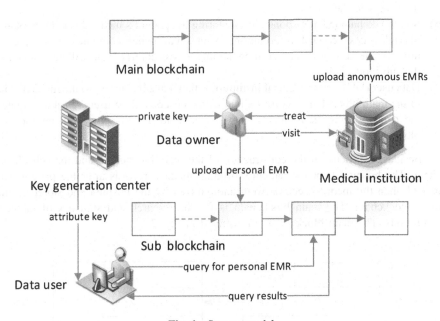

Fig. 1. System model

And the characteristics and functions of each entity are described as follows:

(1) Key generation center (KGC): responsible for generating system public parameter and creating master system key MSK. Meanwhile, the key generation center generates key pairs for patients and data users.

(2) Medical institutions (MI): they are composed of various hospitals with medical capacity. A medical institution manages its staff and provides medical services to patients. After the registration of a medical institution, KGC generates public private key pairs for the medical institution, and securely transmits the private key to the medical institution. A medical institution generates a set of attributes for its medical staff to describe their data access characteristics and generate an attribute key for them.

(3) Data owner (DO): mainly composed of the patients themselves, who can manage their EMRs stored in the sub blockchain directly. When a new medical record is inserted in the main blockchain, the data owner will compute the sub blockchain sequence number according to the patient's unique identity, private key and the main blockchain's sequence number (which block the new record is stored in the blockchain). Meanwhile, the data owner will give the record a set of attributes for authorized access, (such as hospitals, categories of doctor and title of doctor).

(4) Main blockchain (MB): used for storing medical data which is processed anonymously and every EMR's corresponding sub blockchain sequence number. EMRs are all stored in chronological order. Only medical and statistical institutions have the access to the data.

(5) Sub blockchain (SB): responsible for storing the patient's medical data by personal id and access control information. After data owner computes the sequence number, the new medical record containing access control logic will be updated in the sub blockchain.

(6) Data user (DU): as for medical institutions, they have the access to the medical data of the main blockchain. As for individual data users, only if users' attributes match the data owner's requirements, they are able to access some data on the sub blockchain.

Besides, we focus on the construction of the main blockchain and the sub blockchain. The hospital kept the patient's original medical records after the patient was treated, then the medical records were anonymized by the hospital and stored in the main blockchain. The main blockchain is for the research and statistics of medical institutions. The main blockchain is illustrated in Fig. 2.

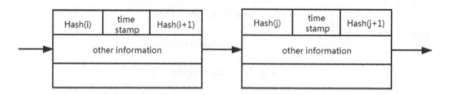

Fig. 2. Main blockchain

Every block has its own sequence number, and sub blockchain's sequence number of the same medical record is computed by the hash function on the basis of patient's id, private key and main blockchain's sequence number. What is different from the main blockchain is that the medical record is stored not by time, but by patient's id. As for patient Alice, her first block on the sub blockchian is B0. B0 is Alice's head block, which stores Alice's id, authorized attribute set for each medical record and the block sequence number corresponding to the attribute set. The sub blockchain is illustrated in Fig. 3. Therefore, when a data user makes a query, only one comparison is needed to get the result of the query. Compared to the traditional way of data retrieval, the query efficiency is greatly improved.

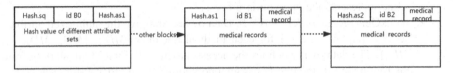

Fig. 3. Sub blockchain

3.2 Description of Proposed PSBM Scheme

We build an efficient retrieval scheme based on blockchain with support for fine-grained access control. The scheme is composed of four polynomial time algorithms, as described below (Table 1).

Table 1. Symbols used in this paper.

Symbols	Meaning of the symbols
PK	The public key
MSK	The system master key
MD	The plaintext medical document
A	The authorized attribute set A
Γ	The access strategy
K	The encryption key
C	The ciphertext medical document
I	The safe index
SK_U	The attribute key

(1) Setup(1^λ) → (*PK, MSK*). The key generation center operates the Setup algorithm. This is a randomized algorithm that takes no input other than the implicit security parameter λ.
The key generation center defines G_0, $G_1 \in Z_p^*$, which are two multiplicative cyclic groups of $\mathbb{G}$. And p is a big secure prime number, let g be the generator of group G_0, and define bilinear mapping e: $G_0 \times G_0 \to G_1$. Select the random hash

function H_1: $\{0,1\}^* \rightarrow G_0$, H_2: $\{0,1\}^k \times \{0,1\}^* \rightarrow \{0,1\}^l$. Key generation center sets the random number α, $\beta \in Z_p^*$ and then computes bilinear pairs $Y = e(g_1, g_2)^{\alpha}$. Finally, it outputs system common parameter PK = G_0, G_1, p, $h = g^{\beta}$, Y, H_1, H_2) and master secret key MSK = (α, β).

(2) Enc(*PK, MD, A, Γ, K*) $\rightarrow$ (*C, I*). The data user public key *PK*, the clear personal medical document *MD*, authorized attribute set *A*, access strategy Γ and encryption key *K*. The outputs are encrypted medical data *C* and safe index *I*.

1. FileEnc(*MD, K*) $\rightarrow$ *C*. Data owner randomly selects $K \leftarrow \{0,1\}^K$ as a clear medical document symmetric encryption key. And the data owner uses *K* to encrypt medical document MD_i(i $\in$ [1, *n*]), then gets the $C_i = \{\varepsilon.Enc_k(D_i)|$ i [1, *n*]}. In the formula, ε represents a secure symmetric encryption scheme, and ε.Enc represents the encryption process.

2. IndexGen(*PK, A, Γ, K*) $\rightarrow$ (*C, I*). Data owner first give the authorized attribute set $A = \{A_1, A_2, ..., A_n\}$ for each medical document $MD = \{MD_1, MD_2, ..., MD_n\}$. The data owner defines the access structure Γ. First of all, starting from the root node *r* of the tree Γ, assign a polynomial q_x of order d_x for each node *x* (the q_x of the leaf node is a constant number). Let k_x be the threshold value of node *x*, and set $\deg(q_x) = d_x = k_x - 1$. The data owner randomly selects $s \leftarrow$ from the root node *r* and make $q_r(0) = s$, and selects $\deg(q_r)$ random coefficients to determine the polynomial q_r. As for other nodes *x*, set $q_x(0) = q_{parent(x)}(index(x))$ and selects $\deg(q_x)$ random coefficients to determine the polynomial q_x. Let Y represents all the leaf nodes of tree Γ, data owner gets the encrypted safe medical data *C* and index *I*.

$$C = \left\{ \Gamma, \bar{C} = Ke(g,g)^{\alpha s}, C = h^s, \left\{ C_y = g^{q_{y(0)}}, C_y' = H_1(att(y))^{q_{y(0)}} \right\} \right\}$$

(1)

In the above expression, *s* is a random number, *att(y)* means attribute value.

(3) UserRegister(*PK, MSK, S*) $\rightarrow$ (*SK_U*). The data user offers the attribute set *S* to the key generation center, the key generation center inputs the public parameter *PK*, master key *MSK* and data user's attribute set *S*. Then it outputs the attribute key SK_U and gives it to the data user. When $\forall j \in S$, the key generation center randomly selects $r \in Z_p^*$. Evaluate the corresponding attribute private key:

$$SK_U = \left\{ E = g^{\frac{\alpha+r}{\beta}}, \left\{ E_j = gH_1(j)^{r_j}, E_j' = g^{r_j} \right\} \right\}$$

(2)

(4) Dec(*SK_U, C*) $\rightarrow$ *MD*/$\perp$. Data user inputs the attribute key SK_U and encrypted medical data *C*. If the attribute key SK_U satisfies the access structure Γ defined by the data owner, the data user can restore symmetric key K, decrypt ciphertext medical document collection *C*. And output contains the plaintext medical document collection *MD*. Otherwise, output $\perp$. K is computed as below:

$$K = \frac{\bar{C}}{\frac{e(C,E)}{A}} = \frac{Ke(g, g)^{\alpha s}}{e\left(h^s, g^{\frac{\alpha+r}{\beta}}\right)} = \frac{Ke(g, g)^{\alpha s}e(g, g)^{rs}}{e(g, g)^{s(\alpha+r)}} \qquad (3)$$

3.3 Insert Algorithm of PMBS

While a patient p visits the hospital, the hospital will anonymize personal medical data and upload it to the main blockchain in chronological order. At the same time, p is supposed to update his or her own personal EMR on the sub blockchain.

(1) If it is the first EMR to store, the patient will compute the sub blockchain's sequence number according to patient's id, private key and main blockchain's sequence number. The head block of the patient will be used to store the hash value of the authorized attribute set.

(2) If not, the patient can directly store the hash value of the authorized attribute set in his or her head block, and the EMR will be stored in the block which is linked to the head block. The EMR of same authorized attribute set is stored in one block.

(3) If the block used to store the EMR of same authorized attribute set is full, the next block will be used for new storage. And all the EMR block's sequence number is stored in the patient's head block. The main-sub blockchain system is shown in Fig. 4.

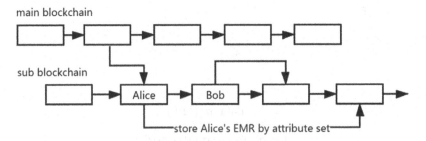

Fig. 4. The main-sub blockchain system

3.4 Query Algorithm of PMBS

While a doctor d attempts to retrieve the certain patient's EMR, d needs to register first according to UserRegister algorithm. Then d generates a query request Q (SK_U, PID) and asks the patient to compare the doctor's attributes with his or her authorized attribute set.

(1) If they are not the same, the doctor cannot visit the patient's EMR.

(2) If the doctor d's attributes match the authorized attribute set, d can visit the patient's head block in the sub blockchain and get the blockchain sequence to

another block which d can get the medical records. In the case of that block is full, d will continue to visit the next block via obtained sequence number.

(3) After d gains the encrypted medical records, d can easily decrypt them by Decryption algorithm.

4   Analysis of Performance

In this section, we verify the efficiency of the efficient patient-friendly medical blockchain system based on attribute-based encryption (PBMS). The experimental environment is on a Windows 10 (64-bit) operating system with an Intel Core i7 4 GHz processor with 8 GB RAM. The encoding is implemented by using the JPBC 2.0 library. Three groups of comparative experiments were conducted to compare the efficiency of this scheme with the existing MedRec scheme [5], MedShare scheme [14] and MedBlock scheme [7] in encryption algorithm, query algorithm and memory consumption. The experimental comparison is shown below.

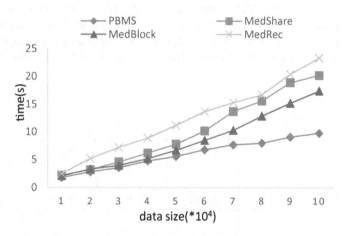

Fig. 5. Time cost for encryption

As illustrated in Fig. 5, the time cost between the other three schemes and our scheme in the encryption phase is positively correlated with the data size. And our scheme has more excellent performance in terms of time consumption than the other three schemes. This is because we use the main blockchain and the sub blockchain to work together, which can significantly reduce time overhead and achieve load balance. By contrast, MedRec scheme, MedShare scheme and MedBlock scheme show a significant increase in time consumption with the increase of data size. In general, our scheme is more effective under the condition of huge data size.

The results in Fig. 6 shows that the efficiency of data query is greatly improved. In our scheme, we adopt main-sub blockchain structure to enhance information retrieval efficiency. If a data user wants to retrieve someone's EMR information, he can directly find the corresponding block according to the block sequence number and the hash value of attribute set. The original search method needs to traverse the data on the block until finding the useful data. Compared with the traditional way of data retrieval, the efficiency of sub-blockchain increases too much. And the sub-blockchain can balance the query pressure of the main blockchain.

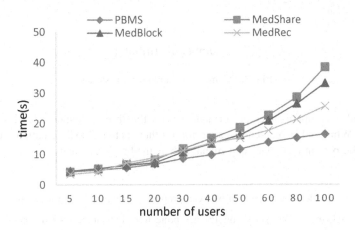

Fig. 6. Time cost for query

When the number of users is small, the time cost of query in our scheme, MedRec scheme, MedShare scheme and MedBlock scheme has small difference. The original search method can also find the EMR information in a relatively short time. However, as the number of users' increases, the advantages of PBMS over the other three schemes become more and more obvious. The records of sub blockchain can directly guide the users to find the corresponding blocks. Even if the number of users is quite large, it will not be a constraint on efficiency.

As shown in Fig. 7, memory consumption increases almost linearly as the number of attributes increases. MedRec scheme uses the least memory consumption, and our scheme costs slightly more memory than MedShare scheme and MedBlock scheme. The increase in memory consumption is due to the introduction of sub blockchain, but the storage overhead is acceptable. When the number of attributes was 100, our scheme memory consumption was 10.80 KB.

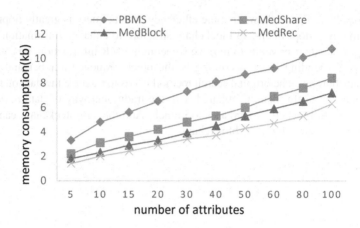

Fig. 7. Memory consumption for storage

We also compare other performances with the three schemes, it is illustrated in Table 2. What makes our scheme superior to others is that PBMS is patient-authorized, so that the patient can share and manage their medical data easily.

Table 2. Comparison between proposed system and other systems

Scheme	Tamper proof	Privacy protection	Patient-authorized access
PBMS	Y	Y	Y
MedRec	Y	N	N
MedShare	Y	N	N
MedBlock	Y	Y	N

5 Conclusion

In this paper, we focused on the practical problem of patient's privacy preserving and management over the personal electronic medical data in the EMR sharing system. We further discussed different stakeholder's practical requirements for this problem. To fulfill these requirements, we proposed a novel patient-friendly medical blockchain sharing system based on attribute-based encryption. For the convenience of sharing and managing EMRs, we designed our system by introducing a sub blockchain to reduce the overheads of query and balance the load. We have carried out several experiments to show that the query processes are efficient and our scheme is appropriate for using in the blockchain-based EMR sharing systems.

Acknowledgement. This work is supported by the key projects of the National Natural Science of Foundation of China (No. U1811263, 61572378), the major technical innovation project of Hubei Province (No. 2019AAA072), the Science and Technology Project of State Grid Corporation of China (No. 5700-202072180A-0-0-00), the Teaching Research Project of Wuhan University (No. 2018JG052). We also thank anonymous reviewers for the helpful reports.

References

1. Marwan, M., Kartit, A., Ouahmane, H.: A cloud based solution for collaborative and secure sharing of medical data. Int. J. Enterp. Inf. Syst. (IJEIS) **14**, 128–145 (2018)
2. Yang, J.J., Li, J.Q., Niu, Y.: A hybrid solution for privacy preserving medical data sharing in the cloud environment. Future Gener. Comput. Syst. (FGCS) **43**, 74–86 (2015)
3. Wu, Y., et al.: Adaptive authorization access method for medical cloud data based on attribute encryption. In: Ni, W., Wang, X., Song, W., Li, Y. (eds.) WISA 2019. LNCS, vol. 11817, pp. 361–367. Springer, Cham (2019). https://doi.org/10.1007/978-3-030-30952-7_36
4. Ding, Y., Song, W., Sheng, Y., Yan, S.: Enabling efficient multi-keyword search over fine-grained authorized healthcare blockchain system. In: The Asia Pacific Web (APWeb) (2020)
5. Goyal, V., Pandey, O., Sahai, A.: Attribute-based encryption for fine-grained access control of encrypted data. In: Proceedings of the 13th ACM Conference on Computer and Communications Security (CCS), pp. 89–98 (2006)
6. Zyskind, G., Nathan, O., Pentland, A.: Decentralizing privacy: using blockchain to protect personal data. In: Proceedings of IEEE Symposium on Security and Privacy (S&P) Workshop, pp. 180–184 (2015)
7. Kish, L.J., Topol, E.J.: Unpatients-why patients should own their medical data. Nat. Biotechnol. **33**, 921–924 (2015)
8. Nakamoto, S.: Bitcoin: a peer-to-peer electronic cash system, Bitcoin White Paper (2008)
9. Azaria, A., Ekblaw, A., Vieira, T., Lippman, A.: MedRec: using blockchain for medical data access and permission management. In: 2016 2nd International Conference on Open and Big Data (OBD), pp. 25–30 (2016)
10. Ajtai, M.: Generating hard instances of lattice problems. In: Proceedings of ACM Symposium on Theory of Computing, pp. 99–108 (1996)
11. Fan, K., Wang, S., Ren, Y., Li, H., Yang, Y.: MedBlock: efficient and secure medical data sharing via blockchain. J. Med. Syst. **42**(8), 1–11 (2018). https://doi.org/10.1007/s10916-018-0993-7
12. Tep, K.S., Martini, B., Hunt, R., Choo, K.K.R.: A taxonomy of cloud attack consequences and mitigation strategies: the role of access control and privileged access management. In: Proceedings of the 2015 IEEE Trustcom/BigDataSE/ISPA, pp. 1073–1080 (2015)
13. Sahai, A., Waters, B.: Fuzzy identity-based encryption. In: Cramer, R. (ed.) EUROCRYPT 2005. LNCS, vol. 3494, pp. 457–473. Springer, Heidelberg (2005). https://doi.org/10.1007/11426639_27
14. Xia, Q., Sifah, E.: MeDShare: trust-less medical data sharing among cloud service providers via blockchain. IEEE Access **5**, 14757–14767 (2017). ISSN 2169-3536

Serving at the Edge: A Redactable Blockchain with Fixed Storage

Jingning Zhang, Youshui Lu$^{(\boxtimes)}$, Yuhao Liu, Xu Yang, Yong Qi, Xinpei Dong, and Haoming Wang

School of Electrical and Information Engineering,
Xi'an Jiaotong University, Xi'an 710071, China
{jingning,lyuhao}@stu.xjtu.edu.cn, lucienlu@me.com,
sunshine561@163.com, qiy@mail.xjtu.edu.cn, dongxinpei@gmail.com,
wanghaomingwj@126.com

Abstract. As a promising approach to extend cloud resource and service on the Internet-of-Things (IoT), edge computing has attracted significant attention. However, edge computing faces challenges in its decentralized management and data reliability. To meet this gap, many approaches propose to use blockchain technology to enable distributed storage and computation at the edge nodes, thus guaranteeing reliable access and control of the network. However, the resource-constraint nature of the edge node makes it difficult to store the entire chain as the data volume increases. To address this issue, we propose Re-chain, a re-writable blockchain *with fixed storage*. Re-chain supports re-writing of the onchain historical transactions in chronological order without changing the block hash, as a result, the total size of the blockchain will not increase. Our protocol is consensus-based and uses the proposed threshold trapdoor chameleon hash (TTCH) to constraint re-write operations. With this regards, Re-chain achieves both decentralized re-writing design and fault-tolerance at the same time. We provide security analysis and evaluation experiments to demonstrate the feasibility of Re-chain, the results show that the performance of Re-chain is acceptable when it is executed at a medium scale.

Keywords: Internet of Things · Chameleon hash · Redactable blockchain · Edge computing

1 Introduction

As the rapid advancement in computing technologies has enabled a wide range of applications, edge computing proposes a novel model for providing computational resources close to billions of end devices at the edge of the network. Edge computing has numerous applications in the Internet of Things (IoT), including healthcare, smart grids, manufacturing, etc. [7]. Edge computing that scales to a large number of sites is a cheaper way to achieve scalability than servers in the corporate center. However, its heterogeneity and resource-constraint nature will

© Springer Nature Switzerland AG 2020
G. Wang et al. (Eds.): WISA 2020, LNCS 12432, pp. 654–667, 2020.
https://doi.org/10.1007/978-3-030-60029-7_58

bring security challenges. During data transmissions, some attacks (e.g., jamming attacks, sniffer attacks) could disable the links by congesting the network. Further, the data in edge networks are separate into many parts and stored in different storage locations, which may cause data reliability issues [8,20].

To address the inherent drawbacks above, many scholars use blockchain as a building block to integrate with edge computing in IoT systems [22,23,25]. With the blockchain technology, it is possible to build a distributed control at dozens of edge nodes. Thanks to the chain structure and consensus process, blockchain can protect the accuracy, consistency and validity of the collected IoT data transparently. The integration of blockchain and edge computing seems to be a win-win solution which can provide secure and reliable services.

However, the integration of blockchain and edge computing still encounters data storage capacity problem. Although the edge node could offer relatively large storage, as the collected data and transactions increase, the storage required for blockchain is ever-growing. As a result, edge nodes will eventually consume the entire storage. Current Bitcoin chain is more than 225 GB large, in the industrial IoT settings, the size of the chain could be even more significant. To address this problem, approaches such as Ethereum differentiate full node and light node. Only the full node stores the entire chain while the light node only stores the state.

Nevertheless, it brings centralization risk, since the full nodes may be malicious and the light nodes have no way to detect this [14,21]. Also, the edge node must be able to verify transactions and blocks, therefore it should store the full chain. Another solution is to overwrite the original chain directly, but it would break tamper-resistance property of the blockchain. The existing approaches cannot effectively solve storage issues.

To address the storage issue, we propose a redactable and reusable blockchain architecture *with fixed storage* called Re-chain. In collaborative edges, Re-chain allows the re-write operations from the earliest block seamlessly when the edge node reaches the maximum storage size, thus mitigating the storage limitation of the edge node. Moreover, the re-writing process is controlled by a consortium of edge nodes. Only approved by a sufficient number of edge nodes can the new transactions be re-written to the chain. Such a consensus mechanism brings trust to the system and increases the attacking overheads for the adversaries.

Our observation is that data and transactions in IoT scenario are time-sensitive. In specific, data generated by the IoT devices (such as sensor measurements, device logs, monitoring data and environmental data) may lose its value after some time. Based on this observation, Re-chain safely re-writes the new transactions in the earliest blocks in a seamless way. We can further use the cloud as a backup node to store the overwritten blocks, which ensures that the historical data and previous transactions are still accessible and verifiable, but it beyond the scope of this paper. Edge nodes in the near-end stores blocks generated in more recent periods and allow transaction query and verification.

Re-chain uses chameleon hash [15] to enable the re-write operation. The concept of chameleon hash was first proposed by Krawczyk and Rabin [3]. It is a one-way hash function that contains public key and trapdoor, where hashing is parametrized by public key pk. As long as the trapdoor sk is not known, it is hard to find a collision. Conversely, if the trapdoor sk is known, the arbitrary collision can be efficiently found. However, the chameleon hash cannot be directly used in our scenario since the trapdoor needs to be managed by a centralized party. Instead, we require a consortium of edge nodes to reach consensus before re-writing a block, and the consensus process should be able to tolerate a considerable number of faulty nodes. To meet this gap, we propose threshold trapdoor chameleon hash (TTCH) by using multiple secret keys instead of a single fixed secret key to finding collisions. We incorporate TTCH into our consensus mechanism. Moreover, to achieve a higher performance in the throughput and reduce the large cryptographic overhead incurred during the consensus process, we construct two different hashes in the Merkel tree, TTCH hash function and SHA256, which only needs to calculate once of the TTCH collision during the block re-write process.

The main contributions of this work are summarized as follows:

- We propose a reusable and redactable blockchain called Re-chain, which can re-write the historical data and transaction of the earlier blocks without affecting the integrity of the original chain. Re-chain allows the most recent data and transactions on the blockchain accessible with a fixed size in space, addressing the storage issues for the large scale edge-based IoT system.
- We propose TTCH to achieve a consensus-based re-write operations, which allows a t-out-of-n edge nodes to compute a hash collision collaboratively in order to re-write block transactions.
- We instantiate a prototype implementation of Re-chain and TTCH, and evaluate the performance of the different operations through comprehensive experiments. The results demonstrate that Re-chain is practical when applying to a medium-scale IoT system.

2 Related Work

2.1 Integrated Blockchain and Edge Computing Systems

The study of [16] proposes a permissioned blockchain edge model to address privacy protections and energy security in smart grid, and they use the voting functionality of blockchain to validate the users' identities and smart contract to achieve optimal energy resource management. Sharma et al. [17] presented distributed blockchain cloud architecture with Software-Defined Networking (SDN) enabled edge computing, the SDN controllers based on blockchain is used for a low-latency service of computing resources. Qi et al. [24] use blockchain technology to build Cpds to prevent the participants from acquiring trusted traceability of products in industrial IoT data sharing arena.

In edge computing environments with a massive amount of data collected from IoT devices, the scalability issues hinder the practical feasibility of blockchain-based solutions. In the following, we summarized works on massive data storage empowered by blockchain. InterPlanetary File System (IPFS) [18] is a decentralized and distributed file system with high integrity and robustness. IPFS runs over a peer-to-peer network to store and share data over the network nodes.

However, none of those above works has explicitly addressed the storage issue of ever-growing blockchain and the reuse of blockchain, existing approaches can only enhance the limited scalability, and security level will decrease accordingly. By contrast, Re-chain can gain storage space by constantly re-writing outdated historical transactions to ensure edge node acting as a full node.

2.2 Redactable Blockchain

Ateniese et al. [15] first proposed redactable blockchain by using chameleon hash (CH) to replace traditional hash function, so that block content can be re-written without causing hard forks. Huang et al. [10] proposed a threshold chameleon hash (TCH) for Industrial Internet-of-Things (IIoT) environment, and it allows a group of authorized sensors to re-write blockchain. However, in the re-write process, TCH requires a ring of k authorized sensors to compute collision one by one, which means, k sensors must behave correctly, and the system cannot tolerate malicious sensors, once a sensor being compromised, the entire system can fail.

In Re-chain, the re-write operation is governed by a consortium of edge nodes, which makes the process more controllable. Further, TTCH find collision through a threshold number of trapdoor keys rather than relies on a single trapdoor key, hence, has better resilience to key compromise. The entire re-write process relies on a consensus protocol similar to Proof of Authority (PoA), which can tolerate faulty nodes. When an edge node leading the re-write process, other edge nodes remain unlinkable, this will not bring higher latency and cost. We believe Re-chain achieves a higher decentralized design and acts more efficient.

3 Background

3.1 System Model

The existing multi-layer edge-IoT system mainly consists of four entities as identified below.

1. *Cloud*: Central cloud in the cloud layer can provide gigantic data storage and computational power. The cloud can back up the blocks which have been overwritten at the edge node in an off-chain manner, therefore it can still recover the overwritten data through the cloud and make the data accessible to the users.

2. *Edge node*: Edge nodes provide fixed storage and computational power which is higher than which of the end device's, but smaller than the cloud. We notice that Re-chain stores at the edge node which has fixed storage. Re-chain is used to record the transactions and data collected from end devices, along with resource management and data processing operations.

3. *Key authority*: Key authority is a trusted entity in the edge layer which is responsible for publishing the genesis block and generating secret keys for edge nodes.
4. *End device*: End devices (e.g., sensors, actuators) are used for collecting data in different circumstances. The storage and computational power are constrained, therefore the end device will send and store the collected data in the edge node.

3.2 Design Goals

Re-chain has the following design goals:

- **Liveness:** Re-chain must guarantee the liveness as long as the minimum number of edge nodes maintains liveness and honest, thus weak synchrony assumptions hold for the key distribution [2].
- **Collision-resistance:** Without knowing the trapdoor of TTCH, no adversary can efficiently find a collision for any pair of $(\hbar^*, R^*, m^*)$ and $(\hbar^*, R'^*, m'^*)$ with PPT algorithm.
- **Unforgeability:** Unforgeability requires that it is even intractable for the adversary who possesses secret keys to find collisions, this definition is stronger than and key-exposure freeness [4,5].
- **Efficiency:** When the space used for Re-chain does not reach the edge node's storage limit, Re-chain uses baseline consensus to write new blocks to the chain, while it reaches the storage limit, Re-chain re-write the historical blocks with our proposed proof-of-concept consensus protocol. The performance of re-writing consensus must be as efficient as basic consensus.

3.3 Threat Model

Our adversary's goal is to find arbitrary collision to break the security of the re-write process. The adversary may hold part of secret keys (less than the threshold) and try to compute collision by invalid credentials. The adversary may take any action within enough secret keys to obtain the hash trapdoor among all edge nodes it controls. We assume that the adversary cannot break standard cryptographic primitives and assumption, such as finding hash collisions or forging digital credentials. The adversary also cannot compromise the private keys of arbitrary domains. In our instantiation, we further assume that the adversary cannot control a majority of edge nodes (more than the threshold) in the blockchain network. We do not consider the privacy and access control of the data stored in the cloud in this paper, as works [9,11,19] are orthogonal with our work and they can integrated into Re-chain.

3.4 Preliminaries

Notations. Let g be a generator of a cyclic group $\mathbb{G}$ of order p a for a λ-bit prime p, an algorithm is efficient if it runs in probabilistic polynomial time (PPT) in the length of its input. We write the integer modulo p as $\mathbb{Z}_p$ and $r \xleftarrow{R} \mathbb{Z}_p$ denote that r is chosen uniformly at random from $\mathbb{Z}_p$, ab represent the multiplication of two integers $a \in \mathbb{Z}_p$ and $b \in \mathbb{Z}_p$. Let BGGen be a PPT algorithm that returns BG $= (\mathbb{G}_1, \mathbb{G}_2, \mathbb{G}_T, p, g_1, g_2, e)$ of asymmetric pairing groups where $\mathbb{G}_1$, $\mathbb{G}_2$, $\mathbb{G}_T$ are cyclic groups of order p, g_1 and g_2 are generators of $\mathbb{G}_1$ and $\mathbb{G}_2$, $e: \mathbb{G}_1 \times \mathbb{G}_2 \to \mathbb{G}_T$ is a computable bilinear map.

Chameleon Hash

Definition 1 (Chameleon Hash). A chameleon hash with message space $\mathcal{M}$ contains five algorithms (HGen, Hash, HVer, HCol) specified as follows:

- HGen(1^λ). The algorithm HGen, on input a parameter λ output a public hash key hk and a secret key sk.
- Hash(m, hk). The algorithm Hash, on input the public hash key hk and a message $m \in \mathcal{M}$ and output the hash value and randomness r.
- HVer$(m, hk, (\hbar, r))$. The algorithm HVer, on input $m, \hbar, r$ verify that $(\hbar, r)$ is a valid hash pair for message m.
- HCol$(sk, (\hbar, m, r), m')$. The algorithm HCol, on input the secret key and $(\hbar, m, r)$, for a new message $m' \in \mathcal{M}$ return new randomness r' to satisfies

$$\text{HVer}(m, hk, (\hbar, r)) = \text{HVer}(m', hk, (\hbar, r'))$$

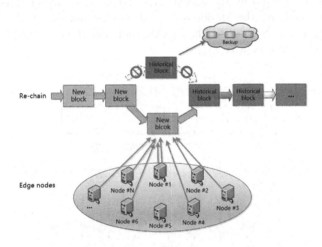

Fig. 1. Re-chain architecture

4   Re-chain Design

4.1   Re-chain Architecture

The core property of Re-chain is the re-writing of the historical transactions when reaching a consensus. We assume that the data collected by the end device is time-sensitive, and the value of data is reducing as the time passing by. Meanwhile, the data collected earlier is less likely to be accessed by the user. After the IoTs system run a period of time, the storage of the edge node will eventually use up by the Re-chain as the collected data volume increases. The edge node will back up all the data to the cloud in case that the data may be requested in later time if the requested data has been overwritten at the edge node. As shown in Fig. 1, in the re-writing phase, Re-chain re-writes from the earliest block seamlessly, which will maintain the structure of the original chain without affecting the accessibility of the most recent transactions and data. The re-writing phase is controlled by the consortium of edge nodes while not by a single entity, it means that the transactions are validated in the same way as which for the ordinary transactions.

As for the block structure in Re-chain, like what is for most blockchains, each block consists of data records (e.g., transactions), block headers, and the source device signed individual transactions. Blocks are chained together by referencing previous blocks via the inclusion of the hash of the previous block header into the header of the current block. Also, block headers include a timestamp which corresponds to the time window of the current block, and integrity measurements, a Merkle tree [12]. In the Merkle Tree, each leaf node contains the hash of a transaction, while each non-leaf node carries with the hash of the concatenation of its child nodes' hashes. To construct our consensus protocol, we modified the hash used by the Merkle tree, for all intermediate tree nodes, we use SHA256, while for the Merkle root hash, we use our designed TTCH which will be discussed in Sect. 4-B. The Merkle root will then be used to verify the integrity of the data records.

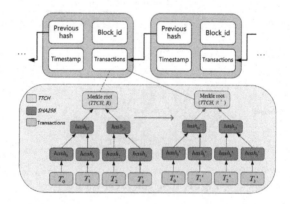

Fig. 2. The structure of blocks used by Re-chain and the Merkle tree update process

4.2   Threshold Trapdoor Chameleon Hash

Building Block of TTCH Scheme. Before giving the full scheme construction, we first recall the public coin chameleon hash proposed by Mojtaba Khalili et al.[1], their hash function satisfies DCDH assumption in the standard model. The scheme works in a billinear group $\mathbb{G}_1, \mathbb{G}_2, \mathbb{G}_T$ as described in Sect. 3-B.

- **K.Setup(1^λ)**: Choose a billinear group $(\mathbb{G}_1, \mathbb{G}_2, \mathbb{G}_T)$ with order p, where p is a λ-bit prime number. Let g_1 be a generator of $\mathbb{G}_1$ and g_2 be a generator of $\mathbb{G}_2$. The system parameters are $params = (\mathbb{G}_1, \mathbb{G}_2, p, g_1, g_2)$.
- **K.KeyGen($params$)**: Choose $x \xleftarrow{R} \mathbb{Z}_p$ and $\widehat{h} \xleftarrow{R} \mathbb{G}_1$, set $h1 = g_1^x$ and $h2 = g_2^x$. Then set hk $= (h_1, h_2, \widehat{h})$ and tk $= x$.
- **K.Hash(m, hk)**: Select a random number $r \xleftarrow{R} \mathbb{Z}_p$. compute $h = h_1^r \widehat{h}^m$ and $R = h_1^r$.
- **K.HVerify(m, hk, R)**: Output $true$ if $e(\hbar/\widehat{h}^m, g^2) = e(R, h_2)$, otherwise output $false$.
- **K.Hcol(tk, m, m')**: For message m', computes new R' as follows:

$$R' = (\frac{\hbar}{\widehat{h}^{m'}})^{\frac{1}{x}}$$

The randomness R in this scheme belongs to a source group of a bilinear pairing and correctness of it can be verified by a pairing product equation. Based on the above scheme, we construct a scheme to support threshold trapdoor aggregate to achieve our design goals as the next section describes.

Construction of TTCH Scheme

- **TT.Setup(1^λ)** $\rightarrow$ ($params$): Choose a billinear group $(\mathbb{G}_1, \mathbb{G}_2, \mathbb{G}_T)$ with order p, where p is a λ-bit prime number. Let g_1 be a generator of $\mathbb{G}_1$ and g_2 be a generator of $\mathbb{G}_2$. The system parameters are $params = (\mathbb{G}_1, \mathbb{G}_2, p, g_1, g_2)$
- **TT.KeyGen($params$, t, n)** $\rightarrow$ (hk, ($sk_1, .., sk_n$)): Pick $x \xleftarrow{R} \mathbb{Z}_p$ and $\widehat{h} \xleftarrow{R} \mathbb{G}_1$, set $h_1 = g_1^x$ and $h_2 = g_2^x$. Compute d s.t. $xd \equiv 1$ mod p. Pick a polynomial v of degree $t - 1$ with coefficients in $\mathbb{Z}_p$, and set the constant term of v to d, which means $v(0) = d$. Issue to each edge node $i \in [1, ..., n]$ a secret key $sk_i = v(i)$, and public the hash key hk $= (h_1, h_2, \widehat{h})$.
- **TT.Hash(m, hk)** $\rightarrow$ ($\hbar$, R): On input a hash key hk and message m. Select a random number $r \xleftarrow{R} \mathbb{Z}_p$, compute $h = h_1^r \widehat{h}^m$ and $R = h_1^r$.
- **TT.HVerify(m, $\hbar$, hk, R)** $\rightarrow$ ($true$ or $false$): On input a committed message m, chameleon hash $\hbar$, hash key hk and randomness R, check whether $e(h/\widehat{h}^m, g_2) = e(R, h_2)$, if yes, output $true$, otherwise output $false$.
- **TT.Sign(sk_i, m', $\hbar$)** $\rightarrow$ (σ_i): The edge node i parses its secret key sk_i and a new message m', output:

$$\sigma_i = (\frac{\hbar}{\widehat{h}^{m'}})^{sk_i}$$

as a credential of message m'.

- **TT.Hcol**$((\sigma_1, .., \sigma_t), R, hk, \hbar, m) \to (\bot$ or $R')$: First make a check by running **Hverify**$(m, \hbar, hk, R)$, if check failed then return $\bot$. Otherwise, parse each σ_i for $i \in [1, .., t]$. compute Lagrange coefficient l:

$$l_i = \left[\prod_{j=1, j\neq i}^{t} (0-j) \right] \left[\prod_{j=1, j\neq i}^{t} (i-j) \right]^{-1} \mod p$$

Then compute a new randomness $R' = \prod_{i=1}^{t} \sigma_i^{l_i}$.

Next, check whether equation: $e(\hbar/\widehat{\hbar}^{m'}, g_2) = e(R', h_2)$ holds. If yes, output R', otherwise output $\bot$.

4.3 Consensus Protocol

In order to put the block re-writing operation to the hands of the consortium of edge nodes, we propose a consensus protocol inspired by Proof of Authority (PoA) [6]. Our consensus protocol can tolerate a number of malicious or failure edge nodes. We assume the majority of N edge nodes are honest, which means at least $N/2 + 1$ edge nodes are honest. The edge nodes are responsible for processing data and transactions from the end devices, and also executing the block re-writing operations on Re-chain. The protocol will run by round where in each round an edge node will be selected as proposal node. To prevent a single Byzantine node from attacking the network by imposing a large number of blocks, each edge node is allowed to propose only one block every $N/2 + 1$ blocks.

To run the consensus protocol, the key authority first generates system parameters during the system initialization. Then it sets the threshold parameter of TTCH function as $t = N/2+1$, and runs **TT.KeyGen** to generate N threshold secret keys $(sk_1, ..., sk_N)$ and a hash key hk. The key authority assigns the secret key to the corresponding edge nodes. Each edge node has its own secret key for issuing credential in consensus. We assume that this consensus protocol only applies to the re-writing operation when the storage space of the Re-chain is used up. The detailed process of the consensus protocol is listed in the following.

- *Step 1: Block proposal.* At the very beginning of the process, an edge node generates transactions with the collected data assembled and proposes a block as proposal node. The proposal node will verify the *Block_id* of the re-written block, and then re-write the earliest transactions. To form a new Merkle Tree, the proposed node calculates the hash corresponding to each new transaction and the hash of the intermediate node with SHA256, while the chameleon hash of the Merkle root remains the same as the original block. Therefore, the Merkle root hash cannot be verified by an intermediate node hash at this time. The proposal node then digitally signs the block and use the timestamp to guarantee authenticity and accuracy. Lastly, it broadcasts the proposed block to other edge nodes.

- *Step 2: Credential issuance.* Upon receiving the proposed block, other edge nodes which act as validators will first validate the signature and the *Block_id*, if verification fails, the proposed block will be dropped by the validators. Upon successfully authenticated, the validator node j will run the **TT.Sign** algorithm to sign the Merkle root's child nodes' hashes with its secret key sk_j, as Fig. 2 shows, the credential $\sigma_j = \textbf{TT.Sign}(sk_j, \hbar'_{01}, \hbar'_{23}, \hbar_{root})$ will be issued by node j and sent to the proposal node, while the validated block will be temporarily saved by node j.
- *Step 3: Computing collision.* After collecting t, i.e. $(N/2+1)$ credentials (Including self-signed credential) $(\sigma_1, ..., \sigma_t)$ from the edge nodes, proposal node enters collision computation phase. By running $\textbf{TT.Hcol}((\sigma_1, ..., \sigma_t), R, hk, \hbar_{root})$, while keeping the chameleon hash of Merkle root unchanged, the proposal node will calculate a new randomness R', which enables the $\hbar_{root}$ to be verifiable by its child nodes' hash. After that, the proposal node broadcasts the *commit* message with the new randomness R' to all the edge nodes in the network. If the proposal node does not collect t credentials within the pre-defined time limit, this consensus round will be terminated and enter into the next round of consensus.
- *Step 4: Commit.* When a validator receives the *commit* message with randomness R', it will first validate the Merkle root hash by running **TT.Hverify**, if verified, the validator will re-write the original block with the new validated block, and update the randomness corresponding to the Merkle root hash to R', and the consensus on re-writing is reached.

4.4   Security Analysis

The Security of TTCH Collision Resistance. We argue that if an adversary $\mathcal{A}$ can break collision resistance. Adversary $\mathcal{A}$ receive a divisible tuple (g_1^x, g_2^x, g_1^y), set $h_1 = g_1^x$, $h_2 = g_2^x$ and $\hat{h} = g_1^y$, without knowing x, the $\mathcal{A}$ can find $(\hbar^*, R^*, m^*)$ and $(\hbar^*, R'^*, m'^*)$ as collision where $R^* = g_1^r$ and $R'^* = g_1^{r'}$, obtain the equation:

$$h_1^r \hat{h}^{m^*} = h_1^{r'} \hat{h}^{m'^*}$$
$$\Rightarrow g_1^{xr} g_1^{ym^*} = g_1^{xr'} g_1^{ym'^*}$$
$$\Rightarrow g_1^{xr} g_1^{\frac{y}{x}xm^*} = g_1^{xr'} g_1^{\frac{y}{x}xm'^*}$$
$$\Rightarrow R^* g_1^{\frac{y}{x}m^*} = R'^* g_1^{\frac{y}{x}m'^*}$$

$\mathcal{A}$ divide both sides of the equation and obtain $g_1^{\frac{y}{x}} = (\frac{R'^*}{R^*})^{(\frac{1}{m^*-m'^*})}$. Since divisible CDH in billinear group is hard, so the adversary $\mathcal{A}$ cannot compute x from two collisions, our proposed TTCH is collision-resistant. **Unforgeability:** We consider two possible ways for adversary $\mathcal{A}$ to forge collision.

(1) The adversary $\mathcal{A}$ without valid credentials manages to compute collision. While forged or wrong credentials will not be able to get new random numbers through **TT.Hcol**.

(2) An adversary $\mathcal{A}$ that has successfully collected fewer than t credentials. While running **TT.Hcol** involves performing Lagrange interpolation. If $\mathcal{A}$ has fewer than t partial credentials, then they have fewer than t points, which makes the resulting the $t - 1$ degree polynomial undetermined and impossible to compute collision.

The Security of Re-chain. The security of Re-chain is based on the security of TTCH.

(1) A malicious edge node $\mathcal{A}$ may attempt to calculate the hash trapdoor by continuous collecting the re-write operation records. While according to the property of Collision Resistance of TTCH, $\mathcal{A}$ cannot get trapdoor through hash tuples.
(2) A small subset of malicious edge node cannot corrupt re-write operation, since the threshold property of Re-chain implies that the adversary needs to corrupt at least t authorities for this attack to be possible.

5   Implementation and Experiments

In this section, we first implement our construction and present the evaluation result of the concrete TTCH. Then we instantiate a proof-of-concept prototype of Re-chain and evaluate its performance through experiments.

5.1   TTCH Implementation

We implement the construction described in Sect. 4 in python 3.6 using petlib and bplib. The bilinear pairing is defined over the Barreto-Naehrig [13] curve, using OpenSSL as arithmetic backend. All simulations are run on desktop computer with Intel i5-3210M CPU and 2 core processors running at 2.3 GHz and 4-GB RAM with 64-bits Linux system.

We first evaluate **TT.KeyGen**, **TT.Hash**, **TT.HVerify** and **TT.Sign**, we fix the message size to 1 KB, each of our results is taken by a mean of 1000 executions. As shown in Fig. 3, we can see that as the threshold parameter increases, the time spent in Hash and HVerify phases are constant, this is because the calculation process of these algorithms does not involve threshold parameter t. The average time cost of **Sign** increases linearly, it is reasonable as in Sign phase it has to get t credentials, so that the algorithm will run t times with different secret keys. However, the time spent of the KeyGen phase increases as the threshold parameter increases, this is due to the **KeyGen** algorithm has to generate t secret keys.

The performance of **TT.Hcol** algorithm is depicted in Fig. 4(i), we demonstrate the linear relationship between threshold parameter and computing cost through the experiments by setting t value from 5 to 30. As it is shown in the figure, the time spent on running **Hcol** increases linearly when t increases. It is because as the t increases, the algorithm has to compute more parameters when finding the collision.

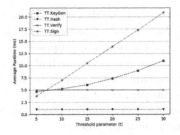

Fig. 3. Average time cost of **KeyGen, Hash, HVerify** and **Sign** in TTCH

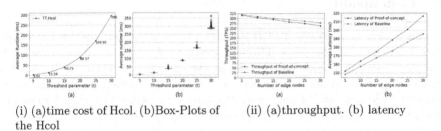

(i) (a)time cost of Hcol. (b)Box-Plots of   (ii) (a)throughput. (b) latency
the Hcol

Fig. 4. Performance of Re-chain

5.2   Re-chain Implementation

We instantiate a prototype of Re-chain with python 3.6. To demonstrate the performance of our proposed consensus protocol, We use the following baseline consensus protocol which is used when the space used for Re-chain is not reach the edge node's storage limit. The baseline consensus procedure are as follows: *1) step 1:* The proposal node propose a block and broadcasts it to the rest of edge nodes. *2) step 2:* The edge node will first validate signatures and *Block_id* of the proposed block, if verified, it will accept the proposal block and broadcasts the preset message to all the other edge nodes. *3) step 3:* If the proposal node receive over $N/2$ correct preset messages, the edge node will broadcast a commit message. *4) step 4:* A consensus is reached if the proposal node accepts $N/2 + 1$ (possibly including its own) commit messages.

We use Aliyun CES as the experimental platform, all nodes are running in the Docker containers on four servers, each server is equipped with two Intel(R) Xeon(R) Platinum 8269CY CPU at 2.50 GHz and 4 GB of RAM, the operating system is 64-bit Ubuntu 18.04.4 LTS. Each block contains 100 transactions. We test the performance of our proposed consensus protocol through the comparison experiments with the baseline consensus protocol. The experiments record the throughput and latency of both protocols with different numbers of nodes varying from 5 to 30. We use the HTTP protocol for the communication between different nodes. And we record the average result out of six tests for each experiment. As shown in Fig. 4(ii)(a), the results show that the throughput for our proposed consensus protocol is about 5% less than which of the baseline when relatively fewer nodes in the system. However, the difference will become more significant when the number of nodes increases. Figure 4(ii)(b) shows that the latency gap is slightly larger between the proposed consensus and baseline consensus. It is reasonable as our proposed method requires more computational overhead in finding collisions and other cryptographic calculations such as digital signing. Moreover, as the number of nodes increases, the threshold t will also increase, therefore it requires more computational power to execute **TT.Hcol** to reach consensus during re-writing process.

6   Conclusion

In this paper, we proposed TTCH to build a reductable and reusable blockchain called Re-chain at the edge. Through Re-chain we address the storage problem of conventional blockchain caused by ever-growing information chunks. The proposed proof-of-concept consensus is used when Re-chain reaches the maximum storage size, and it empowers Re-chain to re-write historical blocks in a controllable and secure way while maintaining the connectivity of the chain, meanwhile, the consensus process can guarantee the liveness even if there are $N/2 - 1$ faulty edge nodes in the network. Re-chain can be applied to the consortium blockchain-based industrial IoT systems which have storage limitation issues. The experimental results showed that the re-write operation is efficient, and the performance of Re-chain is acceptable at a medium scale (under 30 edge nodes).

Acknowledgment. This research is supported by the National key R&D Program of China under Grant No. 2018YFB1402700. It is also partially supported by the National Natural Science Foundation of China under Grant No. 61672421.

References

1. Khalili, M., Dakhilalian, M., Susilo, W.: Efficient chameleon hash functions in the enhanced collision resistant model. Inf. Sci. **510**, 155–164 (2020)
2. Kate, A., Huang, Y., Goldberg, I.: Distributed key generation in the wild. Cryptology ePrint Archive, Report 2012/377 (2012). https://eprint.iacr.org/2012/377
3. Krawczyk, H., Rabin, T.: Chameleon signatures. In: Proceedings of the Network and Distributed System Security Symposium, NDSS 2000, San Diego, California, USA (2000)
4. Ateniese, G., de Medeiros, B.: On the key exposure problem in chameleon hashes. In: Blundo, C., Cimato, S. (eds.) SCN 2004. LNCS, vol. 3352, pp. 165–179. Springer, Heidelberg (2005). https://doi.org/10.1007/978-3-540-30598-9_12
5. Chen, X., Zhang, F., Kim, K.: Chameleon hashing without key exposure. In: Zhang, K., Zheng, Y. (eds.) ISC 2004. LNCS, vol. 3225, pp. 87–98. Springer, Heidelberg (2004). https://doi.org/10.1007/978-3-540-30144-8_8
6. De Angelis, S., Aniello, L., Baldoni, R., Lombardi, F., Margheri, A., Sassone, V.: PBFT vs proof-of-authority: applying the CAP theorem to permissioned blockchain. In: Proceedings of ITASEC, pp. 1–11 (2018)

7. Stanciu, A.: Blockchain based distributed control system for edge computing. In: 21st International Conference on Control Systems and Computer Science (2017)
8. Yang, R., Yu, F.R., Si, P., Yang, Z., Zhang, Y.: Integrated blockchain and edge computing systems: a survey some research issues and challenges. IEEE Commun. Surv. Tutor. **21**, 1–25 (2019)
9. Qi, S., Zheng, Y., Li, M., Liu, Y., Qiu, J.: Scalable industry data access control in RFID-enabled supply chain. IEEE/ACM Trans. Netw. **24**(6), 3551–3564 (2016)
10. Huang, K., et al.: Building redactable consortium blockchain for industrial Internet-of-Things. IEEE Trans. Ind. Inform. **15**(6), 3670–3679 (2019)
11. Qi, S., Zheng, Y., Li, M., Lu, L., Liu, Y.: Secure and private RFID-enabled third-party supply chain systems. IEEE Trans. Comput. **65**(11), 3413–3426 (2016)
12. Merkle, R.C.: A certified digital signature. In: Brassard, G. (ed.) CRYPTO 1989. LNCS, vol. 435, pp. 218–238. Springer, New York (1990). https://doi.org/10.1007/0-387-34805-0_21
13. Kasamatsu, K.: Barreto-Naehrig curves (2014). https://tools.ietf.org/id/draft-kasamatsu-bncurves-01.html. Accessed 14 Aug 2014
14. Buterin, V.: A next generation smart contract and decentralized application platform. https://github.com/ethereum/wiki/wiki/White-Paper
15. Ateniese, G., Magri, B., Venturi, D., Andrade, E.: Redactable blockchain - or - rewriting history in bitcoin and friends. In: 2017 IEEE European Symposium on Security and Privacy, EuroS&P (2017)
16. Gai, K., Wu, Y., Zhu, L., Xu, L., Zhang, Y.: Permissioned blockchain and edge computing empowered privacy-preserving smart grid networks. IEEE Internet Things J. **6**(5), 7992–8004 (2019)
17. Sharma, P.K., Chen, M.-Y., Park, J.H.: A software defined fog node based distributed blockchain cloud architecture for IoT. IEEE Access **6**, 115–124 (2017)
18. Benet, J.: IPFS - content addressed, versioned, p2p file system (draft 3). https://github.com/ipfs/papers/raw/master/ipfs-cap2pfs/ipfs-p2p-file-system.pdf
19. Qi, S., Zheng, Y.: Crypt-DAC: cryptographically enforced dynamic access control in the Cloud. IEEE Trans. Dependable Secure Comput. **16**, 1 (2019)
20. Lu, Y., Qi, Y., Qi, S., Li, Y., Song, H., Liu, Y.: Say no to price discrimination: decentralized and automated incentives for price auditing in ride-hailing services. IEEE Trans. Mob. Comput. (2020). https://doi.org/10.1109/TMC.2020.3008315
21. Xue, S., Zhao, X., Li, X., Zhang, G., Xing, C.: A trusted system framework for electronic records management based on blockchain. In: Ni, W., Wang, X., Song, W., Li, Y. (eds.) WISA 2019. LNCS, vol. 11817, pp. 548–559. Springer, Cham (2019). https://doi.org/10.1007/978-3-030-30952-7_55
22. Tuli, S., Mahmud, R., Tuli, S., Buyya, R.: FogBus: a Blockchain-based lightweight framework for edge and fog computing. J. Syst. Softw. **154**, 22–36 (2019)
23. Ayoade, G., Karande, V., Khan, L., Hamlen, K.: Decentralized IoT data management using blockchain and trusted execution environment. In: 2018 IEEE International Conference on Information Reuse and Integration (IRI), pp. 15–22. IEEE (2018)
24. Qi, S., Lu, Y., Zheng, Y., Li, Y., Chen, X.: Cpds: enabling compressed and private data sharing for industrial IoT over blockchain. IEEE Trans. Ind. Inform. (2020). https://doi.org/10.1109/TII.2020.2998166
25. Samaniego, M., Deters, R.: Virtual resources & blockchain for configuration management in IoT. J. Ubiquit. Syst. Pervasive Netw. **9**(2), 1–13 (2017)

Author Index

Printed in the United States
By Bookmasters